Birkhäuser Advanced Texts Basler Lehrbücher

Series editors
Steven G. Krantz, Washington University, St. Louis, USA
Shrawan Kumar, University of North Carolina at Chapel Hill, Chapel Hill, USA
Jan Nekovář, Sorbonne Université, Paris, France

More information about this series at http://www.springer.com/series/4842

Andreas Rosén

Geometric Multivector Analysis

From Grassmann to Dirac

 Birkhäuser

Andreas Rosén
Department of Mathematical Sciences
Chalmers University of Technology
and the University of Gothenburg
Gothenburg, Sweden

ISSN 1019-6242   ISSN 2296-4894   (electronic)
Birkhäuser Advanced Texts Basler Lehrbücher
ISBN 978-3-030-31413-2   ISBN 978-3-030-31411-8   (eBook)
https://doi.org/10.1007/978-3-030-31411-8

Mathematics Subject Classification (2010): 15-01, 15A72, 15A66, 35-01, 35F45, 45E05, 53-01, 58A10, 58A12, 58J20

This book is published under the imprint Birkhäuser, www.birkhauser-science.com by the registered company Springer Nature Switzerland AG
The registered company address is: Gewerbestrasse 11, 6330 Cham, Switzerland

Contents

Preface

I guess all mathematicians have had their defining moments, some events that led them to devote much of their lives and energy to mathematics. Myself, I vividly recall the spring and summer of 1997, spending my days reading about Clifford algebras in David Hestenes's inspirational books and listening to the Beatles. Don't misunderstand me. To a Swede, there is nothing that beats ABBA, but that summer it happened that the Clifford algebras were enjoyed in this particular way. I was a fourth-year undergraduate student at Linköping University studying the civil engineering program of applied physics and electrical engineering, and the very last course I took there came to change my life in a way that no one could have anticipated. The course was on "applied mathematics", and we were supposed to pursue a math project of our choice, typically to solve some differential equation. One odd topic proposed was learning Clifford algebras, and it appealed to me. I fell deeply in love with the beauty of it all, and I read and I read. I found the biographies [34, 37] about Hermann Grassmann, and I learned what an unfortunate turn mathematics had taken since the 1800s. During my university studies I had had a sense of something missing in the vector calculus that we were taught. I remember students asking me in the linear algebra sessions that I taught how the vector product could have area as dimension while at the same time being a vector. I discovered that Grassmann had figured it all out more than 150 years ago, and now it was all strangely hidden to us students of mathematics, all but the one-dimensional vectors. No one had told me anything about vector products in dimensions other than three, or about determinants of rectangular matrices.

My personal relations with the vector product had in fact begun some five years earlier, when I borrowed a telescope from my high school for a science project on satellites. Using Kepler's laws, I calculated a formula for the altitude of a satellite's orbit, using as input two observations of the satellite's position and the time elapsed between the two observations. Of course you don't need a telescope for this, it's just to look for a slow falling star, but I did other things as well. As you may guess, I stumbled upon a curious expression involving three mixed products, for the plane of rotation of the satellite. It was only the following year, when I had started my university studies, that I learned in the linear algebra lectures that this intriguing formula was called a vector product.

A second defining moment occurred two years later, around May 1999. I was spending a Saturday or Sunday in the library at the mathematics department in Lund, and stumbled upon a friend. We started a discussion that led to a search on this rather new thing called the internet, where I found the perfect PhD supervisor, Alan McIntosh, from Australia, one of the giants in harmonic analysis and operator theory. It was a perfect match, since he was doing real analysis, singular integrals and operator theory, as well as mixing in the algebras of Clifford and Grassmann when needed. And so I ended up down under in Canberra, and spent three years applying singular integrals and Clifford algebra to solve Maxwell boundary value problems on Lipschitz domains with Alan McIntosh. The publications [11, 8, 9, 7, 14, 10] related to my thesis work are perhaps the real starting point for this book. To shed light on the confusion: Axelsson = Rosén before 2011.

The reason for telling this story is not that I think the reader is more interested in my personal story than in the subject of the book. I certainly hope not. But nothing is without context, and it may help to know the background to understand this book. The basic algebra is not new; it goes back to the pioneering works of Hermann Grassmann, first published in 1843, whose exterior algebra of multivectors is the topic of Chapter 2, and of William Kingdon Clifford from 1878, whose geometric algebra is the topic of Chapter 3. Although these algebras are geometric and useful enough that one would expect them to fit into the mainstream mathematics curriculum at a not too advanced level, this has not really happened. But over the last century, they have been rediscovered over and over again. Inspired by the Grassmann algebra, Élie Cartan developed his calculus of differential forms in the early 1900s. He was also the first to discover spinors in general in 1913, which is the topic of Chapter 5. In 1928, Paul Dirac formulated his famous equation that describes massive spin $1/2$ particles in relativistic quantum mechanics, which we discuss in Section 9.2, and which makes use of spacetime spinors and matrix representations of Clifford's algebra. In 1963, Michael Atiyah and Isadore Singer rediscovered and generalized the Dirac operator to Riemannian manifolds in connection with their celebrated index theorem, which is the topic of Chapter 12.

There are also works by Marcel Riesz from 1958 on spacetime isometries and by Lars Ahlfors from 1985 on Möbius maps, using Clifford algebra, which is the topic of Chapter 4. Mentioned above, David Hestenes has been advocating the use of Clifford algebra, in particular in mathematical physics, since the 1960s. There is also a research field of Clifford analysis, where a higher-dimensional complex analysis using Clifford algebras has been developed, starting from around 1980 and which is the topic of Chapter 8.

Included in this book are also some more recent results related to my own research. The material in Sections 9.3 to 10.4 on Dirac integral equations and Hodge decompositions originates with my early thesis work with Alan McIntosh in 2000–2002, and most of the key ideas there are an inheritance from him. Since then, the material covered in this book has been a continued source of inspiration for my research. The following publications of mine in particular make use, explicitly or implicitly, of the algebras of Grassmann and Clifford in real analysis: Axelsson, Keith, and McIntosh [12]; Auscher, Axelsson, and Hofmann [4]; Auscher, Axelsson, and McIntosh [5]; Axelsson, Kou, and Qian [13]; Rosén [82, 83]; Bandara, McIntosh, and Rosén [17]; Bandara and Rosén [18]; and Rosén [84, 80].

This book was written in four stages. The first part, on the algebras of Grassmann and Clifford, was written around 2008 at Stockholm University and was used as material for a graduate course given there. In the second stage I wrote basically Chapters 7, 8, and 10 for a graduate course given in Linköping in 2010. In the third stage I wrote Chapters 11 and 12 for a graduate course in Gothenburg 2014. In between and after these writing periods, the manuscript was collecting dust until I decided, upon returning to mathematics after an extended period of parental leave in 2018, to prepare this final version for publication. Having been away from math for a while gave me new perspectives on things, and this final preparation turned into a major rewriting of the whole book, which I hope will benefit the reader.

A number of mathematicians and friends deserve a sincere thanks for being helpful, directly or indirectly, in the creation of this book. Those who untimely have passed away by now, Peetre, McIntosh, and Passare, will always be remembered fondly by me. In mainly chronological order the following people come to mind. Hans Lundmark, who was my mentor for that very first Clifford algebra project in Linköping. I wonder whether and where I would have discovered this mathematics had he not proposed this project to me. Mats Aigner in Linköping, whom I first met in Lund and with whom I have had uncountably many interesting discussions about the algebras of Clifford and Grassmann. Jaak Peetre, who encouraged me and provided interesting discussions on the subject. Wulf Staubach at Uppsala University, that friend from the library who changed my life by being well read and knowing about Alan McIntosh. Alan Mcintosh at the Australian National University, my mathematical father from whom I have learned so much. I doubt very much that I will ever again meet someone with as deep an understanding of life and mathematics as he possessed. Mikael Passare at Stockholm University, who supported me at a critical stage. Erik Duse, who was a student attending that first course that I gave in Stockholm, and who more recently himself gave a course based on the third version of this book in Helsinki, and who has given me valuable feedback, including some exercises contained in this book.

The book is organized so that the reader finds in the introduction to each chapter a description of and a road map to the material in that chapter. Comments and references are collected in the final section of each chapter. There are two parts of the book. In the first part, the affine multivector and spinor algebra and geometry are explained. A key idea here is the principle of abstract algebra, as explained in the introduction to Chapter 1. In the second part, we use multivectors and spinors in analysis, first in affine space and later on manifolds. A key idea here is that of splittings of function spaces, as explained in the introduction to Chapter 6. My intention is that the material covered should be accessible to basically anyone with mathematical maturity corresponding to that of an advanced undergraduate student, with a solid understanding of standard linear algebra, multi-variable and vector calculus, and complex analysis. My hope is that you will find this beautiful mathematics as useful and inspiring as I have.

Andreas Rosén
Göteborg, August 2019

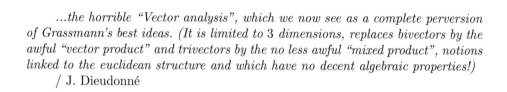

...the horrible "Vector analysis", which we now see as a complete perversion of Grassmann's best ideas. (It is limited to 3 dimensions, replaces bivectors by the awful "vector product" and trivectors by the no less awful "mixed product", notions linked to the euclidean structure and which have no decent algebraic properties!)
/ J. Dieudonné

Chapter 1

Prelude: Linear Algebra

Road map:

This chapter is **not** where to start reading this book, which rather is Chapter 2. The material in the present chapter is meant to be used as a reference for some background material and ideas from linear algebra, which are essential to this book, in particular to the first part of it on algebra and geometry consisting of Chapters 2 through 5. The main idea in this part of the book is what may be called *the principle of abstract algebra*:

> It is not important *what* you calculate with, it is only important *how* you calculate.

Let us explain by examples. Consider for example the complex numbers $x + iy$, where you of course ask what is $i = \sqrt{-1}$ when you first encounter this mathematical construction. But that uncomfortable feeling of what this strange imaginary unit really is fades away as you get more experienced and learn that $\mathbf{C}$ is a field of numbers that is extremely useful, to say the least. You no longer care what kind of object i is but are satisfied only to know that $i^2 = -1$, which is how you calculate with i.

It is this principle of abstract algebra that one needs to bear in mind for all our algebraic constructions in this book: the exterior algebra of multivectors in Chapter 2, Clifford algebras in Chapter 3, and spinors in Chapter 5. In all cases the construction starts by specifying how we want to calculate. Then we prove that there exist objects that obey these rules of calculation, and that any two constructions are isomorphic. Whenever we know the existence and uniqueness up to isomorphism of the objects, we can regard them as geometric objects with an invariant meaning. Which concrete representation of the objects we have becomes irrelevant.

In this chapter, Sections 1.1, 1.2, and 1.4 contain background material for Chapter 2, whereas Sections 1.3 and 1.5 are mainly relevant for Chapters 4 and 5 respectively.

© Springer Nature Switzerland AG 2019
A. Rosén, *Geometric Multivector Analysis*, Birkhäuser Advanced Texts Basler Lehrbücher,
https://doi.org/10.1007/978-3-030-31411-8_1

1.1   Vector Spaces

Two general notations which we use throughout this book are the following. By $X := Y$ or $Y =: X$ we mean that X is defined to be / is assigned the value of Y. By $A \leftrightarrow B$ we denote a one-to-one correspondence, or an isomorphism between A and B, depending on context.

We shall distinguish the concept of a vector space from the more general concept of a linear space. Except for function spaces, which we use later in part two of the book, we shall assume that our linear spaces are finite-dimensional. The difference between linear spaces and vector spaces is only a conceptual one, though. Indeed, any linear space V is naturally an affine space (V, V), where V acts on itself through the addition in V; see below. Thus, strictly mathematically speaking, a linear space is the same thing as a vector space. The difference between linear and vector spaces lies in the geometric interpretation of their objects, and we want to make this distinction clear to start with, since we are going to work with linear spaces whose objects are not to be interpreted as geometric vectors.

Definition 1.1.1 (Linear space). A real *linear space* $(L, +, \cdot)$ is an abelian group $(L, +)$ together with a scalar multiplication $\mathbf{R} \times L \to L$ that is bilinear with respect to addition and a group action of the multiplicative group $\mathbf{R}^* = \mathbf{R} \setminus \{0\}$ on L.

We recall that a group is a set equipped with a binary associative multiplication, containing an identity element and an inverse to each element. For an abelian group, we assume commutativity and write the binary operation as addition.

In a linear space, we sometimes write a product xv of $x \in \mathbf{R}$ and $v \in L$ as vx. Since the product of real numbers is commutative, this presents no problem.

On the other hand, by a vector space V we mean a linear space consisting of geometric vectors, that is, "one-dimensional directed objects", which we refer to as vectors. More precisely, this means that V is the space of translations in some affine space X as follows.

Definition 1.1.2 (Vector space). An *affine space* (X, V) is a set X on which a real linear space V, the space of *translations/vectors* in X, acts freely and transitively by addition, that is, there exists an addition by vectors map $X \times V \to X$ that is a (left or right) action of $(V, +)$ on X such that for all $x, y \in X$ there exists a unique $v \in V$, the vector denoted by $y - x$, for which $x + v = y$.

If $x, y \in X$, then the vector $v = y - x$ has the interpretation of a one-dimensional arrow starting at x and ending at y. Starting at a different point $x' \in X$, the same vector v also appears as the arrow from x' to $x' + v$. Thus a vector v is characterized by its orientation and length, but not its position in X. In general affine spaces, the notion of lengths, and more generally k-volumes, have only a relative meaning when we do not have access to an inner product on the space to measure angles and absolute lengths. Thus in general affine spaces, only the relative lengths of two parallel vectors v_1 and v_2 can be compared: if $v_1 = \lambda v_2$, then v_1 is λ times longer than v_2.

In practice, one often identifies the affine space X and its vector space V. The difference is the origin 0: X is V, but where we have "forgotten" the origin. Given an origin point $x_0 \in X$, we can identify the vector $v \in V$ with the point $x_0 + v \in X$. In particular, $x_0 \in X$ is identified with $0 \in V$. The reader will notice that in Chapters 2 and 7 we carefully distinguish between X and its vector space V, but that in the later chapters, we become more pragmatic and often identify $X = V$.

Definition 1.1.3 ($\mathbf{R}^n$). The vector space $\mathbf{R}^n$ is the set of n-tuples

$$\mathbf{R}^n := \{(x_1, \ldots, x_n) \; ; \; x_i \in \mathbf{R}\},$$

with the usual addition and multiplication by scalars. This linear space has a distinguished basis, the *standard basis* $\{e_i\}$, where

$$e_i := (0, \ldots, 0, 1, 0, \ldots, 0),$$

with coordinate 1 at the ith position.

We adopt the practical convention that we identify row vectors with column vectors, as is often done in doing analysis in $\mathbf{R}^n$. More precisely, $\mathbf{R}^n$ should be the space of column vectors, since matrix multiplication is adapted to this convention. However, whenever no matrix multiplication is involved, it is more convenient to write $\begin{bmatrix} x_1 & \cdots & x_n \end{bmatrix}$ than $\begin{bmatrix} x_1 \\ \vdots \\ x_n \end{bmatrix} = \begin{bmatrix} x_1 & \cdots & x_n \end{bmatrix}^t$, where $\cdot^t$ denotes matrix transpose. We will not distinguish between parentheses $(\cdot)$ and brackets $[\cdot]$.

Note the decreasing generality of the notions: an affine space is homogeneous and isotropic, that is, without any distinguished points or directions. A linear space is isotropic, but has a distinguished point: the origin 0. The linear space $\mathbf{R}^n$ is neither homogeneous or isotropic: it has an origin and a distinguished basis, the standard basis. Whenever we have fixed a basis $\{e_i\}$ in a vector space V, there is a natural identification between V and $\mathbf{R}^n$, where a vector $v = \sum x_i e_i$ corresponds to the coordinate tuple $x = (x_1, \ldots, x_n) \in \mathbf{R}^n$.

Recall the notion of direct sums of linear spaces. Define the sum of subspaces

$$V_1 + V_2 := \{v_1 + v_2 \; ; \; v_1 \in V_1, v_2 \in V_2\}$$

when V_1 and V_2 are two subspaces of a linear space V. When $V_1 \cap V_2 = \{0\}$, we write $V_1 \oplus V_2$ and call the sum a *direct sum*. This is an intrinsic direct sum. In contrast, suppose that we are given two linear spaces V_1 and V_2, without any common embedding space V. In this case we define the (extrinsic) *direct sum* of these spaces as

$$V_1 \oplus V_2 := \{(v_1, v_2) \in V_1 \times V_2 \; ; \; v_1 \in V_1, v_2 \in V_2\}.$$

In a natural way, $V_1 \oplus V_2$ is a linear space that contains both spaces V_1, V_2, under suitable identifications. As an example, $\mathbf{R}^n$ is the exterior direct sum of n copies of the one-dimensional linear space $\mathbf{R}$.

Recall the notions of linear independence of a set $S \subset V$ and its linear span $\mathrm{span}(S) \subset V$. For concrete calculations in a given linear space V, it is often needed to fix a basis

$$\{e_1, \ldots, e_n\} \subset V,$$

with $n = \dim V$ being the dimension of V. It is conceptually important to understand that a basis in general is an unordered set. But often bases for vector spaces are linearly ordered $e_1, e_2, e_3, \ldots$ by the positive integers and considered as ordered sets. In particular, this is needed in order to represent $v \in V$,

$$v = x_1 e_1 + \cdots + x_n e_n = \sum_{i=1}^{n} x_i e_i = \begin{bmatrix} e_1 & \cdots & e_n \end{bmatrix} \begin{bmatrix} x_1 \\ \vdots \\ x_n \end{bmatrix},$$

by its *coordinates* $(x_1, \ldots, x_n) \in \mathbf{R}^n$, and in order to represent a linear map $T : V_1 \to V_2$ between linear spaces V_1, V_2,

$$T(x_1 e_1 + \cdots + x_n e_n) = \sum_{i=1}^{m} \sum_{j=1}^{n} e'_i a_{i,j} x_j = \begin{bmatrix} e'_1 & \cdots & e'_m \end{bmatrix} \begin{bmatrix} a_{1,1} & \cdots & a_{1,n} \\ \vdots & \ddots & \vdots \\ a_{m,1} & \cdots & a_{m,n} \end{bmatrix} \begin{bmatrix} x_1 \\ \vdots \\ x_n \end{bmatrix},$$

by its matrix $A = (a_{i,j})$ relative to the bases $\{e_j\}$ for V_1 and $\{e'_i\}$ for V_2.

However, many fundamental types of bases used in mathematics do not come with any natural linear order. Indeed, this will be the usual situation in this book, where the basic linear spaces of multivectors, tensors, and spinors have standard bases that are not linearly ordered but rather have some sort of lattice ordering, meaning that the basis elements naturally are indexed by subsets of integers or tuples of integers.

Another central theme in this book is that many basic linear spaces that appear are not only linear spaces, but associative algebras in the sense that they come equipped with an associative, but in general noncommutative, product.

Definition 1.1.4 (Associative algebra). A real *associative algebra* $(A, +, *, 1)$, with identity, is a linear space over $\mathbf{R}$ equipped with a bilinear and associative product $*$, with identity element 1. Scalars $\lambda \in \mathbf{R}$ are identified with multiples $\lambda 1 \in A$ of the identity, and it is assumed that $(\lambda 1) * v = \lambda v = v * (\lambda 1)$ for all $v \in A$.

Let $(A_1, +_1, *_1, 1_1)$ and $(A_2, +_2, *_2, 1_2)$ be two algebras. Then a map $T : A_1 \to A_2$ is said to be an algebra *homomorphism* if it is linear and satisfies $T(v_1 *_1 v_2) = T(v_1) *_2 T(v_2)$ for all $v_1, v_2 \in A_1$ and if $T(1_1) = 1_2$. An invertible homomorphism is called an algebra *isomorphism*.

Exercise 1.1.5. Let A be an associative algebra. Define the exponential function

$$\exp(x) := \sum_{k=0}^{\infty} \frac{1}{k!} x^k, \quad x \in A.$$

Show that $\exp(x+y) = \exp(x) \exp(y)$, provided that x and y commute, that is, if $xy = yx$. If $\phi \in \mathbf{R}$, show that

$$\exp(\phi j) = \begin{cases} \cos\phi + j\sin\phi, & \text{if } j^2 = -1, \\ \cosh\phi + j\sinh\phi, & \text{if } j^2 = 1, \\ 1 + \phi j, & \text{if } j^2 = 0. \end{cases}$$

1.2 Duality

There are several reasons for us to consider inner products and dualities more general than Euclidean ones. A first reason is that we want to study the geometry of multivectors in Minkowski spacetimes, the closest relative to Euclidean spaces among inner product spaces, which are modeled by an indefinite inner product as in Section 1.3. A second reason is that we want to study real Clifford algebras where the fundamental representation Theorem 3.4.2 involves inner product spaces of signature zero. A third reason is that we want to study spinor spaces, where more general nonsymmetric dualities may appear.

Definition 1.2.1 (Duality and inner product). A *duality* of two linear spaces V_1 and V_2 is a bilinear map

$$V_1 \times V_2 \to \mathbf{R} : (v_1, v_2) \mapsto \langle v_1, v_2 \rangle$$

that is non-degenerate in the sense that $\langle v_1, v_2 \rangle = 0$ for all $v_1 \in V_1$ only if $v_2 = 0$, and $\langle v_1, v_2 \rangle = 0$ for all $v_2 \in V_2$ only if $v_1 = 0$. In the case $V_1 = V_2 = V$, we speak of a duality on V.

If a duality on V is symmetric in the sense that $\langle v_1, v_2 \rangle = \langle v_2, v_1 \rangle$ for all $v_1, v_2 \in V$, then we call the duality an *inner product* and V an *inner product space*. We use the notation

$$\langle v \rangle^2 := \langle v, v \rangle \in \mathbf{R}.$$

A vector v such that $\langle v \rangle^2 = 0$ is called *singular*. If an inner product has the additional property that $\langle v \rangle^2 > 0$ for all $0 \neq v \in V$, then we call it a *Euclidean inner product*, and V is called a *Euclidean space*. In this case, we define the *norm* $|v| := \sqrt{\langle v \rangle^2} \geq 0$, so that $\langle v \rangle^2 = |v|^2$.

If a duality on V is skew-symmetric in the sense that $\langle v_1, v_2 \rangle = -\langle v_2, v_1 \rangle$ for all $v_1, v_2 \in V$, then we call the duality a *symplectic form* and V a *symplectic space*.

Note carefully that in general, $\langle v \rangle^2$ may be negative, as compared to the square of a real number. We do not define any quantity $\langle v \rangle$, and the square in the notation $\langle v \rangle^2$ is only formal.

Exercise 1.2.2. Show that an inner product is Euclidean if $\langle v \rangle^2 \geq 0$ for all $v \in V$.

Let V be a linear space. There is a canonical linear space V^* and duality $\langle V^*, V \rangle$, namely the *dual space* of V defined as

$$V^* := \{\text{linear functionals } \theta : V \to \mathbf{R}\}.$$

Given such a scalar-valued linear function $\theta \in V^*$, its value $\theta(v) \in \mathbf{R}$ at $v \in V$ will be denoted by

$$\langle \theta, v \rangle := \theta(v) \in \mathbf{R}.$$

Note that this is indeed a duality: if $\theta(v) = 0$ for all $v \in V$, then $\theta = 0$ by definition. On the other hand, if $\theta(v) = 0$ for all θ, then it follows that $v = 0$, since otherwise, we can take a complementary subspace $V' \subset V$ so that $V = \text{span}\{v\} \oplus V'$ and define the linear functional $\theta(\alpha v + v') := \alpha$, $\alpha \in \mathbf{R}$, $v' \in V'$ for which $\theta(v) \neq 0$.

If V is a vector space with a geometric interpretation of $v \in V$ as in Section 1.1, then $\theta \in V^*$, which we refer to as a *covector*, is best described in V by its level sets

$$\{v \in V \; ; \; \langle \theta, v \rangle = C\},$$

for different fixed values of $C \in \mathbf{R}$. Since θ is linear, these level sets are parallel hyperplanes.

The following observation is fundamental in understanding dualities.

Proposition 1.2.3 (Representation of dual space). *Fix a linear space V. Then there is a one-to-one correspondence between dualities*

$$\langle V', V \rangle$$

and invertible linear maps

$$g : V' \to V^* : v \mapsto \theta,$$

given by

$$\langle g(v'), v \rangle := \langle v', v \rangle, \quad v \in V.$$

Here the pairing on the left is the functional value $g(v')v$, whereas the pairing on the right is as in Definition 1.2.1.

If $V' = V$, then V is an inner product/symplectic space if and only if $g : V \to V^*$ is a symmetric/antisymmetric linear map.

With Proposition 1.2.3 in mind, we write a duality between two linear spaces as $\langle V^*, V \rangle$, where V^* not necessarily is *the* dual space of V, but rather *a* linear space dual to V in the sense of Definition 1.2.1. By Proposition 1.2.3 this abuse of notation presents no problem. In particular, when we have a duality or inner product on V, we shall write

$$\theta = v$$

to mean $\theta = g(v)$.

Definition 1.2.4 (Orthogonal complement). Consider a linear space V and a duality $\langle V^*, V \rangle$. If $\langle v', v \rangle = 0$, then we say that $v' \in V^*$ and $v \in V$ are *orthogonal*. The *orthogonal complement* of a set $S' \subset V^*$ is the subspace

$$(S')^\perp := \{v \in V \; ; \; \langle v', v \rangle = 0, \text{ for all } v' \in S'\} \subset V.$$

For $S \subset V$ we similarly define the orthogonal complement

$$S^\perp := \{v' \in V^* \; ; \; \langle v', v \rangle = 0, \text{ for all } v \in S\} \subset V^*.$$

Definition 1.2.5 (Dual basis). Let $\{e_1, \ldots, e_n\}$ be a basis for V. Then each $v \in V$ can be uniquely written $v = \sum_j x_j e_j$, and we define covectors e_j^* by

$$\langle e_j^*, v \rangle := x_j = \text{ the } j\text{th coordinate of } v.$$

We call $\{e_1^*, \ldots, e_n^*\} \subset V^*$ the *dual basis* of $\{e_1, \ldots, e_n\} \subset V$.

Note that the dual basis $\{e_1^*, \ldots, e_n^*\}$ is indeed a basis for V^* whenever $\{e_1, \ldots, e_n\}$ is a basis for V, and is characterized by the property

$$\langle e_i^*, e_j \rangle = \begin{cases} 1, & i = j, \\ 0, & i \neq j. \end{cases}$$

When we have a duality on V, then the dual basis is another basis for V.

Exercise 1.2.6. Consider $V = \mathbf{R}^2$, the Euclidean plane with its standard inner product. Find the dual basis to $\{(3/2, 0), (1/4, 1/2)\}$ and draw the two bases.

Example 1.2.7 (Crystal lattices). Let $\{e_1, e_2, e_3\}$ be the standard basis for $\mathbf{R}^3$. In solid-state physics one studies crystal structures. These have the atoms arranged/packed in a regular pattern that repeats itself, a lattice, which may be different for different crystals. Mathematically a crystal lattice is described by a basis $\{v_1, v_2, v_3\}$, which is such that the atoms in the crystal are located at the lattice points

$$\{n_1 v_1 + n_2 v_2 + n_3 v_3 \; ; \; n_1, n_2, n_3 \in \mathbf{Z}\}.$$

Two commonly occurring crystal structures are the *body-centered cubic* lattice, which has basis

$$\{\tfrac{1}{2}(-e_1 + e_2 + e_3), \tfrac{1}{2}(e_1 - e_2 + e_3), \tfrac{1}{2}(e_1 + e_2 - e_3)\},$$

and the *face-centered cubic* lattice, which has basis

$$\{\tfrac{1}{2}(e_2 + e_3), \tfrac{1}{2}(e_1 + e_3), \tfrac{1}{2}(e_1 + e_2)\}.$$

Except for a factor 2, these two bases are seen to be dual bases: one speaks of *reciprocal lattices* for crystal lattices. The names of these lattices are clear if one draws the basis vectors in relation to the unit cube $\{0 \leq x_1, x_2, x_3 \leq 1\}$ and its integer translates.

Example 1.2.8 (Basis FEM functions). When solving partial differential equations numerically using the finite element method (FEM), the following problem appears. For a three-dimensional computation we consider simplices D, the closed convex hull of four points. Using one corner as the origin 0, and vectors $\{v_1, v_2, v_3\}$ along the edges to the other three corners, we wish to construct linear functions $f_k : D \to \mathbf{R}$ such that $f_k(v_k) = 1$ and $f_k = 0$ on the opposite face of D, for $k = 1, 2, 3$. Using the dual basis $\{v_1^*, v_2^*, v_3^*\}$, we immediately obtain

$$f_k(x) = \langle v_k^*, x \rangle.$$

For practical calculations in an inner product space, we prefer to use the simplest bases: the ON-bases.

Definition 1.2.9 (ON-bases). Let $\langle \cdot, \cdot \rangle$ be a duality on V. Then $\{e_i\}$ is called an *ON-basis* if $\langle e_i, e_j \rangle = 0$ when $i \neq j$ and if $\langle e_i \rangle^2 = \pm 1$ for all i.

In terms of dual bases, a basis $\{e_i\}$ is an ON-basis if and only if

$$e_i^* = \pm e_i, \quad i = 1, \ldots, n.$$

In particular, for a Euclidean space, a basis is an ON-basis if and only if it coincides with its dual basis.

Proposition 1.2.10 (Existence of ON-bases). *Consider a linear space V with a duality $\langle V, V \rangle$. Then V is an inner product space if and only if there exists an ON-basis for V.*

Proof. Clearly V is an inner product space if an ON-basis exists. Conversely, fix any basis $\{v_i\}$ for V, and define the matrix $A = (a_{i,j})$ of $\langle V, V \rangle$ in this basis by $a_{i,j} := \langle v_i, v_j \rangle$. If V is an inner product space, then A is a symmetric matrix. Using the spectral theorem, we can write $D = M^* A M$, for some invertible matrix $M = (m_{i,j})$ and diagonal matrix D with ± 1 as diagonal elements. The basis $\{e_i\}$ defined by $e_i := \sum_j v_j m_{j,i}$ is seen to be an ON-basis. $\qquad \square$

For symplectic spaces, the following is the analogue of ON-bases. Let $\langle \cdot, \cdot \rangle$ be a duality on V, with $\dim V = 2k$. Then $\{e_i\}_{i=1}^k \cup \{e_i'\}_{i=1}^k$ is called a *Darboux basis* if

$$\begin{cases} \langle e_i, e_j \rangle = 0 = \langle e_i', e_j' \rangle, & 1 \leq i, j \leq k, \\ \langle e_i, e_j' \rangle = 0 = \langle e_i', e_j \rangle, & i \neq j, 1 \leq i, j \leq k, \\ \langle e_i', e_i \rangle = 1 = -\langle e_i, e_i' \rangle, & 1 \leq i \leq k. \end{cases}$$

In terms of dual bases, a basis is clearly a Darboux basis if and only if

$$e_i^* = e_i', \quad (e_i')^* = -e_i, \quad \text{for each } i = 1, \ldots, n.$$

Exercise 1.2.11 (Existence of Darboux bases). Consider a linear space V with a duality $\langle V, V \rangle$. Adapt the proof of Proposition 1.2.10 and prove that V is a symplectic space if and only if there exists a Darboux basis for V. Hint: The spectral theorem for normal complex linear operators applies.

1.3 Inner Products and Spacetime

In this section we consider non-Euclidean inner product spaces, and in particular Minkowski spacetimes, the mathematical model for special relativity theory.

Definition 1.3.1. Let V be an inner product space.

Let n_+ be the maximal dimension of a subspace $V_+ \subset V$ such that $\langle v \rangle^2 > 0$ for all $v \in V_+ \setminus \{0\}$, and let n_- be the maximal dimension of a subspace $V_- \subset V$ such that $\langle v \rangle^2 < 0$ for all $v \in V_- \setminus \{0\}$. The *signature* of V is the integer $n_+ - n_-$.

We say that a subspace $V_1 \subset V$ is *degenerate* if there exists $0 \neq v_1 \in V_1$ such that $\langle v_1, v \rangle = 0$ for all $v \in V_1$. Otherwise, V_1 is called *nondegenerate*. If $\langle u, v \rangle = 0$ for all $u, v \in V_1$, then V_1 is called *totally degenerate*.

Note that a subspace of an inner product space is itself an inner product space if and only if the subspace is nondegenerate. Also, a subspace of an inner product space is totally degenerate if and only if all its vectors are singular, as is seen through polarization, that is, the identity

$$\langle u + v \rangle^2 - \langle u - v \rangle^2 = 4\langle u, v \rangle.$$

A nonzero singular vector spans a one-dimensional totally degenerate subspace.

Proposition 1.3.2 (Sylvester's law of inertia). *Let $\langle \cdot, \cdot \rangle$ be an inner product on an n-dimensional vector space V, and let n_+ and n_- be as in Definition 1.3.1. For every ON basis $\{e_i\}$ for V, the number of basis vectors with $\langle e_i \rangle^2 = 1$ equals n_+, and the number of basis vectors with $\langle e_i \rangle^2 = -1$ equals n_-. If n_0 denotes the maximal dimension of a totally degenerate subspace $V_0 \subset V$, then*

$$n_+ + n_- = n, \quad \min(n_+, n_-) = n_0.$$

Proof. Let V_+, V_-, and V_0 be any Euclidean, anti-Euclidean and totally degenerate subspaces, respectively. Then clearly $V_+ \cap V_- = V_+ \cap V_0 = V_- \cap V_0 = \{0\}$, and it follows that $n_+ + n_- \leq n$, $n_+ + n_0 \leq n$, and $n_- + n_0 \leq n$.

Fix an ON-basis $\{e_i\}$ for V and choose $V_\pm := \mathrm{span}\{e_i \; ; \; \langle e_i \rangle^2 = \pm 1\}$. Then $\dim V_+ + \dim V_- = n$ and $\dim V_\pm \leq n_\pm$. It follows that $n_\pm = \dim V_\pm$ and $n_+ + n_- = n$. From $n_+ + n_- = n$, it follows that $n_0 \leq \min(n - n_+, n - n_-) = \min(n_-, n_+) =: m$. To see that equality is attained, let $V_0 := \{e_{i_1} - e_{j_1}, \ldots, e_{i_m} - e_{j_m}\}$, where $\langle e_{i_k} \rangle^2 = 1$ and $\langle e_{j_k} \rangle^2 = -1$. Then V_0 is seen to be totally degenerate. □

Exercise 1.3.3. Generalize Proposition 1.3.2 to degenerate bilinear and symmetric forms $B(\cdot, \cdot)$. Let

$$\mathrm{Rad}(V) := \{v \in V \; ; \; B(v, v') = 0 \text{ for all } v' \in V\}$$

be the *radical* of V, and let $n_{00} := \dim \mathrm{Rad}(V)$. Show that $n_+ + n_- + n_{00} = n$ and $n_0 = n_{00} + \min(n_+, n_-)$.

Geometrically, the most important difference between a general inner product space and Euclidean spaces concerns orthogonal complements. For any subspace V_1 of a Euclidean space V, we always have a direct sum decomposition $V = V_1 \oplus V_1^\perp$, since $V_1 \cap V_1^\perp = \{0\}$, because there are no singular vectors. This is not always true in general inner product spaces, but we have the following general result.

Proposition 1.3.4 (Orthogonal sums). *Let V_1 be a k-dimensional subspace in an n-dimensional inner product space V. Then $\dim V_1^\perp = n - k$ and $(V_1^\perp)^\perp = V_1$, and V_1 is a nondegenerate subspace if and only if $V_1 \cap V_1^\perp = \{0\}$, or equivalently,*

$$V = V_1 \oplus V_1^\perp.$$

In particular, if V_1 is one-dimensional and is spanned by a vector v, then $V = \mathrm{span}\{v\} \oplus \mathrm{span}\{v\}^\perp$ if and only if v is a nonsingular vector.

For the remainder of this section, we study the following non-Euclidean inner product spaces.

Definition 1.3.5 (Spacetime). An inner product space $(W, \langle \cdot, \cdot \rangle)$ is said to be a *Minkowski spacetime*, or *spacetime* for short, with n space dimensions if $\dim W = 1 + n$ and the signature is $n - 1$.

We always index spacetime ON-bases as $\{e_0, e_1, \ldots, e_n\}$, where $\langle e_0 \rangle^2 = -1$.

Note that in spacetime coordinates,

$$\langle x_0 e_0 + x_1 e_1 + \cdots + x_n e_n \rangle^2 = -x_0^2 + x_1^2 + \cdots + x_n^2.$$

To describe the geometry given by such an inner product, we use the following terminology. See Figure 1.1.

- The double cone

$$W_l := \{v \in W \; ; \; \langle v \rangle^2 = 0\}$$

  consisting of all singular vectors v is referred to as the *light cone* in spacetime. Vectors $v \in W_l$ are called *light-like*. We make a choice and declare one of these two cones to be the *future light cone* W_{l+}, and the other cone W_{l-} is the *past light cone*. Thus $W_l = W_{l+} \cup W_{l-}$ and $W_{l+} \cap W_{l-} = \{0\}$.

- We denote the interior of the light cone by $W_t := \{v \in W \; ; \; \langle v \rangle^2 < 0\}$, and it contains the *time-like* vectors. Since W_t is disconnected, we write it as the disjoint union of the *future time-like vectors* W_{t+}, which is the interior of the future light cone, and the *past time-like vectors* W_{t-}, which is the interior of the past light cone. We always assume that $e_0 \in W_{t+}$, that is, that e_0 is a future-pointing time-like vector.

- We denote the exterior of the light cone by $W_s := \{v \in W \; ; \; \langle v \rangle^2 > 0\}$, and it contains the *space-like* vectors. Except when the space dimension is $n = 1$,

W_s is connected. The whole spacetime thus can be written as the disjoint union

$$W = W_{t+} \cup W_{t-} \cup W_s \cup W_{l+} \cup W_{l-},$$

except for the origin.

- The analogue of the Euclidean unit sphere is the spacetime unit hyperboloid

$$H(W) := \{v \in W \; ; \; \langle v \rangle^2 = \pm 1\}.$$

Except for space dimension $n = 1$, this hyperboloid has three connected components:

  the future time-like part $H(W_{t+}) := H(W) \cap W_{t+}$,

  the past time-like part $H(W_{t-}) := H(W) \cap W_{t-}$, and

  the space-like part $H(W_s) := H(W) \cap W_s = \{v \in W \; ; \; \langle v \rangle^2 = +1\}$.

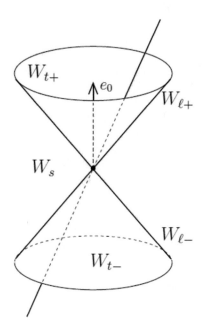

Figure 1.1: The lightcone partition of spacetime, and the straight line representing an inertial observer.

Exercise 1.3.6. Let $\{e_0, e_1, e_2\}$ be an ON-basis for a Minkowski spacetime W. Calculate the dual basis $\{v_1, v_2, v_3\} \subset W$ to $\{e_0 + e_1, e_2, e_0 - e_1\}$. If instead $\{e_0, e_1, e_2\}$ were an ON-basis for a Euclidean space V, what would this dual basis be?

A main reason for considering Minkowski spacetime is that it is the mathematical model for Einstein's special relativity theory, when $n = 3$. Fix an ON-basis $\{e_0, e_1, e_2, e_3\}$ with $\langle e_0 \rangle^2 = -1$. Once an origin is fixed, points in W are identified with vectors

$$x_0 e_0 + x_1 e_1 + x_2 e_2 + x_3 e_3.$$

The coordinates x_i are lengths, and we shall use the meter [m] as the unit of length. We shall write the time coordinate x_0 as

$$x_0 = ct,$$

where t is time measured in seconds [s] and

$$c = 299792458 \, [\text{m/s}]$$

is the exact speed of light. In relativity theory, the points in spacetime are referred to as *events*, at time t and position x. The entire life of an *observer* forms a curve $\gamma(s) \in W$, $s \in \mathbf{R}$, containing all the events that he is present at, at least if he has lived and will live forever. For each $s \in \mathbf{R}$, the tangent vector $\gamma'(s) \in W_{t+}$ will be future-pointing and time-like, since the observer always moves at a speed less than that of light. An observer moving without acceleration is called an *inertial observer*, and is described by a straight line in spacetime W spanned by a time-like vector. The quantity $\sqrt{-\langle v \rangle^2}/c$ for a time-like vector v has the meaning of time elapsed as measured by an inertial observer present at two events separated by v in spacetime. We refer to the physics literature for further details on relativity theory. See Section 1.6.

In the literature, one often models spacetime as an inner product space with signature $1 - 3$, as opposed to the signature convention $3 - 1$ used here. An advantage is that the important time-like vectors then have $\langle v \rangle^2 > 0$. A disadvantage is that in this case, spacetimes are close relatives to the anti-Euclidean space, rather than the Euclidean spaces. Of course, these differences are minor technical ones rather than real geometrical or physical ones.

A geometric result about spacetime subspaces that we need is the following.

Proposition 1.3.7. *Let W be a spacetime and let $V \subset W$ be a subspace. Then V is of exactly one of the following types.*

(i) *A space-like subspace. In this case V is nondegenerate and is a Euclidean space, whereas $V^\perp$ is a spacetime.*

(ii) *A time-like subspace. In this case V is nondegenerate and is a spacetime, whereas $V^\perp$ is a Euclidean space.*

(iii) *A light-like subspace. In this case V is a degenerate subspace and contains a unique one-dimensional subspace V_0 spanned by a light-like vector. The hyperplane $V_0^\perp$ in W is the tangent space to the light cone W_l along the line V_0 and $V_0 \subset V \subset V_0^\perp$. If V' is a complement of V_0 in V, so that $V = V_0 \oplus V'$, then V' is space-like.*

Proof. Consider first the case that V is nondegenerate, and let $n'_\pm$ be the signature indices for V as in Proposition 1.3.2. If $n_+ = n$ and $n_- = 1$ are the indices for W, then clearly $n'_- \leq n_- = 1$ and $n'_+ \leq n_+$. Thus two cases are possible. Either $n'_- = 0$, in which case V is a Euclidean space, or $n'_- = 1$, in which case V is a spacetime. Furthermore, if $n''_\pm$ are the indices for $V^\perp$, then $n'_- + n''_- = n_-$, which proves the statement about $V^\perp$.

On the other hand, if V is a degenerate subspace, write n'_{00} and n'_0 for the dimensions of the radical and a maximal totally degenerate subspace in V as in Exercise 1.3.3. Then

$$1 \leq n'_{00} \leq n'_0 \leq n_0 = \min(n_-, n_+) = 1.$$

Therefore $\min(n'_+, n'_-) = n'_0 - n'_{00} = 1 - 1 = 0$, and also $n'_- \leq n_- = 1$. We claim that $n'_- = 0$. To prove this, assume on the contrary that $n'_- = 1$. Then $n'_+ = 0$, so that $\dim V = n'_{00} + n'_+ + n'_- = 1 + 0 + 1 = 2$. Let $v_- \in V$ be a time-like vector, and consider the splitting $W = \text{span}\{v_-\} \oplus \text{span}\{v_-\}^\perp$. If $v_0 \in \text{Rad}(V) \setminus \{0\}$, then $v_0 = \alpha v_- + v_+$, which shows that V contains a space-like vector $v_+ = v_0 - \alpha v_-$ by (ii). This contradicts $n'_+ = 0$. We have proved that

$$n'_- = 0, n'_{00} = n'_0 = 1, n'_+ = \dim V - 1.$$

Write $V_0 := \text{Rad}(V)$. Then $V_0 \subset V \subset V_0^\perp$. Let $t \mapsto v(t) \in W_l$ be a curve on the light cone such that $v(0) \in V_0 \setminus \{0\}$. Then $0 = \partial_t \langle v(t), v(t) \rangle|_{t=0} = 2 \langle v'(0), v(0) \rangle$. This shows that the hyperplane $V_0^\perp$ must contain the tangent space to W_l along V_0. Since the dimensions are equal, this proves the proposition. $\square$

1.4 Linear Maps and Tensors

We denote the set of linear operators between two given linear spaces V_1 and V_2 by

$$\mathcal{L}(V_1; V_2) := \{T : V_1 \to V_2 \; ; \; T \text{ is linear}\},$$

which itself forms a linear space of dimension $\dim V_1 \times \dim V_2$. For $V_1 = V_2 = V$, we write $\mathcal{L}(V)$. The null space of a linear map T is denoted by $\mathsf{N}(T)$, and its range is denoted by $\mathsf{R}(T) = TV_1$. In this section we discuss a less well known generalization that is essential to this book: the *tensor product* of linear spaces. Just as a linear operator can be represented by its matrix, a two-dimensional rectangular scheme of numbers, general tensor products can be represented by k-dimensional schemes of numbers. However, we shall restrict ourselves to $k = 2$ and the relation between operators and tensors. The construction of tensors uses the following maps.

Definition 1.4.1 (Multilinearity). A map $M : V_1 \times \cdots \times V_k \to V$, where $V_1, \ldots, V_k$ and V are linear spaces, is called *multilinear*, or more precisely *k-linear*, if for each $1 \leq j \leq k$, the restricted map

$$V_j \ni v_j \mapsto M(v_1, \ldots, v_j, \ldots, v_k) \in V$$

is linear for every fixed $v_i \in V_i$, $i \neq j$. When $k = 2$, we use the name *bilinear*.

The construction of tensors is very similar to that of multivectors in Section 2.1, but is less geometrically transparent. Following the principle of abstract algebra, we proceed as follows to construct the tensor product $V \otimes V'$ of two given linear spaces V and V'.

- We first note that there exist a linear space V_M and a bilinear map $M : V \times V' \to V_M$ such that for two given bases $\{e_i\}_{1 \leq i \leq n}$ and $\{e'_j\}_{1 \leq j \leq n'}$ for V and V' respectively, the set

$$\{M(e_i, e'_j)\}_{1 \leq i \leq n, 1 \leq j \leq n'}$$

  forms a basis for V_M. To see this, just let V_M be any linear space of dimension nn' and define $M(e_i, e_j)$ to be some basis for V_M. Then extend M to a bilinear map.

- We next note that if $\{M(e_i, e'_j)\}_{ij}$ is a basis, then $\{M(f_i, f'_j)\}_{ij}$ is also a basis for V_M, for any other choice of bases $\{f_i\}_i$ and $\{f'_j\}_j$ for V and V' respectively. Indeed, using the bilinearity one checks that $\{M(f_i, f'_j)\}_{ij}$ is a linearly independent set in V_M.

- If $M : V \times V' \to V_M$ maps bases onto bases as above, we note the following. If $N : V \times V' \to V_N$ is any other bilinear map, then since $\{M(e_i, e'_j)\}_{ij}$ is a basis, setting

$$T(M(e_i, e'_j)) := N(e_i, e'_j), \quad 1 \leq i \leq n, 1 \leq j \leq n',$$

  we have the existence of a unique linear map $T : V_M \to V_N$ such that $N = T \circ M$. If M has the property that every other bilinear map factors through it in this way, we say that M has the *universal property* (U). We shall encounter universal properties for other constructions, so more precisely, this is the universal property for tensor products.

  Conversely, if a given bilinear map M satisfies (U), then it must map bases onto bases as above. Indeed, take any bilinear map $N : V \times V' \to V_N$ such that $\{N(e_i, e'_j)\}_{ij}$ is a basis. We now have a unique linear map $T : V_M \to V_N$ mapping $\{M(e_i, e'_j)\}_{ij}$ onto a basis. This is possible only if $\{M(e_i, e'_j)\}_{ij}$ is a basis.

Definition 1.4.2 (Tensor product). Let V and V' be linear spaces. Fix any bilinear map $M : V \times V' \to V_M$ satisfying (U). The *tensor product* of V and V' is the linear space

$$V \otimes V' := V_M.$$

We call elements in $V \otimes V'$ *tensors* and we write $u \otimes v := M(u, v)$.

Note that if some other bilinear map $N : V \times V' \to V_N$ satisfies (U), then the linear map $T : V_M \to V_N$ given by the universal property for M has inverse $T^{-1} : V_N \to V_M$ given by the universal property for N. Therefore, T provides a unique identification of V_M and V_N. By the principle of abstract algebra, our definition of $V \otimes V'$ makes sense.

If $\{e_i\}$ and $\{e'_j\}$ are bases for V and V', then a general tensor in $V \otimes V'$ is of the form

$$\sum_i \sum_j \alpha_{ij} e_i \otimes e'_j,$$

for some $\alpha_{ij} \in \mathbf{R}$.

Proposition 1.4.3 (Operator=tensor). *Let V_1 and V_2 be linear spaces and consider a duality $\langle V_1^*, V_1 \rangle$. Then there is a unique invertible linear map*

$$V_2 \otimes V_1^* \to \mathcal{L}(V_1; V_2)$$

such that $v \otimes \theta \mapsto T$, where $Tx := \langle \theta, x \rangle v$, $x \in V_1$.

Proof. Consider the bilinear map

$$V_2 \times V_1^* \to \mathcal{L}(V_1; V_2) : (v, \theta) \mapsto T,$$

where $T(x) := \langle \theta, x \rangle v$ for all $x \in V_1$. According to the universal property for $V_2 \otimes V_1^*$, there exist a unique linear map $V_2 \otimes V_1^* \to \mathcal{L}(V_1; V_2)$ such that $v \otimes \theta \mapsto T$. Let $\{e'_i\}$ be a basis for V_2, and let $\{e_j\}$ be a basis for V_1 with dual basis $\{e_i^*\}$ for V_1^*. Then we see that the tensor

$$\sum_{ij} \alpha_{ij} e'_i \otimes e_j^*$$

maps onto the linear operator with matrix $\{\alpha_{ij}\}_{ij}$. This proves the invertibility. $\square$

The following shows how this translation between tensors and linear operators works.

- If $T = v \otimes \theta : V_1 \to V_2$ and $T' = v' \otimes \theta' : V_2 \to V_3$, then the composed operator $T' \circ T : V_1 \to V_3$ corresponds to the tensor

$$(v' \otimes \theta')(v \otimes \theta) = \langle \theta', v \rangle v' \otimes \theta.$$

This yields a multiplication of tensors, which is referred to as a *contraction*.

- Let $T : V \to V$ be a linear operator on a linear space V. Applying the universal property to the pairing

$$V \times V^* \to \mathbf{R} : (v, \theta) \mapsto \langle \theta, v \rangle,$$

we get a canonical linear map

$$\text{Tr} : \mathcal{L}(V) = V \otimes V^* \to \mathbf{R}.$$

The obtained number $\text{Tr}(T) \in \mathbf{R}$ is called the *trace* of the operator T. If $\{e_i\}$ is a basis for V, then $\text{Tr}(T) = \sum_i \alpha_{ii}$ if $\{\alpha_{ij}\}$ is the matrix for T.

- If V_1 and V_2 are two linear spaces, then there is a natural swapping map

$$S : V_2 \otimes V_1^* \to V_1^* \otimes V_2 : v \otimes \theta \mapsto \theta \otimes v,$$

defined using the universal property. Identifying $V_2 \otimes V_1^* = \mathcal{L}(V_1; V_2)$ and $V_1^* \otimes V_2 = \mathcal{L}(V_2^*; V_1^*)$, this map S of tensors corresponds to the operation of taking adjoints of linear operators. Recall that the *adjoint*, or *dual*, of a linear operator $T \in \mathcal{L}(V_1; V_2)$ is $T^* \in \mathcal{L}(V_2^*; V_1^*)$ given by

$$\langle T^*\theta, v \rangle = \langle \theta, Tv \rangle, \quad \theta \in V_2^*, v \in V_1.$$

- Let V be a Euclidean space, and let $T = T^*$ be a symmetric operator. By the spectral theorem, there exists an ON-basis $\{e_i\}$ for V in which T has a diagonal matrix. Translated to tensors, this result means that if a tensor $w \in V \otimes V$ is fixed by the above swapping map S, then there is an ON-basis in which

$$w = \sum_i \alpha_i e_i \otimes e_i,$$

where we as usual identify V and V^* through the inner product.

- Let V and V' be two Euclidean spaces, and $w \in V \otimes V'$. Then there exist ON-bases $\{e_j\}$ for V and $\{e_j'\}$ for V', and $\mu_j \in \mathbf{R}$ such that

$$w = \sum_{j=1}^{n} \mu_j e_j \otimes e_j'.$$

This follows, by translation to tensors, from the spectral theorem and Proposition 1.4.4 for operators, where μ_j are the singular values of the corresponding operator.

Proposition 1.4.4 (Polar decomposition). *Let V_1, V_2 be Euclidean spaces, and consider an invertible linear map $T \in \mathcal{L}(V_1, V_2)$. Then there exists a unique symmetric map $S \in \mathcal{L}(V_1)$ such that $\langle Su, u \rangle > 0$ for all $u \in V_1 \setminus \{0\}$, and a unique isometric map $U \in \mathcal{L}(V_1, V_2)$ such that*

$$T = US.$$

Similarly, there exists a unique factorization $T = S'U'$ of T, where S' is positive symmetric on V_2 and $U' : V_1 \to V_2$ is isometric. We have $U' = U$ and $S' = USU^*$.

Proof. For such S, U we have

$$T^*T = S(U^*U)S = S^2.$$

Thus $S = (T^*T)^{1/2}$, so S and U are uniquely determined by T. To show existence, define $S := (T^*T)^{1/2}$ and $U := T(T^*T)^{-1/2}$. Then S is positive, $T = US$, and

$$\langle Ux, Uy \rangle = \langle TS^{-1}x, TS^{-1}y \rangle = \langle S^2 S^{-1}x, S^{-1}y \rangle = \langle x, y \rangle.$$

Similarly, $U' = (TT^*)^{-1/2}T = T(T^*T)^{-1/2} = U$, since $VA^{-1/2}V^{-1} = (VAV^{-1})^{-1/2}$ for every positive A and invertible V. $\qquad\square$

1.5   Complex Linear Spaces

The fundamental constructions of exterior algebras and Clifford algebras in this book can be made for linear spaces over more general fields than the real numbers **R**. We will consider only the field of complex numbers **C** besides **R**, which is particularly useful in analysis. We write the complex conjugate of $z \in \mathbf{C}$ as z^c. Given a complex matrix $A = (a_{ij})_{ij}$, its *conjugate transpose* is $A^* := (a_{ji}^c)_{ij}$, as compared to its transpose $A^t := (a_{ji})_{ij}$.

Definition 1.5.1. A complex *linear space* $(\mathcal{V}, +, \cdot)$ is an abelian group $(\mathcal{V}, +)$ together with a scalar multiplication $\mathbf{C} \times \mathcal{V} \to \mathcal{V}$ that is bilinear with respect to the addition operations and a group action of the multiplicative group $\mathbf{C}^* = \mathbf{C} \setminus \{0\}$ on $\mathcal{V}$.

By a complex vector space we shall mean simply a complex linear space, without any interpretation like that in Definition 1.1.2, since this concerns the additive structure of the vector space.

Before proceeding with the algebra, an example is in order, to show why complex linear spaces are natural and very useful in analysis.

Example 1.5.2 (Time-harmonic oscillations). Consider a quantity $f(t, x)$ that depends on time $t \in \mathbf{R}$ and position x in some space X. We assume that f takes values in some real linear space. Fixing a basis there, we can assume that $f(t, x) \in \mathbf{R}^N$. One example is the electromagnetic field in which case $N = 6$, since it consists of a three-dimensional electric field and a three-dimensional magnetic field. The most convenient way to represent f oscillating at a fixed frequency $\omega \in \mathbf{R}$ is to write

$$f(t, x) = \mathrm{Re}(F(x)e^{-i\omega t}),$$

for a function $F : X \to \mathbf{C}^N$, where the real part is taken componentwise. In this way, each component $f_k(t, x)$, $k = 1, \ldots, N$, at each point x will oscillate at frequency ω. The complex-valued function F has a very concrete meaning: the absolute value $|F_k(x)|$ is the amplitude of the oscillation of component k at the point x, and the argument $\arg F_k(x)$ is the phase of this oscillation. Note that we do not assume that the oscillations at different points have the same phase; this happens only for standing waves.

Since the complex field has two automorphisms, the identity and complex conjugation, there are two types of dualities that are natural to consider. These correspond to linear and antilinear identification of $\mathcal{V}'$ and the dual space

$$V^* = \{\theta : V \to \mathbf{C} \; ; \; \theta \text{ is complex linear}\}$$

of $\mathcal{V}$.

- A complex *bilinear duality* of two complex linear spaces $\mathcal{V}'$ and $\mathcal{V}$ is a complex bilinear map

$$\mathcal{V}' \times \mathcal{V} \to \mathbf{C} : (v', v) \mapsto \langle v', v \rangle$$

  that is nondegenerate.

    When $\mathcal{V}' = \mathcal{V}$, we refer to a bilinear duality as a complex *bilinear inner product* if it is symmetric, that is, if $\langle x, y \rangle = \langle y, x \rangle$. A main difference is that notions like signature are not present in the complex bilinear case since we can normalize $-\langle x, x \rangle = \langle ix, ix \rangle$.

- A complex *sesquilinear duality* of $\mathcal{V}'$ and $\mathcal{V}$, is a nondegenerate pairing $(\cdot, \cdot)$ such that $(v', \cdot)$ is complex linear for each $v' \in \mathcal{V}'$ and $(\cdot, v)$ is complex antilinear for each $v \in V$. Note the difference in left and right parantheses, which we use to indicate the sesquilinearity.

    When $\mathcal{V}' = \mathcal{V}$, we refer to a sesquilinear duality as a complex *inner product* if it is symmetric, that is, if $(x, y)^c = (y, x)$. A complex inner product is called *Hermitian* if it is positive definite, that is $(u, u) > 0$ for all $u \in \mathcal{V} \backslash \{0\}$. The *norm* associated with a Hermitian inner product is $|u| := \sqrt{(u, u)}$.

The existence of the following types of canonical bases can be derived from the spectral theorem for normal complex linear operators.

Proposition 1.5.3 (Complex ON-bases). *Let $\mathcal{V}$ be a complex linear space.*

(i) *A sesquilinear duality $(\cdot, \cdot)$ is symmetric if and only if there exists a basis $\{e_i\}$ that is ON in the sense that $(e_i, e_j) = 0$ when $i \neq j$ and $(e_i, e_i) = \pm 1$.*

(ii) *A bilinear duality $\langle \cdot, \cdot \rangle$ is symmetric in the sense that $\langle v_1, v_2 \rangle = \langle v_2, v_1 \rangle$ if and only if there exists a basis $\{e_i\}$ that is ON in the sense that $(e_i, e_j) = 0$ when $i \neq j$ and $(e_i, e_i) = 1$.*

Exercise 1.5.4. (i) Prove that a sesquilinear duality (x, y) is skew-symmetric, that is, $(x, y)^c = -(y, x)$, if and only if $i(x, y)$ is an inner product.

(ii) Prove that a bilinear duality $\langle \cdot, \cdot \rangle$ is skew-symmetric in the sense that $\langle v_1, v_2 \rangle = -\langle v_2, v_1 \rangle$ if and only if $\dim \mathcal{V} = 2k$ and there exists a Darboux basis, that is, a basis $\{e_i\}_{i=1}^k \cup \{e_i'\}_{i=1}^k$ in which the only nonzero pairings are $\langle e_i', e_i \rangle = 1$, $\langle e_i, e_i' \rangle = -1$, $i = 1, \ldots, k$.

We next consider the relation between real and complex linear spaces. We first consider how any complex linear space can be turned into a real linear space, and how to reverse this process.

- Let $\mathcal{V}$ be a complex linear space. Simply forgetting about the possibility of scalar multiplication by nonreal numbers, $\mathcal{V}$ becomes a real linear space, which we denote by $V = \mathcal{V}$. Note that $\dim_{\mathbf{C}} \mathcal{V} = 2 \dim_{\mathbf{R}} V$. Besides this real linear structure, V also is equipped with the real linear operator

$$J : V \to V : v \mapsto iv,$$

which has the property that $J^2 = -I$.

A complex linear map $T : \mathcal{V}_1 \to \mathcal{V}_2$ is the same as a real linear map $T : V_1 \to V_2$ between these spaces regarded as real linear spaces, for which $T J_1 = J_2 T$.

Given a complex functional $\theta \in \mathcal{V}^*$, the real linear functional

$$V \ni v \mapsto \operatorname{Re} \theta(v) \in \mathbf{R}$$

belongs to V^*. This gives a real linear one-to-one correspondence between $\mathcal{V}^*$ and V^*. In particular, if $\langle \cdot, \cdot \rangle$ is a complex inner product on $\mathcal{V}$, taking the real part of the antilinear identification $\mathcal{V} \to \mathcal{V}^*$, we obtain a real inner product

$$\langle v', v \rangle_{\mathbf{R}} := \operatorname{Re}\langle v', v \rangle$$

on V, and $\langle \cdot, \cdot \rangle_{\mathbf{R}}$ is a Euclidean inner product if and only if $\langle \cdot, \cdot \rangle$ is a Hermitian inner product. It is possible but less useful to start with a complex bilinear inner product, since this always leads to a real inner product with signature zero.

- We can reverse the above argument. Let V be a real linear space equipped with a *complex structure*, that is, a real linear operator $J : V \to V$ such that $J^2 = -I$. Then

$$(\alpha + \beta i)v := \alpha v + \beta J(v), \quad v \in V, \ \alpha, \beta \in \mathbf{R},$$

defines a complex scalar multiplication, which turns V into a complex linear space $\mathcal{V}$. If $\dim V$ is odd, then no such J exists, since we would then have $(\det J)^2 = \det(-I) = (-1)^n = -1$, which is unsolvable over $\mathbf{R}$. If $\dim V$ is even, there are infinitely many complex structures among which to choose. Indeed, if $\{e_1, \ldots, e_{2k}\}$ is any basis, then

$$J\left(\sum_{j=1}^{k} (\alpha_{2j-1} e_{2j-1} + \alpha_{2j} e_{2j}) \right) = \sum_{j=1}^{k} (-\alpha_{2j} e_{2j-1} + \alpha_{2j-1} e_{2j})$$

is one such complex structure.

If furthermore the complex structure J on V is an isometry $J^*J = I$, or equivalently skew-adjoint, then polarizing $\langle v', v \rangle_{\mathbf{R}} = \mathrm{Re}(v', v)$ recovers the sesquilinear duality

$$(v', v) = \langle v', v \rangle_{\mathbf{R}} - i\langle v', Jv \rangle_{\mathbf{R}}.$$

We next consider how any real linear space can be embedded in a complex linear space, and how to reverse this process.

- Let V be a real linear space. Define the real linear space $V \oplus V$, and consider V as a subspace of $V \oplus V$ by identifying $v \in V$ and $(v, 0) \in V \oplus V$. Define the standard complex structure

$$J(v_1, v_2) := (-v_2, v_1), \quad (v_1, v_2) \in V \oplus V.$$

Then the complex linear space $V_c := (V \oplus V, J)$ is called the *complexification* of V. The complex vector (v_1, v_2) is usually written as the formal sum $v_1 + iv_2$, so that complex scalar multiplication becomes $(\alpha + \beta i)(v_1 + iv_2) = (\alpha v_1 - \beta v_2) + i(\alpha v_2 + \beta v_1)$.

The complexification V_c of a real linear space V is a complex linear space $\mathcal{V}$, with $\dim_{\mathbf{C}} V_c = \dim_{\mathbf{R}} V$, which comes with two canonical real linear subspaces. Defining a complex conjugation operator $(x + iy)^c := x - iy$, this is a complex antilinear operation that fixes $V \subset V_c$ and squares to the identity.

A real linear map $T : V \to V'$ extends to a complex linear map $T_c : V_c \to V'_c$ by complexification: $T_c(v_1 + iv_2) := Tv_1 + iTv_2$.

The complexification $(V^*)_c$ of the real dual can in a natural way be identified with the complex dual $(V_c)^*$ of the complexification, through the complex linear invertible map given by $\langle \theta_1 + i\theta_2, v_1 + iv_2 \rangle := \langle \theta_1, v_1 \rangle - \langle \theta_2, v_2 \rangle + i(\langle \theta_1, v_2 \rangle + \langle \theta_2, v_1 \rangle)$. In particular, if $\langle \cdot, \cdot \rangle$ is a duality on V, by complexifying the linear identification $V \to V^*$, we obtain a complex bilinear inner product $\langle \cdot, \cdot \rangle_{\mathbf{C}}$ on V_c, described by $V_c \mapsto (V^*)_c = (V_c)^*$. Concretely,

$$\langle u' + iv', u + iv \rangle_{\mathbf{C}} := \langle u', u \rangle - \langle v', v \rangle + i(\langle v', u \rangle + \langle u', v \rangle).$$

Alternatively, we may equip V_c with the complex (sesquilinear) inner product

$$(u' + iv', u + iv)_{\mathbf{C}} := \langle u', u \rangle + \langle v', v \rangle + i(-\langle v', u \rangle + \langle u', v \rangle),$$

which is Hermitian if $\langle \cdot, \cdot \rangle$ is Euclidean.

We can also complexify a real associative algebra $(A, +, *, 1)$, by complexifying the linear space A as well as the bilinear product $*$, to obtain an associative algebra A_c over the complex field.

- We can reverse the above argument. Let $\mathcal{V}$ be any complex linear space equipped with a *real structure*, that is, a complex antilinear operator

$$\mathcal{V} \to \mathcal{V} : z \mapsto z^c$$

such that $(z^c)^c = z$. Then $\mathcal{V}$ is isomorphic to the complexification V_c of the *real subspace* $V := \{z \in \mathcal{V} \; ; \; z^c = z\}$ through

$$\mathcal{V} \ni z = x + iy \longleftrightarrow (x, y) = \left(\tfrac{1}{2}(z + z^c), \tfrac{1}{2i}(z - z^c)\right) \in V_c.$$

Clearly, on any complex linear space there are infinitely many real structures.

A important advantage over the real theory is that every complex linear operator has an eigenvector, by the fundamental theorem of algebra. For a normal operator, that is, if $T^*T = TT^*$ on a Hermitian space, we can iterate this result on the orthogonal complement, yielding an ON-basis of eigenvectors. If we apply these results to the complexification of a real linear operator, we obtain the following real result.

- Every real linear map $T : V \to V$ has either an eigenvector or an invariant two-dimensional subspace. More precisely, in the latter case there exist $\alpha, \beta \in \mathbf{R}$, with $\beta \neq 0$, and linearly independent vectors $v_1, v_2 \in V$ such that

$$T(v_1) = \alpha v_1 - \beta v_2, \quad T(v_2) = \beta v_1 + \alpha v_2.$$

- Let $T : V \to V$ be a real linear normal operator, that is, $T^*T = TT^*$, on a Euclidean space. Then, there exists an ON-basis in which the matrix for T is block diagonal, with 2×2 and 1×1 blocks along the diagonal. Examples include isometries and skew-symmetric maps.

1.6 Comments and References

1.1 A reference for basic algebraic structures such as groups, rings, fields, vector spaces, and algebras is Nicholson [73].

1.2 I thank Mats Aigner, Linköping University, for suggesting the notation for dualities used in this book, which incorporates *the* dual space of linear functionals as a special case.

1.3 Spacetime in the sense of Definition 1.3.5 was first constructed by Hermann Minkowski (1864–1909), for Maxwell's equations. He had Albert Einstein as a student and realized later when Einstein created his special theory of relativity that this could be modeled mathematically by a four-dimensional spacetime. A reference for the theory of relativity is Rindler [79]. The most common sign convention for spacetime in the literature is $+ - --$, that is, opposite to the sign $- + ++$ used in this book.

1.4 Tensors and tensor products appear in the work of J.W. Gibbs (1839–1903), although some specific examples of tensors such as the Cauchy stress tensor and the Riemann curvature tensor had been found earlier.

   A reference for our construction of tensor products, using the universal property, is Greub [46].

1.5 We use of the word *Hermitian* as the complex analogue of *Euclidean*, with a meaning of positivity. However, in many contexts in the literature, Hermitian refers to the conjugate-symmetry, without any implied positivity.

   The proof of Proposition 1.5.3(ii) uses a variant of the spectral theorem known as the Autonne–Takagi factorization.

   An equivalent way to define the complexification V_c of a real linear space V, which is standard but not used in this book, is as the tensor product $V_c := V \otimes \mathbf{C}$ of real linear spaces.

Chapter 2

Exterior Algebra

Prerequisites:

This chapter is where this book starts, and everything else in the book depends on it, except for Section 2.9, which is not needed elsewhere. Chapter 1 is meant to be used as a reference while reading this and later chapters. Otherwise, a solid background in linear algebra should suffice. Section 2.4 requires a small amount of analysis.

Road map:

We all know the algebra of vectors, the one-dimensional oriented/directed arrows. Here we construct and develop the algebra for bivectors, the two-dimensional oriented objects, and 3-vectors, the three-dimensional oriented objects, and so on, which live in n-dimensional affine space. In total we obtain a linear space of dimension 2^n containing all the multivectors in the space, referred to as the exterior algebra. Algebraically, multivectors are in some sense nothing but rectangular determinants, but it is important to understand the geometry to be able to use the theory. Sections 2.2 and 2.4 aim to convey the geometric meaning of multivectors to the reader.

Most applications use Euclidean space, but for a number of practical reasons, including applications to Minkowski spacetime, we allow for more general inner product spaces and dualities. The exterior product $u \wedge v$ can be seen as a higher-dimensional generalization of the vector product, but in a more fundamental way, so that it corresponds to the direct sum $[u] \oplus [v]$ of subspaces $[u]$ and $[v]$. Since $\wedge$ is noncommutative, two different but closely related dual products come into play, the right and left interior products $v \llcorner u$ and $u \lrcorner v$, which geometrically correspond to the orthogonal complement $[u]^\perp \cap [v]$ of subspace $[u]$ in a larger subspace $[v]$. When the larger space is the whole space, we have the Hodge star map, which corresponds to taking orthogonal complements of subspaces.

© Springer Nature Switzerland AG 2019

A. Rosén, *Geometric Multivector Analysis*, Birkhäuser Advanced Texts Basler Lehrbücher,
https://doi.org/10.1007/978-3-030-31411-8_2

Developing the algebra of these products of multivectors, we obtain a geometric birds-eye view on various algebraic results in linear algebra such as identities for the vector product, Cramer's rule, and the cofactor formula for inverses of linear maps, and expansion rules for determinants.

Highlights:

- Simple k-vectors $\leftrightarrow$ k-dimensional subspaces: 2.2.3
- Factorization algorithm for k-vectors: 2.2.8
- Geometry of Cramer's rule: 2.3.6
- Algebra for interior product: 2.6.3
- Geometry of cofactor formula: 2.7.1
- Anticommutation relation between exterior and interior products: 2.8.1

2.1 Multivectors

Let us fix an affine space (X, V) of dimension $1 \leq n < \infty$. The letter n will be the standard notation for the dimension of the vector space V. We set out to construct, for any $0 \leq k \leq n$, a linear space $\wedge^k V$ of k-*vectors* in X. A k-vector $w \in \wedge^k V$ is to be interpreted as an affine k-dimensional object in X determined by its orientation and k-volume. When $k = 1$, then $\wedge^1 V := V$ and 1-vectors are simply vectors in X, or oriented 1-volumes. We build k-vectors from vectors using certain multilinear maps. See Definition 1.4.1.

Lemma 2.1.1. *For a multilinear map $M : V \times \cdots \times V \to L$, the following are equivalent:*

(i) $M(v_1, \ldots, v_k) = 0$ *whenever* $\{v_1, \ldots, v_k\}$ *are linearly dependent.*

(ii) $M(v_1, \ldots, v_k) = 0$ *whenever* $v_i = v_j$ *for some* $i \neq j$.

(iii) M *is alternating, that is, for all* $1 \leq i < j \leq k$ *and vectors* $\{v_m\}$, *we have*

$$M(v_1, \ldots, v_i, \ldots, v_j, \ldots, v_k) = -M(v_1, \ldots, v_j, \ldots, v_i, \ldots, v_k).$$

Proof. That (i) implies (ii) is clear, as is (iii) implies (ii). For (ii) implies (i), recall that if $\{v_1, \ldots, v_k\}$ are linearly dependent, then $v_j = \sum_{i \neq j} x_i v_i$ for some j. Doing this substitution and expanding with multilinearity shows that all terms have two identical factors. This proves (i), using (ii). Finally, to prove (ii) implies (iii), note that

$$0 = M(v_1, \ldots, v_i + v_j, \ldots, v_i + v_j, \ldots, v_k)$$
$$= M(v_1, \ldots, v_i, \ldots, v_j, \ldots, v_k) + M(v_1, \ldots, v_j, \ldots, v_i, \ldots, v_k),$$

from which (iii) follows. $\square$

The theory of k-vectors can be thought of as a theory of rectangular determinants. Let us start with a definition of the usual concept of a (quadratic) determinant from linear algebra.

Proposition 2.1.2 (Determinant). *There exists a unique multilinear map*

$$\det : \mathbf{R}^n \times \cdots \times \mathbf{R}^n \to \mathbf{R},$$

where the number of copies of $\mathbf{R}^n$ is n, with the following properties.

(A) *If the vectors $\{v_1, \ldots, v_n\}$ are linearly dependent, then $\det(v_1, \ldots, v_n) = 0$.*

(B) *If $\{e_i\}$ is the standard basis, then $\det(e_1, \ldots, e_n) = 1$.*

Let us sketch the proof of this well-known fact. If det exists, then (A), (B), and multilinearity show that for any vectors $v_j = \sum_i \alpha_{i,j} e_i$, we must have

$$\det(v_1, \ldots, v_n) = \sum_{s_1=1}^{n} \cdots \sum_{s_n=1}^{n} \alpha_{s_1,1} \cdots \alpha_{s_n,n}\, \epsilon(s_1, \ldots, s_n), \qquad (2.1)$$

where $\epsilon(s_1, \ldots, s_n)$ is zero if an index is repeated and otherwise denote the sign of the permutation $(s_1, \ldots, s_n) \mapsto (1, \ldots, n)$. Hence uniqueness is clear. Note now that if such det exists, then necessarily it must satisfy (2.1). Thus all that remains is to take (2.1) as the definition and verify properties (A) and (B). Note carefully this frequently useful technique to prove existence, using inspiration from a uniqueness proof.

If $v_j = \sum_i \alpha_{i,j} e_i$ and $A = (\alpha_{i,j})$, then we use the standard notation

$$\det(v_1, \ldots, v_n) = \det(A) = \begin{vmatrix} \alpha_{1,1} & \cdots & \alpha_{1,n} \\ \vdots & \ddots & \vdots \\ \alpha_{n,1} & \cdots & \alpha_{n,n} \end{vmatrix}.$$

We now generalize this construction to fewer than n vectors, replacing the range $\mathbf{R}$ by a more general linear space L.

Proposition 2.1.3. *Let $2 \le k \le n$ and let $\{e_1, \ldots, e_n\}$ be a basis for V. Then there exist a linear space L and a multilinear map*

$$\wedge^k : V \times \cdots \times V \to L,$$

where the number of copies of V is k, that satisfy the following properties.

(A) *If the $\{v_1, \ldots, v_k\}$ are linearly dependent, then $\wedge^k(v_1, \ldots, v_k) = 0$.*

(B) *The set $\{\wedge^k(e_{s_1}, \ldots, e_{s_k})\}_{s_1 < \cdots < s_k}$ is a basis for L.*

Proof. Fix a basis $\{e_1, \ldots, e_n\}$ for V and write vectors $v_j = \sum_{s_j=1}^{n} \alpha_{s_j,j} e_{s_j}$, $j = 1, \ldots, k$, in this basis. If $\wedge^k$ is a multilinear and alternating map, then we must have

$$\wedge^k(v_1, \ldots, v_k) = \sum_{i_1=1}^{n} \cdots \sum_{i_k=1}^{n} \alpha_{i_1,1} \cdots \alpha_{i_k,k} \wedge^k (e_{i_1}, \ldots, e_{i_k})$$

$$= \sum_{s_1 < \cdots < s_k} \begin{vmatrix} \alpha_{s_1,1} & \cdots & \alpha_{s_1,k} \\ \vdots & \ddots & \vdots \\ \alpha_{s_k,1} & \cdots & \alpha_{s_k,k} \end{vmatrix} \wedge^k (e_{s_1}, \ldots, e_{s_k}), \qquad (2.2)$$

where the set $\{i_1, \ldots, i_k\}$ has been discarded unless all elements are distinct, and in that case reordered to $\{s_1, \ldots, s_k\}$. Thus $\wedge^k$ is uniquely determined by the $\binom{n}{k}$ values $\wedge^k(e_{s_1}, \ldots, e_{s_k})$, $s_1 < \cdots < s_k$.

As in the proof of Proposition 2.1.2 above, we can now turn this into an existence proof. Note that the range space L is not given, but that we are free to choose/construct this. We let L be the linear space $\mathbf{R}^d$ of dimension $d = \binom{n}{k}$ and write the standard basis as $\{e_s\}_{|s|=k}$, indexed by all subsets s of $\{1, \ldots, n\}$ of size $|s| = k$. (Just fix any ordering of the subsets s.) Then define

$$\wedge^k(e_{s_1}, \ldots, e_{s_k}) := e_s, \quad s = \{s_1, \ldots, s_k\}, \ s_1 < \cdots < s_k,$$

and take (2.2) as a definition. This construction yields a well-defined multilinear map $\wedge^k$ that clearly satisfies (A) and (B). $\qquad\square$

The basis property (B) is easy to understand. We next show that it is equivalent to a property (U), referred to as a *universal property*, which is surprisingly useful.

Proposition 2.1.4 (Universal property). *Let $\wedge^k : V \times \cdots \times V \to L$ be a multilinear map satisfying* (A). *Then the following are equivalent.*

(B+) *For every basis $\{e_1, \ldots, e_n\}$ in V, the set $\{\wedge^k(e_{s_1}, \ldots, e_{s_k})\}_{s_1 < \cdots < s_k}$ is a basis for L.*

(B) *There is a basis $\{e_1, \ldots, e_n\}$ in V, so that the set $\{\wedge^k(e_{s_1}, \ldots, e_{s_k})\}_{s_1 < \cdots < s_k}$ is a basis for L.*

(U) *Whenever $\wedge_1^k : V \times \cdots \times V \to L_1$ is another multilinear map satisfying* (A), *there exists a unique linear map $T : L \to L_1$ such that $T\wedge^k = \wedge_1^k$.*

Proof. Clearly (B+) implies (B). To verify (U), given (B), consider the given basis $\{e_1, \ldots, e_n\}$ and define $T : L \to L_1$ to be the unique linear map such that

$$Te_s = \wedge_1^k(e_{s_1}, \ldots, e_{s_k}), \quad s = \{s_1, \ldots, s_k\}, \quad s_1 < \cdots < s_k.$$

By multilinearity and (A), it follows that $T \wedge^k (v_1, \ldots, v_k) = \wedge_1^k(v_1, \ldots, v_k)$ for any vectors.

Assume finally that (U) holds for $\wedge^k$. Fix any basis $\{e_1, \ldots, e_n\}$ for V and let $\wedge_1^k : V \times \cdots \times V \to L_1$ be the multilinear map constructed in Proposition 2.1.3, based on the basis $\{e_i\}$. This $\wedge_1^k$ satisfies (A) and (B). By assumption, there exists a linear map $T : L \to L_1$, which maps $\{\wedge^k(e_{s_1}, \ldots, e_{s_k})\}_{s_1 < \cdots < s_k}$ onto a basis in L_1. Since $\dim L_1 = \binom{n}{k}$, this proves (B+) for $\wedge^k$. $\qquad\square$

Proposition 2.1.5. *Let* $\wedge_i^k : V \times \cdots \times V \to L_i$, $i = 1, 2$, *be any multilinear maps with properties (A) and (U). Then there exists a unique invertible linear map* $T : L_1 \to L_2$ *such that* $T\wedge_1^k = \wedge_2^k$.

Proof. From property (U), we get the existence of unique maps T_i such that $T_1\wedge_1^k = \wedge_2^k$ and $T_2\wedge_2^k = \wedge_1^k$. In particular, $T_2T_1\wedge_1^k = \wedge_1^k$, and the uniqueness statement in (U) for $\wedge_1^k$ shows that $T_2T_1 = I$. Similarly, $T_1T_2 = I$, so $T := T_1$ is invertible. $\qquad\square$

Proposition 2.1.5 shows that the only difference between different maps $\wedge^k : V \times \cdots \times V \to L$ as in Proposition 2.1.3 is the names of the objects in L. It is important here to note that the map T identifying objects in two different spaces L is unique. With the philosophy of abstract algebra that only *how* we calculate is important, not with *what*, the following definition makes sense.

Definition 2.1.6 (*k*-vectors). Let $\wedge^1 V := V$ and $\wedge^0 V := \mathbf{R}$. For each $2 \leq k \leq n$, fix a multilinear map $\wedge^k : V^k \to L$ with properties (A) and (U) and call L the *k*th *exterior power* $\wedge^k V$ of V. Call elements in this linear space *k-vectors* and write

$$v_1 \wedge \cdots \wedge v_k := \wedge^k(v_1, \ldots, v_k) \in \wedge^k V, \quad v_1, \ldots, v_k \in V,$$

for the *k*-vector determined by the vectors $v_1, \ldots, v_k$. We refer to 0-vectors, 1-vectors, and 2-vectors as *scalars*, *vectors*, and *bivectors* respectively. When $k \notin \{0, 1, \ldots, n\}$, we let $\wedge^k V := \{0\}$.

Lemma 2.1.7. *Assume that* $\wedge^k V$ *is defined by a map* $M : V \times \cdots \times V \to L$. *Let* V_1 *be a subspace of the vector space* V, *and assume that* $\wedge^k V_1$ *is defined by a map* $M_1 : V_1 \times \cdots \times V_1 \to L_1$. *Then* $\wedge^k V_1 = L_1$ *is, in a canonical way, isomorphic to* $\mathrm{span}(M(V_1 \times \cdots \times V_1)) \subset L$.

This means that we can, and will, view $\wedge^k V_1$ as a subspace of $\wedge^k V$, whenever $V_1 \subset V$.

Proof. It suffices to show that $\{M(e_{s_1}, \ldots, e_{s_k})\}_{1 \leq s_1 < \cdots < s_k \leq m}$ is a basis for $\mathrm{span}(M(V_1 \times \cdots \times V_1))$ whenever $\{e_1, \ldots, e_m\}$ is a basis for V_1. The set clearly spans this space and is linearly independent, since M satisfies (U). Property (U) for $\wedge_1^k$ now yields a unique invertible map $L_1 \to \mathrm{span}(M(V_1 \times \cdots \times V_1))$. $\qquad\square$

Definition 2.1.8. If $s \subset \overline{n} := \{1, 2, \ldots, n\}$, write $|s|$ for the number of elements in s. If $\{e_1, \ldots, e_n\}$ is a basis for V, then $\{e_s\}_{|s|=k}$ is called the *induced basis* for $\wedge^k V$, where $e_s := e_{s_1} \wedge \cdots \wedge e_{s_k}$ if $s = \{s_1, \ldots, s_k\}$ with $s_1 < \cdots < s_k$. A useful scheme

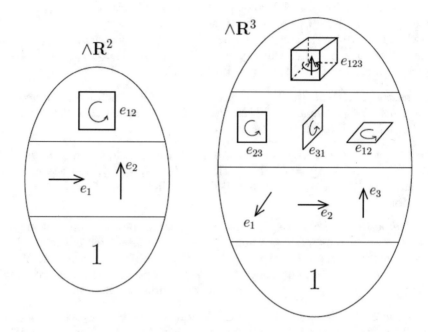

Figure 2.1: Basis multivectors in two and three dimensions.

for calculating the *exterior product* $v_1 \wedge \cdots \wedge v_k$ of vectors $v_j = \sum_i \alpha_{i,j} e_i$ is the *rectangular determinant*. This is calculated as

$$
v_1 \wedge \cdots \wedge v_k =
\begin{vmatrix}
e_1 & \alpha_{1,1} & \cdots & \alpha_{1,k} \\
e_2 & \alpha_{2,1} & \cdots & \alpha_{2,k} \\
\vdots & \vdots & \ddots & \vdots \\
e_n & \alpha_{n,1} & \cdots & \alpha_{n,k}
\end{vmatrix}
:=
\sum_{s_1 < \cdots < s_k}
\begin{vmatrix}
\alpha_{s_1,1} & \cdots & \alpha_{s_1,k} \\
\vdots & \ddots & \vdots \\
\alpha_{s_k,1} & \cdots & \alpha_{s_k,k}
\end{vmatrix}
e_{s_1} \wedge \cdots \wedge e_{s_k},
$$

where the $\binom{n}{k}$ determinants on the right-hand side are the usual quadratic ones.

Note that the rectangular determinant is a definition: it is just a shorthand notation for the k-vector on the right-hand side, with coordinates the usual quadratic determinants. Thus the rectangular determinant cannot, for example, be expanded along some column or row as usual, not even if $k = n - 1$. Vectors can be written as rows rather than columns in a rectangular determinant, just like usual determinants, if this is more convenient.

Note also that $s = \{s_1, \ldots, s_k\}$ is considered an unordered set as compared to the ordered k-tuple $(s_1, \ldots, s_k)$, so that a different order of elements in s gives the same basis k-vector e_s. For example,

$$
e_{\{1,2\}} = e_{\{2,1\}} = e_1 \wedge e_2 = -e_2 \wedge e_1.
$$

We also use the shorthand notation $e_{s_1,\ldots,s_k} := e_{s_1} \wedge \cdots \wedge e_{s_k}$ (sometimes omitting commas when $n \leq 9$) for any ordered index set $(s_1, \ldots, s_k)$ of distinct elements. For example,

$$e_{12} = -e_{21} = e_1 \wedge e_2 = -e_2 \wedge e_1.$$

More generally, for an ordered index set $(s_1, \ldots, s_k)$ of nondistinct elements, we will write $e_{s_1,\ldots,s_k} := e_{s_1} \vartriangle \cdots \vartriangle e_{s_k}$, where $\vartriangle$ denotes a Clifford product as in Section 3.1.

The k-vector map $\wedge^k$ is a *measure of linear independence* in the following sense.

Proposition 2.1.9. *Vectors $v_1, \ldots, v_k$ are linearly dependent if and only if the k-vector $v_1 \wedge \cdots \wedge v_k$ is the zero vector.*

Proof. If they are linearly dependent, then property (A) shows that $v_1 \wedge \cdots \wedge v_k = 0$. On the other hand, if they are linearly independent, then there exist vectors $v_{k+1}, \ldots, v_n$, such that $\{v_1, \ldots, v_n\}$ is a basis for V. Therefore $v_1 \wedge \cdots \wedge v_k$ is an element of a basis and hence nonzero. $\square$

Example 2.1.10. If $\dim V = 3$ and $k = 2$, then the exterior product is closely related to the vector product. Indeed,

$$(\alpha_1 e_1 + \alpha_2 e_2 + \alpha_3 e_3) \wedge (\beta_1 e_1 + \beta_2 e_2 + \beta_3 e_3)$$

$$= \begin{vmatrix} e_1 & \alpha_1 & \beta_1 \\ e_2 & \alpha_2 & \beta_2 \\ e_3 & \alpha_3 & \beta_3 \end{vmatrix}$$

$$= \begin{vmatrix} \alpha_1 & \beta_1 \\ \alpha_2 & \beta_2 \end{vmatrix} e_{12} + \begin{vmatrix} \alpha_1 & \beta_1 \\ \alpha_3 & \beta_3 \end{vmatrix} e_{13} + \begin{vmatrix} \alpha_2 & \beta_2 \\ \alpha_3 & \beta_3 \end{vmatrix} e_{23}$$

$$= (\alpha_1\beta_2 - \alpha_2\beta_1)e_{12} + (\alpha_1\beta_3 - \alpha_3\beta_1)e_{13} + (\alpha_2\beta_3 - \alpha_3\beta_2)e_{23} \in \wedge^2 V.$$

The exact relation between this bivector and the vector product, which is a vector, will be studied in Section 2.6.

Example 2.1.11. Rectangular determinants generalize usual (quadratic) determinants in the following way. Let $k = n = \dim V$ and consider $\wedge^n V$, which is 1-dimensional and spanned by $e_{\overline{n}} = e_{1,2,\ldots,n}$. Then

$$\begin{vmatrix} e_1 & \alpha_{1,1} & \cdots & \alpha_{1,n} \\ e_2 & \alpha_{2,1} & \cdots & \alpha_{2,n} \\ \vdots & \vdots & \ddots & \vdots \\ e_n & \alpha_{n,1} & \cdots & \alpha_{n,n} \end{vmatrix} = \begin{vmatrix} \alpha_{1,1} & \cdots & \alpha_{1,n} \\ \alpha_{2,1} & \cdots & \alpha_{2,n} \\ \vdots & \ddots & \vdots \\ \alpha_{n,1} & \cdots & \alpha_{n,n} \end{vmatrix} e_{\overline{n}}.$$

Instead of having $n+1$ different linear spaces $\wedge^k V$ associated to X, it is convenient to collect these as an extrinsic direct sum, the (full) *exterior algebra*

$$\wedge^0 V \oplus \wedge^1 V \oplus \wedge^2 V \oplus \cdots \oplus \wedge^n V.$$

We show below that this is not only a linear space, but an associative algebra with the exterior product $* = \wedge$ and identity element $1 \in \wedge^0 V$. See Definition 1.1.4.

To define the exterior product between any two objects in $\wedge V$, we need the following bilinear version of the universal property.

Lemma 2.1.12 (Bilinear universal property). *Let V_1 and V_2 be vector spaces and let*

$$M : (V_1 \times \cdots \times V_1) \times (V_2 \times \cdots \times V_2) \to L$$

be a multilinear map with range in some linear space L. Assume that $M(u_1, \ldots, u_k, v_1, \ldots, v_l) = 0$ whenever $\{u_1, \ldots, u_k\}$ or $\{v_1, \ldots, v_l\}$ are linearly dependent. Then there exists a unique bilinear map $B : \wedge^k V_1 \times \wedge^l V_2 \to L$ such that

$$B(u_1 \wedge \cdots \wedge u_k, v_1 \wedge \cdots \wedge v_l) = M(u_1, \ldots, u_k, v_1, \ldots, v_l)$$

for all vectors.

Proof. Fix a basis $\{e_i\}$ for V_1 and a basis $\{e'_j\}$ for V_2. If B exists, it must verify $B(e_{s_1} \wedge \cdots \wedge e_{s_k}, e_{t_1} \wedge \cdots \wedge e_{t_l}) = M(e_{s_1}, \ldots, e_{s_k}, e_{t_1}, \ldots, e_{t_l})$ for all $1 \le s_1 < \cdots < s_k \le n$ and $1 \le t_1 < \cdots < t_l \le m$, where $n = \dim V_1$ and $m = \dim V_2$. This uniquely determines B, since these k-vectors and l-vectors form bases for $\wedge^k V_1$ and $\wedge^l V_2$ respectively. Using the partial alternating property for M, this can be turned into an existence proof in the obvious way. $\qquad\square$

Definition 2.1.13 (Multivectors). Let the *exterior algebra* of V be

$$\wedge V := \wedge^0 V \oplus \wedge^1 V \oplus \wedge^2 V \oplus \cdots \oplus \wedge^n V.$$

An element $w = w_0 + w_1 + w_2 + \cdots + w_n \in \wedge V$ is called a *multivector*, and the j-vector w_j is called the *j-vector part* of w. If $w \in \wedge^k V$ for some k, then w is called a *homogeneous* multivector, of *degree* k. If w is not homogeneous, that is, if $w \notin \bigcup_{k=0}^n \wedge^k V$, then w is called an *inhomogeneous* multivector. Define the spaces of *even* and *odd* multivectors

$$\wedge^{\mathrm{ev}} V := \wedge^0 V \oplus \wedge^2 V \oplus \wedge^4 V \oplus \cdots, \quad \wedge^{\mathrm{od}} V := \wedge^1 V \oplus \wedge^3 V \oplus \wedge^5 V \oplus \cdots,$$

so that $\wedge V = \wedge^{\mathrm{ev}} V \oplus \wedge^{\mathrm{od}} V$.

The *exterior product* of two multivectors is defined as

$$w \wedge w' = \left(\sum_{k=0}^n w_k \right) \wedge \left(\sum_{l=0}^n w'_l \right) := \sum \sum_{k+l \le n} B_{k,l}(w_k, w'_l),$$

where the products $B_{0,l}(w_0, w'_l) := w_0 w'_l$ and $B_{k,0}(w_k, w'_0) := w'_0 w_k$ are defined to be multiplication by scalars, and $B_{k,l} : \wedge^k V \times \wedge^l V \to \wedge^{k+l} V$ is the unique bilinear map such that

$$B_{k,l}(u_1 \wedge \cdots \wedge u_k, v_1 \wedge \cdots \wedge v_l) = u_1 \wedge \cdots \wedge u_k \wedge v_1 \wedge \cdots \wedge v_l,$$

when $k, l \ge 1$.

Proposition 2.1.14 (Exterior algebra). *The exterior algebra* $(\wedge V, +, \wedge, 1)$, *equipped with the exterior product*

$$\wedge V \times \wedge V \to \wedge V : (w, w') \mapsto w \wedge w',$$

is an associative algebra of dimension 2^n. *Indeed, the exterior product is bilinear, associative, and satisfies the commutation relation*

$$w \wedge w' = (-1)^{kl} w' \wedge w, \quad w \in \wedge^k V, \ w' \in \wedge^l V.$$

If $w \in \wedge^k V$ *and* $w' \in \wedge^l V$, *then* $w \wedge w' \in \wedge^{k+l} V$. *In particular,* $(\wedge^{\mathrm{ev}}, \wedge)$ *is a commutative subalgebra.*

Proof. We have

$$\dim(\wedge V) = 1 + n + \binom{n}{2} + \binom{n}{3} + \cdots + \binom{n}{n-1} + \binom{n}{n} = 2^n.$$

Bilinearity of the exterior product is clear from the construction. To verify $(w \wedge w') \wedge w'' = w \wedge (w' \wedge w'')$, by bilinearity it suffices to consider exterior products of vectors. By definition of the exterior product, we have

$$\begin{aligned}
&\big((v_1 \wedge \cdots \wedge v_k) \wedge (v_1' \wedge \cdots \wedge v_l')\big) \wedge (v_1'' \wedge \cdots \wedge v_m'') \\
&= (v_1 \wedge \cdots \wedge v_k \wedge v_1' \wedge \cdots \wedge v_l') \wedge (v_1'' \wedge \cdots \wedge v_m'') \\
&= v_1 \wedge \cdots \wedge v_k \wedge v_1' \wedge \cdots \wedge v_l' \wedge v_1'' \wedge \cdots \wedge v_m'' \\
&= (v_1 \wedge \cdots \wedge v_k) \wedge (v_1' \wedge \cdots \wedge v_l' \wedge v_1'' \wedge \cdots \wedge v_m'') \\
&= (v_1 \wedge \cdots \wedge v_k) \wedge \big((v_1' \wedge \cdots \wedge v_l') \wedge (v_1'' \wedge \cdots \wedge v_m'')\big).
\end{aligned}$$

To prove the commutation relation, note that for vectors $v, v' \in V$, Lemma 2.1.1 shows that $v \wedge v' = -v' \wedge v$. The general case follows from iterated use of this, bilinearity, and associativity. $\qquad\square$

Exercise 2.1.15. Show that the only invertible homogeneous multivectors in the exterior algebra are the nonzero scalars $\wedge^0 V \setminus \{0\}$. Give examples of inhomogeneous multivectors that are invertible in $\wedge V$.

We next consider the calculation of exterior products in a basis $\{e_s\}$ for $\wedge V$ induced by $\{e_1, \ldots, e_n\}$. From the alternating property of the exterior product, we see that the product of two basis multivectors e_s and e_t is

$$e_s \wedge e_t = e_{s_1} \wedge \cdots \wedge e_{s_k} \wedge e_{t_1} \wedge \cdots \wedge e_{t_l} = \begin{cases} \epsilon(s,t)\, e_{s \cup t}, & s \cap t = \varnothing, \\ 0, & s \cap t \neq \varnothing, \end{cases}$$

where $s_1 < \cdots < s_k$, $t_1 < \cdots < t_l$ and the sign $\epsilon(s,t)$ is the following.

Definition 2.1.16 (Permutation sign). For sets $s, t \subset \overline{n} := \{1, 2, \ldots, n\}$, let

$$\epsilon(s, t) := (-1)^{|\{(s_i, t_j) \in s \times t \,;\, s_i > t_j\}|}$$

be the sign of the permutation that rearranges $s \cup t$ in increasing order.

Note that this definition is made for all index sets, not only disjoint ones. The strict inequality $s_i > t_j$ will be the appropriate choice later for Clifford products. By the above definition and bilinearity, we have

$$\left(\sum_{s \subset \overline{n}} x_s e_s \right) \wedge \left(\sum_{t \subset \overline{n}} y_t e_t \right) = \sum_{s \cap t = \varnothing} x_s y_t \, \epsilon(s, t) e_{s \cup t}.$$

In linear algebra it is often told that vectors cannot be multiplied by each other. (The standard inner product and the vector product do not qualify here as "products", since they are not associative algebra products.) The problem is that only the space V of 1-dimensional objects in X is considered. By enlarging the vector space to the complete space $\wedge V$ of all k-dimensional objects in X, $k = 0, 1, \ldots, n$, we get access to the exterior product, which is an associative product.

Example 2.1.17. A concrete example of an exterior product in the two-dimensional plane is

$$(4 - 7e_1 + 3e_2 + 2e_{12}) \wedge (1 + 5e_1 - 6e_2 - 8e_{12})$$
$$= (4 + 20e_1 - 24e_2 - 32e_{12}) + (-7e_1 + 0 + 42e_{12} + 0)$$
$$+ (3e_2 + 15e_{21} + 0 + 0) + (2e_{12} + 0 + 0 + 0)$$
$$= 4 + 13e_1 - 21e_2 - 3e_{12}.$$

Note that $15e_{21} = -15e_{12}$. A concrete example of an exterior product in three-dimensional space is

$$(3 - e_1 - e_2 + 2e_3 + e_{12} + 5e_{13} - 2e_{23} - 7e_{123})$$
$$\wedge (-2 + 4e_1 - 3e_2 + e_3 - 6e_{12} + 3e_{13} - e_{23} + 2e_{123})$$
$$= (-6 + 12e_1 - 9e_2 + 3e_3 - 18e_{12} + 9e_{13} - 3e_{23} + 6e_{123})$$
$$+ (2e_1 + 0 + 3e_{12} - e_{13} + 0 + 0 + e_{123} + 0)$$
$$+ (2e_2 - 4e_{21} + 0 - e_{23} + 0 - 3e_{213} + 0 + 0)$$
$$+ (-4e_3 + 8e_{31} - 6e_{32} + 0 - 12e_{312} + 0 + 0 + 0)$$
$$+ (-2e_{12} + 0 + 0 + e_{123} + 0 + 0 + 0 + 0)$$
$$+ (-10e_{13} + 0 - 15e_{132} + 0 + 0 + 0 + 0 + 0)$$
$$+ (4e_{23} - 8e_{231} + 0 + 0 + 0 + 0 + 0 + 0) + (14e_{123} + 0 + 0 + 0 + 0 + 0 + 0 + 0)$$
$$= -6 + 14e_1 - 7e_2 - e_3 - 13e_{12} - 10e_{13} + 6e_{23} + 20e_{123}.$$

Note, for example, that the e_{123} coordinate is $6 + 1 + 3 - 12 + 1 + 15 - 8 + 14 = 20$ by checking the permutation signs.

Let us also check the three-dimensional calculation by computing the bivector part of the exterior product as

$$3(-6e_{12} + 3e_{13} - e_{23}) + (-2)(e_{12} + 5e_{13} - 2e_{23}) + \begin{vmatrix} e_1 & -1 & 4 \\ e_2 & -1 & -3 \\ e_3 & 2 & 1 \end{vmatrix},$$

again giving $-13e_{12} - 10e_{13} + 6e_{23}$.

For calculations in the large inhomogeneous noncommutative exterior algebra $\wedge V$, the following operations turn out to be convenient.

Definition 2.1.18. The *involution* $\widehat{w}$ of a multivector $w \in \wedge V$ is given by

$$\widehat{w} = \sum_j \widehat{w_j} := \sum_j (-1)^j w_j = w_0 - w_1 + w_2 - w_3 + w_4 - w_5 + \cdots, \quad w_j \in \wedge^j V.$$

The *reversion* $\overline{w}$ of a multivector $w \in \wedge V$ is given by

$$\overline{w} = \sum_j \overline{w_j} := \sum_j (-1)^{j(j-1)/2} w_j = w_0 + w_1 - w_2 - w_3 + w_4 + w_5 - \cdots, \quad w_j \in \wedge^j V.$$

The main motivations for these operations are as follows.

Exercise 2.1.19. Let $w = v_1 \wedge \cdots \wedge v_k$, for vectors v_j. Show that $\widehat{w}$ is the result of negating the vectors:

$$\widehat{w} = (-v_1) \wedge \cdots \wedge (-v_k).$$

Show that $\overline{w}$ is the result of reversing the order of the vectors:

$$\overline{w} = v_k \wedge \cdots \wedge v_1.$$

Show also that the involution is an algebra automorphism, i.e., $\widehat{w_1 \wedge w_2} = \widehat{w_1} \wedge \widehat{w_2}$, and that the reversion is an algebra antiautomorphism, i.e., $\overline{w_1 \wedge w_2} = \overline{w_2} \wedge \overline{w_1}$. The involution is helpful in commuting through a vector $v \in V$: show that $v \wedge w = \widehat{w} \wedge v$.

Exercise 2.1.20 (Cartan's lemma). Let $v_1, \ldots, v_k$ be linearly independent vectors. Show that vectors $u_1, \ldots, u_k$ satisfy

$$u_1 \wedge v_1 + \cdots + u_k \wedge v_k = 0$$

if and only if there exists a symmetric matrix $(a_{ij})_{i,j=1}^k$ such that $u_i = \sum_{j=1}^k a_{ij} v_j$.

2.2   The Grassmann Cone

This section concerns the geometric interpretation of k-vectors. The main result is the fundamental correspondence between simple k-vectors and k-dimensional subspaces in V described in Proposition 2.2.3. Namely, the k-vectors in a one-dimensional subspace in the range of $\wedge^k$ correspond to one k-dimensional subspace in V. The two half-lines separated by 0 correspond to the two different orientations of the subspace. Just as a vector is interpreted as a one-dimensional object in the affine space X with length and orientation characterizing it, a k-vector is interpreted as a k-dimensional object in X with its k-volume and orientation characterizing it. Unfortunately, as we shall see, in dimension four or greater, there exist composite k-vectors that do not have a straightforward geometric interpretation. However, these composite k-vectors will be used in Section 2.4 to represent the "oriented measure" of curved k-surfaces.

Consider the alternating multilinear k-vector map

$$\wedge^k : V \times \cdots \times V \to \wedge^k V,$$

for $2 \le k \le n$. In contrast to linear maps, the null space is not in general a linear subspace. In fact, the domain of definition, $V \times \cdots \times V$, is not even considered a linear space, so the notion of linear subspace is not well defined. Rather, Proposition 2.1.9 characterizes the null space as the set of k-sets of linearly dependent vectors.

As for the range, this is a subset of the linear space $\wedge^k V$. However, since the map $\wedge^k$ is not linear, we shall see that in general, the range is not a linear subspace either.

Definition 2.2.1 (Grassmann cone). Let $1 \le k \le n = \dim V$. Denote the range of $\wedge^k$ by

$$\widehat{\wedge}^k V := \{ w \in \wedge^k V \ ; \ w = v_1 \wedge \cdots \wedge v_k \text{ for some vectors } v_1, \ldots, v_k \in V \}.$$

The, in general nonlinear, subset $\widehat{\wedge}^k V \subset \wedge^k V$ is called the *Grassmann cone*. A k-vector $w \in \wedge^k V$ is called *simple* if $w \in \widehat{\wedge}^k V$. Otherwise, it is called *composite*. Let $\widehat{\wedge}^0 V := \wedge^0 V = \mathbf{R}$.

We stress that this notion of simple k-vectors is different from, and more nontrivial than, that of homogeneous multivectors from Definition 2.1.13. Being homogeneous is a necessary condition for a multivector to be simple, but not a sufficient condition.

The range $\widehat{\wedge}^k V$ is not in general closed under addition. However, it is closed under multiplication by scalars, since, for example,

$$\lambda(v_1 \wedge \cdots \wedge v_k) = (\lambda v_1) \wedge v_2 \wedge \cdots \wedge v_k.$$

This motivates calling $\widehat{\wedge}^k V$ a *cone*. In fact, we shall see in Section 2.9 that $\widehat{\wedge}^k V$ is described by a certain set of homogeneous quadratic equations. We have

$$\widehat{\wedge}^k V = \wedge^k V$$

when $k = 0$, $k = 1$, and $k = n$, since $\dim \wedge^n V = 1$, but also when $k = n - 1$. See Proposition 2.2.11.

Exercise 2.2.2. Prove that $w \wedge w = 0$ if $w \in \wedge^k V$ is simple. Let $\{e_1, e_2, e_3, e_4\}$ be a basis for a four-dimensional vector space V. Show that the bivector $e_1 \wedge e_2 + e_3 \wedge e_4$ is composite.

Proposition 2.2.3 (Geometry of simple k-vectors). *The k-vector map*

$$\wedge^k : (v_1, \dots, v_k) \mapsto v_1 \wedge \cdots \wedge v_k \in \widehat{\wedge}^k V$$

induces a one-to-one correspondence between k-dimensional subspaces $W \subset V$ and one-dimensional subspaces/lines through 0 in $\widehat{\wedge}^k V$. If $\{v_1, \dots, v_k\}$ is a basis for W, then the correspondence is

$$W = \operatorname{span}\{v_1, \dots, v_k\} \leftrightarrow \operatorname{span}\{v_1 \wedge \cdots \wedge v_k\}.$$

Proof. First assume that $\{v_1, \dots, v_k\}$ and $\{v_1', \dots, v_k'\}$ both are bases for W. Then we may write $v_j = \sum_i a_{ij} v_i'$, and the multilinearity and alternating property of $\wedge^k$ show that

$$v_1 \wedge \cdots \wedge v_k = \det(a_{ij}) v_1' \wedge \cdots \wedge v_k'.$$

Hence the line spanned by $v_1 \wedge \cdots \wedge v_k$ in $\widehat{\wedge}^k V$ depends only on W, not the exact choice of basis for W.

Conversely, assume that $\{v_1'', \dots, v_k''\}$ is a basis for some k-dimensional subspace W'' and that $v_1'' \wedge \cdots \wedge v_k''$ is parallel to $v_1 \wedge \cdots \wedge v_k$. Then

$$v_j'' \wedge v_1 \wedge \cdots \wedge v_k = \lambda v_j'' \wedge v_1'' \wedge \cdots \wedge v_k'' = 0,$$

so $v_j'' \in W$ by Proposition 2.1.9, for all j. It follows that $W'' = W$. $\qquad\square$

Given a simple nonzero k-vector

$$w = v_1 \wedge \cdots \wedge v_k,$$

we write

$$[w] := W = \operatorname{span}\{v_1, \dots, v_k\}$$

for the corresponding k-dimensional subspace of V as in Proposition 2.2.3, and we call $[w]$ simply the *space* of w. It is not clear what the space of a composite k-vector should be. The following two generalizations turn out to be useful.

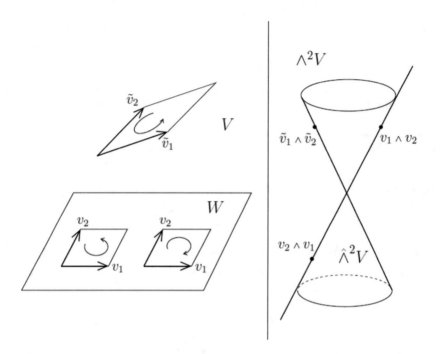

Figure 2.2: Left: Two-dimensional subspaces of V. Right: Lines on the Grassmann cone in $\wedge^2 V$.

Definition 2.2.4. Let $w \in \wedge^k V$. The *inner space* and *outer space* of w are

$$\lfloor w \rfloor := \{v \in V \; ; \; v \wedge w = 0\},$$
$$\lceil w \rceil := \bigcap \{V' \subset V \; ; \; w \in \wedge^k V'\},$$

respectively.

Note that we do not define spaces associated to inhomogeneous multivectors, since this is not natural geometrically. We note that when $w = v_1 \wedge \cdots \wedge v_k$ is simple and nonzero, or equivalently $\{v_1, \ldots, v_k\}$ is linearly independent, then

$$\lfloor w \rfloor = \lceil w \rceil = [w] = \operatorname{span}\{v_1, \ldots, v_k\}.$$

This is clear for the outer space. Indeed, $V' = \lceil w \rceil$ is clearly the smallest subspace such that $w \in \wedge^k V'$. Recall from Lemma 2.1.7 how $\wedge^k V' \subset \wedge^k V$. For the inner space, $\lfloor w \rfloor = [w]$ is a direct consequence of Proposition 2.1.9.

Under the correspondence in Proposition 2.2.3, the exterior product of multivectors corresponds to direct sums of subspaces. More precisely, we have the following.

Proposition 2.2.5 (Geometry of $\wedge$). *Let $w_1 \in \wedge^k V$ and $w_2 \in \wedge^l V$. Then*

$$\lfloor w_1 \wedge w_2 \rfloor \supset \lfloor w_1 \rfloor + \lfloor w_2 \rfloor,$$
$$\lceil w_1 \wedge w_2 \rceil \subset \lceil w_1 \rceil + \lceil w_2 \rceil.$$

If w_1 and w_2 are simple, so is $w_1 + \wedge w_2$, and if $w_1 \wedge w_2 \neq 0$, then

$$[w_1 \wedge w_2] = [w_1] \oplus [w_2].$$

Proof. Assume $v = v_1 + v_2$, where $v_i \in \lfloor w_i \rfloor$. Then

$$v \wedge (w_1 \wedge w_2) = (v_1 \wedge w_1) \wedge w_2 + (-1)^k w_1 \wedge (v_2 \wedge w_2) = 0 + 0 = 0,$$

and therefore $v \in \lfloor w_1 \wedge w_2 \rfloor$.

For the outer spaces, let $V' := \lceil w_1 \rceil + \lceil w_2 \rceil$. Then $w_1 \in \wedge^k \lceil w_1 \rceil \subset \wedge^k V'$ and $w_2 \in \wedge^l \lceil w_2 \rceil \subset \wedge^l V'$. Since $\wedge V'$ is closed under exterior multiplication, it follows that $w_1 \wedge w_2 \in \wedge^{k+l} V'$, so $\lceil w_1 \wedge w_2 \rceil \subset V'$.

The last claim for simple multivectors is clear, since

$$\lfloor w_1 \rfloor + \lfloor w_2 \rfloor \subset \lfloor w_1 \wedge w_2 \rfloor = \lceil w_1 \wedge w_2 \rceil \subset \lceil w_1 \rceil + \lceil w_2 \rceil,$$

where the left- and right-hand sides are equal. That $w_1 \wedge w_2$ is simple is clear from the definition. $\qquad\square$

We continue by studying the inner and outer spaces for composite multivectors, for which we shall see that these two subspaces will always be distinct.

Proposition 2.2.6 (Outer simplicity criterion). *Let $0 \neq w \in \wedge^k V$. Then $\dim \lceil w \rceil \geq k$ with equality if and only if $w \in \widehat{\wedge}^k V$.*

Proof. Consider the vector space $V' := \lceil w \rceil$. Since $\wedge^k V' \subset \wedge^k V$ is isomorphic to the kth exterior power of V' by Lemma 2.1.7, and $0 \neq w \in \wedge^k V'$, it follows that $\dim \lceil w \rceil = \dim V' \geq k$. When equality holds, $\dim \wedge^k V' = 1$, and it is clear that w must be simple. Conversely, we have already observed that $\dim \lceil w \rceil = k$ when w is simple. $\qquad\square$

For the corresponding result for the inner space, we require the following lemma. Later, when we have access to the full algebra, we shall see this as a simple example of a Hodge decomposition as in Example 10.1.7.

Lemma 2.2.7. *Let $V' \subset V$ be a subspace of codimension 1 and $v \in V \setminus V'$ so that $V = \operatorname{span}\{v\} \oplus V'$. Then the map*

$$\wedge^k V' \to \{w \in \wedge^{k+1} V \; ; \; v \wedge w = 0\} : w' \mapsto v \wedge w'$$

is an invertible linear map.

Proof. Since $v \wedge (v \wedge w') = (v \wedge v) \wedge w' = 0$, the map is well defined. Pick a basis $\{e_1, \ldots, e_{n-1}\}$ for V' and add $e_n := v$ to make a basis for V. Then clearly $\wedge^k V' = \text{span}\{e_s\}_{|s|=k, n \notin s}$, and since exterior multiplication by e_n acts by adding n to index sets, modulo signs, injectivity is clear. To prove surjectivity, write $w = w_1 + v_n \wedge w_2$, where $w_1 \in \wedge^{k+1} V'$, $w_2 \in \wedge^k V'$, by checking $n \notin t$ or $n \in t$ for basis multivectors e_t. Then $0 = e_n \wedge w = e_n \wedge w_1$. Applying injectivity, but with k replaced by $k+1$, shows that $w_1 = 0$ and thus $w = v \wedge w_2$. Thus the map is invertible.   □

Proposition 2.2.8 (Inner simplicity criterion). *Let* $0 \neq w \in \wedge^k V$. *Then* $\dim \lfloor w \rfloor \leq k$ *with equality if and only if* $w \in \hat{\wedge}^k V$. *Moreover, if* $\{v_1, \ldots, v_l\}$ *is a basis for* $\lfloor w \rfloor$, *then there exists* $w' \in \wedge^{k-l} V$ *such that* $w = v_1 \wedge \cdots \wedge v_l \wedge w'$ *and* $\lfloor w' \rfloor = \{0\}$.

Note from Proposition 2.2.8 that either a k-vector has a k-dimensional inner space, if it is simple, or else the dimension of its inner space is at most $k - 2$, since all vectors $w' \neq 0$ are simple. From this and Proposition 2.6.9 we will also be able to deduce that no k-vector can have a $(k+1)$-dimensional outer space.

Proof. Extend $\{v_1, \ldots, v_l\}$ to a basis $\{v_1, \ldots, v_n\}$ for V. From Lemma 2.2.7 with $v := v_1$ and $V' := \text{span}\{v_2, \ldots, v_n\}$, it follows that $w = v_1 \wedge w_1$ for a unique $0 \neq w_1 \in \wedge^{k-1} V'$. Clearly $\lfloor w_1 \rfloor \subset \lfloor w \rfloor$ and $v_1 \notin \lfloor w_1 \rfloor$. If $2 \leq j \leq l$, then

$$0 = v_j \wedge w = -v_1 \wedge (v_j \wedge w_1).$$

Since $v_j \wedge w_1 \in \wedge^k V'$, it follows from Lemma 2.2.7 that $v_j \wedge w_1 = 0$, that is, $v_j \in \lfloor w_1 \rfloor$. Thus $\lfloor w_1 \rfloor = \text{span}\{v_2, \ldots, v_l\}$. Since $\lceil w_1 \rceil \subset V'$, we can iterate this argument, replacing V by V', to prove the factorization of w. In particular, $l \leq k$ with equality if and only if w is simple.   □

Corollary 2.2.9. *For every* $0 \neq w \in \wedge^k V$ *we have*

$$\lfloor w \rfloor \subset \lceil w \rceil,$$

with equality if and only if w is simple. However, we have $\lfloor 0 \rfloor = V$ *and* $\lceil 0 \rceil = \{0\}$.

Proof. If $v \notin \lceil w \rceil$, then there exists a subspace V' of codimension 1 such that $v \notin V' \supset \lceil w \rceil$. Since $w \in \wedge^k V'$, it follows from Lemma 2.2.7 that $v \wedge w \neq 0$, that is, $v \notin \lfloor w \rfloor$. That equality holds if and only if w is simple follows from Propositions 2.2.6 and 2.2.8.   □

Proposition 2.2.8 gives an algorithm for factorizing simple k-vectors. Below in Proposition 2.6.9 we shall prove a duality between the outer and inner spaces that permits us to calculate the outer space as well. That such a duality should exist can be anticipated from Proposition 2.2.8, which characterizes the inner space $\lfloor w \rfloor$ as the maximal space corresponding to a factor $v_1 \wedge \cdots \wedge v_l$ in w.

Example 2.2.10. We wish to factorize

$$w = e_{12} + 2e_{13} + 3e_{14} + e_{23} + 2e_{24} + e_{34} \in \wedge^2 V.$$

To do this, we determine $\lfloor w \rfloor$, that is, the null space of the map $V \ni v \mapsto v \wedge w \in \wedge^3 V$. Its matrix with respect to bases $\{e_1, e_2, e_3, e_4\}$ and $\{e_{123}, e_{124}, e_{134}, e_{234}\}$ is

$$\begin{bmatrix} 1 & -2 & 1 & 0 \\ 2 & -3 & 0 & 1 \\ 1 & 0 & -3 & 2 \\ 0 & 1 & -2 & 1 \end{bmatrix}.$$

We find that the null space is spanned by $(1, 1, 1, 1)^t$ and $(3, 2, 1, 0)^t$. Thus w is simple, since the null space is two-dimensional. According to Proposition 2.2.8, the exterior product

$$\begin{vmatrix} e_1 & 1 & 3 \\ e_2 & 1 & 2 \\ e_3 & 1 & 1 \\ e_4 & 1 & 0 \end{vmatrix} = -e_{12} - 2e_{13} - 3e_{14} - e_{23} - 2e_{24} - e_{34}$$

must be parallel to w. Checking the proportionality constant, we find that one possible (among many) factorization is

$$w = (3e_1 + 2e_2 + e_3) \wedge (e_1 + e_2 + e_3 + e_4).$$

Since the Grassmann cone it not closed under addition, composite k-vectors are generally the result of an addition of simple k-vectors. The following gives a geometric criterion for when the sum of two simple k-vectors happens to be simple.

Proposition 2.2.11. *If* $\dim V = n$, *then* $\widehat{\wedge}^k V = \wedge^k V$ *if and only if* $k = 0, 1, n - 1$, *or* n.

Let $2 \leq k \leq n - 2$ *and* $w_1, w_2 \in \widehat{\wedge}^k V \setminus \{0\}$. *Then* $w_1 + w_2$ *is composite if and only* $\dim([w_1] \cap [w_2]) \leq k - 2$. *In this case* $\lfloor w_1 + w_2 \rfloor = [w_1] \cap [w_2]$.

Note that when $w_1 + w_2$ is simple, then $\dim \lfloor w_1 + w_2 \rfloor = k$ (or n if $w_1 + w_2 = 0$). Even though $\dim \lfloor w_1 + w_2 \rfloor = k - 1$ is not possible, it may happen that $\dim(\lfloor w_1 \rfloor \cap \lfloor w_2 \rfloor) = k - 1$.

Proof. It is clear that $\widehat{\wedge}^k V = \wedge^k V$ if $k = 0, 1, n$. For $k = n - 1$, let $0 \neq w \in \wedge^{n-1} V$ and consider the map

$$V \ni v \mapsto v \wedge w \in \wedge^n V.$$

Since $\dim \wedge^n V = 1$, the dimension theorem shows that $\dim \lfloor w \rfloor \geq n - 1$. By the remark after Proposition 2.2.8, w is simple.

Next consider $2 \leq k \leq n - 2$. Note that $\lfloor w_1 \rfloor \cap \lfloor w_2 \rfloor \subset \lfloor w_1 + w_2 \rfloor$. Thus, if $w_1 + w_2$ is composite, then the intersection is at most $(k - 2)$-dimensional, since $\lfloor w_1 + w_2 \rfloor$ cannot be $(k-1)$-dimensional. Conversely, assume that $m :=$ $\dim(\lfloor w_1 \rfloor \cap \lfloor w_2 \rfloor) \le k - 2$. By extending a basis from $\lfloor w_1 \rfloor \cap \lfloor w_2 \rfloor$, we see that there is a basis $\{e_i\}$ such that $w_1 = e_{1,\dots,k}$ and $w_1 = e_{k-m+1,\dots,2k-m}$. In particular, $2k - m \le n$. For the k-vector

$$w_1 + w_2 = e_{1,\dots,k} + e_{k-m+1,\dots,2k-m}, \tag{2.3}$$

we can calculate its inner space as in Example 2.2.10, which is seen to be $\mathrm{span}\{e_{k-m+1}, \dots, e_k\}$. Thus $\dim\lfloor w_1 + w_2 \rfloor = m$ and therefore $\lfloor w_1 + w_2 \rfloor = \lfloor w_1 \rfloor \cap \lfloor w_2 \rfloor$ and $w_1 + w_2$ is composite. In particular, we conclude that whenever $2 \le k \le n-2$, there are composite k-vectors, since k-vectors of the form (2.3) exist, for example, with $m = \max(2k - n, 0)$. $\square$

Remark 2.2.12. Proposition 2.2.11 sheds some light on the universal property (U), or equivalently the basis property (B). It is natural to ask whether it is possible to replace conditions (A) and (U) in the definition of $\wedge^k V$ by the conditions that the range of $\wedge^k$ spans L and $\wedge^k(v_1, \dots, v_k) = 0$ if and only if the vectors are linearly dependent. However, this would not determine $\wedge^k$ uniquely up to isomorphism. To see this, let $2 \le k \le n-2$ and fix a one-dimensional subspace $L \subset \wedge^k V$ such that $L \cap (\widehat{\wedge}^k V) = \{0\}$. Then one can verify that $M : V^k \to (\wedge^k V)/L$, defined as $\wedge^k$ followed by projection down to the quotient space, satisfies the above conditions, although the range space of M has dimension $\binom{n}{k} - 1$.

2.3   Mapping Multivectors

Consider a linear map $T : V_1 \to V_2$ between vector spaces V_1 and V_2. We wish to extend this to a linear map $T : \wedge V_1 \to \wedge V_2$. To do this, consider the alternating multilinear map

$$(v_1, \dots, v_k) \mapsto T(v_1) \wedge \cdots \wedge T(v_k) \in \wedge^k V_2.$$

According to the universal property of $\wedge^k V_1$, there exists a unique linear map $T_\wedge : \wedge^k V_1 \to \wedge^k V_2$ such that

$$T_\wedge(v_1 \wedge \cdots \wedge v_k) = T(v_1) \wedge \cdots \wedge T(v_k).$$

Definition 2.3.1. Given $T : V_1 \to V_2$, the linear map $T = T_\wedge : \wedge V_1 \to \wedge V_2$ is called the *induced* map between exterior algebras, where $T_\wedge(1) := 1$, $T_\wedge : \wedge^k V_1 \to \wedge^k V_2$ is as above, and $T_\wedge(\sum_k w_k) := \sum_k T_\wedge(w_k)$, for $w_k \in \wedge^k V_1$.

With some abuse of notation we keep the notation T for $T_\wedge$.

Proposition 2.3.2 (Geometry of induced maps). *Let $T : \wedge V_1 \to \wedge V_2$ be an induced linear map as above. Then T is an algebra homomorphism with respect to the exterior product, that is,*

$$T(w_1 \wedge w_2) = T(w_1) \wedge T(w_2), \quad w_1, w_2 \in \wedge V_1,$$

and T maps the inner and outer spaces as

$$\lfloor T(w) \rfloor \supset T(\lfloor w \rfloor) \quad and \quad \lceil T(w) \rceil \subset T(\lceil w \rceil), \quad w \in \wedge^k V_1.$$

If T is invertible, then we have $\lfloor T(w) \rfloor = T(\lfloor w \rfloor)$ and $\lceil T(w) \rceil = T(\lceil w \rceil)$. In particular, if w is simple, then $T(w)$ is simple, and $\lceil T(w) \rceil = T(\lceil w \rceil)$ whenever T is invertible.

Proof. To verify that T is a homomorphism, by linearity we may assume that the w_i are simple. In this case, the identity follows directly from the definition of T.

To show that $\lfloor T(w) \rfloor \supset T(\lfloor w \rfloor)$, let $v \in \lfloor w \rfloor$. Then

$$T(v) \wedge T(w) = T(v \wedge w) = T(0) = 0,$$

that is, $T(v) \in \lfloor T(w) \rfloor$ as claimed. To show that $\lceil T(w) \rceil \subset T(\lceil w \rceil)$, note that $w \in \wedge^k \lceil w \rceil$. It follows that $T(w) \in \wedge^k (T(\lceil w \rceil))$, so that $\lceil T(w) \rceil \subset T(\lceil w \rceil)$.

If T is invertible, then replacing T by T^{-1} and w by $T(w)$, we get $\lfloor w \rfloor \supset T^{-1}(\lfloor T(w) \rfloor)$, which proves that $\lfloor T(w) \rfloor = T(\lfloor w \rfloor)$. Equality for the outer space is proved similarly. $\square$

For obvious practical reasons, we choose to keep the notation T for the map $T_\wedge$ of exterior algebras induced by $T : V \to V$. However, note carefully that the operation $T \mapsto T_\wedge$ of extending T from V to $\wedge V$ is not linear. Indeed, it is not even homogeneous, since the extension $(\alpha T)_\wedge$ of $\alpha T : V \to V$, $\alpha \in \mathbf{R}$, equals $\alpha^k T_\wedge$ on $\wedge^k V$. However, for compositions the following result holds. The proof is left as an exercise.

Proposition 2.3.3. *Let $T_1 : V_1 \to V_2$ and $T_2 : V_2 \to V_3$ be linear maps of vector spaces. Then the map $\wedge V_1 \to \wedge V_3$ induced by $T_2 T_1 : V_1 \to V_3$ equals the composition $T_2 T_1$ of the induced maps $T_1 : \wedge V_1 \to \wedge V_2$ and $T_2 : \wedge V_2 \to \wedge V_3$.*

Let $\{e_1, \ldots, e_n\}$ and $\{e_1', \ldots, e_m'\}$ be bases for V_1 and V_2, and write $Te_i = \sum_j e_j' \alpha_{j,i}$, where $(\alpha_{j,i})$ is the matrix of T with respect to these bases. Then

$$Te_s = \begin{vmatrix} e_1' & \alpha_{1,s_1} & \cdots & \alpha_{1,s_k} \\ e_2' & \alpha_{2,s_1} & \cdots & \alpha_{2,s_k} \\ \vdots & \vdots & \ddots & \vdots \\ e_m' & \alpha_{m,s_1} & \cdots & \alpha_{m,s_k} \end{vmatrix}, \quad s = \{s_1, \ldots, s_k\}, s_1 < \cdots < s_k.$$

We see that the matrix of $T : \wedge^k V_1 \to \wedge^k V_2$ consists of all $k \times k$ subdeterminants of the matrix for $T : V_1 \to V_2$. In particular, if $V_1 = V_2 = V$, then the induced linear map $T : \wedge^n V \to \wedge^n V$ is simply multiplication by a number, since $\dim \wedge^n V = 1$. It is clear that

$$T(w) = (\det T) \, w, \quad \text{for } w \in \wedge^n V.$$

Example 2.3.4. If $\dim V = 3$ and $T : V \to V$ has matrix A in the basis $\{e_1, e_2, e_3\}$, then the induced map $T : \wedge V \to \wedge V$ has matrix B in the induced basis $\{1, e_1, e_2, e_3, e_{12}, e_{13}, e_{23}, e_{123}\}$, where

$$A = \begin{bmatrix} 1 & 1 & 0 \\ -1 & 1 & 2 \\ 2 & 0 & -1 \end{bmatrix} \quad \text{and} \quad B = \begin{bmatrix} 1 & 0 & 0 & 0 & 0 & 0 & 0 & 0 \\ 0 & 1 & 1 & 0 & 0 & 0 & 0 & 0 \\ 0 & -1 & 1 & 2 & 0 & 0 & 0 & 0 \\ 0 & 2 & 0 & -1 & 0 & 0 & 0 & 0 \\ 0 & 0 & 0 & 0 & 2 & 2 & 2 & 0 \\ 0 & 0 & 0 & 0 & -2 & -1 & -1 & 0 \\ 0 & 0 & 0 & 0 & -2 & -3 & -1 & 0 \\ 0 & 0 & 0 & 0 & 0 & 0 & 0 & 2 \end{bmatrix}.$$

Exercise 2.3.5. Consider the operations from Definition 2.1.18. Show that involution $\wedge V \to \wedge V : w \mapsto \widehat{w}$ is the map induced by $V \to V : v \mapsto -v$. On the other hand, show that reversion $\wedge V \to \wedge V : w \mapsto \overline{w}$ is not induced by any map $V \to V$ in the sense of Definition 2.3.1.

A formula for the solution of a linear system of equations is *Cramer's rule*.

Proposition 2.3.6 (Cramer's rule). *Assume that $T : V \to V$ is an invertible linear map and consider the equation $Tu = v$. Fix a basis $\{e_i\}$ and write $u = \sum_i x_i e_i$ and $v = \sum_i y_i e_i$. Then given v, we have*

$$x_j = \begin{vmatrix} \alpha_{1,1} & \cdots & y_1 & \cdots & \alpha_{1,n} \\ \vdots & \ddots & \vdots & \ddots & \vdots \\ \alpha_{n,1} & \cdots & y_n & \cdots & \alpha_{n,n} \end{vmatrix} \Bigg/ \begin{vmatrix} \alpha_{1,1} & \cdots & \alpha_{1,n} \\ \vdots & \ddots & \vdots \\ \alpha_{n,1} & \cdots & \alpha_{n,n} \end{vmatrix},$$

where v replaces the jth column and $(\alpha_{i,j})_{i,j}$ is the matrix for T.

The proof by multivector algebra is indeed very simple.

Proof. Given $1 \le j \le n$ and

$$Tu = x_1 Te_1 + \cdots + x_n Te_n = y,$$

multiply this equation from the left by $Te_1 \wedge \cdots \wedge Te_{j-1}$ and from the right by $Te_{j+1} \wedge \cdots \wedge Te_n$, with the exterior product. Only term j on the left survives, since $Te_i \wedge Te_i = 0$, giving

$$x_j Te_1 \wedge \cdots \wedge Te_n = Te_1 \wedge \cdots \wedge Te_{j-1} \wedge y \wedge Te_{j+1} \wedge \cdots \wedge Te_n.$$

Calculating the n-vectors using (rectangular=quadratic) determinants yields Cramer's rule. □

As is well known, $T : V \to V$ is invertible if and only if $\det T$ is nonzero. More generally, we have the following invertibility criterion.

Proposition 2.3.7 (Generalized determinant criterion). *Let* $T : \wedge V \to \wedge V$ *be an induced linear map, and let* $2 \le k \le n$. *Then*

$$T : \wedge^k V \to \wedge^k V$$

is invertible if and only if $T : V \to V$ *is invertible.*

Proof. Assume that $T : \wedge^{k-1} V \to \wedge^{k-1} V$ is not invertible. Then there exists $0 \ne w \in \wedge^{k-1} V$ such that $T(w) = 0$. Consider the inner space $\lfloor w \rfloor$. Proposition 2.2.8 shows that $\lfloor w \rfloor \subsetneq V$, so there exists $v \in V$ such that $v \wedge w \ne 0$. Since

$$T(v \wedge w) = T(v) \wedge T(w) = T(v) \wedge 0 = 0,$$

$T : \wedge^k V \to \wedge^k V$ cannot be invertible.

To complete the proof, it suffices to prove that if $T : V \to V$ is invertible, then so is $T : \wedge^n V \to \wedge^n V$. But this is clear, since

$$T(e_1 \wedge \cdots \wedge e_n) = T(e_1) \wedge \cdots \wedge T(e_n),$$

and the image vectors form a basis. $\qquad \square$

Example 2.3.8 (Cauchy–Binet formula). Let $V_1 \subset V$, and consider linear maps $T_1 : V_1 \to V$ and $T_2 : V \to V_1$, where $\dim V_1 =: m < n =: \dim V$. Fix a basis $\{e_1, \dots, e_m\}$ for V_1, and extend it to a basis $\{e_1, \dots, e_n\}$ for V. Note that $\wedge^m V_1$ is one-dimensional and spanned by $e_{1,\dots,m}$, and that $\dim \wedge^m V = \binom{n}{m}$ with basis $\{e_s\}_{|s|=m}$. We see that

$$T_1(e_{1,\dots,m}) = \sum_{|s|=m} D_s^1 e_s \in \wedge^m V,$$

where D_s^1 is the $m \times m$ subdeterminant of the matrix for T_1 obtained by deleting rows, keeping only those determined by s. Next consider the induced map $T_2 : \wedge^m V \to \wedge^m V_1$. Here

$$T_2(e_s) = D_s^2 e_{1,\dots,m},$$

where D_s^2 is the $m \times m$ subdeterminant of the matrix for T_2 obtained by deleting columns, keeping only those determined by s. Applying Proposition 2.3.3, we deduce that

$$\det(T_2 T_1) = \sum_s D_s^2 D_s^1.$$

This is the *Cauchy–Binet* formula, which is a generalization of the multiplicativity of the determinant to rectangular matrices.

Recall that a one-dimensional subspace, a line $V_1 \subset V$, is invariant under the action of T, that is, $T V_1 \subset V_1$ if and only if every $0 \ne v \in V_1$ is an eigenvector for T. In this case $T : V_1 \to V_1$ is invertible, or equivalently $T V_1 = V_1$, if and only if the eigenvalue λ is nonzero. This readily generalizes to higher-dimensional subspaces with Proposition 2.2.3 and Proposition 2.3.2.

Proposition 2.3.9 (Invariant subspaces). *Let* $T : \wedge^k V \to \wedge^k V$ *be an induced linear map, let* $V_k \subset V$ *be a* k-*dimensional subspace, and let* $w \in \widehat{\wedge}^k V \setminus \{0\}$ *be such that* $[w] = V_k$. *Then* w *is an eigen-*k*-vector with nonzero eigenvalue if and only if* $TV_k = V_k$.

If $TV_k \subsetneqq V_k$, *then* $Tw = 0$. *However, if* $Tw = 0$, *then* V_k *need not be invariant if* $k \geq 2$.

Proof. If $Tw = \lambda w$ and $\lambda \neq 0$, then $Tw \neq 0$, so Proposition 2.3.2 shows that $TV_k = [Tw] = [w] = V_k$. Conversely, assume $TV_k = V_k$ and write $w = v_1 \wedge \cdots \wedge v_k$, where $\{v_1, \ldots, v_k\}$ is a basis for V_k. If $Tw = Tv_1 \wedge \cdots \wedge Tv_k = 0$, then the images of the basis vectors are linearly dependent and thus $\dim TV_k < k$. Hence $Tw \neq 0$, and Proposition 2.3.2 shows that $[Tw] = TV_k = V_k = [w]$. By Proposition 2.2.3 we must have $Tw = \lambda w$ for some $\lambda \neq 0$.

If $TV_k \subsetneqq V_k$, then in particular $\dim TV_k < k$. Then we cannot have $Tw \neq 0$, since this would give a k-dimensional space $TV_k = [Tw]$. Finally, consider a generic map T with one-dimensional range. Then all induced maps $T : \wedge^k V \to \wedge^k V$, $k \geq 2$, are zero. This proves the last statement. $\qquad\square$

2.4 Oriented Measure

In this section we give composite k-vectors a certain geometric meaning as oriented measures of curved k-surfaces in an affine space X. For this, we require the notion of orientation of affine spaces. As a special case of Proposition 2.2.3, the whole n-dimensional affine space X corresponds to the one-dimensional line $\wedge^n V$.

- A choice of one of the half-lines of $\wedge^n V$, starting at 0, corresponds to choosing an orientation of the affine space X.

- A choice of norm $|\cdot|$ for $\wedge^n V$ corresponds to defining the n-volume of subsets of X.

If we choose a nonzero n-vector $e_{\overline{n}} \in \wedge^n V$, then we specify both an orientation of X, namely the half-line of $\wedge^n V$ containing $e_{\overline{n}}$, and a volume norm, the one having $|e_{\overline{n}}| = 1$. To simplify notation, instead of talking about oriented and/or volume-normed spaces we shall always assume that we have fixed $e_{\overline{n}} \in \wedge^n V$ in this way, although sometimes the volume-norm is irrelevant, and sometimes the orientation is irrelevant.

Definition 2.4.1. Let (X, V) be an affine space of dimension n, and consider the one-dimensional linear space $\wedge^n V$. Fixing a nonzero n-vector $e_{\overline{n}} \in \wedge^n V$, we refer to $(X, V, e_{\overline{n}})$ as an *oriented affine space*. The *orientation* of X is the half-line $\{\lambda e_{\overline{n}} \; ; \; \lambda > 0\}$, and the *volume-norm* is the norm $|\cdot|$ on $\wedge^n V$ such that $|e_{\overline{n}}| = 1$. Whenever we use a basis $\{e_1, \ldots, e_n\}$ in an oriented space, we shall require that

$$e_1 \wedge \cdots \wedge e_n = e_{\overline{n}}.$$

In particular, this means that the basis is positively oriented.

Orientation and volume-norm for a k-dimensional subspace $V_1 \subset V$ are defined similarly using the one-dimensional linear space $\wedge^k V_1 \subset \wedge^k V$.

We first consider the notion of oriented measure of a flat k-surface.

Example 2.4.2 (Flat oriented measure). Let $\{v_1, \ldots, v_k\}$ be linearly independent vectors in a given affine space (X, V) and let $x_0 \in X$. Consider a subset M of the affine k-dimensional subspace

$$x_0 + \text{span}\{v_1, \ldots, v_k\} \subset X$$

that is parametrized by $\rho : \mathbf{R}^k \to X : (y_1, \ldots, y_k) \mapsto x_0 + y_1 v_1 + \cdots + y_k v_k$. Let $M_0 \subset \mathbf{R}^k$ be the parameter domain so that $\rho(M_0) = M$. We define the oriented measure of M as

$$\wedge^k(M) := |M_0| v_1 \wedge \cdots \wedge v_k \in \wedge^k V,$$

where $|M_0|$ denotes the k-volume of M_0 in $\mathbf{R}^k$. This k-vector is independent of choice of basis/parametrization as long as we do not change its orientation. Indeed, consider a second basis $\{v_1', \ldots, v_k'\}$ that spans the same subspace, and let $A = (\alpha_{i,j})$ be the matrix connecting the bases, so that $v_j' = \sum_i \alpha_{i,j} v_i$. If M has parameter domain $M_0' \subset \mathbf{R}^k$ in coordinates y_i', then $|M_0| = |\det A| |M_0'|$ by a change of variables in the integral. One the other hand,

$$v_1' \wedge \cdots \wedge v_k' = (\det A) v_1 \wedge \cdots \wedge v_k.$$

This shows the stated independence of basis.

We note from Example 2.4.2 that the oriented measure of a flat k-surface is actually a simple k-vector. Extending this notion additively, we soon encounter composite oriented measures when we add those of different flat surfaces. A continuous setup is as follows: We consider a curved k-dimensional surface $M \subset X$ with varying tangent spaces $T_x M$. We expect the oriented measure of M to be a composite k-vector in general, since the orientations of $T_x M$ vary over the surface. In general, we let M be an oriented compact C^1-regular k-surface, that is a manifold embedded in the affine space X, as in Section 6.1. For the purpose of this section, we limit ourselves to the case in which M is covered by only one chart

$$\rho : M_\rho \to M : y \mapsto x.$$

Let $\{e_1, \ldots, e_k\}$ denote the standard basis in $\mathbf{R}^k$. We map the oriented volume element $\underline{dy} := e_1 \wedge \cdots \wedge e_k \, dy$ in $\mathbf{R}^k$, where $dy = dy_1 \cdots dy_k$ is the usual scalar Lebesgue volume element, onto the oriented volume element

$$\underline{dx} := \rho_{\underline{y}}(e_1 \wedge \cdots \wedge e_k) dy_1 \cdots dy_k$$

for M, which we formally regard as an infinitesimal k-vector. Here $\rho_{\underline{y}}$ denotes the total derivative of ρ as in Definition 6.1.2, or more precisely the map of k-vectors that it induces as in Definition 2.3.1. Note that the space of $\underline{\rho}_y(e_1 \wedge \cdots \wedge e_k)$ is the tangent space $T_x M$ at $x = \rho(y)$ by Proposition 2.3.2. This motivates the following definition.

Definition 2.4.3 (Curved oriented measure). Let M be a compact oriented C^1-regular k-surface in X with parametrization $\rho : M_\rho \to M$. Then the *oriented measure* of M is defined as

$$\wedge^k(M) := \int_M \underline{dx} = \int_{M_\rho} \underline{\rho}_y(e_1 \wedge \cdots \wedge e_k)\, dy_1 \cdots dy_k \in \wedge^k V,$$

where y_i denote coordinates with respect to the standard basis $\{e_i\}$ in $\mathbf{R}^k$. The integral on the right-hand side is calculated componentwise.

Example 2.4.4. Let M be the paraboloid

$$M := \{(x_1, x_2, x_3)\ ;\ x_3 = x_1^2 + x_2^2 \le 1\} \subset \mathbf{R}^3.$$

With parametrization $\rho : (y_1, y_2) \mapsto (y_1, y_2, y_1^2 + y_2^2)$ and M_ρ the unit disk, the matrix for the derivative $\underline{\rho}_y$ is $\begin{bmatrix} 1 & 0 \\ 0 & 1 \\ 2y_1 & 2y_2 \end{bmatrix}$. This gives the oriented area element $\underline{dx} = w(y)dy_1 dy_2$ on M, where the orientation is

$$w = \begin{vmatrix} e_1 & 1 & 0 \\ e_2 & 0 & 1 \\ e_3 & 2y_1 & 2y_2 \end{vmatrix} = \begin{vmatrix} 1 & 0 \\ 0 & 1 \end{vmatrix} e_{12} + \begin{vmatrix} 1 & 0 \\ 2y_1 & 2y_2 \end{vmatrix} e_{13} + \begin{vmatrix} 0 & 1 \\ 2y_1 & 2y_2 \end{vmatrix} e_{23}$$

$$= e_{12} + 2y_2 e_{13} - 2y_1 e_{23},$$

since $w = \underline{\rho}(e_1 \wedge e_2) = \underline{\rho}(e_1) \wedge \underline{\rho}(e_2) = \partial_1 \rho \wedge \partial_2 \rho$. The oriented measure of M is

$$\wedge^2(M) = \int_{M_\rho} (e_{12} + 2y_2 e_{13} - 2y_1 e_{23})\, dy_1 dy_2 = \pi e_{1,2}.$$

Note that the flat surface $M_0 := \{x_3 = 1,\ x_1^2 + x_2^2 \le 1\}$ has the same oriented measure as M. This is always the case for two surfaces with the same boundary, and compatible orientations, as is shown in Proposition 7.3.15.

Since there are no composite bivectors in three-dimensional spaces, we need to go to higher dimensions for a generic example.

Example 2.4.5. Let M be a piece of the graph of the complex analytic function $w = e^z$:

$$M := \{(x_1, x_2, x_3, x_4)\ ; x_3 = e^{x_1} \cos x_2,\ x_4 = e^{x_1} \sin x_2,$$
$$0 \le x_1 \le 1,\ 0 \le x_2 \le 2\pi\} \subset \mathbf{R}^4.$$

Parametrizing with $(y_1, y_2) = (x_1, x_2)$, that is,

$$\rho(y_1, y_2) = (y_1, y_2, e^{y_1} \cos y_2, e^{y_1} \sin y_2),$$

the oriented area element on M is $\underline{dx} = w(y) dy_1 dy_2$, where the orientation is

$$w = \begin{vmatrix} e_1 & 1 & 0 \\ e_2 & 0 & 1 \\ e_3 & e^{y_1} \cos y_2 & -e^{y_1} \sin y_2 \\ e_4 & e^{y_1} \sin y_2 & e^{y_1} \cos y_2 \end{vmatrix}$$

$$= e_{12} - e^{y_1} \sin y_2 e_{13} + e^{y_1} \cos y_2 e_{14} - e^{y_1} \cos y_2 e_{23} - e^{y_1} \sin y_2 e_{24} + e^{2y_1} e_{34}.$$

The oriented measure of M is the composite bivector

$$\wedge^2(M) = \int_0^{2\pi} \int_0^1 w(y_1, y_2) dy_1 dy_2 = 2\pi e_{12} + \pi(e^2 - 1) e_{34}.$$

Note that $\pi(e^2 - 1)$ equals the area of the image, the projection of the graph M onto the $x_3 x_4$-plane, of the rectangle under $w = e^z$, that is, the annulus $\{1 \le x_3^2 + x_4^2 \le e^2\}$.

2.5 Multicovectors

Let (X, V) be an affine space, and fix throughout this section a duality $\langle V^*, V \rangle$. Recall that even though V^* may be any linear space in duality with V, we can always identify V^* with the dual space of V by Proposition 1.2.3.

Clearly, exterior algebras of any linear space can be built not only on vector spaces, since it uses only the notion of linear dependence. The assumption that the linear space was a vector space was imposed only to clarify the geometric interpretation. Replacing V with V^* in the previous sections, we obtain the exterior algebra of V^*.

Definition 2.5.1 (Dual exterior algebra). Let (X, V) be an n-dimensional affine space. The dual exterior algebra $\wedge V^*$ is the 2^n dimensional linear space $\wedge V^* = \wedge^0 V^* \oplus \wedge^1 V^* \oplus \wedge^2 V^* \oplus \cdots \oplus \wedge^n V^*$, where the space $\wedge^k V^*$ of k-*covectors* is obtained by fixing a multilinear map

$$\wedge^k : V^* \times \cdots \times V^* \to \wedge^k V^*$$

satisfying the properties (A) and (U) as in Definition 2.1.6. An element Θ in $\wedge V^*$ is referred to as a *multicovector*.

We consider the geometric interpretation of multicovectors from the point of view of X and the vector space V.

- Let $\theta \in V^*$ be a nonzero covector. Since it defines a scalar linear functional

$$V \to \mathbf{R} : v \mapsto \langle \theta, v \rangle,$$

  the best way we can visualize it inside V is to consider its level sets. These will consist of a density/stack of parallel hyperplanes in V. So a covector can be thought of as a density of $(n-1)$-dimensional subspaces in V.

- Let $\Theta = \theta_1 \wedge \theta_2 \in \wedge^2 V^*$. We note that the $[\theta_i] \subset V^*$ have orthogonal complements $[\theta_i]^\perp \subset V$, being one of the level sets by which we visualize θ_i. Since

$$[\Theta]^\perp = ([\theta_1] \oplus [\theta_2])^\perp = [\theta_1]^\perp \cap [\theta_2]^\perp$$

  by Proposition 2.2.5, we see that the best way to visualize Θ is as a density of $(n-2)$-dimensional subspaces, the intersections between the level sets of θ_1 and θ_2.

- Similarly, a k-covector is best visualized in V as a density of $(n-k)$-dimensional subspaces. At the end of the scale, an n-covector is viewed as a density of points.

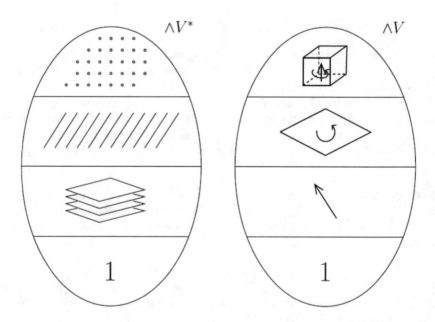

Figure 2.3: Geometry of multicovectors and multivectors in a three-dimensional space X.

We would like to extend the pairing $\langle \theta, v \rangle$ between vectors and covectors to a pairing $\langle \Theta, w \rangle$ between k-covectors and k-vectors. To see what this pairing should be, we argue geometrically as follows. Interpret a simple k-vector w as the oriented measure of some flat k-surface lying in the subspace $W := [w]$, and interpret a simple k-covector Θ as a density of $(n-k)$-dimensional subspaces parallel to P. If W and P span V together, or equivalently $W \cap P = \{0\}$, then the intersections of W with the subspaces parallel to P consist of points in W. We want the number of these lying in the k-surface w, with appropriate sign depending on the orientations of W and P to be the value of $\langle \Theta, w \rangle$. If W and P do not span V together, then we want $\langle \Theta, w \rangle = 0$.

Now fix dual bases $\{e_1^*, \ldots, e_n^*\}$ and $\{e_1, \ldots, e_n\}$ and consider the induced basis vectors $\{e_s^*\}_{|s|=k}$ and $\{e_t\}_{|t|=k}$. From the geometric argument above, we require that $\langle e_s^*, e_t \rangle = 0$ if $s \neq t$ since $W \cap P \neq \{0\}$. Note that W is spanned by e_j, $j \in t$, while P is spanned by e_i, $i \notin s$. In this way, we are also led to the condition that

the induced bases $\{e_s^*\}_{|s|=k}$ and $\{e_t\}_{|t|=k}$ are required to be dual whenever $\{e_1^*, \ldots, e_n^*\}$ and $\{e_1, \ldots, e_n\}$ are dual.

This uniquely determines the pairing $\langle \Theta, w \rangle$. Algebraically, it can be defined as follows.

Definition 2.5.2 (Duality for multivectors). Let

$$\wedge^k V^* \times \wedge^k V \to \mathbf{R} : (\Theta, w) \mapsto \langle \Theta, w \rangle$$

be the unique bilinear pairing given by Lemma 2.1.12, with $V_1 = V^*$ and $V_2 = V$, which extends the multilinear map

$$M(\theta_1, \ldots, \theta_k, v_1, \ldots, v_k) := \begin{vmatrix} \langle \theta_1, v_1 \rangle & \cdots & \langle \theta_1, v_k \rangle \\ \vdots & \ddots & \vdots \\ \langle \theta_k, v_1 \rangle & \cdots & \langle \theta_k, v_k \rangle \end{vmatrix},$$

so that $\langle \theta_1 \wedge \cdots \wedge \theta_k, v_1 \wedge \cdots \wedge v_k \rangle = \det((\langle \theta_i, v_j \rangle)_{i,j}$. We define a pairing between the full exterior algebras $\wedge V^*$ and $\wedge V$ by

$$\left\langle \sum_{l=0}^n \Theta_l, \sum_{k=0}^n w_k \right\rangle := \sum_{k=0}^n \langle \Theta_k, w_k \rangle, \quad w_k \in \wedge^k V, \ \Theta_l \in \wedge^l V^*,$$

where $\langle \Theta_0, w_0 \rangle := \Theta_0 w_0$ denotes real multiplication.

In practice, the pairing $\langle \wedge V^*, \wedge V \rangle$ is calculated using dual bases. Starting from a basis $\{e_1, \ldots, e_n\} \subset V$, we calculate the dual basis $\{e_1^*, \ldots, e_n^*\} \subset V^*$ and the induced bases

$$\{e_s^*\} \subset \wedge V^* \quad \text{and} \quad \{e_t\} \subset \wedge V,$$

which also are dual. Then a multicovector $\Theta = \sum_s x_s e_s^*$ and a multivector $w = \sum_t y_t e_t$ pair to $\langle \Theta, w \rangle = \sum_s x_s y_s$. In particular, it is clear that $\langle \wedge V^*, \wedge V \rangle$ is a duality.

Proposition 2.5.3 (Alternating forms). *Let $\langle V^*, V \rangle$ be a duality. Then $\langle \wedge V^*, \wedge V \rangle$ is also a duality. In particular, there is a natural one-to-one correspondence between*

(i) *k-covectors $\Theta \in \wedge^k V^*$,*

(ii) *linear functionals $f \in (\wedge^k V)^*$, and*

(iii) *real-valued alternating k-linear maps*

$$\omega : V^k \to \mathbf{R} : (v_1, \ldots, v_k) \mapsto \omega(v_1, \ldots, v_k).$$

Proof. We have seen that $\langle \wedge V^*, \wedge V \rangle$ is a duality, and Proposition 1.2.3 therefore establishes the correspondence between (i) and (ii). For (iii), each linear functional $f \in (\wedge^k V)^*$ gives an alternating k-linear map

$$\omega(v_1, \ldots, v_k) := f(v_1 \wedge \cdots \wedge v_k),$$

by restriction to simple k-vectors. The converse is a consequence of the universal property for $\wedge^k$. $\qquad\square$

Let us now summarize our constructions. We are given an n-dimensional affine space (X, V), with space of translations/vectors V. From this we build the exterior algebra

$$\wedge V = \mathbf{R} \oplus V \oplus \wedge^2 V \oplus \wedge^3 V \oplus \cdots \oplus \wedge^{n-2} V \oplus \wedge^{n-1} V \oplus \wedge^n V$$

of multivectors in X, which is a linear space of dimension 2^n. The linear space $\wedge^k V$, $k \in \{0, 1, 2, \ldots, n\}$, has dimension $\binom{n}{k}$ and consists of all k-vectors in X. For simple k-vectors $0 \neq w \in \widehat{\wedge}^k V$, we have a geometric interpretation of w as an oriented k-volume lying in the k-dimensional subspace $[w] \subset V$. More generally, $w \in \wedge^k V$ can be interpreted as the oriented measure of some curved k-surface in X.

We next consider a duality $\langle V^*, V \rangle$, where V^* is another n-dimensional linear space. We can always identify V^* in a natural way with the dual space of V, consisting of covectors, that is linear functionals on V. From V^*, we build the dual exterior algebra

$$\wedge V^* = \mathbf{R} \oplus V^* \oplus \wedge^2 V^* \oplus \wedge^3 V^* \oplus \cdots \oplus \wedge^{n-2} V^* \oplus \wedge^{n-1} V^* \oplus \wedge^n V^*$$

of multicovectors in X. Since V^* is also an n-dimensional linear space, $\wedge V^*$ has the same algebraic structure as $\wedge V$, the only difference being the geometric interpretation. A simple k-covector $0 \neq \Theta = \theta_1 \wedge \cdots \wedge \theta_k \in \widehat{\wedge}^k V^*$ can be interpreted, from the point of view of X, as a density of parallel $(n-k)$-dimensional subspaces in X, the intersections of level sets of the functionals θ_j.

For each integer $0 \leq k \leq n$, we thus have four linear spaces,

$$\wedge^k V, \ \wedge^{n-k} V, \ \wedge^k V^*, \text{ and } \wedge^{n-k} V^*,$$

all of which have the same dimension $\binom{n}{k} = \binom{n}{n-k}$. Hence it would be natural to identify these spaces. There are two kinds of identifications.

- The identification of $\wedge^k V$ with $\wedge^k V^*$ (and $\wedge^{n-k} V$ with $\wedge^{n-k} V^*$). This requires a duality $\langle V, V \rangle$ on V, which gives a one-to-one correspondence between V and V^*.

- The identification of $\wedge^k V$ with $\wedge^{n-k} V^*$ (and $\wedge^{n-k} V$ with $\wedge^k V^*$). This requires only that an orientation $e_{\overline{n}} \in \wedge^n V \setminus \{0\}$ has been fixed as in Definition 2.4.1 and the discussion below it.

The latter identification is furnished by the Hodge star maps, which we construct in Section 2.6. We end this section with a few remarks concerning the first identification by duality. We consider the case of an inner product space.

Lemma 2.5.4. *Let V be an inner product space. Then $\wedge V$ is also an inner product space. If $\{e_1, \ldots, e_n\}$ is an ON-basis for V, then the induced basis $\{e_s\}$ for the exterior algebra is also an ON-basis. The sign of the basis multivector e_s equals*

$$\langle e_s \rangle^2 = \langle e_{s_1} \rangle^2 \cdots \langle e_{s_k} \rangle^2, \quad s = \{s_1, \ldots, s_k\}.$$

In particular, if V is a Euclidean space, then so is $\wedge V$, considered as an inner product space.

Proof. It is straightforward to show from Definition 2.5.2 that $\langle \wedge V, \wedge V \rangle$ is symmetric if $\langle V, V \rangle$ is so. For the second claim, we write $e_i^* = \langle e_i \rangle^2 e_i$, and note that $\{\langle e_i \rangle^2 e_i\}$ and $\{e_i\}$ are dual bases for V. Therefore the corresponding two induced bases for $\wedge V$ are dual, which is seen to translate to the stated claim. $\square$

Exercise 2.5.5. Given an inner product space V with signature $n_+ - n_-$, find a formula for the signature of $\wedge V$.

2.6 Interior Products and Hodge Stars

Let (X, V) be an affine space, and fix throughout this section a duality $\langle V^*, V \rangle$. We define in this section another fundamental bilinear, but nonassociative, product on the exterior algebra $\wedge V$ besides the exterior product: the interior product. The interior product on $\wedge V$ is the operation dual to the exterior product on $\wedge V^*$. In fact, since the exterior product is noncommutative, there will be two interior products on $\wedge V$. Swapping the roles of V and V^*, there are also interior products on $\wedge V^*$, being dual to the exterior products on $\wedge V$, but we shall leave the statement of these analogous results to the reader.

Note that we have a well-defined exterior product

$$\wedge V^* \times \wedge V^* \to \wedge V^* : (\Theta, \Theta') \mapsto \Theta \wedge \Theta'$$

also on $\wedge V^*$, with properties as in Proposition 2.1.14.

Definition 2.6.1 (Interior products). The *left and right interior products* on $\wedge V$ are defined as follows.

- Let $\Theta_0 \in \wedge V^*$. *Left interior multiplication* $w \mapsto \Theta_0 \lrcorner w$ by Θ_0 on $\wedge V$ is the operation adjoint to left exterior multiplication $\Theta \mapsto \Theta_0 \wedge \Theta$ by Θ_0 on $\wedge V^*$, that is,

$$\langle \Theta, \Theta_0 \lrcorner w \rangle = \langle \Theta_0 \wedge \Theta, w \rangle, \quad \text{for all } \Theta \in \wedge V^*, \ w \in \wedge V.$$

- Let $\Theta_0 \in \wedge V^*$. *Right interior multiplication* $w \mapsto w \llcorner \Theta_0$ by Θ_0 on $\wedge V$ is the operation adjoint to right exterior multiplication $\Theta \mapsto \Theta \wedge \Theta_0$ by Θ_0 on $\wedge V^*$, that is,

$$\langle \Theta, w \llcorner \Theta_0 \rangle = \langle \Theta \wedge \Theta_0, w \rangle, \quad \text{for all } \Theta \in \wedge V^*, \ w \in \wedge V.$$

We remark that in most applications V is an inner product space, in which case $V^* = V$ and the interior products are then bilinear products on $\wedge V$.

Concrete calculation of an interior product $\Theta \lrcorner w$ is best performed in a pair of dual bases $\{e_s^*\}$ and $\{e_t\}$. By bilinearity, we have

$$\Theta \lrcorner w = \Big(\sum_{s \subset \overline{n}} x_s e_s^* \Big) \lrcorner \Big(\sum_{t \subset \overline{n}} y_t e_t \Big) = \sum_{s \subset t} x_s y_t \, e_s^* \lrcorner e_t,$$

if Θ and w have coordinates x_s and y_t respectively. Thus it remains to find $e_s^* \lrcorner e_t$. Note that

$$\langle e_{s'}^*, e_s^* \lrcorner e_t \rangle = \langle e_s^* \wedge e_{s'}^*, e_t \rangle = \langle \epsilon(s,s') e_{s \cup s'}^*, e_t \rangle = \begin{cases} \epsilon(s,s'), & t = s' \cup s, \\ 0, & t \neq s' \cup s, \end{cases}$$

using the notation from Definition 2.1.16. This vanishes for all s' unless $s \subset t$. If $s \subset t$, then all $e_{s'}$ coordinates of $e_s^* \lrcorner e_t$, except for $s' = t \setminus s$, are zero and $\langle e_{t \setminus s}^*, e_s^* \lrcorner e_t \rangle = \epsilon(s, t \setminus s)$. We have shown that

$$e_s^* \lrcorner e_t = \begin{cases} \epsilon(s, t \setminus s) \, e_{t \setminus s}, & s \subset t, \\ 0, & s \not\subset t. \end{cases}$$

Thus, modulo signs, just as exterior products of basis multivectors $e_s \wedge e_t$ correspond to taking unions $s \cup t$ of index sets (with zero result if $s \cap t \neq \varnothing$), left interior products $e_s^* \lrcorner e_t$ correspond to taking set differences $t \setminus s$ of index sets (with zero result if $s \not\subset t$). The same holds for the right interior product $e_t \llcorner e_s^*$, the only difference to $e_s^* \lrcorner e_t$ is possibly the sign. Indeed, as above, one verifies that

$$e_t \llcorner e_s^* = \begin{cases} \epsilon(t \setminus s, s) \, e_{t \setminus s}, & s \subset t, \\ 0, & s \not\subset t. \end{cases}$$

Exercise 2.6.2. Show that if $w \in \wedge^k V$ and $\Theta \in \wedge^l V^*$, then

$$\Theta \lrcorner w, \quad w \llcorner \Theta \in \wedge^{k-l} V.$$

In particular, if $k = l$, show that

$$\Theta \lrcorner w = w \llcorner \Theta = \langle \Theta, w \rangle,$$

where as usual $\mathbf{R}$ and $\wedge^0 V$ are identified.

At first it may seem unnecessary to have two different interior products $\lrcorner$ and $\llcorner$, rather than just one such product. However, to obtain good algebraic properties, the notation needs to convey the asymmetry of the interior products. Thus $\Theta \lrcorner w$ indicates that "Θ is removed from w from the left side", whereas $w \llcorner \Theta$ indicates that "Θ is removed from w from the right side". In some cases it does not matter whether $\Theta \lrcorner w$ or $w \llcorner \Theta$ is used, and in such cases we shall prefer the left interior product $\Theta \lrcorner w$ for the reason that we view $\Theta \lrcorner (\cdot)$ as operating on w, using standard functional notation $f(\cdot)$.

To complete the algorithm of calculating interior products $\Theta \lrcorner w$, which boils down to $e_s^* \lrcorner e_t$ as above, we need a practical method for finding the signs $\epsilon(s, t \setminus s)$. The usual method is to use associativty properties. Consider first the associative exterior product. For example, to calculate $e_{135} \wedge e_{246}$, we would calculate

$$e_{13} \wedge (e_5 \wedge e_{246}) = e_{13} \wedge (-1)^2 e_{2456} = e_1 \wedge (e_3 \wedge e_{2456}) = e_1 \wedge (-1)^1 e_{23456} = -e_{123456},$$

using associativity and the alternating property. Associativity like "$\Theta_1 \lrcorner (\Theta_2 \lrcorner w) = (\Theta_1 \lrcorner \Theta_2) \lrcorner w$" does not hold for the interior product. In fact, the right-hand side is not even well defined. The kind of associativity that holds for the interior products is the following.

Proposition 2.6.3 (Interior algebra). *The interior products are bilinear and satisfy the associativity relations*

$$(\Theta_1 \wedge \Theta_2) \lrcorner w = \Theta_2 \lrcorner (\Theta_1 \lrcorner w),$$
$$(\Theta_1 \lrcorner w) \llcorner \Theta_2 = \Theta_1 \lrcorner (w \llcorner \Theta_2),$$
$$w \llcorner (\Theta_1 \wedge \Theta_2) = (w \llcorner \Theta_2) \llcorner \Theta_1,$$

and the commutativity relation

$$\Theta \lrcorner w = (-1)^{l(k-l)} w \llcorner \Theta, \quad w \in \wedge^k V, \ \Theta \in \wedge^l V^*.$$

Proof. The proposition is a dual version of Proposition 2.1.14. Bilinearity is clear. For $\Theta' \in \wedge V^*$, we have

$$\langle \Theta', (\Theta_1 \wedge \Theta_2) \lrcorner w \rangle = \langle (\Theta_1 \wedge \Theta_2) \wedge \Theta', w \rangle = \langle \Theta_1 \wedge (\Theta_2 \wedge \Theta'), w \rangle$$
$$= \langle \Theta_2 \wedge \Theta', \Theta_1 \lrcorner w \rangle = \langle \Theta', \Theta_2 \lrcorner (\Theta_1 \lrcorner w) \rangle,$$

which proves the first associativity relation, since Θ' is arbitrary. The other two are proved similarly. To prove the commutativity relation, we argue similarly for $\Theta' \in \wedge^{k-l} V^*$ that

$$\langle \Theta', \Theta \lrcorner w \rangle = \langle \Theta \wedge \Theta', w \rangle = (-1)^{l(k-l)} \langle \Theta' \wedge \Theta, w \rangle = \langle \Theta', (-1)^{l(k-l)} w \llcorner \Theta \rangle. \qquad \square$$

Example 2.6.4. Consider the left interior product of a covector e_j^* and a multivector e_s, where $j \in s$. In this case, the sign $\epsilon(j, s \setminus j)$ is simply $(-1)^k$, where k is the number of indices in s strictly smaller than j. With this observation and the first associativity relation above, we can calculate a more general interior product like $e_{245}^* \lrcorner e_{12345}$ as

$$e_{245}^* \lrcorner e_{12345} = e_{45}^* \lrcorner (e_2^* \lrcorner e_{12345}) = e_{45}^* \lrcorner (-1)^1 e_{1345}$$
$$= -e_5^* \lrcorner (e_4^* \lrcorner e_{1345}) = -e_5^* \lrcorner (-1)^2 e_{135} = -e_{13}.$$

For the following two examples, compare Example 2.1.17. A concrete example of an interior product in the plane is

$$(4 - 7e_1^* + 3e_2^* + 2e_{12}^*) \lrcorner (1 + 5e_1 - 6e_2 - 8e_{12})$$
$$= (4 + 20e_1 - 24e_2 - 32e_{12}) + (0 - 35 + 0 + 56e_2) + (0 + 0 - 18 + 24e_1)$$
$$+ (0 + 0 + 0 - 16)$$
$$= -65 + 44e_1 + 32e_2 - 32e_{12}.$$

A concrete example of an interior product in three-dimensional space is

$$(3 - e_1^* - e_2^* + 2e_3^* + e_{12}^* + 5e_{13}^* - 2e_{23}^* - 7e_{123}^*)$$
$$\lrcorner (-2 + 4e_1 - 3e_2 + e_3 - 6e_{12} + 3e_{13} - e_{23} + 2e_{123})$$
$$= (-6 + 12e_1 - 9e_2 + 3e_3 - 18e_{12} + 9e_{13} - 3e_{23} + 6e_{123})$$
$$+ (0 - 4 + 0 + 0 + 6e_2 - 3e_3 + 0 - 2e_{23})$$
$$+ (0 + 0 + 3 + 0 - 6e_1 + 0 + e_3 + 2e_{13})$$
$$+ (0 + 0 + 0 + 2 + 0 - 6e_1 + 2e_2 + 4e_{12})$$
$$+ (0 + 0 + 0 + 0 - 6 + 0 + 0 + 2e_3) + (0 + 0 + 0 + 0 + 0 + 15 + 0 - 10e_2)$$
$$+ (0 + 0 + 0 + 0 + 0 + 0 + 2 - 4e_1) + (0 + 0 + 0 + 0 + 0 + 0 + 0 - 14)$$
$$= -8 - 4e_1 - 11e_2 + 3e_3 - 14e_{12} + 11e_{13} - 5e_{23} + 6e_{123}.$$

For the remainder of this section, we also assume that the space V is oriented and an n-vector $e_{\overline{n}} \in \wedge^n V$ is chosen. We write $e_{\overline{n}}^*$ for the unique nonzero n-covector such that $\langle e_{\overline{n}}^*, e_{\overline{n}} \rangle = 1$, which specifies the corresponding orientation of V^*. From the duality $\langle V^*, V \rangle$, we have defined interior products on $\wedge V$. Similarly, by swapping the roles of V and V^*, we construct interior products on $\wedge V^*$.

Definition 2.6.5 (Hodge stars). Let (X, V) be an oriented affine space, and consider a duality $\langle V^*, V \rangle$. Then the *Hodge star maps* are

$$\wedge V \to \wedge V^* : w \mapsto *w := e_{\overline{n}}^* \llcorner w,$$
$$\wedge V^* \to \wedge V : \Theta \mapsto \Theta* := \Theta \lrcorner e_{\overline{n}}.$$

Note that the Hodge star maps depend on the choice of $e_{\overline{n}}$. However, making another choice $e_{\overline{n}}'$ with $\langle e_{\overline{n}}^{*\prime}, e_{\overline{n}}' \rangle = 1$, we must have $e_{\overline{n}}' = \lambda e_{\overline{n}}$ and $e_{\overline{n}}^{*\prime} = \lambda^{-1} e_{\overline{n}}^*$ for some $0 \neq \lambda \in \mathbf{R}$. Thus the only change is that the Hodge star map $w \mapsto *w$ is scaled with the factor λ^{-1} and $\Theta \mapsto \Theta*$ is scaled by λ.

Remark 2.6.6. The side on which the Hodge star is placed is chosen to agree with the side the multi(co)vector is removed from the n-(co)vector (and the order used in the duality $\langle \Theta, w \rangle$). Even if we will not need it, we could also consider Hodge star maps

$$\wedge V \to \wedge V^* : w \mapsto w* := w \lrcorner\, e_{\overline{n}}^*,$$
$$\wedge V^* \to \wedge V : \Theta \mapsto *\Theta := e_{\overline{n}} \llcorner \Theta.$$

Note that if $w \in \wedge^k V$ and $\Theta \in \wedge^l V^*$, then from the commutativity properties of the interior products, it follows that $w* = (-1)^{k(n-k)} *w$ and $*\Theta = (-1)^{l(n-l)} \Theta*$. In particular, when $\dim V$ is odd, we have $w* = *w$ for all $w \in \wedge V$ and $*\Theta = \Theta*$ for all $\Theta \in \wedge V^*$.

As a special case of the algebra for interior products, we see that in dual induced bases $\{e_s^*\} \subset \wedge V^*$ and $\{e_t\} \subset \wedge V$, the formulas

$$*\left(\sum_s x_s e_s \right) = \sum_s x_s\, \epsilon(\overline{n} \setminus s, s) e_{\overline{n}\setminus s}^*,$$
$$\left(\sum_t y_t e_t^* \right)* = \sum_t y_t\, \epsilon(t, \overline{n} \setminus t) e_{\overline{n}\setminus t},$$

hold for the Hodge star maps. Note that both correspond to taking complements of index sets relative to $\overline{n}$, modulo signs.

Example 2.6.7. In the oriented plane, one has

$$*w = *(1 + 5e_1 - 6e_2 - 8e_{12}) = e_{12}^* - 5e_2^* - 6e_1^* - 8 = \Theta.$$

One checks that $\Theta* = w$, but that $*\Theta$ is the involution $\widehat{w}$ of w. Compare to Proposition 2.6.3.

In oriented 3-dimensional space, one has

$$*w = *(-2 + 4e_1 - 3e_2 + e_3 - 6e_{12} + 3e_{13} - e_{23} + 2e_{123})$$
$$= -2e_{123}^* + 4e_{23}^* + 3e_{13}^* + e_{12}^* - 6e_3^* - 3e_2^* - e_1^* + 2 = \Theta.$$

One checks that $\Theta* = w$, and also $*\Theta = w$.

Note carefully the useful fact that the left and right Hodge star maps $*w$ and $w*$ coincide in 3-dimensional space, and more generally in odd-dimensional spaces, but not in even-dimensional spaces, where they differ in sign on k-vectors, for k odd.

Proposition 2.6.8 (Star algebra). *The Hodge star maps $*w$ and $\Theta*$ have the following properties.*

(i) *They are each other's inverses, that is,*

$$(*w)* = (e_{\overline{n}}^* \llcorner w) \lrcorner\, e_{\overline{n}} = w, \quad \textit{for all } w \in \wedge V,$$
$$*(\Theta*) = e_{\overline{n}}^* \llcorner (\Theta \lrcorner\, e_{\overline{n}}) = \Theta, \quad \textit{for all } \Theta \in \wedge V^*.$$

(ii) *They swap exterior and interior products in the sense that if $w_i \in \wedge V$ and $\Theta_i \in \wedge V^*$, $i = 1, 2$, then*

$$*(w_1 \wedge w_2) = (*w_2) \llcorner w_1, \quad (\Theta_2 \llcorner w_1)* = w_1 \wedge (\Theta_2*),$$
$$(\Theta_1 \wedge \Theta_2)* = \Theta_2 \lrcorner (\Theta_1*), \quad *(\Theta_2 \lrcorner w_1) = (*w_1) \wedge \Theta_2.$$

Proof. To verify $(*w)* = w$, by linearity it suffices to consider $w = e_s$. In this case $(*w)* = \epsilon(\overline{n} \setminus s, s)^2 e_s = w$. A similar basis calculation proves $*(\Theta*) = \Theta$ as well. Part (ii) follows from the associative relations for interior products from Proposition 2.6.3 applied to $e_{\overline{n}}^* \llcorner (w_1 \wedge w_2)$ and $(\Theta_1 \wedge \Theta_2) \lrcorner e_{\overline{n}}$ respectively. The right two identities follow from the left identities by applying Hodge stars on both sides, using (i) and renaming the variables. $\qquad \square$

As noted above, the Hodge star maps set up identifications

$$\wedge^k V \longleftrightarrow \wedge^{n-k} V^*,$$

where both spaces have dimension $\binom{n}{k}$. The geometric interpretation of this for simple k-vectors, is that a k-volume in a k-dimensional subspace W is identified with a density of subspaces parallel to W.

Proposition 2.6.9 (Geometry of $*$). *If $w \in \wedge^k V$, then*

$$\lfloor *w \rfloor = \lceil w \rceil^\perp \quad and \quad \lceil *w \rceil = \lfloor w \rfloor^\perp.$$

Thus $\dim \lfloor *w \rfloor = n - \dim \lceil w \rceil$ and $\dim \lceil *w \rceil = n - \dim \lfloor w \rfloor$. In particular, $*w$ is simple if and only if w is simple.

Proof. We claim that $\lfloor *w \rfloor \supset V_1^\perp$ if and only if $\lceil w \rceil \subset V_1$, for subspaces $V_1 \subset V$. By minimizing V_1, this will prove $\lfloor *w \rfloor = \lceil w \rceil^\perp$. Switching the roles of V and V^*, one similarly proves $\lfloor \Theta* \rfloor = \lceil \Theta \rceil^\perp$. Writing $\Theta = *w$ and taking orthogonal complements, this proves $\lfloor w \rfloor^\perp = (\lceil *w \rceil^\perp)^\perp = \lceil *w \rceil$.

To see the claim, fix dual bases so that $V_1 = \text{span}\{e_{m+1}, \dots, e_n\}$ and $V_1^\perp = \text{span}\{e_1^*, \dots, e_m^*\}$. That $\lceil w \rceil \subset V_1$ means that $w \in \wedge^k V_1$. Writing

$$w = \sum_{s \subset \{m+1, \dots, n\}} w_s e_s$$

in the induced basis, we see that this happens if and only if $*w = e_1^* \wedge \cdots \wedge e_m^* \wedge \Theta$ for some $\Theta \in \wedge V^*$, since $*e_s$ contains this factor $e_1^* \wedge \cdots \wedge e_m^*$ when $s \subset \{m+1, \dots, n\}$. Propositions 2.2.6 and 2.2.8 prove the last statement. $\qquad \square$

Corollary 2.6.10 (Outer space formula). *Let $w \in \wedge^k V$. Then its outer space is*

$$\lceil w \rceil = \{\theta \in V^* \; ; \; \theta \lrcorner w = 0\}^\perp.$$

Proof. By Proposition 2.6.9, $v \in \lceil w \rceil$ if and only if v is orthogonal to $\lfloor *w \rfloor$. By Proposition 2.6.8(ii) we have $\theta \in \lfloor *w \rfloor$ if and only if $\theta \lrcorner w$. This proves the claim. $\qquad \square$

Proposition 2.6.11 (Geometry of $\lrcorner$). *Let* $w \in \wedge^k V$ *and* $\Theta \in \wedge^l V^*$. *Then*

$$\lfloor \Theta \lrcorner w \rfloor \supset \lfloor w \rfloor \cap \lceil \Theta \rceil^\perp,$$
$$\lceil \Theta \lrcorner w \rceil \subset \lceil w \rceil \cap \lfloor \Theta \rfloor^\perp.$$

If w *and* Θ *are simple, then so is* $\Theta \lrcorner w$, *and if* $\Theta \lrcorner w \neq 0$, *then*

$$[\Theta \lrcorner w] = [w] \cap [\Theta]^\perp.$$

Proof. It follows from Propositions 2.6.8(ii), 2.6.9, and 2.2.5 that

$$\lfloor \Theta \lrcorner w \rfloor = \lfloor ((*w) \wedge \Theta)* \rfloor = \lceil (*w) \wedge \Theta \rceil^\perp \supset (\lceil *w \rceil + \lceil \Theta \rceil)^\perp = \lceil *w \rceil^\perp \cap \lceil \Theta \rceil^\perp$$
$$= \lfloor w \rfloor \cap \lceil \Theta \rceil^\perp.$$

The second inclusion is proved similarly, and the claim for simple multi(co)vectors is clear, since

$$\lfloor w \rfloor \cap \lceil \Theta \rceil^\perp \subset \lfloor \Theta \lrcorner w \rfloor \subset \lceil \Theta \lrcorner w \rceil \subset \lceil w \rceil \cap \lfloor \Theta \rfloor^\perp,$$

where the left- and right-hand sides are equal. $\qquad\square$

We summarize the geometric interpretations of the exterior and interior products and the Hodge star maps. Under suitable assumptions they are the algebraic operations corresponding to direct sums, direct differences, and orthogonal complements of subspaces:

$$\lfloor w_1 \wedge w_2 \rfloor = \lfloor w_1 \rfloor \oplus \lfloor w_2 \rfloor,$$
$$\lfloor \Theta \lrcorner w \rfloor = \lfloor w \llcorner \Theta \rfloor = \lfloor w \rfloor \ominus \lfloor \Theta \rfloor := \lfloor w \rfloor \cap \lfloor \Theta \rfloor^\perp,$$
$$\lfloor *w \rfloor = \lfloor w \rfloor^\perp.$$

We end this section by considering the calculation of interior products and Hodge stars in inner product spaces, that is, we consider the special case of an inner product $\langle V, V \rangle$ that identifies V^* and V.

Let $\{e_1, \ldots, e_n\}$ be an ON-basis for V. Since the exterior product does not at all depend on the inner product, products of basis multivectors are as before,

$$e_s \wedge e_t = \begin{cases} \epsilon(s,t)\, e_{s \cup t}, & s \cap t = \varnothing, \\ 0, & s \cap t \neq \varnothing. \end{cases}$$

For the interior products, the formulas are

$$e_s \lrcorner e_t = \begin{cases} 0, & s \not\subset t, \\ \langle e_s \rangle^2\, \epsilon(s, t \setminus s)\, e_{t \setminus s}, & s \subset t, \end{cases} \qquad e_t \llcorner e_s = \begin{cases} 0, & s \not\subset t, \\ \langle e_s \rangle^2\, \epsilon(t \setminus s, s)\, e_{t \setminus s}, & s \subset t. \end{cases}$$

These formulas follow by recalling the general formulas in the affine case, and writing $e_s^* = \langle e_s \rangle^2 e_s$. Note that in a Euclidean space, the signs $\langle e_s \rangle^2 = \pm 1$ are all $+$, but in spacetimes there are equally many basis multivectors with $\langle e_s \rangle^2 = -1$ as with $\langle e_s \rangle^2 = +1$.

For the Hodge star maps, the basis formulas become

$$*e_s = \langle e_{\overline{n}\setminus s} \rangle^2 \epsilon(\overline{n}\setminus s, s)\, e_{\overline{n}\setminus s}, \quad e_s* = \langle e_s \rangle^2 \epsilon(s, \overline{n}\setminus s)\, e_{\overline{n}\setminus s}.$$

Note carefully that the right Hodge star $*w = e_{\overline{n}} \lrcorner w$ uses the dual volume element $e_{\overline{n}}^* = \langle e_{\overline{n}} \rangle^2 e_{\overline{n}}$.

Definition 2.6.12 (Vector product). Let V be an oriented Euclidean space and assume that $\dim V = 3$. Then the *vector product* of two vectors $v_1, v_2 \in V$ is the vector

$$v_1 \times v_2 := *(v_1 \wedge v_2) \in V.$$

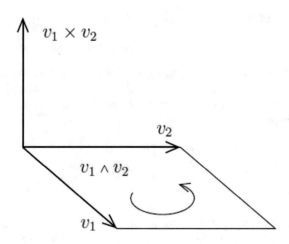

Figure 2.4: Hodge duality between the exterior product and the vector product.

Let $\{e_1, e_2, e_3\}$ be an ON-basis such that $e_1 \wedge e_2 \wedge e_3$ is the orientation of V. Then

$$e_1 \times e_2 = *e_{12} = e_3, \quad e_1 \times e_3 = *e_{13} = -e_2, \quad e_2 \times e_3 = *e_{23} = e_1.$$

Thus this definition of the vector product agrees with the usual one. It should be noted that having access to bivectors, the exterior product and the Hodge star map, it becomes clear that the vector product is a hybrid, a composition of two operations.

Example 2.6.13 (Triple products). Let v_1, v_2, v_3 be vectors in a three-dimensional oriented Euclidean space. Then their triple scalar product is

$$\langle v_1, v_2 \times v_3 \rangle = \big(*(v_2 \wedge v_3) \big) \llcorner v_1 = *(v_1 \wedge v_2 \wedge v_3),$$

and their triple vector product is

$$v_1 \times (v_2 \times v_3) = *(v_1 \wedge (v_2 \times v_3)) = (*(v_2 \times v_3)) \llcorner v_1 = -v_1 \lrcorner (v_2 \wedge v_3),$$

using Proposition 2.6.8. Note the geometry of the last right-hand side: It is a vector orthogonal to v_1 in the plane $\mathrm{span}\{v_2, v_3\}$ according to Propositions 2.2.5 and 2.6.11.

To demonstrate the dependence of the interior products and Hodge star maps on the inner product, we consider the following examples.

Example 2.6.14. Consider the left interior product calculated in three-dimensional affine space in Example 2.6.4. With a Euclidean inner product such that the basis used is an ON-basis, we get the same result

$$(3 - e_1 - e_2 + 2e_3 + e_{12} + 5e_{13} - 2e_{23} - 7e_{123})$$
$$\lrcorner \, (-2 + 4e_1 - 3e_2 + e_3 - 6e_{12} + 3e_{13} - e_{23} + 2e_{123})$$
$$= -8 - 4e_1 - 11e_2 + 3e_3 - 14e_{12} + 11e_{13} - 5e_{23} + 6e_{123}.$$

The only difference is that we write e_i in the left factor instead of $e_i^* = g(e_i)$, with g from Proposition 1.2.3. Assume now instead that an inner product for spacetime is used such that $\{e_i\}$ is an ON-basis with $\langle e_1 \rangle^2 = -1$ and $\langle e_2 \rangle^2 = \langle e_3 \rangle^2 = +1$. Then the interior product instead becomes

$$(3 - e_1 - e_2 + 2e_3 + e_{12} + 5e_{13} - 2e_{23} - 7e_{123})$$
$$\lrcorner \, (-2 + 4e_1 - 3e_2 + e_3 - 6e_{12} + 3e_{13} - e_{23} + 2e_{123})$$
$$= (-6 + 12e_1 - 9e_2 + 3e_3 - 18e_{12} + 9e_{13} - 3e_{23} + 6e_{123})$$
$$- (0 - 4 + 0 + 0 + 6e_2 - 3e_3 + 0 - 2e_{23})$$
$$+ (0 + 0 + 3 + 0 - 6e_1 + 0 + e_3 + 2e_{13})$$
$$+ (0 + 0 + 0 + 2 + 0 - 6e_1 + 2e_2 + 4e_{12}) - (0 + 0 + 0 + 0 - 6 + 0 + 0 + 2e_3)$$
$$- (0 + 0 + 0 + 0 + 0 + 15 + 0 - 10e_2) + (0 + 0 + 0 + 0 + 0 + 0 + 2 - 4e_1)$$
$$- (0 + 0 + 0 + 0 + 0 + 0 + 0 - 14)$$
$$= 10 - 4e_1 - 3e_2 + 5e_3 - 14e_{12} + 11e_{13} - e_{23} + 6e_{123}.$$

Example 2.6.15. In oriented 3-dimensional affine space, one has

$$*w = e_{123}^* \llcorner (-2 + 4e_1 - 3e_2 + e_3 - 6e_{12} + 3e_{13} - e_{23} + 2e_{123})$$
$$= -2e_{123}^* + 4e_{23}^* + 3e_{13}^* + e_{12}^* - 6e_3^* - 3e_2^* - e_1^* + 2.$$

If $\{e_i\}$ is an ON-basis in oriented Euclidean space, then this shows that

$$*w = e_{123} \lrcorner w = -2e_{123} + 4e_{23} + 3e_{13} + e_{12} - 6e_3 - 3e_2 - e_1 + 2.$$

Check that $(*w)* = w$. If on the other hand, $\{e_i\}$ is an ON-basis in oriented spacetime with $\langle e_1 \rangle^2 = -1$ and $\langle e_2 \rangle^2 = \langle e_3 \rangle^2 = +1$, then the result is instead the multivector

$$*w = -e_{123} \lrcorner w = 2e_{123} + 4e_{23} - 3e_{13} - e_{12} - 6e_3 - 3e_2 + e_1 + 2.$$

Check that $(*w)* = w$.

In an inner product space the two Hodge stars act on the same space $\wedge V$, and we have the following additional identities.

Proposition 2.6.16. *Let V be an oriented inner product space. Then*

$$\langle *w_1, *w_2 \rangle = \langle e_{\overline{n}} \rangle^2 \langle w_1, w_2 \rangle = \langle w_1*, w_2* \rangle, \quad \text{for all } w_1, w_2 \in \wedge V,$$

and

$$*w = \begin{cases} \langle e_{\overline{n}} \rangle^2 w*, & \text{if } \dim V \text{ is odd}, \\ \langle e_{\overline{n}} \rangle^2 \widehat{w}*, & \text{if } \dim V \text{ is even}. \end{cases}$$

In particular, in Euclidean spaces, the Hodge star map $w \mapsto *w : \wedge V \to \wedge V$ is isometric, and also symmetric in odd dimension.

Proof. For example, the first formula follows from the calculation

$$\langle *w_1, *w_2 \rangle = \langle e_{\overline{n}} \rangle^2 \langle e_{\overline{n}}, (*w_2) \wedge w_1 \rangle = \langle e_{\overline{n}} \rangle^2 \langle e_{\overline{n}}, *(w_1 \lrcorner w_2) \rangle$$
$$= \langle e_{\overline{n}} \wedge (w_1 \lrcorner w_2), e_{\overline{n}} \rangle = \langle w_1 \lrcorner w_2, e_{\overline{n}} \lrcorner e_{\overline{n}} \rangle = \langle e_{\overline{n}} \rangle^2 \langle w_1, w_2 \rangle.$$

The $*$-commutation formula follows from $w \lrcorner e_{\overline{n}} = (-1)^{k(n-k)} e_{\overline{n}} \lrcorner w$, deduced from Proposition 2.6.3 for $w \in \wedge^k V$. Note that $(-1)^{k(n-k)} = 1$ if n is odd, and $(-1)^{k(n-k)} = (-1)^k$ if n is even. $\square$

Exercise 2.6.17. Prove the relation

$$(w_1 \wedge (*w_2))* = \langle w_1, w_2 \rangle, \quad w_1, w_2 \in \wedge V,$$

between the exterior product, Hodge star maps, and the inner product in an oriented inner product space, and show how it generalizes to oriented affine space.

Exercise 2.6.18. Let V be an oriented Euclidean space of even dimension $n = 2m$, and define the linear map

$$Tw := w*$$

on m-vectors $w \in \wedge^m V$. Show that $T^2 w = (-1)^m w$, for $w \in \wedge^m V$.

2.7   Mappings of Interior Products

From Proposition 2.3.2 we recall that

$$T(w_1 \wedge w_2) = (Tw_1) \wedge (Tw_2)$$

for induced maps. Since

$$\langle \theta_1 \wedge \cdots \wedge \theta_k, Tv_1 \wedge \cdots \wedge Tv_k \rangle = \begin{vmatrix} \langle \theta_1, Tv_1 \rangle & \cdots & \langle \theta_1, Tv_k \rangle \\ \vdots & \ddots & \vdots \\ \langle \theta_k, Tv_1 \rangle & \cdots & \langle \theta_k, Tv_k \rangle \end{vmatrix}$$

$$= \begin{vmatrix} \langle T^*\theta_1, v_1 \rangle & \cdots & \langle T^*\theta_1, v_k \rangle \\ \vdots & \ddots & \vdots \\ \langle T^*\theta_k, v_1 \rangle & \cdots & \langle T^*\theta_k, v_k \rangle \end{vmatrix}$$

$$= \langle T^*\theta_1 \wedge \cdots \wedge T^*\theta_k, v_1 \wedge \cdots \wedge v_k \rangle$$

by Definition 2.5.2, it is clear from Definition 2.3.1 that the induced $\wedge$-homomorphisms $T_\wedge$ and $(T^*)_\wedge$ are dual/adjoint, that is, $(T_\wedge)^* = (T^*)_\wedge$. We consider how this T^* maps interior products.

Proposition 2.7.1 (Mapping of interior products). *For dual maps $T : \wedge V_1 \to \wedge V_2$ and $T^* : \wedge V_2^* \to \wedge V_1^*$ induced by linear $T : V_1 \to V_2$ and $T^* : V_2^* \to V_1^*$, we have the identities*

$$T((T^*\Theta) \lrcorner w) = \Theta \lrcorner T(w),$$
$$T^*((Tw) \lrcorner \Theta) = w \lrcorner T^*(\Theta), \quad w \in \wedge V_1, \ \Theta \in \wedge V_2^*.$$

Swapping the roles of V_i and V_i^* gives the second identity.

Proof. For every $\Theta_2 \in \wedge V_2^*$, we have

$$\langle \Theta_2, T((T^*\Theta) \lrcorner w) \rangle = \langle T^*\Theta_2, (T^*\Theta) \lrcorner w \rangle = \langle (T^*\Theta) \wedge (T^*\Theta_2), w \rangle$$
$$= \langle T^*(\Theta \wedge \Theta_2), w \rangle = \langle \Theta \wedge \Theta_2, T(w) \rangle = \langle \Theta_2, \Theta \lrcorner T(w) \rangle.$$

This proves the first identity, and the second is proved similarly.   □

At first, the formula $T((T^*\Theta) \lrcorner w) = \Theta \lrcorner T(w)$ seems impossible by homogeneity considerations. But as noted, the map $T \mapsto T_\wedge$ is not linear, not even homogeneous, with respect to T. For example, if $T : V \to V$ is multiplied by α, then $T : \wedge^k V \to \wedge^k V$ is scaled by α^k.

Corollary 2.7.2 (Generalized cofactor formula). *Let V be an oriented vector space. For an induced linear map $T : \wedge V \to \wedge V$ we have Hodge star identities*

$$T((T^*\Theta)*) = \det(T)\, \Theta*, \quad \Theta \in \wedge V^*,$$
$$T^*(*(Tw)) = \det(T)\, *w, \quad w \in \wedge V.$$

Proof. Fixing n-(co)vectors $\langle e_{\overline{n}}^*, e_{\overline{n}} \rangle = 1$, we have according to Proposition 2.7.1 that

$$T((T^*\Theta) \lrcorner e_{\overline{n}}) = \Theta \lrcorner T(e_{\overline{n}}) = \det(T) \,\Theta *,$$
$$T^*(e_{\overline{n}}^* \llcorner (Tw)) = (T^* e_{\overline{n}}^*) \llcorner w = \det(T^*) \,*w.$$

Note the multivector proof of a well-known fact: $\det(T^*) = \langle \det(T^*)e_{\overline{n}}^*, e_{\overline{n}} \rangle = \langle T^*(e_{\overline{n}}^*), e_{\overline{n}} \rangle = \langle e_{\overline{n}}^*, T(e_{\overline{n}}) \rangle = \det(T)$. $\qquad\qquad\square$

Example 2.7.3. Evaluating the spaces for both sides of the identities in Corollary 2.7.2 gives relations between the mapping properties of T and T^*. Assume for simplicity that T is invertible and w is simple and nonzero. Then

$$[w]^\perp = [*w] = [T^*(*(Tw))] = T^*[*(Tw)] = T^*([Tw]^\perp) = T^*((T[w])^\perp),$$

using Propositions 2.3.2 and 2.6.9. Thus with $V_1 := [w]$ and $V_2 := T[w]$, we see that if $T : V_1 \to V_2$, then $T^* : V_2^\perp \to V_1^\perp$.

Recall the *cofactor formula* for computation of inverses from linear algebra, which in two dimensions reads

$$\begin{bmatrix} a & b \\ c & d \end{bmatrix} = \frac{1}{ad - bc} \begin{bmatrix} d & -b \\ -c & a \end{bmatrix}.$$

Multivector algebra shows that the inverse map on V is closely related to the dual map on $\wedge^{n-1} V^*$.

Corollary 2.7.4 (Cofactor formula). *Let V be an oriented vector space, and let $T : V \to V$ be an invertible linear map. Consider the adjoint operator $T^* : V^* \to V^*$ and the induced operator $C := T^*|_{\wedge^{n-1}V^*} : \wedge^{n-1}V^* \to \wedge^{n-1}V^*$ on $(n-1)$-covectors. Then the inverse $T^{-1} : V \to V$ is related to C via the* cofactor formula

$$T^{-1}v = \frac{1}{\det T}\Big(C(*v)\Big)*, \quad v \in V. \tag{2.4}$$

Proof. Use Corollary 2.7.2 to get the identity

$$T^*(*(Tu)) = \det T \,(*u), \quad u \in V.$$

Applying the right Hodge star map, we get

$$(T^*(*(Tu)))* = \det(T)\,u.$$

Replacing Tu and u by v and $T^{-1}v$ respectively proves the cofactor formula. $\quad\square$

Example 2.7.5. Consider $T : V \to V$ from Example 2.3.4 with matrix

$$A = \begin{bmatrix} 1 & 1 & 0 \\ -1 & 1 & 2 \\ 2 & 0 & -1 \end{bmatrix}$$

in Euclidean ON-basis $\{e_1, e_2, e_3\}$. To find its inverse T^{-1} using the cofactor formula, we note that $T^* : V \to V$ has matrix A^t, the transpose of A. The induced operator $T^* : \wedge^2 V \to \wedge^2 V$ is seen to have matrix $\begin{bmatrix} 2 & -2 & 2 \\ 2 & -1 & -3 \\ 2 & -1 & -1 \end{bmatrix}$ in the induced basis $\{e_{12}, e_{13}, e_{23}\}$, and the Hodge star map $w \mapsto *w = w* : V \to \wedge^2 V$ has matrix $\begin{bmatrix} 0 & 0 & 1 \\ 0 & -1 & 0 \\ 1 & 0 & 0 \end{bmatrix}$, with respect to bases $\{e_1, e_2, e_3\}$ and $\{e_{12}, e_{13}, e_{23}\}$. Thus the matrix for T^{-1} in basis $\{e_1, e_2, e_3\}$ is

$$\frac{1}{2} \begin{bmatrix} 0 & 0 & 1 \\ 0 & -1 & 0 \\ 1 & 0 & 0 \end{bmatrix} \begin{bmatrix} 2 & -2 & 2 \\ 2 & -1 & -3 \\ 2 & -1 & -1 \end{bmatrix} \begin{bmatrix} 0 & 0 & 1 \\ 0 & -1 & 0 \\ 1 & 0 & 0 \end{bmatrix} = \frac{1}{2} \begin{bmatrix} -1 & 1 & 2 \\ 3 & -1 & -2 \\ -2 & 2 & 2 \end{bmatrix}.$$

Exercise 2.7.6. The cofactor formula (2.4) can be generalized to give a relation between $T^{-1} : \wedge^k V \to \wedge^k V$ and $T^* : \wedge^{n-k} V^* \to \wedge^{n-k} V^*$ for any $0 \le k \le n$. Work out the details of this.

Exercise 2.7.7. Let $T : V \to V$ be an invertible linear map of Euclidean space. Show that the induced map of multivectors $T : \wedge V \to \wedge V$ from Definition 2.3.1 is a homomorphism with respect to the left interior product, that is,

$$T(w_1 \lrcorner w_2) = (Tw_1) \lrcorner (Tw_2), \quad w_1, w_2 \in \wedge V,$$

if and only if $T : V \to V$ is an isometry. Show also that the same statement holds for the right interior product.

2.8 Anticommutation Relations

We know that
$$v_1 \wedge v_2 \wedge w = -v_2 \wedge v_1 \wedge w,$$

for any two vectors, that is, exterior multiplications by vectors always anticommute. There is also a fundamental anticommutation relation between exterior multiplication by a vector v and interior multiplication by a covector θ, but this depends on $\langle \theta, v \rangle$.

Theorem 2.8.1 (Anticommutation). *Let* (X, V) *be an affine space. For a vector* $v \in V$ *and a covector* $\theta \in V^*$, *we have the anticommutation relation*

$$\theta \lrcorner (v \wedge w) + v \wedge (\theta \lrcorner w) = \langle \theta, v \rangle w, \quad w \in \wedge V.$$

We remark that this anticommutation relation lies at the very heart of the theory of Clifford algebras, which we study from Chapter 3. Just as we shall see there that the Clifford product is closely related to isometries, we shall see in this section that the exterior and interior products are closely related to projections. For more general linear operators, the tensor product is the natural multilinear setting. See Proposition 1.4.3.

Proof. To prove the anticommutation relation, by homogeneity and continuity it suffices to consider the case $\langle \theta, v \rangle = 1$. In this case $V = [v] \oplus V_0$, where $V_0 := [\theta]^\perp$. We can choose a basis such that $v = e_1$ and $V_0 = \mathrm{span}\{e_2, \ldots, e_n\}$. Note that $\theta = e_1^*$. Writing w in the induced basis, we see that it can be uniquely decomposed:

$$w = w_0 + e_1 \wedge w_1, \qquad w_0, w_1 \in \wedge V_0.$$

Note from the basis expressions for the exterior and interior products that $e_1^* \lrcorner (e_1 \wedge w') = w'$ for all $w' \in \wedge V_0$. It follows that

$$\theta \lrcorner (v \wedge w) = \theta \lrcorner (v \wedge w_0) + 0 = w_0, \quad \text{and}$$
$$v \wedge (\theta \lrcorner w) = 0 + v \wedge (\theta \lrcorner (v \wedge w_1)) = e_1 \wedge w_1.$$

Summing gives $\theta \lrcorner (v \wedge w) + v \wedge (\theta \lrcorner w) = w = \langle \theta, v \rangle w.$ $\qquad \square$

Example 2.8.2. Consider three vectors v_1, v_2, v_3 in three-dimensional oriented Euclidean space V as in Example 2.6.13. Then the anticommutation relation shows the well-known expansion formula

$$v_1 \times (v_2 \times v_3) = -v_1 \lrcorner (v_2 \wedge v_3) = -\langle v_1, v_2 \rangle v_3 + \langle v_1, v_3 \rangle v_2$$

for the triple vector product. More generally, in any inner product space V, we have the formula

$$v_1 \lrcorner (v_2 \wedge v_3) = \langle v_1, v_2 \rangle v_3 - \langle v_1, v_3 \rangle v_2, \quad \text{for vectors } v_1, v_2, v_3 \in V,$$

as is seen by choosing $\theta = v_1$ and a vector $w = v_3 \in \wedge^1 V$ in Theorem 2.8.1.

Iterated use of the anticommutation relation clearly yields the following general expansion rule.

Corollary 2.8.3 (Expansion). *For a covector $\theta \in V^*$ and vectors $v_1, \ldots, v_k \in V$, we have the expansion rule*

$$\theta \lrcorner (v_1 \wedge \cdots \wedge v_k) = \sum_{j=1}^{k} (-1)^{j-1} \langle \theta, v_j \rangle v_1 \wedge \cdots \check{v}_j \cdots \wedge v_k,$$

where $\check{v}_j$ means omission of this factor from the exterior product.

The expansion rule from Corollary 2.8.3 can be seen as a geometric form of the expansion rules for determinants along rows or columns. To see this, pair the identity with a $(k-1)$-covector $\theta_2 \wedge \cdots \wedge \theta_k$ and write $\theta = \theta_1$. Then according to Definition 2.5.2, we have

$$\langle \theta_2 \wedge \cdots \wedge \theta_k, \theta_1 \lrcorner (v_1 \wedge \cdots \wedge v_k) \rangle = \langle \theta_1 \wedge \cdots \wedge \theta_k, v_1 \wedge \cdots \wedge v_k \rangle$$

$$= \begin{vmatrix} \langle \theta_1, v_1 \rangle & \cdots & \langle \theta_1, v_k \rangle \\ \vdots & \ddots & \vdots \\ \langle \theta_k, v_1 \rangle & \cdots & \langle \theta_k, v_k \rangle \end{vmatrix} \overset{k}{\underset{j=1}{=}} \sum_{j=1}^{k} (-1)^{j-1} \langle \theta_1, v_j \rangle D_{j,1}$$

$$= \sum_{j=1}^{k} (-1)^{j-1} \langle \theta_1, v_j \rangle \langle \theta_2 \wedge \cdots \wedge \theta_k, v_1 \wedge \cdots \check{v}_j \cdots \wedge v_k \rangle,$$

where $D_{j,1}$ is the subdeterminant obtained by deleting row j and column 1. This gives an alternative proof of the expansion rule, based on the well-known expansion rule for determinants. In the spirit of our notion of rectangular determinants, one can think of Corollary 2.8.3 as an expansion rule for rectangular determinants.

Corollary 2.8.4 (Generalized anticommutation). *Let (X, V) be an affine space. For a covector $\theta \in V^*$ and multivectors $w_1, w_2 \in \wedge V$, we have*

$$\theta \lrcorner (w_1 \wedge w_2) = (\theta \lrcorner w_1) \wedge w_2 + \widehat{w}_1 \wedge (\theta \lrcorner w_2).$$

Here $\widehat{w}_1$ denotes the involution from Definition 2.1.18.

Proof. We consider the case $w_1 = v_1 \wedge v_2 \in \wedge^2 V$ and leave the proof of the general case as an exercise. Applying Theorem 2.8.1 three times gives

$$\theta \lrcorner (v_1 \wedge v_2 \wedge w_2) = \langle \theta, v_1 \rangle v_2 \wedge w_2 - v_1 \wedge (\theta \lrcorner (v_2 \wedge w_2))$$

$$= \langle \theta, v_1 \rangle v_2 \wedge w_2 - \langle \theta, v_2 \rangle v_1 \wedge w_2 + v_1 \wedge v_2 \wedge (\theta \lrcorner w_2)$$

$$= (\langle \theta, v_1 \rangle v_2 - \langle \theta, v_2 \rangle v_1) \wedge w_2 + (v_1 \wedge v_2) \wedge (\theta \lrcorner w_2)$$

$$= (\theta \lrcorner w_1) \wedge w_2 + w_1 \wedge (\theta \lrcorner w_2). \qquad \square$$

Example 2.8.5 (Gram–Schmidt). Recall from linear algebra the Gram–Schmidt orthogonalization process to construct an ON-basis $\{e_1, \ldots, e_n\}$ from an arbitrary basis $\{v_1, \ldots, v_n\}$ for a Euclidean space V. Given $\{v_i\}$, the recursive process is to first set $\tilde{v}_1 := v_1$, $e_1 := \tilde{v}_1 / |\tilde{v}_1|$, and then, having defined $\{\tilde{v}_1, \ldots, \tilde{v}_k\}$ and $\{e_1, \ldots, e_k\}$, one sets

$$\tilde{v}_{k+1} := v_{k+1} - \sum_{i=1}^{k} \langle v_{k+1}, e_i \rangle e_i \quad \text{and} \quad e_{k+1} := \tilde{v}_{k+1} / |\tilde{v}_{k+1}|.$$

With exterior algebra, it is straightforward to find $\{e_1, \ldots, e_n\}$. Indeed, since the idea of the Gram–Schmidt process is to define the ON-basis $\{e_i\}$ such that

$$e_{k+1} \in \text{span}\{v_1, \ldots, v_{k+1}\} \cap \text{span}\{v_1, \ldots, v_k\}^\perp,$$

it is clear from Proposition 2.6.11 that we can take

$$\tilde{v}_{k+1} := (v_1 \wedge \cdots \wedge v_k) \lrcorner (v_1 \wedge \cdots \wedge v_{k+1})$$

and then normalize to $e_{k+1} := \tilde{v}_{k+1}/|\tilde{v}_{k+1}|$. As the reader can check using the expansion rule from Corollary 2.8.3, this yields the same ON-basis $\{e_i\}$ as the vector calculus algorithm. Note that the vectors $\tilde{v}_i$ will not be the same as before, but parallel.

Next consider the geometric meaning of the anticommutation relations: As we saw in the proof of Theorem 2.8.1, it is closely related to projections.

Definition 2.8.6 (Tangential and normal multivectors). Let $V_0 \subset V$ be a hyperplane in an affine space (X, V) and fix a covector $\theta \in V^*$ such that $[\theta]^\perp = V_0$. Let $v \in V$ be a vector such that $\langle \theta, v \rangle = 1$ and define the line $V_1 = [v]$, so that $V = V_0 \oplus V_1$. We say that a multivector $w \in \wedge V$ is *tangential* to V_0 if $w \in \wedge V_0$, that is, $\lceil w \rceil \subset V_0$, or equivalently if $\theta \lrcorner w$. We say that it is *v-transversal* to V_0, if $v \wedge w = 0$, or equivalently $v \in \lfloor w \rfloor$. If (X, V) is an inner product space and if $\theta = v$, then we say that w is *normal* to V_0 if it is *v-transversal* to V_0.

Concretely this means that in a basis with v as one basis vector and the others spanning V_0, a multivector w is tangential if it can be written in the induced basis with only basis multivectors containing no factor v, and it is is normal if it can be written with only basis multivectors containing a factor v.

Proposition 2.8.7 (Hyperplane projection). *In an affine space (X, V), let $V_0 = [\theta]^\perp$ and $V_1 = [v]$ be as in Definition 2.8.6. Consider the linear operators*

$$P_0 : \wedge V \to \wedge V : w \mapsto \theta \lrcorner (v \wedge w),$$
$$P_1 : \wedge V \to \wedge V : w \mapsto v \wedge (\theta \lrcorner w), \quad w \in \wedge V.$$

Then P_0 is a projection onto the multivectors tangential to V_0 and P_1 is a projection onto those that are v-transversal to V_0. Furthermore, $P_0 + P_1 = 1$, so that $R(P_0) \oplus R(P_1) = \wedge V$.

Note that if $\theta = v$ is a unit vector in Euclidean space and $w \in V$ is a vector, then we have the well-known formula $P_1 w = \langle w, v \rangle v$ for its projection onto the line $[v]$. With multivector algebra, we see that we can write its projection onto the hyperplane $[v]^\perp$ as $P_0 w = w - \langle w, v \rangle v = v \lrcorner (v \wedge w)$.

Proof. In the proof of Theorem 2.8.1, it was proved that by decomposing $w = w_0 + v \wedge w_1$, with $w_0, w_1 \in \wedge V_0$, then

$$P_0(w) = w_0 \quad \text{and} \quad P_1(w) = v \wedge w_1.$$

The claims follow immediately. $\qquad \square$

Exercise 2.8.8. In Proposition 2.8.7, show that P_0 is the map on $\wedge V$ induced by $V \to V : u \mapsto \theta \lrcorner (v \wedge u)$. Show also that P_1 is not the map induced by $V \to V : u \mapsto v \wedge (\theta \lrcorner u)$. In fact P_1 is not induced by any map $V \to V$.

Exercise 2.8.9. Let $V_0 \subset V$ be a k-dimensional subspace in a Euclidean space V, with orthogonal complement V_1 so that $V = V_0 \oplus V_1$. Let $w \in \widehat{\wedge}^k V$ be such that $[w] = V_1$ and $|w| = 1$. Define maps $P_0, P_1 : V \to V$ by

$$P_0(x) = w \lrcorner (w \wedge x), \quad P_1(x) = (w \llcorner x) \lrcorner w, \quad x \in V.$$

Show that P_0 is the orthogonal projection of vectors onto V_0 and that $P_1 = I - P_0$ is the orthogonal projection onto V_1.

2.9 The Plücker Relations

Recall that Propositions 2.2.6 and 2.2.8 give two different criteria for checking whether a given k-vector is simple, and moreover, the latter gives an algorithm for factorizing it into vectors. We now combine these results and derive a set of quadratic equations that describe the Grassmann cone of simple k-vectors.

Theorem 2.9.1 (Plücker relations). *The Grassmann cone $\widehat{\wedge}^k V$ of simple k-vectors is given by the Plücker relations*

$$\widehat{\wedge}^k V = \{w \in \wedge^k V \ ; \ \langle w \lrcorner \Theta_1, \Theta_2 \lrcorner w \rangle = 0 \text{ for all } \Theta_1 \in \wedge^{k+1} V^*, \ \Theta_2 \in \wedge^{k-1} V^* \}.$$

Note that by bilinearity, it suffices to check whether $\langle w \lrcorner e_t^*, e_s^* \lrcorner w \rangle = 0$ for $|s| = k - 1$ and $|t| = k + 1$, in a given basis $\{e_i\}$. Thus the Grassmann cone is the intersection of at most $\binom{n}{k-1}\binom{n}{k+1}$ quadratic surfaces in the $\binom{n}{k}$-dimensional space $\wedge^k V$.

Proof. We may assume that $w \neq 0$. By Proposition 2.6.9, w is simple if and only if $*w$ is simple. According to Corollary 2.2.9, this happens if and only if $\lceil *w \rceil \subset \lfloor *w \rfloor$, that is, if

$$\theta \wedge *w = 0 \quad \text{for all } \theta \in \lceil *w \rceil.$$

Note that since $\lfloor w \rfloor$ is the null space of the map $V \to \wedge^{k+1} V : v \mapsto v \wedge w$, the orthogonal complement $\lfloor w \rfloor^\perp = \lceil *w \rceil$ is the range of the map

$$\wedge^{k+1} V^* \to V^* : \Theta_1 \mapsto \Theta_1 \llcorner w.$$

Thus w is simple if and only if

$$(\Theta_1 \llcorner w) \wedge (*w) = 0, \quad \text{for all } \Theta_1 \in \wedge^{k+1} V^*.$$

Next rewrite this obtained formula, using identities $(\Theta_1 \llcorner w) \wedge (*w) = (-1)^{n-k}(*w) \wedge (\Theta_1 \llcorner w)$ and $((*w) \wedge (\Theta_1 \llcorner w))* = (\Theta_1 \llcorner w) \lrcorner w \in \wedge^{k-1} V$. Therefore w is simple if and only if for all $\Theta_1 \in \wedge^{k+1} V^*$ and $\Theta_2 \in \wedge^{k-1} V^*$, we have

$$0 = \langle \Theta_2, (\Theta_1 \llcorner w) \lrcorner w \rangle = \langle (\Theta_1 \llcorner w) \wedge \Theta_2, w \rangle = \langle \Theta_1 \llcorner w, w \llcorner \Theta_2 \rangle = -\langle w \lrcorner \Theta_1, \Theta_2 \lrcorner w \rangle.$$

This proves the Plücker relations.   □

It should be noted that the algebra of interior and exterior products together is in general quite nontrivial. Indeed, great care when applying the associative and commutative properties for these products was needed in the above proof of the Plücker relations. But this is not that surprising, since these relations contain a large amount of geometric information.

That the Plücker relations are necessary for w to be simple can also be seen as follows. Assume $w = v_1 \wedge \cdots \wedge v_k \neq 0$. Extend these linearly independent vectors to a basis $\{v_1, \ldots, v_n\}$ and let $\{v_1^*, \ldots, v_n^*\}$ be the dual basis. Then

$$\Theta_2 \lrcorner\, w \in \operatorname{span}\{v_1, \ldots, v_k\}$$

for all $\Theta_2 \in \wedge^{k-1} V^*$, and

$$w \lrcorner\, \Theta_1 \in \operatorname{span}\{v_{k+1}^*, \ldots, v_n^*\}$$

for all $\Theta_1 \in \wedge^{k+1} V^*$. Therefore the Plücker relations are satisfied whenever w is simple.

Example 2.9.2. The space V with smallest dimension in which composite k-vectors exist is $\wedge^2 V$ when $\dim V = 4$. As noted in Exercise 2.2.2, the bivector $w = e_{12} + e_{34}$ is composite. This also follows from the Plücker relations, since for example,

$$\langle w \lrcorner\, e_{234}^*, e_1^* \lrcorner\, w \rangle = \langle e_2^*, e_2 \rangle = 1 \neq 0.$$

The following proves the converse of Exercise 2.2.2.

Corollary 2.9.3. *A bivector* $w \in \wedge^2 V$ *is simple if and only if* $w \wedge w = 0$.

Proof. Apply Corollary 2.8.4 to obtain

$$\theta \lrcorner\, (w \wedge w) = (\theta \lrcorner\, w) \wedge w + w \wedge (\theta \lrcorner\, w) = 2w \wedge (\theta \lrcorner\, w).$$

Thus, the Plücker relations read

$$0 = \langle w \lrcorner\, \Theta_1, \theta \lrcorner\, w \rangle = \tfrac{1}{2} \langle \Theta_1, \theta \lrcorner\, (w \wedge w) \rangle = \tfrac{1}{2} \langle \theta \wedge \Theta_1, w \wedge w \rangle,$$

since $\Theta_2 =: \theta \in V^*$ is a covector.

Since the 4-covectors $\theta \wedge \Theta_1$ span $\wedge^4 V^*$, the Plücker relations are equivalent to $w \wedge w = 0$.   □

Clearly, this result does not generalize to higher k-vectors. For example, if $2k > n$, then $w \wedge w = 0$ for all $w \in \wedge^k V$.

Exercise 2.9.4. Show that if

$$w = \sum_{i,j} w_{i,j} e_i \wedge e_j \in \wedge^2 V, \quad \text{where } w_{i,j} = -w_{j,i},$$

then the Plücker relations in coordinate form are

$$w_{i,j}w_{k,l} + w_{i,k}w_{l,j} + w_{i,l}w_{j,k} = 0 \qquad \text{for all } 1 \leq i < j < k < l \leq n.$$

In particular, if $\dim V = 4$, there is only one Plücker relation, which is $w_{12}w_{34} + w_{13}w_{42} + w_{14}w_{23} = 0$.

We end with a remark on mappings of the Grassmann cone. Consider a linear map $T : \wedge V_1 \to \wedge V_2$ induced by $T : V_1 \to V_2$. From Definition 2.3.1, it is clear that T preserves simple k-vectors, that is it maps the Grassmann cone into itself $T : \widehat{\wedge}^k V_1 \to \widehat{\wedge}^k V_2$. To verify this from the Plücker relations, we use Proposition 2.7.1 to calculate

$$\langle T(w) \lrcorner \Theta_1, \Theta_2 \lrcorner T(w) \rangle = \langle T(w) \lrcorner \Theta_1, T(T^*(\Theta_2) \lrcorner w) \rangle$$
$$= \langle T^*(T(w) \lrcorner \Theta_1), T^*(\Theta_2) \lrcorner w \rangle = \langle w \lrcorner T^*(\Theta_1), T^*(\Theta_2) \lrcorner w \rangle = 0,$$

for $w \in \wedge^k V_1$, $\Theta_1 \in \wedge^{k+1} V_2^*$ and $\Theta_2 \in \wedge^{k-1} V_2^*$.

2.10 Comments and References

2.1 Exterior algebra was discovered by Hermann Grassmann (1809–1877), starting from around 1832 and presented in the two editions of his book *Die lineale Ausdehnungslehre* (Linear Extension Theory) from 1844 [44] and 1862 [45]. Because he was very much ahead of his time and he presented his ideas in an obscure way, his ideas were not at all recognised until long after his death. More on the extraordinary life and fate of Grassmann can be read in Dieudonné [34] and in Fearnley-Sander [37]. A quote from [37]: "All mathematicians stand, as Newton said he did, on the shoulders of giants, but few have come closer than Hermann Grassmann to creating, single-handedly, a new subject."

A reference for our construction of exterior algebra, building on the universal property, is Greub [46]. The wedge ∧ is standard for denoting the exterior product. However, often a larger wedge is used in the literature.

2.2,2.9 The Grassmann cone and the Grassmannian manifold of all subspaces of a certain dimension are well researched objects that appear in the literature. We have not seen a notion of inner and outer spaces as in Definition 2.2.4 in the literature, and have chosen to explore these constructions for the obvious benefit for the geometric understanding of exterior algebra. These spaces also provide a natural way to derive the Plücker relations for the Grassmann cone. A reference for these is Griffiths and Harris [47].

2.3 The method for solving linear systems using the exterior product as in Proposition 2.3.6 was discovered by Grassmann. This is a geometric derivation of the formula that Cramer had found a century earlier.

Regarding the usefulness of determinants, the paper [15] by Axler deserves attention.

2.4 Oriented measures appear in the standard mathematics curriculum, for example as dz in contour integrals in complex analysis, and as νdS in surface integrals as in Stokes's theorem in vector calculus. These are special cases of our construction. We have chosen to present the continuous version rather than the discrete analogue, even though we shall not use this until the second part of the book, since we judge that it benefits the geometric understanding of exterior algebra.

2.5 Élie Cartan (1869–1951) was inspired by Grassmann's ideas when he created his theory of differential forms, from around 1899 in [25]. His theory, which is the standard presentation of exterior algebra in the modern literature, is based on the dual alternating forms/multicovectors. This approach has advantages, since clearly the objects in Proposition 2.5.3(iii) are easier to define algebraically than k-vectors. However, the geometric visualization of k-covectors is more complicated than that of k-vectors. We would face a similar problem in standard linear algebra if we tried to teach the dual space V^* before mastering the vector space V itself.

A question, which is more important than it may appear, is when one should identify objects and when one should not identify them. When working in a Euclidean space, or more generally in a space on which we have fixed a duality, it is very convenient to identify multivectors and multicovectors. This is natural and should rightly be done. We have also seen that there is a one-to-one correspondence between k-vectors and $(n-k)$-(co)vectors via the Hodge star maps. However, to identify these two objects would be very unnatural geometrically, and we refrain from doing so. In fact, Grassmann himself made such an identification, and this was probably his biggest mistake. See [34]. As noted in [45], it is better for applications to physics not to use the Hodge star maps as an identification as has been standard in linear algebra. Physics texts, in attempting to explain the physical distinction between "polar vectors" (1-vectors) and "axial vectors" (2-vectors), are handicapped by the fact that in the accepted model both have the same representation.

2.6 Our definitions of interior products and Hodge stars do not follow the standard in the literature. It is often only the left interior product that appears in the literature, and often $v \lrcorner w$ is defined only for vectors $v \in \wedge^1 V$. The notation is not standardized in the literature. For example, there appear different notations such as $v \lrcorner w$, $v \llcorner w$, and $i_v(w)$ for the same left interior product, which is also referred to as "contraction". When the interior product is defined for multivectors v, it often incorporates a reversion of v.

It is standard in the literature to use only the Hodge star maps $w \mapsto w \lrcorner e_{\bar{n}}^*$ and $\Theta \mapsto \Theta \lrcorner e_{\bar{n}}$, and contrary to the notation used here, they are denoted by $*w$ and $*\Theta$. These maps are not each others inverses in general, which complicates the algebra.

Chapter 3

Clifford Algebra

Prerequisites:

This chapter builds on Chapter 2. Besides a solid background in linear algebra, the reader is assumed to be familiar with the algebra and geometry of complex numbers.

Road map:

In Chapter 2, we defined the exterior algebra $\wedge V$ for a vector space V. If V is an inner product space, then the interior products $\lrcorner$ and $\llcorner$ are dual to the exterior product $\wedge$, and all these are bilinear products on $\wedge V$. However, only the exterior product is associative. In this chapter, we combine the exterior and interior products to obtain a "hypercomplex" associative product on $\wedge V$: the Clifford product $\triangle$. In the complex plane we know well the complex exponential

$$e^{i\phi} = \cos\phi + i\sin\phi,$$

and how it simplifies algebra by combining the complementary cosine and sine functions. The Clifford product is defined in Section 3.1 in a similar way by combining the complementary interior and exterior products. Unlike the complex numbers, this construction works in any inner product space of any dimension. In three-dimensional Euclidean space we show in Section 3.2 how the Clifford algebra reduces to Hamilton's classical algebra of quaternions. However, life is not perfect, and the Clifford algebra in higher dimensions is not best viewed as a generalization of complex numbers. The Clifford product $\triangle$ is noncommutative, and the standard algebraic way to define the Clifford product is by requiring the anticommutation relation

$$u \triangle v + v \triangle u = 2\langle u, v \rangle, \quad u, v \in V,$$

something we discuss in Section 3.3. But even worse, the Clifford algebra is not a division algebra, that is, many multivectors are not invertible with respect to

© Springer Nature Switzerland AG 2019

A. Rosén, *Geometric Multivector Analysis*, Birkhäuser Advanced Texts Basler Lehrbücher,

https://doi.org/10.1007/978-3-030-31411-8_3

$\triangle$. The best way to view Clifford algebra is rather as an algebra of matrices, seen in a geometric way. Choosing such a matrix representation, each multivector w corresponds to a matrix, in such a way that the Clifford product $\triangle$ corresponds to matrix multiplication. This is the topic of Section 3.4. Having said this, the Clifford product nevertheless provides a highly useful and fundamental algebra with deep applications to geometry and analysis, as we shall see in this book.

Highlights:

- Clifford product $\triangle = \wedge + \lrcorner + \llcorner = $ symmetric difference: 3.1.3

- Three-dimensional quaternion rotation: 3.2.3

- Spin $= \mathbf{Z}_2$ memory in rotations: 3.2.5

- Operators on multivectors $=$ neutral Clifford algebra: 3.4.2

- Periodicity of 8 for representation of real Clifford algebras: 3.4.13

3.1 The Clifford Product

We aim to define an associative Clifford product $w_1 \triangle w_2$ between multivectors $w_1, w_2 \in \wedge V$, and we begin with a discussion about what it should be in the special case of vectors, or at least when one of the factors is a vector. The full definition of the Clifford product comes in Proposition 3.1.3. The reader who wants to start with a concrete example may want to begin with Example 3.1.4. A natural point of departure for the Clifford product is the following.

Proposition 3.1.1 (Lagrange's identity). *Let V be an inner product space. Then vectors v_1, v_2 satisfy Lagrange's identity $\langle v_1, v_2 \rangle^2 + \langle v_1 \wedge v_2 \rangle^2 = \langle v_1 \rangle^2 \langle v_2 \rangle^2$. If V is a Euclidean space, this reads*

$$|\langle v_1, v_2 \rangle|^2 + |v_1 \wedge v_2|^2 = |v_1|^2 |v_2|^2,$$

and in particular, the Cauchy–Schwarz and Hadamard inequalities

$$|\langle v_1, v_2 \rangle| \le |v_1||v_2|, \quad |v_1 \wedge v_2| \le |v_1||v_2|$$

hold, with equality in the Cauchy–Schwarz inequality if and only if $v_1 \wedge v_2 = 0$, that is, v_1 and v_2 are parallel, and equality in Hadamard's inequality if and only if $\langle v_1, v_2 \rangle = 0$, that is, v_1 and v_2 are orthogonal.

Proof. The anticommutation relation from Theorem 2.8.1 shows that

$$\langle v_1 \wedge v_2 \rangle^2 = \langle v_1 \wedge v_2, v_1 \wedge v_2 \rangle = \langle v_2, v_1 \lrcorner (v_1 \wedge v_2) \rangle$$
$$= \langle v_2, \langle v_1 \rangle^2 v_2 - \langle v_1, v_2 \rangle v_1 \rangle = \langle v_1 \rangle^2 \langle v_2 \rangle^2 - |\langle v_1, v_2 \rangle|^2. \qquad \square$$

Exercise 3.1.2. Using Lagrange's identity, prove *Heron's formula* for the Euclidean area $A = \frac{1}{2}|v_1 \wedge v_2|$ of the triangle with edge vectors v_1, v_2, v_3, so that $v_1 + v_2 + v_3 = 0$, which reads

$$A = \tfrac{1}{4}\sqrt{2(a^2b^2 + a^2c^2 + b^2c^2) - a^4 - b^4 - c^4} = \sqrt{p(p-a)(p-b)(p-c)},$$

where $a = |v_1|$, $b = |v_2|$, $c = |v_3|$, and $p := (a + b + c)/2$ is half the perimeter.

In Euclidean space, we recall the definition of the *angle* ϕ between two vectors v_1 and v_2: This is the unique number $0 \leq \phi \leq \pi$ such that

$$\langle v_1, v_2 \rangle = |v_1||v_2|\cos(\phi),$$

guaranteed by the Cauchy–Schwarz inequality. Lagrange's identity then shows that

$$|v_1 \wedge v_2| = |v_1||v_2|\sin(\phi).$$

We hence have two "complementary" products: the inner product vanishes when the vectors are orthogonal and has maximal modulus when they are parallel. Conversely, the exterior product vanishes when the vectors are parallel and has maximal modulus when they are orthogonal. In view of Lagrange's identity, it is almost irresistible to try to define a "hypercomplex" product

$$v_1 \vartriangle v_2 := \langle v_1, v_2 \rangle + v_1 \wedge v_2 \in \wedge^0 V \oplus \wedge^2 V.$$

This is called the *Clifford product* of the two vectors, and Lagrange's identity shows that $|v_1 \vartriangle v_2| = |v_1||v_2|$. In particular, if $v_1 = v_2 = v$, then $v \vartriangle v = \langle v \rangle^2$. Thus, if the space is Euclidean, any nonzero vector is invertible with respect to the Clifford product, with inverse

$$v^{-1} = \tfrac{1}{|v|^2} v$$

being the inversion of the vector v in the unit sphere. Note that unlike inverses of complex numbers, there is no complex conjugation implicit in the Clifford inverse.

Another feature of the Clifford product is that it is a "complex" product in the sense that the product in general will be an inhomogeneous multivector in the sense of Definition 2.1.13, even if the factors are both homogeneous, and we are thus forced to work in the full exterior algebra $\wedge V$. This should be compared to exterior and interior products, where the product of homogeneous multivectors is also a homogeneous multivector.

We extend the Clifford product to more general multivectors. If $v \in V$ is a vector and $w \in \wedge^k V$ is any k-vector, then we have Lagrange-type identities

$$\langle v \lrcorner w \rangle^2 + \langle v \wedge w \rangle^2 = \langle v \rangle^2 \langle w \rangle^2,$$
$$\langle w \llcorner v \rangle^2 + \langle w \wedge v \rangle^2 = \langle v \rangle^2 \langle w \rangle^2,$$

which follows from the anticommutation relation as above. This suggests a definition of Clifford products

$$v \vartriangle w := v \lrcorner w + v \wedge w \in \wedge^{k-1}V \oplus \wedge^{k+1}V, \qquad (3.1)$$

$$w \vartriangle v := w \llcorner v + w \wedge v \in \wedge^{k-1}V \oplus \wedge^{k+1}V. \qquad (3.2)$$

Note very carefully that these formulas for the Clifford product are valid only when one of the factors is a vector. As we will see in Proposition 3.1.9, the expression for the Clifford product between two general multivectors is more complicated.

We want to extend this product $\vartriangle$ to a bilinear product on all $\wedge V$. Clearly we cannot expect a commutative product, since we have the *anticommutation relation*

$$v_1 \vartriangle v_2 + v_2 \vartriangle v_1 = 2\langle v_1, v_2 \rangle, \quad v_1, v_2 \in V,$$

since the inner product is symmetric and the exterior product is skew symmetric. This formula shows how we encode geometry algebraically, via the inner product, as commutation rules for the Clifford product. If vectors are parallel, they will commute. At the other extreme, orthogonal vectors will anticommute. More generally, the commutation relations for the exterior and interior products yield the *Riesz formulas*

$$v \lrcorner w = \tfrac{1}{2}(v \vartriangle w - \widehat{w} \vartriangle v), \qquad (3.3)$$

$$v \wedge w = \tfrac{1}{2}(v \vartriangle w + \widehat{w} \vartriangle v), \quad v \in V, w \in \wedge^k V, \qquad (3.4)$$

as is seen from (3.1), (3.2), and Propositions 2.1.14 and 2.6.3. Even if we will not have a commutative product, it is possible to obtain an associative product, which we now show.

Proposition 3.1.3 (Clifford product). *Let V be an inner product space. Then there exists a unique Clifford product $\vartriangle$ on $\wedge V$ such that $(\wedge V, +, \vartriangle, 1)$ is an associative algebra, in the sense of Definition 1.1.4, and (3.1) and (3.2) hold for all vectors $v \in V$ and all multivectors $w \in \wedge V$. If $\{e_i\}$ is an ON-basis for V, with induced ON-basis $\{e_s\}$ for $\wedge V$, then the Clifford products of basis multivectors e_s and e_t are*

$$e_s \vartriangle e_t = \langle e_{s \cap t} \rangle^2 \, \epsilon(s, t) \, e_{s \vartriangle t}, \qquad (3.5)$$

where $s \vartriangle t := (s \cup t) \setminus (s \cap t)$ denotes the symmetric difference of the index sets, $\epsilon(s, t) = \pm 1$ denotes the permutation sign from Definition 2.1.16, and $\langle e_{s \cap t} \rangle^2 = \pm 1$ is the square sign from Lemma 2.5.4.

Note that in Euclidean space, $\langle e_{s \cap t} \rangle^2 = +1$. Before the general proof, let us consider a concrete example to see why (3.1) and (3.2) and associativity uniquely define a Clifford product.

Example 3.1.4. A concrete example of a Clifford product in the Euclidean plane, with ON-basis $\{e_1, e_2, e_3\}$, is

$$(4 - 7e_1 + 3e_2 + 2e_{12}) \vartriangle (1 + 5e_1 - 6e_2 - 8e_{12})$$
$$= (4 + 20e_1 - 24e_2 - 32e_{12}) + (-7e_1 - 35 + 42e_{12} + 56e_2)$$
$$+ (3e_2 - 15e_{12} - 18 + 24e_1) + (2e_{12} - 10e_2 - 12e_1 + 16)$$
$$= -33 + 25e_1 + 25e_2 - 3e_{12}.$$

We have used $e_1 \vartriangle e_1 = e_2 \vartriangle e_2 = 1$ and $e_1 \vartriangle e_2 = -e_2 \vartriangle e_1$. The most interesting term is $e_{12} \vartriangle e_{12}$, where we needed to use also associativity and that 1 is the identity element:

$$(e_1 \vartriangle e_2) \vartriangle (e_1 \vartriangle e_2) = -(e_2 \vartriangle e_1) \vartriangle (e_1 \vartriangle e_2) = -e_2 \vartriangle (e_1 \vartriangle e_1) \vartriangle e_2 = -e_2 \vartriangle e_2 = -1.$$

Note that unlike the exterior product in Example 2.1.17 or the interior product in Example 2.6.14, no terms are zero here.

A concrete example of a Clifford product in three-dimensional Euclidean space is

$$(3 - e_1 - e_2 + 2e_3 + e_{12} + 5e_{13} - 2e_{23} - 7e_{123})$$
$$\vartriangle (-2 + 4e_1 - 3e_2 + e_3 - 6e_{12} + 3e_{13} - e_{23} + 2e_{123})$$
$$= (-6 + 12e_1 - 9e_2 + 3e_3 - 18e_{12} + 9e_{13} - 3e_{23} + 6e_{123})$$
$$+ (2e_1 - 4 + 3e_{12} - e_{13} + 6e_2 - 3e_3 + e_{123} - 2e_{23})$$
$$+ (2e_2 + 4e_{12} + 3 - e_{23} - 6e_1 + 3e_{123} + e_3 + 2e_{13})$$
$$+ (-4e_3 - 8e_{13} + 6e_{23} + 2 - 12e_{123} - 6e_1 + 2e_2 + 4e_{12})$$
$$+ (-2e_{12} - 4e_2 - 3e_1 + e_{123} + 6 - 3e_{23} - e_{13} - 2e_3)$$
$$+ (-10e_{13} - 20e_3 + 15e_{123} + 5e_1 - 30e_{23} - 15 + 5e_{12} + 10e_2)$$
$$+ (4e_{23} - 8e_{123} - 6e_3 - 2e_2 - 12e_{13} - 6e_{12} - 2 + 4e_1)$$
$$+ (14e_{123} - 28e_{23} - 21e_{13} - 7e_{12} - 42e_3 - 21e_2 - 7e_1 + 14)$$
$$= -2 + e_1 - 16e_2 - 73e_3 - 17e_{12} - 42e_{13} - 57e_{23} + 20e_{123}.$$

For example,

$$e_{13} \vartriangle e_{123} = e_1 \vartriangle (e_3 \vartriangle e_{123}) = e_1 \vartriangle (e_3 \lrcorner e_{123} + 0) = e_1 \vartriangle e_{12} = e_1 \lrcorner e_{12} + 0 = e_2.$$

Note, however, that $e_{13} \lrcorner e_{123} = e_3 \lrcorner (e_1 \lrcorner e_{123}) = e_3 \lrcorner e_{23} = -e_2$. The reason for the difference in sign is that in $w_1 \lrcorner w_2$ there is an implicit reversion of w_1. This also illustrates the important fact that $w_1 \vartriangle w_2 = w_1 \lrcorner w_2 + w_1 \wedge w_2$ is required only when w_1 is a vector. We shall even see that a Clifford product of two homogeneous multivectors in general has more than two homogeneous parts. See Proposition 3.1.9.

Exercise 3.1.5. In oriented three-dimensional Euclidean space V, the Clifford product can be expressed in terms of inner and vector products of vectors, and multiplication by scalars, using the Hodge star duality. Show that the Clifford product of two multivectors

$$w_1 = a_1 + v_1 + *u_1 + *b_1,$$
$$w_2 = a_2 + v_2 + *u_2 + *b_2,$$

where $a_1, b_1, a_2, b_2 \in \mathbf{R}$ and $v_1, u_1, v_2, u_2 \in V$, equals

$$\begin{aligned}
w_1 \vartriangle w_2 = {} & a_1 a_2 + \langle v_1, v_2 \rangle - \langle u_1, u_2 \rangle - b_1 b_2 \\
& + a_1 v_2 + v_1 a_2 - v_1 \times u_2 - u_1 \times v_2 - u_1 b_2 - b_1 u_2 \\
& + *(a_1 u_2 + u_1 a_2 + v_1 \times v_2 - u_1 \times u_2 + v_1 b_2 + b_1 v_2) \\
& + *(a_1 b_2 + \langle v_1, u_2 \rangle + \langle u_1, v_2 \rangle + b_2 a_1).
\end{aligned}$$

The following lemma, the proof of which is left as a combinatorial exercise, shows that the symmetric difference is natural for the permutation signs $\epsilon(s,t)$ from Definition 2.1.16.

Lemma 3.1.6. *For all $s, t, u \subset \bar{n}$, we have*

$$\epsilon(s \vartriangle t, u) = \epsilon(s, u)\epsilon(t, u) \quad and \quad \epsilon(s, t \vartriangle u) = \epsilon(s, t)\epsilon(s, u).$$

Proof of Proposition 3.1.3. To prove uniqueness, assume that $\vartriangle$ is any associative product, with identity 1, such that (3.1) and (3.2) hold. By bilinearity, it suffices to prove (3.5) for all basis multivectors. It suffices to prove it for $|s|, |t| \geq 1$, since 1 is assumed to be the identity. We proceed by induction on $m = \min(|s|, |t|)$. For $m = 1$, (3.5) follows from (3.1) and (3.2). For example, if $s = \{i\}$, then (3.1) shows that

$$e_i \vartriangle e_t = \begin{cases} \langle e_i \rangle^2 \epsilon(\{i\}, t \setminus \{i\}) \, e_{t \setminus \{i\}}, & i \in t, \\ \epsilon(\{i\}, t) \, e_{\{i\} \cup t}, & i \notin t, \end{cases} = \langle e_{\{i\} \cap t} \rangle^2 \epsilon(\{i\}, t) e_{\{i\} \vartriangle t}.$$

To make the induction step, consider $e_{\{i\} \cup s} \vartriangle e_t$, where $i < \min(s)$ and where (3.5) is assumed to hold for $e_s \vartriangle e_t$. Then

$$e_{\{i\} \cup s} \vartriangle e_t = (e_i \vartriangle e_s) \vartriangle e_t = e_i \vartriangle (e_s \vartriangle e_t) = \langle e_{s \cap t} \rangle^2 \epsilon(s, t) \, e_i \vartriangle e_{s \vartriangle t}$$
$$= \langle e_{\{i\} \cap (s \vartriangle t)} \rangle^2 \langle e_{s \cap t} \rangle^2 \epsilon(\{i\}, s \vartriangle t)\epsilon(s, t) \, e_{\{i\} \vartriangle (s \vartriangle t)}.$$

Since $(\{i\} \cup s) \cap t = (\{i\} \cap (s \vartriangle t)) \cup (s \cap t)$ disjointly, we have $\langle e_{\{i\} \cap (s \vartriangle t)} \rangle^2 \langle e_{s \cap t} \rangle^2 = \langle e_{(\{i\} \cup s) \cap t} \rangle^2$. Since the symmetric difference of sets is associative, we have $e_{\{i\} \vartriangle (s \vartriangle t)} = e_{(\{i\} \cup s) \vartriangle t}$. Finally, since $\epsilon(\{i\}, s) = 1$, the lemma shows that $\epsilon(\{i\}, s \vartriangle t)\epsilon(s, t) = \epsilon(\{i\}, t)\epsilon(s, t) = \epsilon(\{i\} \cup s, t)$. This proves (3.5) for $e_{\{i\} \cup s} \vartriangle e_t$.

If instead the second index is smaller, then we swap the roles of s and t, and make use of (3.2) similarly. We conclude by induction that (3.5) holds for all s and t. This proves uniqueness.

To prove existence of an associative bilinear Clifford product, we define the products of basis multivectors by (3.5) and extend to a bilinear product. The formulas (3.1) and (3.2) are straightforward to verify from (3.5). To prove that this yields an associative product it suffices to show that $(e_s \vartriangle e_t) \vartriangle e_u = e_s \vartriangle (e_t \vartriangle e_u)$ for all index sets. By definition, this amounts to showing that

$$\left(\langle e_{s \cap t} \rangle^2 \langle e_{(s \vartriangle t) \cap u} \rangle^2 \right) \left(\epsilon(s,t) \epsilon(s \vartriangle t, u) \right) e_{(s \vartriangle t) \vartriangle u}$$
$$= \left(\langle e_{s \cap (t \vartriangle u)} \rangle^2 \langle e_{t \cap u} \rangle^2 \right) \left(\epsilon(s, t \vartriangle u) \epsilon(t,u) \right) e_{s \vartriangle (t \vartriangle u)}.$$

Drawing a Venn diagram reveals that $(s \cap t) \cup ((s \vartriangle t) \cap u)$ and $(s \cap (t \vartriangle u)) \cup (t \cap u)$ are two different partitions of the set $(s \cap t) \cup (s \cap u) \cup (t \cap u)$. This shows that the norm square products coincide. Equality of the permutation signs follows from the lemma. Since the symmetric difference of sets is associative, this shows associativity. This proves existence. $\qquad\square$

When V is an inner product space, we have various products defined on the exterior algebra $\wedge V$: the associative exterior and Clifford products, as well as the nonassociative interior products. We denote the space of multivectors by $\wedge V$ when using mainly $\wedge, \lrcorner, \llcorner$. When using mainly $\vartriangle$, we use the following notation.

Definition 3.1.7 (Standard Clifford algebra). Let V be an inner product space. We denote by $\triangle V$ *the standard Clifford algebra* $(\wedge V, +, \vartriangle)$ defined by the Clifford product on the space of multivectors in V. A shorthand notation for the Clifford product $w_1 \vartriangle w_2$ is

$$w_1 w_2 := w_1 \vartriangle w_2, \qquad w_1, w_2 \in \triangle V.$$

Write $\triangle^k V := \wedge^k V$ and define the spaces of *even* and *odd* multivectors $\triangle^{\mathrm{ev}} V = \wedge^{\mathrm{ev}} V$ and $\triangle^{\mathrm{od}} V = \wedge^{\mathrm{od}} V$, so that

$$\triangle V = \triangle^{\mathrm{ev}} V \oplus \triangle^{\mathrm{od}} V.$$

It is standard to write $w_1 w_2$ for the Clifford product. We mainly follow this convention from now on, writing $w_1 \vartriangle w_2$ only when we want to emphasize that the product we use is the Clifford product.

Note that if $w_1, w_2 \in \triangle^{\mathrm{ev}} V$ then $w_1 \vartriangle w_2 \in \triangle^{\mathrm{ev}} V$. On the other hand, $\triangle^{\mathrm{od}} V$ is not closed under Clifford multiplication.

Even though $\wedge V = (\wedge V, +, \wedge)$ and $\triangle V = (\wedge V, +, \vartriangle)$ are the same as linear spaces, they are very different as real algebras. In particular, when writing w^{-1} for the inverse of a multivector, we shall always mean the inverse with respect to the Clifford product, and never with respect to the exterior product. Note that since the Clifford product is noncommutative, it is ambiguous to write x/y in $\triangle V$, since one must specify whether xy^{-1} or $y^{-1}x$ is meant. Only when $x \in \mathbf{R} = \wedge^0 V$ shall we write x/y.

Proposition 3.1.8 (Center and anticenter). *Let V be an inner product space of dimension n. Then the center $\mathcal{Z}(\triangle V) := \{w \in \triangle V \ ; \ wx = xw \text{ for all } x \in \triangle V\}$ and anticenter $\mathcal{Z}^-(\triangle V) := \{w \in \triangle V \ ; \ wx = \hat{x}w \text{ for all } x \in \triangle V\}$ of $\triangle V$ are*

$$
\mathcal{Z}(\triangle V) = \begin{cases} \triangle^0 V, & n \text{ even,} \\ \triangle^0 V \oplus \triangle^n V, & n \text{ odd,} \end{cases} \qquad \mathcal{Z}^-(\triangle V) = \begin{cases} \triangle^n V, & n \text{ even,} \\ \{0\}, & n \text{ odd.} \end{cases}
$$

Thus if $\hat{w}v = vw$ for all vectors $v \in V$, then $w \in \triangle^0 V = \mathbf{R}$.

Proof. Let $w \in \triangle V$ and write $w = \sum_s \alpha_s e_s$ in an induced basis. If $s \neq \varnothing$ and $s \neq \bar{n}$, then there exist $j \in s$ and $j' \notin s$, and e_s cannot commute with both, or anticommute with both e_j and $e_{j'}$. Thus $\mathcal{Z}(\triangle V), \mathcal{Z}^-(\triangle V) \subset \triangle^0 V \oplus \triangle^n V$. Clearly $\triangle^0 V \subset \mathcal{Z}(\triangle V)$ and $\triangle^0 V \cap \mathcal{Z}^-(\triangle V) = \varnothing$. For all vectors $e_j \in V$, we have $e_j e_{\bar{n}} = (-1)^{n-1} e_{\bar{n}} e_j$, and we conclude that the center and anticenter are as stated.

If $\hat{w}v = vw$ for all vectors $v \in V$, write $w = w_e + w_o \in \triangle^{\mathrm{ev}} V \oplus \triangle^{\mathrm{od}} V$. It follows that $w_e \in \mathcal{Z}(\triangle V)$ and $w_o \in \mathcal{Z}^-(\triangle V)$, and therefore that $w = w_e \in \triangle^0 V$. $\qquad \square$

We point out again the important fact that the formulas (3.1) and (3.2) hold only when one factor is a vector. The following result shows what happens in the general case: additional homogeneous terms appear in the Clifford product that cannot be expressed using the exterior and interior products.

Proposition 3.1.9 (Homogeneous Clifford products). *The Clifford product of two bivectors $w_1, w_2 \in \triangle^2 V$ is*

$$
w_1 \vartriangle w_2 = -\langle w_1, w_2 \rangle + \tfrac{1}{2}[w_1, w_2] + w_1 \wedge w_2 \in \triangle^0 V \oplus \triangle^2 V \oplus \triangle^4 V,
$$

where the commutator product is $[w_1, w_2] := w_1 \vartriangle w_2 - w_2 \vartriangle w_1 = 2(w_1 \vartriangle w_2)_2$. More generally, if $w_1 \in \triangle^k V$ and $w_2 \in \triangle^l V$, then

$$
w_1 \vartriangle w_2 \in \triangle^{|k-l|} V \oplus \triangle^{|k-l|+2} V \oplus \cdots \oplus \triangle^{k+l-2} V \oplus \triangle^{k+l} V,
$$

where the highest- and lowest-degree terms of the Clifford product are the exterior and interior products in the sense that

$$
(w_1 \vartriangle w_2)_{k+l} = w_1 \wedge w_2, \quad (w_1 \vartriangle w_2)_{l-k} = \overline{w}_1 \lrcorner w_2, \quad (w_1 \vartriangle w_2)_{k-l} = w_1 \llcorner \overline{w}_2.
$$

In particular, the inner product of k-vectors $w_1, w_2 \in \triangle^k V$ is

$$
\langle w_1, w_2 \rangle = (\overline{w}_1 \vartriangle w_2)_0 = (w_1 \vartriangle \overline{w}_2)_0.
$$

Proof. (i) By linearity, it suffices to consider a simple bivector $w_1 = v_1 \wedge v_2 = v_1 \vartriangle v_2$, with $\langle v_1, v_2 \rangle = 0$. Two applications of (3.1) gives

$$
w_1 w_2 = v_1 \lrcorner (v_2 \lrcorner w_2) + \Big(v_1 \lrcorner (v_2 \wedge w_2) + v_1 \wedge (v_2 \lrcorner w_2) \Big) + v_1 \wedge v_2 \wedge w_2.
$$

The last term is $w_1 \wedge w_2$ and the first term is $v_1 \lrcorner (v_2 \lrcorner w_2) = (v_2 \wedge v_1) \lrcorner w_2 = -w_1 \lrcorner w_2 = -\langle w_1, w_2 \rangle$. Note that $\langle w_1, w_2 \rangle = \langle w_2, w_1 \rangle$ as well as $w_1 \wedge w_2 = w_2 \wedge w_1$. A similar application of (3.2) to $w_2 \vartriangle w_1$ and subtraction gives

$$w_1 w_2 - w_2 w_1 = v_1 \lrcorner (v_2 \wedge w_2) + v_1 \wedge (v_2 \lrcorner w_2) - (w_2 \wedge v_1) \llcorner v_2 - (w_2 \llcorner v_1) \wedge v_2$$
$$= v_1 \lrcorner (v_2 \wedge w_2) + v_1 \wedge (v_2 \lrcorner w_2) - v_2 \lrcorner (v_1 \wedge w_2) - v_2 \wedge (v_1 \lrcorner w_2)$$
$$= 2(v_1 \lrcorner (v_2 \wedge w_2) + v_1 \wedge (v_2 \lrcorner w_2)) = 2(w_1 \vartriangle w_2)_2.$$

(ii) Next consider $w_1 \in \triangle^k V$ and $w_2 \in \triangle^l V$, and in particular $w_1 = e_s$ and $w_2 = e_t$, where $|s| = k$ and $|t| = l$. Then $e_s \vartriangle e_t = \pm e_{s \vartriangle t}$. The mapping property of the Clifford product is now a consequence of linearity and the identity

$$|s \vartriangle t| = |s| + |t| - 2|s \cap t|.$$

That the highest and lowest terms can be expressed in terms of exterior and interior products can be shown by generalizing the argument in (i). ☐

The Hodge star maps can be computed using the Clifford product and reversion as follows.

Corollary 3.1.10 (Clifford Hodge stars). *Let V be an oriented inner product space with orientation $e_{\overline{n}} \in \wedge^n V$. Then the Hodge star maps can be expressed in terms of the Clifford product as*

$$*w = \langle e_{\overline{n}} \rangle^2 e_{\overline{n}} \vartriangle \overline{w}, \quad w* = \overline{w} \vartriangle e_{\overline{n}}, \quad w \in \wedge V.$$

If V is three-dimensional oriented Euclidean space, then the commutator of two bivectors $w_1, w_2 \in \triangle^2 V$ is

$$\tfrac{1}{2}[w_1, w_2] = -(*w_1) \wedge (*w_2).$$

Proof. (i) To prove the formula for $w*$, by linearity it suffices to consider $w \in \wedge^k V$. Since $\overline{w} \vartriangle e_{\overline{n}} \in \wedge^{n-k} V$, as $s \vartriangle \overline{n} = \overline{n} \setminus s$, Proposition 3.1.9 shows that

$$w* = w \lrcorner e_{\overline{n}} = (\overline{w} \vartriangle e_{\overline{n}})_{n-k} = \overline{w} \vartriangle e_{\overline{n}}.$$

The proof for $*w$ is similar. Note that this uses the dual orientation $e_{\overline{n}}^* = \langle e_{\overline{n}} \rangle^2 e_{\overline{n}}$.
(ii) For the commutator formula, we calculate

$$[w_1, w_2] = w_1 w_2 - w_2 w_1 = -e_{123}^2 (w_1 w_2 - w_2 w_1)$$
$$= -(e_{123} \overline{w}_1)(e_{123} \overline{w}_2) + (e_{123} \overline{w}_2)(e_{123} \overline{w}_1)$$
$$= -(*w_1)(*w_2) + (*w_2)(*w_1) = -2(*w_1) \wedge (*w_2). \qquad ☐$$

Exercise 3.1.11. For $w_1, w_2, w_3 \in \triangle V$ in an inner product space V, prove formulas

$$\langle w_1 \vartriangle w_2, w_3 \rangle = \langle w_2, \overline{w}_1 \vartriangle w_3 \rangle = \langle w_1, w_3 \vartriangle \overline{w}_2 \rangle,$$
$$\widehat{w_1 \vartriangle w_2} = \widehat{w}_1 \vartriangle \widehat{w}_2, \quad \text{and} \quad \overline{w_1 \vartriangle w_2} = \overline{w}_2 \vartriangle \overline{w}_1.$$

Exercise 3.1.12. Consider a duality $\langle V, V \rangle$, which is not an inner product, on a vector space V. What kind of generalization of the Clifford product, based on (3.1) and (3.2), can be achieved? Note that no ON-bases exist in this setting, by Proposition 1.2.10.

3.2   Complex Numbers and Quaternions

A disadvantage of the Clifford product is that its geometric interpretation is somewhat more difficult than for exterior and interior products, since we are forced to work with inhomogeneous multivectors. A big advantage, though, is the gain of associativity, as compared to the interior products. Moreover, as we shall see in this section, the Clifford product is intimately related to isometries of the space, just as the exterior and interior products are closely related to orthogonal projections, as seen in Section 2.8.

One of the most fundamental constructions in mathematics is the complex number field $\mathbf{C} = \{a + ib \; ; \; a, b \in \mathbf{R}\}$ and the imaginary unit i, where $i^2 = -1$. This field can be constructed from the reals in various way. For example, a construction of $\mathbf{C}$ from commutative algebra is to form the ring of real polynomials $\mathbf{R}[x]$, and define $\mathbf{C}$ to be the quotient ring modulo the principal ideal generated by the polynomial $x^2 + 1$. Since this is an irreducible polynomial, it follows that $\mathbf{C}$ is a field.

Another fundamental construction of $\mathbf{C}$, through noncommutative algebra, geometry, and Clifford algebra, is to identify $\mathbf{C}$ and the even subalgebra $\triangle^{\mathrm{ev}}V$ of an oriented Euclidean plane V with $\dim V = 2$. To see what this means, let $\{e_1, e_2\}$ be a positive ON-basis and let

$$j := e_{12} = e_1 \wedge e_2 = e_1 e_2 = e_1 \vartriangle e_2$$

be the orientation of the plane. Then $\{1, e_1, e_2, j\}$ is the induced ON basis for $\triangle V$ and the Clifford products of basis multivectors are

$$e_1^2 = e_2^2 = 1 = -j^2, \quad e_1 e_2 = -e_2 e_1 = j, \quad e_1 j = e_2 = -je_1, \quad e_2 j = -e_1 = -je_2.$$

We see that all $\triangle V$ is noncommutative, but that the even subalgebra

$$\triangle^{\mathrm{ev}}V = \triangle^0 V \oplus \triangle^2 V = \{a + bj = a + be_{12} \; ; \; a, b \in \mathbf{R}\}$$

is a commutative algebra that is isomorphic to $\mathbf{C}$ when equipped with the Clifford product. Note also that the full Clifford algebra $\triangle V$ contains noninvertible elements. For example, $(1 + e_1)(1 - e_1) = 0$. However, the even subalgebra $\mathbf{C}$ is of course a field.

Definition 3.2.1. If V is an oriented two-dimensional Euclidean plane, then we refer to $(\triangle^{\mathrm{ev}}V, +, \vartriangle)$ as the *standard geometric representation* of the complex numbers $\mathbf{C}$. When using this geometric representation of $\mathbf{C}$, we write j for the imaginary unit.

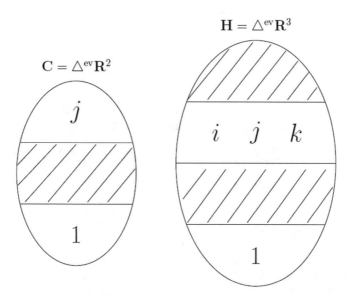

Figure 3.1: Standard geometric representations of $\mathbf{C}$ and $\mathbf{H}$ as even parts of Clifford algebras.

Another observation from the basis products above is that the map $V \ni v \mapsto vj = -jv \in V$ is rotation by $90°$ counterclockwise in V. More generally,

$$v \mapsto v \exp(\phi j) = \exp(-\phi j)v = \exp(-\phi j/2)v \exp(\phi j/2)$$

is rotation counterclockwise through the angle $\phi \in \mathbf{R}$, where the exponential function is $\exp(\phi j) = \cos\phi + j\sin\phi$. As we shall see below, it is the latter formula $\exp(-\phi j/2)v \exp(\phi j/2)$ for rotation that generalizes to higher dimension, where the algebra is much more noncommutative.

Note that so far we have distinguished vectors in V from complex numbers in $\mathbf{C} = \triangle^{\mathrm{ev}}V$. There is a difference between these spaces: the first one is an isotropic plane, whereas the complex plane $\mathbf{C} = \triangle^0 V \oplus \triangle^2 V$ is anisotropic, with fixed real and imaginary axes. If we fix an ON-basis in V, then we may identify the two spaces. In this case we may assume that $V = \mathbf{R}^2$ and $\{e_1, e_2\}$ is the standard basis. Declaring e_1 as the real direction, we shall refer to

$$V \ni v \longleftrightarrow e_1 v = \overline{ve_1} \in \mathbf{C}$$

as the standard identification of $\mathbf{R}^2$ and $\mathbf{C}$, where $e_1 \leftrightarrow 1$ and $e_2 \leftrightarrow j$. Note that the reversion operation in $\triangle V$ restricts to usual complex conjugation in $\mathbf{C}$.

Next we increase the dimension to three and consider an oriented Euclidean space V with $\dim V = 3$. Let $\{e_1, e_2, e_3\}$ be an ON-basis for V, where $J := e_{123}$ is the orientation of V, and define

$$j_1 := *e_1 = e_{23} = e_2 \wedge e_3 = e_2 \,\vartriangle\, e_3,$$
$$j_2 := *e_2 = e_{31} = e_3 \wedge e_1 = e_3 \,\vartriangle\, e_1,$$
$$j_3 := *e_3 = e_{12} = e_1 \wedge e_2 = e_1 \,\vartriangle\, e_2.$$

Then $\{1, e_1, e_2, e_3, j_1, j_2, j_3, J\}$ is (essentially) the induced ON-basis, and the Clifford products of basis multivectors in the even subalgebra

$$\triangle^{\mathrm{ev}} V = \triangle^0 V \oplus \triangle^2 V = \{a + bj_1 + cj_2 + dj_3 \; ; \; a, b, c, d \in \mathbf{R}\}$$

are

$$j_1^2 = j_2^2 = j_3^2 = -1, \quad j_1 j_2 j_3 = 1,$$
$$j_1 j_2 = -j_2 j_1 = -j_3, \quad j_3 j_1 = -j_1 j_3 = -j_2, \quad j_2 j_3 = -j_3 j_2 = -j_1.$$

This is nearly the product rules for Hamilton's *quaternions* $\mathbf{H}$. If we set $i = -j_1$, $j = -j_2$, and $k = -j_3$, then i, j, k are exactly isomorphic to the quaternion units. Since we only changed basis, we see that $\triangle^{\mathrm{ev}} V$ is isomorphic to the quaternions.

Definition 3.2.2. If V is an oriented Euclidean 3-dimensional space, then we refer to $(\triangle^{\mathrm{ev}} V, +, \vartriangle)$ as the *standard geometric representation* of the quaternions $\mathbf{H}$.

From the basis products above, we observe that in contrast to $\mathbf{C}$, the quaternions do not form a commutative algebra. However, the strength of the quaternions is that they form an associative division algebra, that is, every nonzero quaternion is invertible. Indeed,

$$\bar{q}q = (a - bj_1 - cj_2 - dj_3)(a + bj_1 + cj_2 + dj_3) = a^2 + b^2 + c^2 + d^2 = |q|^2 \in \mathbf{R},$$

so that $q^{-1} = \bar{q}/|q|^2$, for all $q \in \mathbf{H}$. Note that the reversion operation in $\triangle V$ restricts to the standard quaternion conjugation in $\mathbf{H}$.

We would like to rotate vectors in V using $\mathbf{H}$, just as we rotated plane vectors with $\mathbf{C}$. Let $T : V \to V$ be a rotation in V. In particular, $\langle Tv_1, Tv_2 \rangle = \langle v_1, v_2 \rangle$ for all $v_1, v_2 \in V$, so that $T^{-1} = T^*$. This also holds for the induced map $T : \triangle^3 V \to \triangle^3 V$, so $T(J) = \pm J$. However, by definition a rotation is orientation preserving, so $T(J) = J$.

Since $\dim V = 3$ and therefore the degree of the characteristic equation is odd, T has a real eigenvalue, with eigenvector v_1, and we normalize so that $|v_1| = 1$. We claim that

$$T(*v) = *T(v), \quad \text{for all } v \in V,$$

where on the left-hand side we apply the induced operator $T : \triangle^2 V \to \triangle^2 V$. Indeed, this follows from the cofactor formula (2.4), replacing T by T^{-1}, which also is a rotation. Let $Tv_1 = \lambda v_1$. Then $T(*v_1) = \lambda(*v_1)$, so that $j := *v_1$ is an eigen-bivector of T. Since $\triangle^2 V = \wedge^2 V = \hat{\wedge}^2 V$, there is an ON-basis $\{v_2, v_3\}$ for the plane $V' := [j]$ such that $j = v_2 \wedge v_3$, and T leaves V' invariant according to Proposition 2.3.2. It is clear that $\{v_1, v_2, v_3\}$ is an ON-basis for V and $J = v_1 \wedge j = v_1 \wedge v_2 \wedge v_3$. Thus $J = T(J) = T(v_1) \wedge T(j) = \lambda v_1 \wedge \lambda(j) = \lambda^2 J$, so that $\lambda = \pm 1$.

If $\lambda = 1$, then $T(v_1) = v_1$ and $T(j) = j$, so that T restricts to a rotation in the plane V'. From the discussion above, if ϕ denotes the rotation angle in V' counterclockwise according to the orientation j, then

$$T(v) = v \exp(\phi j) = \exp(-\phi j)v, \quad v \in V'.$$

To extend this expression to all V, note that

$$T(v) = v \exp(\phi j/2) \exp(\phi j/2) = \exp(-\phi j/2)v \exp(\phi j/2), \quad v \in V'.$$

Since $jv_1 = (v_2 v_3)v_1 = v_1(v_2 v_3) = v_1 j$, we have $\exp(-\phi j/2)v_1 \exp(\phi j/2) = v_1 = T(v_1)$, so in fact the expression for T is valid for all $v \in V$.

If $\lambda = -1$, then $T(v_1) = -v_1$ and $T(j) = -j$. Then $T : V' \to V'$ is a plane orientation-reversing isometry, and all such have an eigenvector $v_1' \in V'$ with eigenvalue 1. Thus replacing v_1 with v_1', we are back at the case $\lambda = 1$.

Let us summarize the discussion.

Proposition 3.2.3 (Rotations by quaternions). *Assume that V is an oriented 3-dimensional Euclidean space. Let $j \in \Delta^2 V$, $|j| = 1$, and $\phi \in \mathbf{R}$. Then rotation through the angle ϕ counterclockwise with rotation plane $[j]$ and rotation axis $[*j] = [j]^\perp$, that is, seen counterclockwise from the side of $[j]$ to which $*j$ points, is given by*

$$v \mapsto \exp(b/2)v \exp(-b/2) = qvq^{-1}, \quad v \in V, \tag{3.6}$$

where $b := -\phi j \in \Delta^2 V$. Here $q = \exp(b/2) = \cos(\phi/2) - j \sin(\phi/2) \in \mathbf{H}$ is a unit quaternion, i.e., $|q| = 1$. Two unit quaternions q_1 and q_2 represent the same rotation if and only if $q_1 = \pm q_2$.

The reason for not choosing the simpler relation $q = e^{\phi j/2}$ is that we want $q \mapsto T$ to be a homomorphism, and then convention unfortunately dictates a minus sign.

Proof. It remains to see the last result. Assume that $q_1 v q_1^{-1} = q_2 v q_2^{-1}$ for all $v \in V$. Then $q_1^{-1} q_2 v = v q_1^{-1} q_2$. Thus $q_1^{-1} q_2$ commutes with all vectors, and therefore with every multivector. But in three dimensions, the center of $\triangle V$ is $\triangle^0 V \oplus \triangle^3 V$ by Proposition 3.1.8. However, $q_1^{-1} q_2 \in \mathbf{H}$, so $q_1^{-1} q_2 \in \mathbf{R}$. Since $|q_1| = 1 = |q_2|$, we must have $q_1 = \pm q_2$. $\qquad \square$

This represents a rotation by three independent real parameters for the unit quaternion q. Compare this to a matrix representation, where nine parameters are needed. Here the parameters are dependent, since the matrix satisfies the orthogonality condition $A^* A = I$. This yields six equations, leaving three independent parameters.

Example 3.2.4. Since $\mathbf{H}$ is noncommutative, three-dimensional rotations will not in general commute. Let $q_1 := \exp(-\frac{\pi}{4}(*e_3)) = \frac{1}{\sqrt{2}}(1 - e_{12})$ and $q_2 := \exp(-\frac{\pi}{4}(*e_1)) = \frac{1}{\sqrt{2}}(1 - e_{23})$ be the unit quaternions representing rotations through the angle $\pi/2$ counterclockwise around e_3 and e_1 respectively. Then

$$q_1 q_2 = \tfrac{1}{2}(1 - e_{12} - e_{23} + e_{13}) = \tfrac{1}{2} - \tfrac{\sqrt{3}}{2}(*(e_1 + e_2 + e_3))/\sqrt{3},$$

and $q_2 q_1 = \frac{1}{2} - \frac{\sqrt{3}}{2}(*(e_1 - e_2 + e_3))/\sqrt{3}$. Thus $q_1 q_2$ and $q_2 q_1$ represent rotations through the angle $2\pi/3$ counterclockwise around $e_1 + e_2 + e_3$ and $e_1 - e_2 + e_3$ respectively.

Formula (3.6) gives a good global understanding of the space of all three-dimensional Euclidean rotations

$$\mathrm{SO}(3) = \{T : V \to V \ ; \ T^*T = I, \ \det T = 1\}.$$

Indeed, each $T \in \mathrm{SO}(3)$ corresponds to two antipodal unit quaternions $\pm q$ on the three-sphere $S^3 \subset \mathbf{H}$. Thus $\mathrm{SO}(3)$ coincides with the so-called real projective space

$$\mathbf{P}^3 := S^3/\{\pm 1\} = \{\text{lines through origin in } \mathbf{H}\}.$$

Example 3.2.5 (Spin). Consider a full $360°$ rotation around a fixed axis $*j$. Thus we consider a family of rotations $T_\phi \in \mathrm{SO}(3)$ depending continuously on the angle $\phi \in [0, 2\pi]$ and $T_0 = I = T_{2\pi}$, whereas T_π is rotation by $180°$. Viewing $\mathrm{SO}(3)$ as $\mathbf{P}^3$ we equivalently have a family of lines through the origin in $\mathbf{H}$, turning along a great circle on S^3 through an angle of $180°$ (not $360°$!). What becomes apparent here is that rotations have an "intrinsic memory". To explain this, put an arrow on the line representing the rotation, that is, orient this one-dimensional subspace, thus distinguishing the two quaternions $\pm q$ representing the same rotation. See Figure 3.2. Having oriented the line, we see that after completing the rotation, the situation is still different from what we had at the beginning: the line is the same, but it has changed orientation, since it turned only $180°$. However, performing yet another full rotation, we have again the same situation as at the beginning.

This sort of intrinsic state is traditionally referred to as *spin* and is not only a mathematical construction but a reality. It appears in quantum mechanics, where, for example, electrons exhibit this sort of intrinsic memory. Mathematically, we say that the fundamental group for $\mathrm{SO}(3)$ is $\mathbf{Z}_2$. See Theorem 4.3.4 for more details.

Just as the exponential function parametrizes the unit circle S^1 in the plane with $\mathbf{R}$, the exponential map $\exp : \triangle^2 V \to S^3 : w \mapsto q = \exp(b/2)$ parametrizes the three-sphere S^3 of unit quaternions with bivectors and enables us to visualize S^3 within the three-dimensional space $\triangle^2 V$. The origin $b = 0$, as well as the spheres $|b| = 4\pi k$, corresponds to the "north pole" $q = 1$, the spheres $|b| = \pi + 2\pi k$, $k = 0, 1, 2, \ldots$, correspond to the "equator" $q \in S^3 \cap \triangle^2 V$, and the spheres $|b| = 2\pi + 4\pi k$, $k = 0, 1, 2, \ldots$, correspond to the "south pole" $q = -1$.

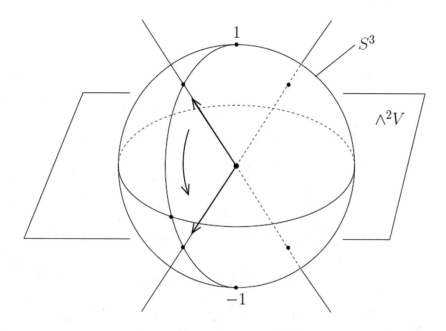

Figure 3.2: SO(3) as the real projective 3-space $\mathbf{P}^3$, and the quaternion unit sphere S^3.

Thus rotations of $180°$, that is, $I \neq T \in SO(3)$ such that $T^2 = I$, correspond to two opposite points $\pm b$ on the sphere $|b| = \pi$. Every other rotation corresponds to a unique bivector b in the ball

$$B(0; \pi) := \{b \in \triangle^2 V \; ; \; |b| < \pi\},$$

where $|b|$ is the clockwise rotation angle around the rotation axis $*b$. Note that if we view the rotation from the opposite direction, we see that it is equivalently a rotation through the angle $2\pi - |b|$ clockwise around $-(*b)$. Thus the point $(1 - 2\pi/|b|)b$ in the annulus $\pi < |b| < 2\pi$ also represents the same rotation.

If we distinguish the two different spin states of a rotation, then each unit quaternion, except $q = -1$, corresponds to a unique bivector in the ball $B(0, 2\pi)$ of radius 2π. All points on the boundary sphere correspond to $q = -1$.

Example 3.2.6 (Clifford trace). Let $\{e_1, e_2, e_3\}$ be a positive ON-basis. We would like to find the matrix of the rotation through the angle $\arccos(1/3)$ counterclockwise around $e_1 + e_2$.

The bivector for the rotation plane is $j = *\frac{1}{\sqrt{2}}(e_1 + e_2) = \frac{1}{\sqrt{2}}(e_{23} + e_{31})$ and the angle of rotation counter clockwise is $\phi = \arccos(1/3)$, so we should use the quaternion

$$q = \cos(\phi/2) - \tfrac{1}{\sqrt{2}}(e_{23} + e_{31})\sin(\phi/2) = (2 - e_{23} - e_{31})/\sqrt{6}.$$

Thus the rotation is

$$v \mapsto \tfrac{1}{6}(2 - e_{23} - e_{31})v(2 + e_{23} + e_{31}).$$

Calculating the rotated basis vectors yields the matrix

$$A = \frac{1}{3}\begin{bmatrix} 2 & 1 & 2 \\ 1 & 2 & -2 \\ -2 & 2 & 1 \end{bmatrix}.$$

Conversely, Clifford algebra proves useful when we want to find the rotation angle and rotation plane for this orthogonal matrix. Consider the *Clifford trace*

$$w = \sum_{i=1}^{3} e_i \vartriangle T(e_i) \in \mathbf{H},$$

where T is the rotation with matrix A. This trace can be seen as the linear map $\mathcal{L}(V) = V \otimes V \to \wedge V$ obtained by lifting the bilinear map $B : V \times V \to \mathbf{H} :$ $(v, v') \mapsto v \vartriangle v'$ by the universal property to the tensor product as in Section 1.4. In particular, it does not depend on the choice of ON-basis. Choosing an ON-basis $\{e_1', e_2', e_3'\}$, where e_3' is the rotation axis and $j = e_1' \wedge e_2'$ is the orientation of the rotation plane, we get

$$w = 2(\cos\phi + j\sin\phi) + 1.$$

On the other hand, from the matrix in the first basis, we get

$$w = \tfrac{1}{3}((2 + e_{12} - 2e_{13}) + (e_{21} + 2 + 2e_{23}) + (2e_{31} - 2e_{32} + 1)) = \tfrac{5}{3} + \tfrac{4}{3}(e_{23} + e_{31}).$$

Comparing the two expressions, we deduce $\phi = \arccos(1/3)$ and rotation axis $e_1 + e_2$, and we are back where we started.

Exercise 3.2.7. Let $v_1 = \tfrac{1}{\sqrt{2}}(e_1 + e_2)$ and $v_2 = \tfrac{1}{3}(2e_1 - e_2 + 2e_3)$ be two unit vectors in a three-dimensional Euclidean space V with ON-basis $\{e_1, e_2, e_3\}$. Find all bivectors $b \in \wedge^2 V$ such that $q := v_1 v_2$ equals $q = e^{b/2}$. Find the rotation axis, angle, and sense of rotation of the corresponding rotation $v \mapsto qvq^{-1}$.

Similar to the situation for $\mathbf{C}$, we can identify vectors in V with "pure quaternions", that is, the bivectors $\triangle^2 V$. For $\mathbf{H}$, this identification is more canonical in that it does not require a "real" vector e_1 to be chosen, and it is furnished by the Hodge star map. Recall that in three-dimensional Euclidean space $*w = w* = \overline{w}J = J\overline{w}$. Thus we identify

$$V \ni v = -\tilde{v}J = -J\tilde{v} \leftrightarrow \tilde{v} = vJ = Jv, \quad v \in V, \ \tilde{v} \in \triangle^2 V \subset \mathbf{H},$$

so that $e_1 \leftrightarrow j_1$, $e_2 \leftrightarrow j_2$, and $e_3 \leftrightarrow j_3$. Since J belongs to the center of $\triangle V$, simply multiplying (3.6) by J shows that in the pure quaternion representation of vectors, rotation is given by the formula

$$\tilde{v} \mapsto \exp(b/2)\tilde{v}\exp(-b/2) = q\tilde{v}q^{-1}, \quad \tilde{v} \in \triangle^2 V.$$

Finally, we remark that these standard representations of $\mathbf{C}$ and $\mathbf{H}$ are by no means the only place within Clifford algebras where $\mathbf{C}$ and $\mathbf{H}$ appear. In fact, isomorphic copies of these appear in many places in Clifford algebras, and for this reason, Clifford algebras are often thought of as algebras of "hypercomplex numbers". For example, if $\dim V = 1$ and V has signature $0 - 1$, then $(\triangle V, \triangle)$ is isomorphic to $\mathbf{C}$; if $\dim V = 2$ and V has signature $0 - 2$, then $(\triangle V, \triangle)$ is isomorphic to $\mathbf{H}$; if V is three-dimensional Euclidean space, then $\mathbf{C} = \triangle^0 V \oplus \triangle^3 V$; and so on.

3.3 Abstract Clifford Algebras

We have constructed a Clifford product $\triangle$ on the exterior algebra $\wedge V$ of an inner product space V, giving us the standard Clifford algebra $\triangle V$. We have chosen to recycle the exterior algebra that we already have constructed and turn this into a Clifford algebra by modifying the product. However, associative algebras isomorphic to $\triangle V$ appear naturally in many other ways. Analogously to exterior algebra, in this section we study Clifford algebras from the point of view of a universal property. For $\triangle V$, the central identity is the anticommutation relation

$$v_1 v_2 + v_2 v_1 = 2\langle v_1, v_2 \rangle, \quad v_1, v_2 \in V. \tag{3.7}$$

From this and associativity, all products of basis vectors can be computed. Indeed, it follows that $e_i e_j = -e_j e_i$ if $i \neq j$ and that $e_i^2 = \langle e_i \rangle^2$. Putting $v_1 = v_2 = v$, we obtain

$$v^2 = \langle v \rangle^2, \quad v \in V, \tag{3.8}$$

which states that if $\langle v \rangle^2 \neq 0$, then v is invertible and $v^{-1} = v/\langle v \rangle^2$. In fact, (3.8) is equivalent to (3.7), as is seen by polarization:

$$2\langle v_1, v_2 \rangle + \langle v_1 \rangle^2 + \langle v_2 \rangle^2 = \langle v_1 + v_2 \rangle^2 = (v_1 + v_2)(v_1 + v_2) = v_1 v_2 + v_2 v_1 + v_1^2 + v_2^2.$$

Indeed, if $v^2 = \langle v \rangle^2$ for $v = v_1$, v_2, and $v_1 + v_2$, then this identity shows that $v_1 v_2 + v_2 v_1 = 2\langle v_1, v_2 \rangle$. We refer to general Clifford algebras defined in the following way as abstract Clifford algebras, as compared to the standard Clifford algebra $(\triangle V, \triangle)$.

Definition 3.3.1 (Clifford algebras). Let $(V, \langle \cdot, \cdot \rangle)$ be an inner product space. An associative algebra A with identity $1 \in A$ and containing V, or more precisely A together with a linear map $C : V \to A$ that identifies V and the linear subspace $C(V) \subset A$, is called a (universal) *Clifford algebra* for $(V, \langle \cdot, \cdot \rangle)$ if

(C) $(C(v))^2 = \langle v \rangle^2 1$ for all $v \in V$, or equivalently $C(v_1)C(v_2) + C(v_2)C(v_1) = 2\langle v_1, v_2 \rangle 1$ for all $v_1, v_2 \in V$, and

(U) $C : V \to A$ satisfies the following universal property. Whenever A' is an algebra and $C' : V \to A'$ satisfies (C), there exists a unique algebra homomorphism $T : A \to A'$ such that $C'(v) = T(C(v))$ for all $v \in V$.

Existence and uniqueness of Clifford algebras is immediate.

Proposition 3.3.2 (Existence and uniqueness). *Let $(V, \langle \cdot, \cdot \rangle)$ be an inner product space.*

(i) *The exterior algebra $\triangle V = (\wedge V, \triangle)$ equipped with the Clifford product is a Clifford algebra, that is, it satisfies (C) and (U).*

(ii) *If (A, C) and (A', C') are two Clifford algebras for V, then there exists a unique algebra isomorphism $T : A \to A'$ such that $C' = T \circ C$. Thus, if $\dim V = n$, then every Clifford algebra for V has dimension 2^n.*

Proof. (i) We already know that the Clifford condition (C) holds for $\triangle$, where we take C as the usual inclusion of $V = \triangle^1 V$ into $\triangle V$. To verify (U), note that if the homomorphism T exists, it must satisfy $T(e_{s_1} \triangle \ldots \triangle e_{s_k}) = C'(e_{s_1}) \cdots C'(e_{s_k})$ and $T(1) = 1$ if $C' = T \circ C$ is to hold. Such a T is unique. To show the existence of T, we define T in this way on basis multivectors and extend to a well-defined linear map $T : \triangle V \to A'$. To verify that T is indeed a homomorphism, note that by linearity it suffices to consider the action of T on products of basis elements. Write $C'(e_s) := C'(e_{s_1}) \cdots C'(e_{s_k})$ for short, and consider a product $e_s \triangle e_t$. Through iterated use of (C), $e_s \triangle e_t$ is reduced to ± 1 times the basis element $e_{s \triangle t}$. The same permutation and squaring procedure reduces $C'(e_s)C'(e_t)$ to $\pm C'(e_{s \triangle t})$. Thus

$$T(e_s \triangle e_t) = T(\pm e_{s \triangle t}) = \pm C'(e_{s \triangle t}) = C'(e_s)C'(e_t) = T(e_s)T(e_t),$$

which proves that T is a homomorphism, and we conclude that (U) holds.

(ii) From the universal properties for A and A', we get unique homomorphisms $T : A \to A'$ and $T' : A' \to A$ such that $TC = C'$ and $T'C' = C$. But then $T'T : A \to A$ is a homomorphism such that $(T'T)C = T'C' = C$, and again by the uniqueness hypothesis in (U) we must have $T'T = I$. Similarly, it follows that $TT' = I$. This shows that T is an algebra isomorphism. $\qquad\square$

Let A be any Clifford algebra for V. According to Proposition 3.3.2(ii) there exists a unique algebra isomorphism $T : \triangle V \to A$ that identifies A and the standard Clifford algebra $\triangle V$ and respects the embeddings of V in the two algebras. Writing $A^k := T(\triangle^k V)$ shows that

$$A = A^0 \oplus A^1 \oplus A^2 \oplus \cdots \oplus A^{n-1} \oplus A^n.$$

Translating from $\triangle V$, we see that each product $C(v_1) \cdots C(v_k)$ of k orthogonal vectors is in A^k, and that A^k is the linear span of all such products. However, unlike the situation for the exterior product, if the vectors v_j are not orthogonal, the best we can say is that

$$C(v_1)\cdots C(v_k) \in A^k \oplus A^{k-2} \oplus A^{k-4} \oplus \cdots.$$

As for $\triangle V$, we may also consider the even subalgebra $A^{\mathrm{ev}} := T(\triangle^{\mathrm{ev}}V)$ and the odd part $A^{\mathrm{od}} := T(\triangle^{\mathrm{od}}V)$ of A, which is not a subalgebra. The even subalgebra is seen to be closed under multiplication in A.

In contrast to the alternating multilinear maps $\wedge_1^k$ in Proposition 2.1.4, which very well may be 0, the maps C satisfying (C) almost automatically satisfy (U), due to the highly noncommutative properties of A implied by (C), as the following fundamental result for Clifford algebras shows.

Proposition 3.3.3 (Universality). *Let $(V, \langle \cdot, \cdot \rangle)$ be an inner product space of dimension $n = n_+ + n_-$ and signature $n_+ - n_-$. Let A be an algebra and $C : V \to A$ any map satisfying (C). Denote by $A_0 \subset A$ the subalgebra generated by $C(V)$, that is, the smallest subalgebra containing $C(V)$.*

(i) *If $n_+ - n_- \not\equiv 1 \,(\mathrm{mod}\ 4)$, then $\dim A_0 = 2^n$ and (A_0, C) is a Clifford algebra for V.*

(ii) *If $n_+ - n_- \equiv 1 \,(\mathrm{mod}\ 4)$, then $\dim A_0$ is either 2^n or 2^{n-1}. Here (A_0, C) is a Clifford algebra for V if and only if $\dim A_0 = 2^n$, which is the case if and only if $C(e_1)\cdots C(e_n)$ is not scalar, that is, if it is linearly independent with 1.*

Proof. Consider an ON-basis $\{e_i\}$ for V. To verify the universal property, it suffices to show that

$$\{C(e_{s_1})\cdots C(e_{s_k}) \ ; \ 1 \le s_1 < \cdots < s_k \le n, 0 \le k \le n\}$$

is linearly independent, since the proof of Proposition 3.3.2(i) then can be repeated. Write $C(e_s) = C(e_{s_1})\cdots C(e_{s_k})$ and $C(1) = 1$, and assume that

$$\sum_s \alpha_s C(e_s) = 0$$

for some scalars α_s. Conjugating with a basis vector e_j, it follows that

$$\sum_s \alpha_s C(e_j)^{-1} C(e_s) C(e_j) = 0.$$

Note that (C) implies that either $C(e_j)C(e_s) = C(e_s)C(e_j)$ or $C(e_j)C(e_s) = -C(e_s)C(e_j)$. Adding and subtracting these two linear combinations therefore shows that

$$\sum_{\{s\,;\,C(e_j)C(e_s)=C(e_s)C(e_j)\}} \alpha_s C(e_s) = 0 = \sum_{\{s\,;\,C(e_j)C(e_s)=-C(e_s)C(e_j)\}} \alpha_s C(e_s).$$

Further splitting of the linear combination according to the commutation pattern with all e_j produces 2^n equations. Write $s \sim t$ if both e_s and e_t either commute or anticommute with all e_j, $1 \le j \le n$. Then the equations are

$$\sum_{s \sim s_0} \alpha_s C(e_s) = 0.$$

Note that $s \sim t$ if and only if $C(e_s)C(e_t) = \pm C(e_{s \triangle t})$ commutes with all $C(e_j)$, that is if $s \triangle t \sim \varnothing$. If $u \ne \varnothing$ and $u \ne \overline{n}$, then there exist $j \in u$ and $j' \notin u$, and $C(e_u)$ cannot commute with both $C(e_j)$ and $C(e_{j'})$. Thus $u = \varnothing$ or $u = \overline{n}$ if $u \sim \varnothing$, that is, $s = t$ or $s = \overline{n} \setminus t$ if $s \sim t$, so all equivalence classes contains either one or two index sets. If they are singletons, then $\dim A_0 = 2^n$. If $C(1) = \lambda C(e_{\overline{n}})$ are parallel, then multiplying by $C(e_s)$ shows that $C(e_s) = \lambda C(e_{s \triangle \overline{n}})$, and it follows that $\dim A_0 = 2^{n-1}$.

If n is even, then $\mathcal{C}(e_{\overline{n}})$ anticommutes with vectors and the first case holds. Assume now that $n = 2k+1$. If $C(1) = \lambda C(e_{\overline{n}})$, then we must have $(C(e_{\overline{n}}))^2 > 0$. But

$$(C(e_{\overline{n}}))^2 = (-1)^{n(n-1)/2 + n_-} 1 = (-1)^{k+n_-} 1,$$

so $k - n_-$ is even, that is, $n - 1 - 2n_- = n_+ - n_- - 1 \equiv 0 \, (\mathrm{mod}\ 4)$. Thus, if $n_+ - n_- \not\equiv 1 \, (\mathrm{mod}\ 4)$, then $\{C(e_s)\}$ are necessarily linearly independent. □

Example 3.3.4 (Nonuniversal Clifford algebras). To see that algebras of dimension 2^{n-1} satisfying (C) do exist in case (ii), assume that $n_+ - n_- \equiv 1 \, (\mathrm{mod}\ 4)$ and consider $\triangle V$. Let $e_{\overline{n}}$ be a unit n-vector. From the signature condition it follows that $e_{\overline{n}}^2 = 1$ and $e_{\overline{n}} w = w e_{\overline{n}}$ for all $w \in \triangle V$. Therefore the two multivectors $(1 + e_{\overline{n}})/2$ and $(1 - e_{\overline{n}})/2$ are central complementary projections, that is, $((1 \pm e_{\overline{n}})/2)^2 = (1 \pm e_{\overline{n}})/2$ and $((1 + e_{\overline{n}})/2)((1 - e_{\overline{n}})/2) = 0$. This means that the corresponding multiplication operators are complementary projections that split $\triangle V$ into two ideals. Consider, for example, the algebra

$$A = (1 + e_{\overline{n}})(\triangle V) = \{(1 + e_{\overline{n}})w \; ; \; w \in \triangle V\},$$

where $(1 + e_{\overline{n}})/2$ is the identity in A, and the injection $C : v \mapsto v(1 + e_{\overline{n}})/2 \in A$. This is seen to be an algebra of dimension 2^{n-1} satisfying (C).

A first example of an abstract Clifford algebra is the following.

Proposition 3.3.5 (Even subalgebra). *Let* $(V, \langle \cdot, \cdot \rangle)$ *be an inner product space. Fix a nonsingular vector* $e_0 \in V$ *and define the orthogonal complement* $V' := [e_0]^{\perp}$, *and equip this with the inner product* $\langle \cdot, \cdot \rangle' := -\langle e_0 \rangle^2 \langle \cdot, \cdot \rangle$. *Then the even subalgebra* $\triangle^{\mathrm{ev}} V$ *is a Clifford algebra for* $(V', \langle \cdot, \cdot \rangle')$, *with vector subspace defined by*

$$C : V' \to \triangle^{\mathrm{ev}} V : v \mapsto e_0 v.$$

Proof. Condition (C) for Clifford algebras is satisfied, since

$$(C(v))^2 = (e_0 v)^2 = e_0 v e_0 v = -e_0 e_0 v v = -\langle e_0 \rangle^2 \langle v \rangle^2 1 = \langle v, v \rangle' 1, \quad v \in V'.$$

If $\{e_1, \ldots, e_n\}$ is an ON-basis for V', then

$$C(e_1) \cdots C(e_n) = ((-1)^{n(n-1)/2} e_0^n) e_1 \cdots e_n.$$

Since this volume element is not scalar, and since $\dim \triangle^{\mathrm{ev}} V = 2^n$, where $n = \dim V'$, it follows from Proposition 3.3.3 that $\triangle^{\mathrm{ev}} V$ is a Clifford algebra for $(V', \langle \cdot, \cdot \rangle')$. $\square$

Note that if W is a spacetime and e_0 is a time-like vector such that $\langle e_0 \rangle^2 = -1$, then Proposition 3.3.5 shows that the even subalgebra $\triangle^{\mathrm{ev}} W$ is an algebra isomorphic to the Clifford algebra $\triangle V$ for the Euclidean space $V = [e_0]^\perp$.

Example 3.3.6. The standard geometric representation of $\mathbf{C}$ is as the even subalgebra $\triangle^{\mathrm{ev}} V$ of the Clifford algebra for the Euclidean plane V. Fix an ON-basis $\{e_1, e_2\}$ for V and define on $V' := [e_1]^\perp$ an inner product by $\langle e_2 \rangle^{2'} = -1$. Then $\triangle^{\mathrm{ev}} V$ is a Clifford algebra for V', isomorphic to $\triangle V' = \mathbf{C}$.

Example 3.3.7. The standard geometric representation of $\mathbf{H}$ is as the even subalgebra $\triangle^{\mathrm{ev}} V$ of the Clifford algebra for the Euclidean three-dimensional space V.

Fix an ON-basis $\{e_1, e_2, e_3\}$ for V and define on $V' := [e_1]^\perp$ an inner product by $\langle e_2 \rangle^{2'} = \langle e_3 \rangle^{2'} = -1$, $\langle e_2, e_3 \rangle' = 0$. Then $\triangle^{\mathrm{ev}} V$ is a Clifford algebra for V', isomorphic to $\triangle V' = \mathbf{H}$.

Exercise 3.3.8. The notion of a Clifford algebra does not really need the inner product to be nondegenerate. Generalize Proposition 3.3.2 and Proposition 3.3.3 to Clifford algebras for general symmetric bilinear forms $B(\cdot, \cdot)$, and show that for the trivial form $B = 0$, the Clifford algebra is isomorphic to the exterior algebra $(\wedge V, \wedge)$.

3.4 Matrix Representations

Matrix algebras, with the product being matrix multiplication, are the standard associative algebras. We will prove that all Clifford algebras are isomorphic to matrix algebras, but for this, one needs to allow slightly more general matrix elements than real numbers. The situation simplifies if we consider Clifford algebras over the complex field, but this discussion is postponed until Chapter 5.

Although it is tempting and common to view Clifford algebras as generalizations of the complex field, Clifford algebras are rather matrix algebras, seen from the point of view of a geometric basis through an embedding of the vector space satisfying condition (C).

Note that in applications, being able to represent a Clifford algebra by matrices is very useful in implementing geometric computations with Clifford algebras.

Exercise 3.4.1. Verify that the matrix representations

$$x_0 + x_1 e_1 + x_2 e_2 + x_{12} e_{12} = \begin{bmatrix} x_0 + x_2 & x_1 - x_{12} \\ x_1 + x_{12} & x_0 - x_2 \end{bmatrix}$$

for the Clifford algebra of the Euclidean plane and

$$x_0 + x_1 e_1 + x_2 e_2 + x_3 e_3 + x_{12} e_{12} + x_{13} e_{13} + x_{23} e_{23} + x_{123} e_{123}$$

$$= \begin{bmatrix} x_0 + x_2 + i(x_{123} - x_{13}) & x_1 - x_{12} + i(x_{23} + x_3) \\ x_1 + x_{12} + i(x_{23} - x_3) & x_0 - x_2 + i(x_{123} + x_{13}) \end{bmatrix}$$

for the Clifford algebra for Euclidean three-dimensional space are isomorphisms of real algebras. Furthermore, show that the latter is an isomorphism of complex algebras if we use e_{123} as a complex structure. Verify the calculations in Example 3.1.4 using these representations. In both two and three dimensions, use these representations to write equations for the coordinates that determine when a multivector in the Clifford algebra is not invertible.

We derive these representations in Examples 3.4.18 and 3.4.19 below. Before turning to the main representation Theorem 3.4.13 for real Clifford algebras, we consider the most fundamental case: the case of signature 0.

Theorem 3.4.2 (Neutral Clifford algebras). *Let $(V, \langle \cdot, \cdot \rangle)$ be any inner product space. Then the space of all linear operators $\mathcal{L}(\wedge V)$ acting on its exterior algebra is naturally a Clifford algebra of signature zero. More precisely, consider the inner product space $V \oplus V$, equipped with the inner product*

$$\langle v_1 \oplus v_2, v_1' \oplus v_2' \rangle_2 := \langle v_1, v_1' \rangle - \langle v_2, v_2' \rangle,$$

and define the linear map $C : V \oplus V \to \mathcal{L}(\wedge V)$ as

$$C(v \oplus 0)w := v \lrcorner w + v \wedge w,$$
$$C(0 \oplus v)w := -v \lrcorner w + v \wedge w.$$

Then $(\mathcal{L}(\wedge V), C)$ is a Clifford algebra for $(V \oplus V, \langle \cdot, \cdot \rangle_2)$.

Proof. Since the inner product on $V \oplus V$ has signature zero and since $\dim(V \oplus V) = 2n$ and $\dim \mathcal{L}(\wedge V) = (2^n)^2$, by Proposition 3.3.3 it suffices to verify condition (C). Consider $v_1 \oplus v_2 \in V \oplus V$ and $C(v_1 \oplus v_2)w = (v_1 - v_2) \lrcorner w + (v_1 + v_2) \wedge w$. The associativity and anticommutation relations for exterior and interior products show that

$$C(v_1 \oplus v_2)^2 w = (v_1 - v_2) \lrcorner ((v_1 - v_2) \lrcorner w) + (v_1 - v_2) \lrcorner ((v_1 + v_2) \wedge w)$$
$$+ (v_1 + v_2) \wedge ((v_1 - v_2) \lrcorner w) + (v_1 + v_2) \wedge (v_1 + v_2) \wedge w$$
$$= \langle v_1 - v_2, v_1 + v_2 \rangle w = (\langle v_1 \rangle^2 - \langle v_2 \rangle^2)w$$

for all $w \in \wedge V$. This proves that $(C(v_1 \oplus v_2))^2 = \langle v_1 \oplus v_2, v_1 \oplus v_2 \rangle^2 I$ and thus condition (C).   □

Definition 3.4.3. Let V be an inner product space and consider its exterior algebra $\wedge V$. For each vector $v \in V$, we define the corresponding operators of *positive (negative) Clifford multiplication* $v^+(v^-) \in \mathcal{L}(\wedge V)$ to be

$$v^+(w) := v \lrcorner w + v \wedge w,$$
$$v^-(w) := -v \lrcorner w + v \wedge w,$$

for $w \in \wedge V$.

If $\{e_i\}$ is an ON-basis for V, then $\{e_i \oplus 0\}_i \cup \{0 \oplus e_j\}_j$ is an ON-basis for $V \oplus V$. The *induced basis* for $\mathcal{L}(\wedge V)$ is the set of operators

$$\{e_{s_1}^+ \circ \cdots \circ e_{s_k}^+ \circ e_{t_1}^- \circ \cdots \circ e_{t_l}^-\}_{s_1 < \cdots < s_k, t_1 < \cdots < t_l, 0 \leq k, l \leq n}.$$

Furthermore, we let $\mathcal{L}_\pm(\wedge V)$ be the subalgebras generated by all v^+ and all v^- respectively. The subalgebra $\mathcal{L}_+(\wedge V)$ has a basis $\{e_s^+\}_{s \subset \bar{n}}$, whereas $\mathcal{L}_-(\wedge V)$ has a basis $\{e_s^-\}_{s \subset \bar{n}}$. Thus $\dim \mathcal{L}_\pm(\wedge V) = 2^n$, whereas $\dim \mathcal{L}(\wedge V) = 2^{2n}$.

Exercise 3.4.4. Show that $C : V \to \mathcal{L}_+(\wedge V) : v \mapsto v^+$ realizes $\mathcal{L}_+(\wedge V)$ as a Clifford algebra for $(V, \langle \cdot, \cdot \rangle)$, whereas $C : V \to \mathcal{L}_-(\wedge V) : v \mapsto v^-$ realizes $\mathcal{L}_-(\wedge V)$ as a Clifford algebra for $(V, -\langle \cdot, \cdot \rangle)$. Prove, through a basis calculation, that

$$\mathcal{L}_+(\wedge V) \to \wedge V : T \mapsto T(1)$$

is an invertible linear map, and that $T_1(1) \vartriangle T_2(1) := (T_1 \circ T_2)(1)$ gives another way to construct the Clifford product $\vartriangle$ on $\wedge V$.

Theorem 3.4.2 lies at the heart of the theory of Clifford algebras. On the one hand, it gives an "algebraic" point of view of Clifford algebras: multivectors, under Clifford multiplication, can be represented as matrices. On the other hand, it gives a "geometric" point of view of matrix algebras: by writing an operator in the Clifford basis generated by e_j^+ and e_j^- in $\mathcal{L}(\wedge V)$, one obtains a geometric interpretation of the operator.

Example 3.4.5. Let V be the two-dimensional Euclidean plane in Theorem 3.4.2, with induced ON-basis $\{1, e_{12}, e_1, e_2\}$ for $\wedge V$. Then the 16-dimensional space of 4×4 matrices $\mathcal{L}(\wedge V)$ has a basis induced by $\{e_1^+ e_2^+, e_1^-, e_2^-\}$. One verifies that the matrices for these operators, in the basis $\{1, e_{12}, e_1, e_2\}$, are

$$e_1^+ = \begin{bmatrix} 0 & 0 & 1 & 0 \\ 0 & 0 & 0 & 1 \\ 1 & 0 & 0 & 0 \\ 0 & 1 & 0 & 0 \end{bmatrix}, \quad e_2^+ = \begin{bmatrix} 0 & 0 & 0 & 1 \\ 0 & 0 & -1 & 0 \\ 0 & -1 & 0 & 0 \\ 1 & 0 & 0 & 0 \end{bmatrix},$$

$$e_1^- = \begin{bmatrix} 0 & 0 & -1 & 0 \\ 0 & 0 & 0 & 1 \\ 1 & 0 & 0 & 0 \\ 0 & -1 & 0 & 0 \end{bmatrix}, \quad e_2^- = \begin{bmatrix} 0 & 0 & 0 & -1 \\ 0 & 0 & -1 & 0 \\ 0 & 1 & 0 & 0 \\ 1 & 0 & 0 & 0 \end{bmatrix}.$$

Typically, basis multivectors correspond to signed permutation matrices like these.

Exercise 3.4.6 (Trace as scalar part). Let $T \in \mathcal{L}(\wedge V)$ be a linear operator. Show that

$$\mathrm{Tr}\, T = 2^n T_\emptyset,$$

where $T_\emptyset$ is the scalar part of T when written in an induced basis for $\mathcal{L}(\wedge V)$.

The following examples of abstract Clifford algebras will be relevant in particular in Chapter 12.

Example 3.4.7 (Reflection operators). Let $a \in V$ be a unit vector in Euclidean space and define the operator

$$R_a := a^+ a^- \in \mathcal{L}(\wedge V).$$

We see that it acts on a multivector as

$$R_a(w) = \begin{cases} a \wedge (-a \lrcorner w) = -w, & \text{if } a \wedge w = 0, \\ a \lrcorner (a \wedge w) = w, & \text{if } a \lrcorner w = 0. \end{cases}$$

From Proposition 2.8.7 it follows that R_a coincides with the induced map on $\wedge V$, as in Definition 2.3.1, from the reflection of vectors in the hyperplane $[a]^\perp$.
    Since

$$a^- w = -a \lrcorner w + a \wedge w = \widehat{w} \llcorner a + \widehat{w} \wedge a = \widehat{w} \vartriangle a,$$

we can also write R_a as $R_a(w) = a \vartriangle \widehat{w} \vartriangle a = a\widehat{w}a$.
    For example, in Example 3.4.5, $e_1^+ e_1^-$ is diagonal with diagonal $[1, -1, -1, 1]$, and $e_2^+ e_2^-$ is diagonal with diagonal $[1, -1, 1, -1]$.

Example 3.4.8 (Volume element = involution). Let V be a Euclidean space. The involution operator $w \mapsto \widehat{w}$ on $\wedge V$, viewed in the Clifford basis for $\mathcal{L}(\wedge V)$, is a volume element, that is, it belongs to $\mathcal{L}^{2n}(\wedge V)$. More precisely, if $\{e_i\}$ is an ON-basis for V, then

$$e_1^+ e_1^- e_2^+ e_2^- \cdots e_n^+ e_n^-(w) = R_{e_1} \cdots R_{e_n}(w) = \widehat{w}.$$

This follows from Example 3.4.7, since a k-vector w is tangential to all but k of coordinate hyperplanes $[e_j]^\perp$.
    The operator consisting of right Clifford multiplication $w \mapsto w \vartriangle a$ by a vector a, viewed in the Clifford basis for $\mathcal{L}(\wedge V)$, belongs to $\mathcal{L}^{2n-1}(\wedge V)$. Since $w \vartriangle a = a^-(\widehat{w})$, we get

$$w \vartriangle a = a^- e_1^+ e_1^- e_2^+ e_2^- \cdots e_n^+ e_n^-(w).$$

In particular, $w \vartriangle e_j = e_1^+ e_1^- e_2^+ e_2^- \cdots e_j^+ \overset{\smile}{e_j^-} \cdots e_n^+ e_n^-(w)$, where $\overset{\smile}{e_j^-}$ means that e_j^- is missing in the product.

Exercise 3.4.9 (Reversion, Hodge stars). Let $\{e_i\}$ be an ON-basis for Euclidean space V. Show that $\wedge^k V$ are eigenspaces to the operator

$$b := e_1^+ e_1^- + e_2^+ e_2^- + \cdots + e_n^+ e_n^- \in \mathcal{L}(\wedge V),$$

with eigenvalues $n-2k$. Note that each operator $b_k := e_k^+ e_k^-$ is symmetric, and that $b_k b_l = b_l b_k$. Since all the eigenvalues $n - 2k$ are distinct, any operator with $\wedge^k V$ as eigenspaces can be obtained from b through functional calculus. We complexify the matrices in $\mathcal{L}(\wedge V)$ as in Section 1.5 to simplify calculations and make use of Exercise 1.1.5, noting that $b_k^2 = I$.

For example, the involution in Example 3.4.8 equals

$$\exp(i\tfrac{\pi}{2}(n - b)) = i^n \exp(-i\tfrac{\pi}{2}b_1) \cdots \exp(-i\tfrac{\pi}{2}b_n) = b_1 \cdots b_n,$$

since $\exp(i\tfrac{\pi}{2}(n - \lambda)) = (-1)^k$ when $\lambda = n - 2k$. Similarly, reversion $w \mapsto \overline{w}$ is

$$\sqrt{2}\cos(\tfrac{\pi}{4}(n - 1 - b)) = 2^{1-n} \operatorname{Re}\left((1 + i)^{n-1} \prod_{k=1}^{n}(1 - ib_k) \right),$$

using Euler's formula, since $\sqrt{2}\cos(\tfrac{\pi}{4}(n-1-\lambda)) = (-1)^{k(k-1)/2}$ when $\lambda = n - 2k$.

Assume now that $\dim V = 2$. Simplify the above expression for $w \mapsto \overline{w}$ (in particular, eliminate the imaginary unit i), and verify the result in Example 3.4.5. Show that the linear map $w \mapsto e_{\overline{n}} w$ belongs to $(\mathcal{L}_+(\wedge V))^n$, whereas $w \mapsto w e_{\overline{n}}$ belongs to $(\mathcal{L}_-(\wedge V))^n$, that is, the n-vector parts of the sub-Clifford algebras $\mathcal{L}_\pm(\wedge V)$ of $\mathcal{L}(\wedge V)$. Combine this result with that for the reversion, and write the Hodge star maps $*w$ and $w*$ in the Clifford basis $\{e_s^+ e_t^-\}$ for $\mathcal{L}(\wedge V)$.

We next consider the generalization of Theorem 3.4.2 to real Clifford algebras of arbitrary signature. For this we must allow more general coefficients than elements of the fields $\mathbf{R}$ and $\mathbf{C}$. We need to consider coefficients from the skewfield $\mathbf{H}$ as well as from the algebras $\mathbf{R}^2$, $\mathbf{C}^2$, and $\mathbf{H}^2$.

Definition 3.4.10 (Tuples). Let $\mathbf{A}$ denote one of the fields $\mathbf{R}$, $\mathbf{C}$, or the skew field $\mathbf{H}$, or more generally any real algebra. Then $\mathbf{A}^2$ denotes the real algebra consisting of tuples (x_1, x_2), $x_1, x_2 \in \mathbf{A}$, with componentwise addition $(x_1, x_2)+(y_1, y_2) := (x_1+y_1, x_2 + y_2)$, multiplication by scalars $\lambda(x_1, x_2) := (\lambda x_1, \lambda_2)$, and componentwise multiplication

$$(x_1, x_2)(y_1, y_2) := (x_1 y_1, x_2 y_2), \quad x_i, y_i \in \mathbf{A}.$$

The algebra has identity element $(1, 1)$.

Definition 3.4.11 (Matrices). Let $\mathbf{A}$ denote $\mathbf{R}$, $\mathbf{C}$, $\mathbf{H}$, $\mathbf{R}^2$, $\mathbf{C}^2$, or $\mathbf{H}^2$, or more generally any real algebra. Denote by $\mathbf{A}(n)$ the real algebra of $n \times n$ matrices with coefficients in $\mathbf{A}$, and equip $\mathbf{A}(n)$ with componentwise addition and the standard matrix multiplication, that is, $(x_{i,j})_{i,j=1}^n \cdot (y_{i,j})_{i,j=1}^n := (z_{i,j})_{i,j=1}^n$, where

$$z_{i,j} := \sum_{k=1}^{n} x_{i,k} y_{k,j}.$$

Note the order of $x_{i,k}$ and $y_{k,j}$. The algebra has identity element the identity matrix.

Exercise 3.4.12 (Matrix representations). (i) Show that the map

$$(\mathbf{A}^2)(n) \to (\mathbf{A}(n))^2 : ((x_{i,j}, y_{i,j}))_{i,j=1}^n \mapsto ((x_{i,j})_{i,j=1}^n, (y_{i,j})_{i,j=1}^n)$$

is a real algebra isomorphism. Thus $\mathbf{A}^2(n)$ means matrices of tuples or equivalently tuples of matrices.

(ii) Show that the map

$$(\mathbf{A}(m))(n) \to \mathbf{A}(mn) : \left(((x_{i',j'}^{i,j})_{i',j'=1}^m)_{i,j} \right)_{i,j=1}^n \mapsto (y_{\alpha,\beta})_{\alpha,\beta=1}^{nm},$$

where $y_{(i-1)m+i',(j-1)m+j'} := x_{i',j'}^{i,j}$, is a real algebra isomorphism. Thus $\mathbf{A}(m)(n)$ means $n \times n$ matrices of $m \times m$ matrices, or equivalently $nm \times nm$ matrices.

The main result of this section is the following representation theorem for real Clifford algebras.

Theorem 3.4.13. *Let V be a real inner product space of dimension $n = n_+ + n_-$ and signature $\sigma := n_+ - n_-$. Then $\triangle V$ is isomorphic to*

$$\mathbf{H}(2^{(n-2)/2}) \text{ if } \sigma \equiv -2 \mod 8, \qquad \mathbf{R}(2^{n/2}) \text{ if } \sigma \equiv 2 \mod 8,$$
$$\mathbf{C}(2^{(n-1)/2}) \text{ if } \sigma \equiv -1 \mod 8, \qquad \mathbf{C}(2^{(n-1)/2}) \text{ if } \sigma \equiv 3 \mod 8,$$
$$\mathbf{R}(2^{n/2}) \text{ if } \sigma \equiv 0 \mod 8, \qquad \mathbf{H}(2^{(n-2)/2}) \text{ if } \sigma \equiv 4 \mod 8,$$
$$\mathbf{R}^2(2^{(n-1)/2}) \text{ if } \sigma \equiv 1 \mod 8, \qquad \mathbf{H}^2(2^{(n-3)/2}) \text{ if } \sigma \equiv 5 \mod 8.$$

The result for inner product spaces of small dimension can be summarized by a diagram,

$$
\begin{array}{ccccccccc}
 & & & & \mathbf{R} & & & & \\
 & & & \mathbf{C} & & \mathbf{R}^2 & & & \\
 & & \mathbf{H} & & \mathbf{R}(2) & & \mathbf{R}(2) & & \\
 & \mathbf{H}^2 & & \mathbf{C}(2) & & \mathbf{R}^2(2) & & \mathbf{C}(2) & \\
\mathbf{H}(2) & & \mathbf{H}(2) & & \mathbf{R}(4) & & \mathbf{R}(4) & & \mathbf{H}(2)
\end{array}
$$

where in row $n+1$ are the n-dimensional inner product spaces, with the Euclidean spaces along the right edge of the triangle, and the spacetimes next to them, each row ordered by increasing signature.

The proof builds on four lemmas, which give isomorphisms between different algebras. We will use the fact several times that inner product spaces of the same dimension and signature are isometric.

Lemma 3.4.14. *Let V_1 and V_2 be inner product spaces with the same signature and dimensions n_1 and n_2 such that $n_1 = n_2 + 2$. Then $\triangle V_1$ is isomorphic to $\triangle V_2(2)$, the algebra of 2×2 matrices with entries in $\triangle V_2$.*

Proof. Since $|\sigma_1| = |\sigma_2| \leq n_2 = n_1 - 2$, there are orthogonal vectors $e_+, e_- \in V_1$ such that $\langle e_+\rangle^2 = 1$ and $\langle e_-\rangle^2 = -1$. Let $V_0 := [e_+ \wedge e_-]^\perp \subset V_1$, and note that there is an isomorphism $T : V_0 \to V_2$. Define a map

$$C : V_1 \to \triangle V_2(2) : \alpha_+ e_+ + \alpha_- e_- + v \mapsto \begin{bmatrix} -T(v) & \alpha_+ - \alpha_- \\ \alpha_+ + \alpha_- & T(v) \end{bmatrix},$$

$\alpha_+, \alpha_- \in \mathbf{R}$, $v \in V_0$. Since

$$\begin{bmatrix} -T(v) & \alpha_+ - \alpha_- \\ \alpha_+ + \alpha_- & T(v) \end{bmatrix}^2 = (\alpha_+^2 - \alpha_-^2 + \langle v\rangle^2) \begin{bmatrix} 1 & 0 \\ 0 & 1 \end{bmatrix},$$

it is seen that C realizes $\triangle V_2(2)$ as a Clifford algebra for V_1, using Proposition 3.3.3. □

Lemma 3.4.15. *Let V_1 and V_2 be inner product spaces of the same dimension and signatures σ_1 and σ_2 respectively. Then $\triangle V_1$ and $\triangle V_2$ are isomorphic if $\sigma_1 = 2 - \sigma_2$.*

Proof. Without loss of generality we can assume $\sigma_1 \leq 0$ and $\sigma_2 \geq 2$. Since $\sigma_1 = 2 - \sigma_2 \geq 2 - n > -n$, there exists $e_1 \in V_1$ with $\langle e_1\rangle^2 = 1$. Define $V_0 := [e_1]^\perp \subset V_1$ so that $V_1 = [e_1] \oplus V_0$, and the map

$$C : V_1 \to \triangle V_1 : \alpha e_1 + v_0 \mapsto \alpha e_1 + e_1 v_0, \quad \alpha \in \mathbf{R}, v_0 \in V_0.$$

Note that $(\alpha e_1 + e_1 v_0)^2 = \alpha^2 - \langle v_0\rangle^2$, and define a new inner product on V_1 such that

$$\langle \alpha e_1 + v_0, \alpha e_1 + v_0\rangle_2 := \alpha^2 - \langle v_0, v_0\rangle,$$

where $\langle \cdot, \cdot\rangle$ denotes the inner product of V_1. Therefore there is an isometry $T : V_2 \to (V_1, \langle \cdot, \cdot\rangle_2)$, and $C \circ T$ realizes $\triangle V_1$ as a Clifford algebra for V_2, using Proposition 3.3.3. □

Lemma 3.4.16. *Let V_1 and V_2 be inner product spaces of the same dimension and signatures σ_1 and σ_2 respectively. Then $\triangle V_1$ and $\triangle V_2$ are isomorphic if $\sigma_1 \equiv \sigma_2$ mod 8.*

Proof. Without loss of generality, we may assume that $\sigma_1 = \sigma_2 + 8 \geq 5$, using Lemma 3.4.15. Pick a four-dimensional subspace $V_0 \subset V_1$ on which the inner product is positive definite, and split $V_1 = V_0 \oplus V_0^\perp$ orthogonally. Let $w \in \triangle^4 V_0$ be such that $w^2 = 1$. Then w anticommutes with vectors in V_0 and commutes with vectors in $V_0^\perp$. Define a map

$$C : V_1 \to \triangle V_1 : v + v' \mapsto vw + v', \quad v \in V_0, \ v' \in V_0^\perp.$$

Note that $(vw + v')^2 = -\langle v\rangle^2 + \langle v'\rangle^2$, and define a new inner product on V_1 such that

$$\langle vw + v', vw + v'\rangle_2 := -\langle v, v\rangle + \langle v', v'\rangle,$$

where $\langle \cdot, \cdot \rangle$ denotes the inner product of V_1. Then there is an isometry $T : V_2 \to (V_1, \langle \cdot, \cdot \rangle_2)$, and $C \circ T$ realizes $\triangle V_1$ as a Clifford algebra for V_2, using Proposition 3.3.3. $\square$

The last lemma is not really needed for the proof of Theorem 3.4.13, but we give it, since the proof is instructive.

Lemma 3.4.17. *Let V_1 and V_2 be inner product spaces such that $n_1 = n_2 + 1$ and $\sigma_1 = \sigma_2 \pm 1$, and assume that $\sigma_1 \equiv 1 \mod 4$. Then $\triangle V_1$ is isomorphic to $(\triangle V_2)^2$.*

Proof. As in the proof of Proposition 3.3.3, the hypothesis shows that there exists $w \in \triangle^{n_1} V_1$ such that $w^2 = 1$ and $w \in \mathcal{Z}(\triangle V_1)$. Let

$$p_1 := \tfrac{1}{2}(1 + w) \quad \text{and} \quad p_2 := \tfrac{1}{2}(1 - w).$$

Then the p_i are central complementary projections, that is, $p_i \in \mathcal{Z}(\triangle V)$, $p_1 + p_2 = 1$, $p_1 p_2 = p_2 p_1 = 0$, and $p_i^2 = p_i$. These are noninvertible and act by multiplication as projections. Denote the ranges by $A_i := p_i(\triangle V_1) = \{p_i x \; ; \; x \in \triangle V_1\}$. Note that A_i is an algebra, but with identity element p_i, not 1. Thus we do not consider A_i a subalgebra of $\triangle V_1$. It is straightforward to verify that $\triangle V_1 = A_1 \oplus A_2$ and that the involution gives an automorphism $A_1 \to A_2 : w \mapsto \hat{w}$. Hence $\dim A_i = 2^{n_2}$, and we claim that the A_i are Clifford algebras for V_2. To see this, note that there exists a subspace $V_0 \subset V_1$ of codimension 1 and an isometry $T : V_2 \to V_0$. Define

$$C_i : V_2 \to A_i : v \mapsto p_i T(v).$$

Since $(C_i v)^2 = \langle v \rangle^2 p_i$ and n_2 is even, Proposition 3.3.3 shows that A_1 is a Clifford algebra for V_2. This proves that $\triangle V_1 = A_1 \oplus A_2$ is isomorphic to $(\triangle V_2)^2$. $\square$

Proof of Theorem 3.4.13. We can visualize the collection of all (equivalence classes of) inner product spaces as the points in the n, σ plane with $n \geq |\sigma|$ and $n + \sigma$ even.

That the Clifford algebras for $(n, \sigma) = (0, 0)$, $(1, 1)$, $(1, -1)$, and $(2, -2)$ are isomorphic to $\mathbf{R}$, $\mathbf{R}^2$, $\mathbf{C}$, and $\mathbf{H}$ respectively is clear by inspection. See the examples below. By closer inspection or by applying Lemma 3.4.17, the case $(n, \sigma) = (3, -3)$ is taken care of.

The isomorphisms for the five columns in the diagram corresponding to inner product spaces with signature $-3 \leq \sigma \leq 1$ now follow by induction using Lemma 3.4.14. Then the isomorphisms for the inner product spaces in a row corresponding to a fixed dimension n follow from Lemma 3.4.15 and Lemma 3.4.16. This completes the proof. $\square$

Example 3.4.18 (Two dimensions). From Theorem 3.4.13 we get matrix representations for the Euclidean and Minkowski planes. Consider first the Minkowski plane. Since this has signature 0, Theorem 3.4.2 applies. Writing e_1^+, e_1^- for the Minkowski ON-basis, in the basis $\{1, e_1\}$ we have matrix representations

$$e_1^+ = \begin{bmatrix} 0 & 1 \\ 1 & 0 \end{bmatrix}, \quad e_2^- = \begin{bmatrix} 0 & -1 \\ 1 & 0 \end{bmatrix}.$$

Turning to the Euclidean plane, with ON-basis $\{e_1', e_2'\}$, following Lemma 3.4.15 we identify

$$e_1' = e_1^+ = \begin{bmatrix} 0 & 1 \\ 1 & 0 \end{bmatrix}, \quad e_2' = e_1^+ e_1^- = \begin{bmatrix} 1 & 0 \\ 0 & -1 \end{bmatrix}.$$

Note that this agrees with the standard representation of $\mathbf{C}$ by matrices: the complex number $\alpha + j\beta \in \mathbf{C} = \triangle^{\mathrm{ev}} V$ is represented by the real matrix $\begin{bmatrix} \alpha & -\beta \\ \beta & \alpha \end{bmatrix}$, since $j = e_1' e_2' = e_1^-$.

Example 3.4.19 (Three dimensions). Consider the Euclidean three-dimensional space V with ON-basis $\{e_1, e_2, e_3\}$. We first represent $\triangle[e_1 \wedge e_2]$ as $\mathbf{R}(2)$, and then e_{123} as i times the identity matrix. Using Example 3.4.18, this gives

$$e_1 = \begin{bmatrix} 0 & 1 \\ 1 & 0 \end{bmatrix}, \quad e_2 = \begin{bmatrix} 1 & 0 \\ 0 & -1 \end{bmatrix}, \quad e_3 = e_2 e_1 e_{123} = \begin{bmatrix} 0 & 1 \\ -1 & 0 \end{bmatrix} i = \begin{bmatrix} 0 & i \\ -i & 0 \end{bmatrix}.$$

The description of electron spin in nonrelativistic quantum mechanics usually involves the following three *Pauli matrices*:

$$\sigma_1 = \begin{bmatrix} 0 & 1 \\ 1 & 0 \end{bmatrix}, \quad \sigma_2 = \begin{bmatrix} 0 & -i \\ i & 0 \end{bmatrix}, \quad \sigma_3 = \begin{bmatrix} 1 & 0 \\ 0 & -1 \end{bmatrix}.$$

These anticommute, and all square to the identity matrix. According to Proposition 3.3.3 they generate an 8-dimensional real Clifford algebra, where $\{\sigma_1, \sigma_2, \sigma_3\}$ represents a three-dimensional ON-basis. Comparing with Example 3.4.19, we see that $\sigma_1 = e_1$, $\sigma_2 = -e_3$, and $\sigma_3 = e_2$.

Example 3.4.20 (Four dimensions). Consider four-dimensional spacetime V with ON-basis $\{e_0, e_1, e_2, e_3\}$. One way to derive a matrix representation of $\triangle V$ as $\mathbf{R}(4)$ would be to use Lemma 3.4.15 and Example 3.4.5. However, let us instead apply Lemma 3.4.14, by which we identify

$$x_0 e_0 + x_1 e_1 + x_2 e_2 + x_3 e_3 = \begin{bmatrix} -(x_2 e_2 + x_3 e_3) & x_1 - x_0 \\ x_1 + x_0 & x_2 e_2 + x_3 e_3 \end{bmatrix}.$$

In these matrices, we now expand each of the four plane multivectors to a real 2×2 matrix using Example 3.4.18. We get

$$e_0 = \begin{bmatrix} 0 & 0 & -1 & 0 \\ 0 & 0 & 0 & -1 \\ 1 & 0 & 0 & 0 \\ 0 & 1 & 0 & 0 \end{bmatrix}, \quad e_1 = \begin{bmatrix} 0 & 0 & 1 & 0 \\ 0 & 0 & 0 & 1 \\ 1 & 0 & 0 & 0 \\ 0 & 1 & 0 & 0 \end{bmatrix},$$

$$e_2 = \begin{bmatrix} 0 & -1 & 0 & 0 \\ -1 & 0 & 0 & 0 \\ 0 & 0 & 0 & 1 \\ 0 & 0 & 1 & 0 \end{bmatrix}, \quad e_3 = \begin{bmatrix} -1 & 0 & 0 & 0 \\ 0 & 1 & 0 & 0 \\ 0 & 0 & 1 & 0 \\ 0 & 0 & 0 & -1 \end{bmatrix}.$$

In relativity theory it is at least as common to model spacetime as the four-dimensional inner product space V with signature $1 - 3$. Theorem 3.4.13 shows in this case that the real Clifford algebra $\triangle V$ is isomorphic to $\mathbf{H}(2)$. Let $\{e_0, e_1, e_2, e_3\}$ be an ON-basis with $\langle e_0 \rangle^2 = 1$ and apply Lemma 3.4.14, by which we identify

$$x_0 e_0 + x_1 e_1 + x_2 e_2 + x_3 e_3 = \begin{bmatrix} -(x_2 e_2 + x_3 e_3) & x_0 - x_1 \\ x_0 + x_1 & x_2 e_2 + x_3 e_3 \end{bmatrix}.$$

Here e_2, e_3 belong to the anti-Euclidean plane. Identifying $\triangle[e_2 \wedge e_3]$ with the quaternions $\mathbf{H}$ as in Example 3.3.7 gives the matrix representation

$$e_0 = \begin{bmatrix} 0 & 1 \\ 1 & 0 \end{bmatrix}, \quad e_1 = \begin{bmatrix} 0 & -1 \\ 1 & 0 \end{bmatrix}, \quad e_2 = \begin{bmatrix} k & 0 \\ 0 & -k \end{bmatrix}, \quad e_3 = \begin{bmatrix} -j & 0 \\ 0 & j \end{bmatrix}.$$

From this we can also derive a matrix representation for the Euclidean 4-dimensional space, with ON-basis $\{e_1', e_2', e_3', e_4'\}$. Following Lemma 3.4.15, we make the identifications $e_1' = e_0$, $e_2' = e_0 e_1$, $e_3' = e_0 e_2$, and $e_4' = e_0 e_3$, which gives

$$e_1' = \begin{bmatrix} 0 & 1 \\ 1 & 0 \end{bmatrix}, \quad e_2' = \begin{bmatrix} 1 & 0 \\ 0 & -1 \end{bmatrix}, \quad e_3' = \begin{bmatrix} 0 & -k \\ k & 0 \end{bmatrix}, \quad e_4' = \begin{bmatrix} 0 & j \\ -j & 0 \end{bmatrix}.$$

3.5   Comments and References

3.1 Clifford algebras were discovered by William Kingdon Clifford (1845–1879) and are described in his work [27] from 1878. He himself preferred the name *geometric algebra* for his algebra, and he was inspired by the works of Grassmann and Hamilton.

   Clifford was contemporary with James Clerk Maxwell (1831–1879), the physicist famous for the differential equations describing electromagnetism. However, instead of using Clifford algebra for formulating these equations, which is very natural, as we shall see in Section 9.2, history took a different turn. The physicists J. W. Gibbs and O. Heaviside developed vector algebra as a mathematical framework for Maxwell's equation, and this became standard, while the algebras of Grassmann, Hamilton, and Clifford failed to enter mainstream mathematics.

   The notation $\wedge$ for the exterior product is standard. The notations $\lrcorner$ and $\llcorner$ for interior products also appear in the literature, but not in a standardized way. The notation $\triangle$ for the Clifford product does not appear in the literature. Our motivation for introducing it here is its relation to the symmetric difference of index sets. Another reason is a playing with symbols: $\triangle$ is built from its three corners $\wedge$, $\lrcorner$, and $\llcorner$. Just as the notation $\wedge V$ follows the exterior product, our notation $\triangle V$ follows the Clifford product $\triangle$. A common notation in the literature for the Clifford algebra of an inner product space of signature $p - q$ is $\mathbf{R}_{p,q}$.

Let V be a Euclidean space. In the mathematics literature, usually the Clifford algebra for $-\langle \cdot, \cdot \rangle$ is used, so that vectors have negative squares $v^2 = -|v|^2 \leq 0$, whereas in the physics literature, usually the Clifford algebra for $\langle \cdot, \cdot \rangle$ is used, so that vectors have positive squares $v^2 = |v|^2 \geq 0$. The reason for following the physics convention in this book, beyond its obvious simplicity, is, for example, that concrete algebraic computations with the Clifford product involve fewer minus signs this way. Equally important is that, this way, it is clearer that vectors should not be viewed as imaginary numbers. It is the bivectors that naturally generalize the imaginaries from complex algebra, regardless of which Euclidean or anti-Euclidean Clifford algebra is used.

3.2 The quaternions $\mathbf{H}$ were discovered by William Rowan Hamilton (1805–1865) in 1843. A famous anecdote tells how he had been struggling to find a suitable generalization of the complex product to three dimensions. One day, while walking along a canal in Dublin, he realized how this was instead possible in four dimensions, and he carved the basic rules for multiplication $i^2 = j^2 = j^2 = ijk = -1$ into the bridge nearby.

Our use of $j \in \mathbf{C}$ to denote the geometric representation e_{12} of the imaginary unit, is not standard. A standard in physics literature is to denote the imaginary unit j rather than i, since the latter also denotes electric current.

3.4 A reference for representations of Clifford algebras is Porteous [75]. In particular, Proposition 3.3.3 and Theorem 3.4.13 are found there. Further references for this famous representation theorem, with its periodicity of 8, are Lawson and Michelsohn [63], Delanghe [33], and Lounesto [64].

Chapter 4

Rotations and Möbius Maps

Prerequisites:

This chapter builds on Chapter 3. The reader is assumed to know the spectral theorem for normal operators from linear algebra, as well as basic plane complex geometry. Some basic knowledge of Lie groups and algebras, as well as special relativity, is helpful.

Road map:

We recall from complex analysis how complex algebra can be used to represent fundamental mappings of the Euclidean plane. For example, multiplication by the exponential

$$z \mapsto e^{i\phi} z$$

gives rotation counterclockwise through an angle ϕ. The inverse map $z \mapsto 1/z = \overline{z}/|z|^2$ yields inversion in the unit circle followed by reflection in the real line. More general Möbius maps of the complex plane, which maps circles and lines onto circles and lines, have representations as fractional linear transformations

$$z \mapsto \frac{az + b}{cz + d}.$$

In this chapter we show how this generalizes to higher dimensions if we replace the commutative complex product by the noncommutative Clifford product. In Section 4.1 we show how an analogue of the Grassmann cone from Section 2.2 for the Clifford product contains the multivectors needed to represent isometries. In Section 4.2 we show how the argument $i\phi$ of the exponential function generalizes to bivectors $\triangle^2 V$. Recall from Section 3.2 that the standard geometric representation of i is as the bivector $j = e_1 \wedge e_2$. This representation of rotations with Clifford algebra works not only for Euclidean spaces: in Section 4.4 we show how the Lorentz boosts, or hyperbolic rotations, in special relativity are handled with

© Springer Nature Switzerland AG 2019
A. Rosén, *Geometric Multivector Analysis*, Birkhäuser Advanced Texts Basler Lehrbücher,
https://doi.org/10.1007/978-3-030-31411-8_4

Clifford algebra. Out in space, changing your inertial system by a Lorentz boost by increasing your speed comparable to that of light, your celestial sphere, and the stars you see will change, and the stars will move closer to the direction in which you are travelling. Using Clifford algebra, we compute this map of the celestial sphere and discover that it is indeed a Möbius map. We show in Section 4.5 how all Möbius maps are represented by fractional linear transformations using the Clifford product.

Highlights:

- Cartan–Dieudonné's theorem: 4.1.3,

- Spin = universal cover of SO: 4.3.4

- Structure of spacetime rotations: 4.4.7

- Fractional linear = Möbius = global conformal: 4.5.12

- Spacetime isometries act like Möbius: 4.6.4

- Watching a Lorentz boost: 4.6.5

4.1   Isometries and the Clifford Cone

We recall the notion of isometry, which is fundamental in any inner product space.

Definition 4.1.1. Let V be an inner product space. A linear operator $T : V \to V$ is called an *isometry* if

$$\langle Tv \rangle^2 = \langle v \rangle^2 \quad \text{for all } v \in V.$$

In fact, through polarization $\langle u + v \rangle^2 - \langle u - v \rangle^2 = 4\langle u, v \rangle$, an isometry satisfies $\langle Tu, Tv \rangle = \langle u, v \rangle$ for all $u, v \in V$.

Lemma 4.1.2. *Let T be an isometry on an inner product space V. Then T is invertible and $T^* = T^{-1}$ is another isometry. If $V_1 \subset V$ is invariant under T, that is, $TV_1 \subset V_1$, then $TV_1 = V_1$ and $TV_1^\perp = V_1^\perp$, that is, the orthogonal complement of an invariant subspace is also invariant under T.*

Proof. Clearly, T maps an ON-basis onto another ON-basis, and thus is invertible. We have $\langle Tu, v \rangle = \langle Tu, T(T^{-1}v) \rangle = \langle u, T^{-1}v \rangle$, which shows that $T^* = T^{-1}$. If V_1 is invariant, T must be injective on V_1 in particular, and therefore $TV_1 = V_1$. As in Example 2.7.3, it is clear that $V_1^\perp$ is invariant under $T^* = T^{-1}$. Since T^{-1} is injective, we have $T^{-1}V_1^\perp = V_1^\perp$ and therefore $TV_1^\perp = V_1^\perp$.   □

A set of building blocks for the isometries of a given inner product space V is provided by the hyperplane reflections in Example 3.4.7. If $a \in V$ is any nonsingular vector, then a has inverse $a^{-1} = a/\langle a \rangle^2$, and the reflection of a vector $v \in V$ in the hyperplane $[a]^\perp$ is given by

$$R_a : v \mapsto -ava^{-1} = \widehat{a}va^{-1}.$$

Note that in Example 3.4.7 we put the minus sign on v instead of a. These reflections are the simplest isometries in the sense that they leave an $(n-1)$-dimensional hyperplane invariant. Cartan–Dieudonné's theorem shows that every isometry is a composition of hyperplane reflections.

Theorem 4.1.3 (Cartan–Dieudonné). *Let V be an inner product space and let $T :$ $V \to V$ be an isometry. Then there are nonsingular vectors $a_1, \ldots, a_k$, with $k \le n$, such that T is the composition of the corresponding hyperplane reflections, so that*

$$T(v) = (-1)^k (a_1 a_2 \cdots a_k) v (a_1 a_2 \cdots a_k)^{-1}, \quad v \in V.$$

We first prove the theorem for Euclidean spaces, since the proof is simpler here.

Proof of Theorem 4.1.3 for Euclidean spaces. The proof is by induction over the dimension of V. Take any vector $a \ne 0$. If $Ta = a$, then T leaves $[a]$ and the hyperplane $[a]^\perp$ invariant. By the induction hypothesis, there are vectors $a_2, \ldots, a_k$ in $[a]^\perp$, $k \le n$, such that T is the product of the corresponding reflections inside the hyperplane. In the full space all these reflections leave a invariant, since a is orthogonal to $a_2, \ldots, a_n$, and we obtain the desired factorization of T.

If $Ta \ne a$, then we observe that the reflection R_{a_1} in the hyperplane orthogonal to $a_1 := Ta - a$ maps Ta to a. See Figure 4.1. Indeed, since T is an isometry, it follows that $\langle Ta - a, Ta + a \rangle = 0$, so

$$R_{a_1}(Ta + a) = Ta + a.$$

Since clearly $R_{a_1}(Ta - a) = a - Ta$, we get by adding the equations that $R_{a_1}(Ta) = a$. Thus $R_{a_1}T$ leaves $[a]$ and $[a]^\perp$ invariant, and proceeding as above, replacing T by $R_{a_1}T$, we deduce that $R_{a_1}T = R_{a_2} \cdots R_{a_k}$. Multiplying by R_{a_1}, we obtain the desired factorization. $\qquad\square$

The proof for spacetimes is not much more difficult.

Proof of Theorem 4.1.3 for spacetimes. Let $0 \ne a \in V$ be any time-like vector. We first claim that every nonzero vector v orthogonal to a must be space-like. To see this, note that v cannot be parallel to a, and since V is a spacetime, the plane spanned by v and a contains a nonzero singular vector $a + \lambda v$. Then

$$0 = \langle a + \lambda v \rangle^2 = \langle a \rangle^2 + \lambda^2 \langle v \rangle^2.$$

Thus $\langle v \rangle^2 > 0$, so v is space-like.

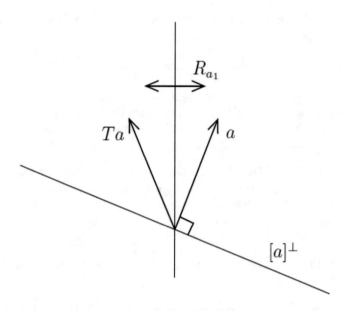

Figure 4.1: Reduction to hyperplane in Euclidean space.

Next we claim that $Ta - a$ is nonsingular or zero. To see this, assume it is singular and nonzero. Then

$$0 = \langle Ta - a \rangle^2 = 2\langle a \rangle^2 - 2\langle Ta, a \rangle = 2\langle a - Ta, a \rangle, \qquad (4.1)$$

so $Ta - a$ must be space-like as above, giving a contradiction.

With this a, we can now proceed as in the Euclidean case, reducing to the case of an isometry in the space-like subspace $[a]^\perp$, in which we can reuse the Euclidean proof.   □

The general proof given below is quite intricate. We give it for completeness, but shall not use it.

Proof of Theorem 4.1.3 for general inner product spaces. Assume that the theorem is not true. Let V be a space of minimal dimension n for which there exists an isometry T that is not the product of at most n reflections. Fix such T. Note that by the above proofs $n \geq 4$, since anti-Euclidean and Euclidean spaces of the same dimension have the same isometries and reflections, and so have anti-spacetimes and spacetimes.

In trying to perform the induction step in the proof for Euclidean spaces in a general inner product space, there are two obstacles. Firstly, the vector $Ta - a$ may be nonzero and singular, so that there is no hyperplane reflection R_a possible.

Secondly, the vector a may be singular, in which case $a \in [a]^{\perp}$, preventing the reduction to the hyperplane.

Define the subspace $V_0 := \mathsf{N}(T - I)$, that is, the set of vectors fixed by T. If there were a nonsingular vector in V_0, then the isometry T would be an isometry in the $(n-1)$-dimensional orthogonal complement. Hence by assumption it would be a product of at most $n-1$ reflections, which contradicts the hypothesis. Hence V_0 is totally degenerate.

Now let $V_1 := \mathsf{R}(T - I)$. If there is a nonsingular vector $v \in V$ such that $Tv - v \in V_1$ is nonsingular, then we would get a contradiction as above, so $Tv - v$ is singular whenever v is nonsingular. We now claim that $Tv - v$ is singular also when v is singular. To prove this, take any singular vector $v_0 \neq 0$ and consider $[v_0]^{\perp}$, to which v_0 belongs. There exists a nonsingular vector $v_1 \in [v_0]^{\perp}$, for if not, then $[v_0]^{\perp}$ is a totally degenerate subspace of V, and Proposition 1.3.2 would give $2(n-1) \leq n$, which contradicts $n \geq 4$. Since v_1 is nonsingular, we have $\langle Tv_1 - v_1 \rangle^2 = 0$. Now take any $\lambda \neq 0$ and define $v_2 := v_0 + \lambda v_1$. Then $\langle v_2 \rangle^2 = \lambda^2 \langle v_1 \rangle^2 \neq 0$, so also v_2 is nonsingular and thus $\langle Tv_2 - v_2 \rangle^2 = 0$. But $v_2 \rightarrow v_0$ when $\lambda \rightarrow 0$, so by continuity we get $\langle Tv_0 - v_0 \rangle^2 = 0$ as claimed.

We have proved that $V_1 = \mathsf{R}(T-I)$ is totally degenerate as well as $V_0 = \mathsf{N}(T-I)$, and thus $V_1 \subset V_1^{\perp}$ and $V_0 \subset V_0^{\perp}$. However, $V_1^{\perp} = \mathsf{N}(T^* - I) = \mathsf{N}(T^{-1} - I) = V_0$, so in fact, $V_0 = V_1 = V_0^{\perp} = V_1^{\perp}$. In particular, $\dim V = n = 2k$ if $\dim V_0 =: k$.

Take any basis $\{e_1, \dots, e_k\}$ for V_0 and extend it to a basis $\{e_1, \dots, e_{2k}\}$ for V. Note that this basis will not be an ON-basis. Then $Te_i = e_i$ when $1 \leq i \leq k$, since V_0 is fixed by T. But also $Te_j = (T - I)e_j + e_j$, where the first term belongs to $V_1 = V_0$ if $k + 1 \leq j \leq 2k$. Thus $\det T = 1$, since its matrix is upper triangular with 1 on the diagonal. Now, if R is any reflection in V, then RT must be the product of at most n reflections, since otherwise, we could run the whole argument for RT and get $\det(RT) = 1$ as well, and thus $\det R = 1$, a contradiction. Thus $T = R(RT)$ is a product of at most $n + 1$ reflections. However, since $n = 2k$ and $\det T = 1$, T is in fact a product of at most n reflections. This proves the Cartan–Dieudonné theorem. □

The Cartan–Dieudonné theorem shows the relevance of the following multiplicative subgroup of the Clifford algebra.

Definition 4.1.4 (Clifford cone). Let V be an inner product space. The *Clifford cone* of V is the multiplicative group $\widehat{\triangle} V \subset \triangle V$ generated by nonsingular vectors, that is, vectors v such that $\langle v \rangle^2 \neq 0$. More precisely, $q \in \widehat{\triangle} V$ if there are finitely many nonsingular vectors $v_1, \dots, v_k \in V$ such that

$$q = v_1 \vartriangle \cdots \vartriangle v_k.$$

The Clifford cone can be viewed as an inhomogeneous analogue of the simple multivectors $\widehat{\wedge}^k V$ for the Clifford product. If $w \in \widehat{\wedge}^k V$ is such that $[w]$ is nondegenerate and $\langle w \rangle^2 = \pm 1$, then $w \in \widehat{\triangle} V$, as is seen by factoring w with an ON-basis for $[w]$. The following result can be viewed as an analogue of the Plücker relations for the Grassmann cone, but instead for the Clifford cone.

Proposition 4.1.5 (Clifford cone equations). *Let $w \in \triangle V$. Then $w \in \widehat{\triangle} V$ if and only if w is invertible and*

$$\widehat{w} v w^{-1} \in V \quad \text{for all } v \in V.$$

In this case w can be written as a product of at most $\dim V$ nonsingular vectors, and $\overline{w} w = w \overline{w} \in \mathbf{R} \setminus \{0\}$.

Proof. Clearly, the condition is necessary for w to be in $\widehat{\triangle} V$. Conversely, assume that $\widehat{w} v w^{-1} \in V$ for all vectors. Then

$$\langle \widehat{w} v w^{-1} \rangle^2 = (\widehat{w} v w^{-1})(\widehat{w} v w^{-1}) = -(\widehat{w} v w^{-1})(\widehat{\widehat{w} v w^{-1}}) = \widehat{w} v w^{-1} w v \widehat{w}^{-1}$$
$$= \widehat{w} \langle v \rangle^2 \widehat{w}^{-1} = \langle v \rangle^2 \widehat{w} \widehat{w}^{-1} = \langle v \rangle^2,$$

so that $v \mapsto \widehat{w} v w^{-1}$ is an isometry. Theorem 4.1.3 shows that there are vectors $v_1, \ldots, v_k$, with $k \leq n$, such that $\widehat{w} v w^{-1} = (\widehat{v_1 \cdots v_k}) v (v_1 \cdots v_k)^{-1}$ for all $v \in V$, that is, $(w^{-1} \widehat{v_1 \cdots v_k}) v = v(w^{-1} v_1 \cdots v_k)$. Proposition 3.1.8 shows that $w = (\lambda v_1) v_2 \cdots v_k$ for some $\lambda \in \mathbf{R} \setminus \{0\}$. $\square$

The following normalized variants of the Clifford cone are standard constructs.

Definition 4.1.6 (Four Lie groups). Let V be an inner product space. Define the *orthogonal, special orthogonal, pin,* and *spin* groups

$$\mathrm{O}(V) := \{\text{isometries } T : V \to V\} \subset \mathcal{L}(V),$$
$$\mathrm{SO}(V) := \{T \in \mathrm{O}(V) \; ; \; \det T = +1\} \subset \mathcal{L}(V),$$
$$\mathrm{Pin}(V) := \{q \in \widehat{\triangle} V \; ; \; \langle q \rangle^2 = \pm 1\} \subset \triangle V,$$
$$\mathrm{Spin}(V) := \{q \in \mathrm{Pin}(V) \; ; \; q \in \triangle^{\mathrm{ev}} V\} \subset \triangle^{\mathrm{ev}} V.$$

We call $T \in \mathrm{SO}(V)$ a *rotation* and we call $q \in \mathrm{Spin}(V)$ a *rotor*.

It is clear that $\mathrm{O}(V)$, $\mathrm{SO}(V)$, $\mathrm{Pin}(V)$, and $\mathrm{Spin}(V)$ all are groups. Note that for $q = v_1 \triangle \cdots \triangle v_k \in \widehat{\triangle} V$, we have

$$\langle q \rangle^2 = \langle v_1 \rangle^2 \cdots \langle v_k \rangle^2.$$

Exercise 4.1.7. Show the following. In two-dimensional Euclidean space, the group $\mathrm{Spin}(V)$ is the unit circle in $\mathbf{C} = \triangle^{\mathrm{ev}} V$. In three-dimensional Euclidean space, $\mathrm{Spin}(V)$ is the unit three-sphere in $\mathbf{H} = \triangle^{\mathrm{ev}} V$.

Example 4.1.8 (Four-dimensional Spin). Let V be a four-dimensional Euclidean space, and consider $\mathrm{Spin}(V)$. A multivector $w \in \triangle^{\mathrm{ev}} V$ belongs to $\mathrm{Spin}(V)$ if and only if $|w| = 1$ and $wvw^{-1} \in V$ for all $v \in V$. Since $wvw^{-1} \in V \oplus \triangle^3 V$ and the reversion negates $\triangle^3 V$, the last condition is equivalent to $\overline{w}^{-1} v \overline{w} = wvw^{-1}$. This means that $v(\overline{w}w) = (\overline{w}w)v$, which holds if and only if $\overline{w}w \in \mathbf{R}$ according to Proposition 3.1.8.

Now note that for every $w' \in \triangle^{\mathrm{ev}} V$, we have $\overline{w'}w' \in \mathbf{R} \oplus \triangle^4 V$, since $\overline{\overline{w'}w'} = \overline{w'}w'$. Note also that if $\overline{w'}w' \in \mathbf{R}$, then $\overline{w'}w' = \langle \overline{w'}w', 1 \rangle = \langle w', w' \rangle = |w'|^2$, so in particular, w' is invertible. To summarize, we have shown that in four dimensions,

$$\mathrm{Spin}(V) = \{ w \in \triangle^{\mathrm{ev}} V \ ; \ \langle w, we_{1234} \rangle = 0, \ |w| = 1 \}.$$

In coordinates, for $w = w_\varnothing + w_{12}e_{12} + w_{13}e_{13} + w_{14}e_{14} + w_{23}e_{23} + w_{24}e_{24} + w_{34}e_{34} + w_{1234}e_{1234}$, this means

$$w_{12}w_{34} - w_{13}w_{24} + w_{14}w_{23} = w_\varnothing w_{1234}.$$

In particular, comparing the right-hand side of this equation to that in Exercise 2.9.4 shows that if $w \in \triangle^2 V$ and $|w| = 1$, then $w \in \mathrm{Spin}(V)$ if and only if w is a simple bivector.

Let V be an n-dimensional inner product space. We record some basic facts. The groups $\mathrm{O}(V)$, $\mathrm{SO}(V) \subset \mathcal{L}(V)$ as well as $\mathrm{Pin}(V)$, $\mathrm{Spin}(V) \subset \triangle V$ are smooth manifolds, so they are all Lie groups. It is well known that the group $\mathrm{SO}(V)$ is a Lie group. Proposition 4.1.9 below shows that locally $\mathrm{Spin}(V) \subset \triangle V$ looks like $\mathrm{SO}(V) \subset \mathcal{L}(V)$, so $\mathrm{Spin}(V)$ is also a Lie group.

The special orthogonal group $\mathrm{SO}(V)$ is a normal subgroup of $\mathrm{O}(V)$, and $\mathrm{O}(V) \setminus \mathrm{SO}(V)$ is the coset of $\mathrm{SO}(V)$. Fix any hyperplane reflection R. Then composition with R yields a one-to-one correspondence between $\mathrm{SO}(V)$ and $\mathrm{O}(V) \setminus \mathrm{SO}(V)$, the two connected components of $\mathrm{O}(V)$.

Similarly, $\mathrm{Spin}(V)$ is a normal subgroup of $\mathrm{Pin}(V)$, and $\mathrm{Pin}(V) \setminus \mathrm{Spin}(V)$ is the coset of $\mathrm{Spin}(V)$. Multiplication by a fixed vector $a \in V$ such that $a^2 = \pm 1$ gives a one-to-one correspondence between $\mathrm{Spin}(V)$ and $\mathrm{Pin}(V) \setminus \mathrm{Spin}(V)$, the two connected components of $\mathrm{Pin}(V)$.

All four Lie groups have dimension $\binom{n}{2}$. Since tangent spaces have the dimension of the manifold, these dimensions follow from Propositions 4.2.2 and 4.2.5 below. Note that for a general Euclidean space, $\mathrm{Spin}(V)$ is contained in the unit sphere in $\triangle^{\mathrm{ev}}(V)$, which has dimension $2^{n-1} - 1 \geq \binom{n}{2}$, with equality only in dimensions 2 and 3. In dimension 4 we have seen in Example 4.1.8 that the codimension is 1, but in general we note that the spin group is only a very small fraction of the unit sphere in the even subalgebra.

Fundamental is the following two-to-one correspondence between the groups from Definition 4.1.6.

Proposition 4.1.9 (Double coverings). *The map*

$$p : \mathrm{Pin}(V) \to \mathrm{O}(V) : q \mapsto T,$$

where $Tv = \widehat{q}vq^{-1}$, is a homomorphism, that is, $p(q_1 q_2) = p(q_1)p(q_2)$. It is surjective and has kernel $\{1, -1\}$. The map p restricts to a surjective homomorphism between subgroups

$$p : \mathrm{Spin}(V) \to \mathrm{SO}(V) : q \mapsto T,$$

where $Tv = qvq^{-1}$, with kernel $\{1, -1\}$.

Proof. First consider $p : \mathrm{Pin}(V) \to \mathrm{O}(V)$. The surjectivity of p follows from Theorem 4.1.3. To prove that the kernel is ± 1, assume that $\widehat{q}vq^{-1} = v$ for all vectors $v \in V$, so that $\widehat{q}v = vq$. Proposition 3.1.8 shows that $q \in \triangle^0 V$, and since $\langle q \rangle^2 = \pm 1$, we have $q = \pm 1$. That p restricts to a homomorphism $p : \mathrm{Spin}(V) \to \mathrm{SO}(V)$ is clear since

$$\det p(a_1 \cdots a_k) = \det p(a_1) \cdots \det p(a_k) = (-1)^k,$$

and we note that $\pm 1 \in \mathrm{Spin}(V)$. $\qquad\square$

We end this section by recording the properties of maps on $\wedge V$ induced, as in Section 2.3, by isometries on V.

Proposition 4.1.10 (Induced isometries). *Let V be an inner product space, let $T : V \to V$ be a linear operator, and let $T : \wedge V \to \wedge V$ be the induced $\wedge$-homomorphism. Then*

$$T(w_1 \vartriangle w_2) = T(w_1) \vartriangle T(w_2) \quad \textit{for all } w_1, w_2 \in \wedge V,$$

if and only if T is an isometry. When T has a representation $Tv = \widehat{q}vq^{-1}$, $v \in V$, for some $q \in \widehat{\triangle}(V)$, then

$$Tw = \begin{cases} q\,w\,q^{-1}, & q \in \triangle^{\mathrm{ev}}V, \\ q\,\widehat{w}\,q^{-1}, & q \in \triangle^{\mathrm{od}}V, \end{cases} \quad w \in \wedge V.$$

Proof. If T is a $\vartriangle$-homomorphism, consider vectors $v_i \in V$. Then

$$\langle Tv_1, Tv_2 \rangle + Tv_1 \wedge Tv_2 = Tv_1 \vartriangle Tv_2$$
$$= T(v_1 \vartriangle v_2) = T(\langle v_1, v_2 \rangle + v_1 \wedge v_2) = \langle v_1, v_2 \rangle + Tv_1 \wedge Tv_2,$$

and it follows that $\langle Tv_1, Tv_2 \rangle = \langle v_1, v_2 \rangle$. Conversely, assume that T is an isometry. Consider the linear map

$$C : V \to \triangle V : v \mapsto Tv.$$

Since $(Cv)^2 = (Tv)^2 = \langle Tv \rangle^2 = \langle v \rangle^2$, the universal property for Clifford algebras shows that there is an algebra homomorphism $\tilde{T} : \triangle V \to \triangle V$ such that $C(v) = \tilde{T}(v)$ for all $v \in V$. We need to show that $\tilde{T}$ coincides with the induced $\wedge$-homomorphism T. To this end, fix an ON-basis $\{e_i\}$. It suffices to show that $\tilde{T}(e_s) = T(e_s)$ for all $s \subset \overline{n}$. But

$$\tilde{T}(e_s) = \tilde{T}(e_{s_1}) \vartriangle \cdots \vartriangle \tilde{T}(e_{s_k}) = T(e_{s_1}) \vartriangle \cdots \vartriangle T(e_{s_k}) = T(e_{s_1}) \wedge \cdots \wedge T(e_{s_k}) = T(e_s),$$

since the Clifford product coincides with the exterior product for orthogonal vectors.

To verify the representation formula for the isometry T, it suffices to consider a basis multivector $w = e_s \in \wedge^k V$. By Theorem 4.1.3,

$$T e_s = (\widehat{q} e_{s_1} q^{-1}) \cdots (\widehat{q} e_{s_k} q^{-1}) = \epsilon (q e_{s_1} q^{-1}) \cdots (q e_{s_k} q^{-1}) = \epsilon q e_s q^{-1},$$

where $\epsilon = 1$ if $q \in \triangle^{\mathrm{ev}} V$ and $\epsilon = (-1)^k$ if $q \in \triangle^{\mathrm{od}} V$. Since $\widehat{w} = (-1)^k w$, this proves the proposition. $\qquad\square$

4.2 Infinitesimal Rotations and Bivectors

In Proposition 3.2.3 we represented Euclidean three-dimensional rotations by unit quaternions, that is, the spin group, and, moreover we wrote the quaternions $q = \exp(b/2)$ in exponential form. This gave a very good understanding of rotations, since b explicitly contains information about the rotation axis and rotation angle. Our next aim is to generalize this to higher-dimensional inner product spaces V through a more detailed analysis of $\mathrm{SO}(V)$.

Definition 4.2.1. By a *one-parameter group* of rotations, we mean a smooth map $\mathbf{R} \ni \phi \mapsto T(\phi) \in \mathrm{SO}(V)$ such that

$$T(\phi_1 + \phi_2) = T(\phi_1) T(\phi_2).$$

Similarly, a one-parameter group in $\mathrm{Spin}(V)$ is a smooth map $\mathbf{R} \ni \phi \mapsto q(\phi) \in \mathrm{Spin}(V)$ such that $q(\phi_1 + \phi_2) = q(\phi_1) q(\phi_2)$.

Denote by $\underline{\mathrm{SO}}(V) \subset \mathcal{L}(V)$ the tangent space to $\mathrm{SO}(V)$ at I. Similarly, let $\underline{\mathrm{Spin}}(V) \subset \triangle V$ be the tangent space to $\mathrm{Spin}(V)$ at 1.

Note that one-parameter groups are commutative subgroups that pass through the identity at $\phi = 0$. A standard example to think of is the rotations around a fixed axis in three-dimensional Euclidean space. Recall from Section 3.2 that this corresponds to a line of bivectors passing through the origin in $\triangle^2 V$.

Proposition 4.2.2 (The Lie algebra for rotations). *Let V be an inner product space. The tangent space $\underline{\mathrm{SO}}(V)$ is the linear space of skew-symmetric maps*

$$\underline{\mathrm{SO}}(V) = \{ A \in \mathcal{L}(V) \ ; \ \langle Au, v \rangle = -\langle u, Av \rangle \text{ for all } u, v \in V \}.$$

There is a one-to-one correspondence between $A \in \underline{\mathrm{SO}}(V)$ and one-parameter groups of rotations $T(\phi)$ given by

$$T'(0) = A \longleftrightarrow T(\phi) = \exp(\phi A).$$

A linear map is skew-symmetric if and only if its matrix $(\alpha_{i,j})$ in an ON-basis $\{e_i\}$ satisfies $\langle e_i \rangle^2 \alpha_{i,j} = -\langle e_j \rangle^2 \alpha_{j,i}$. For the exponential map, see Definition 1.1.5.

Proof. If A is skew-symmetric, then $\exp(\phi A)^* = \exp(\phi A^*) = \exp(-\phi A) = \exp(\phi A)^{-1}$, so that $\exp(\phi A)$ is a one-parameter family of rotations, and clearly $A = \frac{d}{d\phi}\exp(\phi A)|_{\phi=0}$. Thus $A \in \underline{SO}(V)$. Conversely, if $T(\phi)$ is any differentiable family of rotations such that $T(0) = I$, then differentiation of $\langle T(\phi)u, T(\phi)v\rangle = \langle u, v\rangle$ at $\phi = 0$ gives

$$\langle T'(0)u, v\rangle + \langle u, T'(0)v\rangle = 0.$$

Since every tangent at I is of the form $T'(0)$, we have shown that $\underline{SO}(V)$ is the linear space of skew-symmetric maps.

To verify the one-to-one correspondence, it remains to show that $T(\phi) = \exp(\phi A)$ if $T(\phi)$ is a one-parameter group with tangent $A := T'(0)$. But this follows, since

$$T'(\phi) = \lim_{h\to 0}\frac{T(\phi + h) - T(\phi)}{h} = \lim_{h\to 0}\frac{T(h) - T(0)}{h}T(\phi) = AT(\phi),$$

and $\exp(\phi A)$ is the unique solution to this differential equation. $\qquad\square$

Proposition 4.2.3 (Bivector representation). *Let V be an inner product space and let $A \in \underline{SO}(V)$ be a skew-symmetric map. Then there exists a unique bivector $b \in \triangle^2 V$ such that*

$$Av = b \llcorner v, \quad v \in V.$$

Conversely, every $b \in \triangle^2 V$ gives a skew-symmetric map $Av = b \llcorner v$ on V.

Proof. There is a one-to-one correspondence between skew-symmetric maps A and alternating bilinear forms θ:

$$\theta(u, v) = \langle u, Av\rangle, \quad u, v \in V.$$

By Proposition 2.5.3 we can view θ as a bi-covector $\theta \in \wedge^2 V^*$, and through the inner product, we further identify θ with the corresponding bivector $b \in \triangle^2 V$ such that $\theta(u, v) = \langle b, u \wedge v\rangle$. We have

$$\langle u, Av\rangle = \langle b, u \wedge v\rangle = \langle u, b \llcorner v\rangle, \quad \text{for all } u \in V,$$

so that $Av = b \llcorner v$. $\qquad\square$

Exercise 4.2.4. Define

$$\otimes^2_- V := \{w \in V \otimes V \; ; \; Sw = -w\},$$

where S is the map that swaps the factors in a tensor. Show on the one hand that $\otimes^2_- V$ corresponds to skew-symmetric linear maps as in Section 1.4, and on the other hand that $\otimes^2_- V$ corresponds to $\wedge 2V$ by inspection of the definitions. Show that the resulting one-to-one correspondence between skew-symmetric maps and bivectors coincides with that in Proposition 4.2.3.

Concretely, the relation between A and b works as follows. If $\{e_i\}$ is an ON-basis and $b = e_{ij}$, then

$$e_{ij} \llcorner e_i = -\langle e_i \rangle^2 e_j, \quad e_{ij} \llcorner e_j = \langle e_j \rangle^2 e_i,$$

whereas $e_{ij} \llcorner e_k = 0$ for $k \neq i, j$. Hence the matrix elements of A in positions (i, j) and (j, i) are $\pm$ the e_{ij}-coordinate of b.

Proposition 4.2.5 (The Lie algebra $\triangle^2 V$). *Let V be an inner product space. The tangent space* $\underline{\mathrm{Spin}}(V)$ *is the linear space of bivectors $\triangle^2 V$. There is a one-to-one correspondence between $b \in \underline{\mathrm{Spin}}(V) = \triangle^2 V$ and one-parameter groups of rotors $q(\phi) \subset \mathrm{Spin}(V)$ given by*

$$q'(0) = b/2 \longleftrightarrow q(\phi) = \exp(\phi b/2).$$

Corresponding one-parameter groups of rotations $T(\phi) = \exp(\phi b/2) v \exp(-\phi b/2)$ have corresponding tangents $Av = b \llcorner v$ at the identity.

Proof. If $q(\phi) \subset \mathrm{Spin}(V)$ is any differentiable curve and $q(0) = 1$, then $T(\phi)v = q(\phi)vq(\phi)^{-1}$ is a curve of rotations, and

$$T'(0)v = q'(0)v - vq'(0)$$

is skew symmetric. Therefore there exists a unique bivector b such that $T'(0)v = b \llcorner v = \frac{1}{2}(bv - vb)$, according to Proposition 4.2.3 and the reverse version of the Riesz formula (3.3). Therefore $q'(0) - b/2$ commutes with all vectors and belongs to $\triangle^{\mathrm{ev}} V$. By Proposition 3.1.8, there is a scalar $\lambda \in \mathbf{R}$ such that $q'(0) = b/2 + \lambda$. However, $\langle q(\phi) \rangle^2 = 1$ for all ϕ, so differentiation gives $0 = \langle q'(0), q(0) \rangle = \lambda$. Thus $\underline{\mathrm{Spin}}(V) \subset \triangle^2(V)$. Since $\dim \underline{\mathrm{Spin}}(V) = \dim \underline{\mathrm{SO}}(V) = \binom{n}{2}$, we must have equality.

To verify the one-to-one correspondence, assume first that $q(\phi)$ is a one-parameter group. Then $b := 2q'(0)$ is tangent and $q(\phi) = \exp(\phi b/2)$ as in the proof of Proposition 4.2.2. Conversely, assume $b \in \underline{\mathrm{Spin}}(V) = \triangle^2 V$. We claim that $q(\phi) := \exp(\phi b/2)$ is a one-parameter group in $\overline{\mathrm{Spin}}(V)$. The problem is to verify that $q(\phi) \in \mathrm{Spin}(V)$ for all ϕ. By Proposition 4.1.5 it suffices to show that

$$f(\phi) := q(\phi)vq(\phi)^{-1} \in V, \quad \text{for all } v \in V, \phi \in \mathbf{R},$$

since $q(\phi)$ is invertible and $\langle q(\phi) \rangle^2 = \langle 1, \overline{q(\phi)}q(\phi) \rangle = 1$. The function $f(\phi)$ is real analytic, so it suffices to verify that all its derivatives $f^{(j)}(0)$ belong to V. The first derivative is

$$f'(\phi) = q(\phi)(bv/2 - vb/2)q(\phi)^{-1} = q(\phi)(b \llcorner v)q(\phi)^{-1}.$$

Repeating this calculation shows that $f^{(j)}(0) = b \llcorner (b \llcorner \cdots (b \llcorner (b \llcorner v)) \cdots) \in V$. Thus $q(t) = \exp(\phi b/2)$ is a one-parameter group of rotors, and we have $b/2 = \frac{d}{d\phi} \exp(\phi b/2)|_{\phi=0}$. $\qquad \square$

To summarize the results above, we have the following diagram:

$$q = \exp(b/2) \in \mathrm{Spin}(V) \xrightarrow{\;\;p\;\;} \mathrm{SO}(V) \ni T = \exp(A)$$

$$\exp \uparrow \qquad\qquad\qquad\qquad \uparrow \exp$$

$$b \in \triangle^2 V = \underline{\mathrm{Spin}}(V) \xrightarrow{Av = b \llcorner v} \underline{\mathrm{SO}}(V) \ni A$$

Here the map p, which represents rotations $v \mapsto Tv$ by rotors q as $v \mapsto qvq^{-1}$, is a surjective homomorphism and has kernel ± 1. It is a $2-1$ covering map, that is, locally a diffeomorphism. For the tangent spaces $\underline{\mathrm{Spin}}(V)$ and $\underline{\mathrm{SO}}(V)$, we have a one-to-one linear map $b \mapsto A$, where $Av = b \llcorner v$, $v \in V$, between bivectors $b \in \underline{\mathrm{Spin}}(V)$ and skew-symmetric maps $A \in \underline{\mathrm{SO}}(V)$. This is seen to be the derivative map of p at $q = 1$, and moreover, the diagram commutes, that is,

$$\exp(A)v = \exp(b/2)v\exp(-b/2) \quad \text{if } Av = b \llcorner v \text{ for all } v \in V.$$

Consider the smooth map

$$\exp : \underline{\mathrm{SO}}(V) \to \mathrm{SO}(V) : A \mapsto \exp(A)$$

from the linear tangent space of skew-symmetric maps to the smooth manifold of isometries. The origin $0 \in \underline{\mathrm{SO}}(V)$ is mapped onto the identity $I \in \mathrm{SO}(V)$, and a straight line $\phi \mapsto \phi A$ in $\underline{\mathrm{SO}}(V)$ is mapped onto a one-parameter group $\phi \mapsto \exp(\phi A)$ in such a way that the tangent A to the line is mapped onto the tangent A to the curve by the derivative $\exp_0$. This means that the derivative is the identity, so by the inverse function theorem, exp is an invertible map between a neighborhood of 0 in $\underline{\mathrm{SO}}(V)$ and a neighborhood of I in $\mathrm{SO}(V)$, with a smooth inverse. If $\underline{\mathrm{SO}}(V) \ni A \approx 0$, then $\exp(A) \approx I + A$ to first-order approximation. We say that $I + A$ is an *infinitesimal rotation*. Similar considerations apply to the spin group.

In Sections 4.3 and 4.4 we are going to consider the global behavior of exp. More precisely, we ask whether $\exp : \underline{\mathrm{SO}}(V) \to \mathrm{SO}(V)$ is surjective, so that every rotation T can be represented by a skew matrix A. If so, then we have a bivector b representing T through the formula

$$Tv = \exp(b/2)v\exp(-b/2).$$

For Euclidean spaces, we shall indeed show that both exponential maps are surjective. For spacetimes this is essentially true as well, although one needs to consider the connected component of the groups containing the identity. In contrast to the Euclidean case, $\mathrm{SO}(V)$ and $\mathrm{Spin}(V)$ are not connected for spacetime. For the spin group, however, there are exceptional behaviors in dimension ≤ 4. Details are in Sections 4.3 and 4.4.

Surprisingly this result is sharp: if V is not a Euclidean space or Minkowski spacetime, then $\exp : \underline{\mathrm{SO}}(V) \to \mathrm{SO}(V)$ is not surjective, not even onto the connected component of $I \in \mathrm{SO}(V)$. Hence $\exp : \underline{\mathrm{Spin}}(V) \to \mathrm{Spin}(V)$ cannot be surjective either. This follows from the following counterexample.

Example 4.2.6 (Nonsurjective exp). Let V be an *ultrahyperbolic* four-dimensional inner product space, that is, the signature is 2–2. Fix a non-ON-basis $\{f_1, f_2, f_3, f_4\}$ in which the inner product is

$$\langle u_1 f_1 + \cdots + u_4 f_4, v_1 f_2 + \cdots + v_4 f_4 \rangle = u_1 v_4 + u_4 v_1 + u_2 v_3 + u_3 v_2.$$

Let A be the map that in this basis has the matrix $\begin{bmatrix} 1 & 1 & 0 & 0 \\ 0 & 1 & 0 & 0 \\ 0 & 0 & -1 & -1 \\ 0 & 0 & 0 & -1 \end{bmatrix}$. It is verified that A is skew-symmetric, and the one-parameter group of rotations it generates is

$$\exp(\phi A) = T(\phi) = \begin{bmatrix} e^\phi & \phi e^\phi & 0 & 0 \\ 0 & e^\phi & 0 & 0 \\ 0 & 0 & e^{-\phi} & -\phi e^{-\phi} \\ 0 & 0 & 0 & e^{-\phi} \end{bmatrix}.$$

Fix $\phi \neq 0$ and consider the rotation $\tilde{T} := -T(\phi)$. Assume that there exists $\tilde{A} \in \underline{SO}(V)$ such that $\tilde{T} = \exp(\tilde{A})$. Then in particular, $\tilde{T} = T^2$, where $T := \exp(\tilde{A}/2) \in SO(V)$. Clearly $\tilde{T}$ has the two eigenvalues $\{-e^\phi, -e^{-\phi}\}$. Since T is a real matrix, its complexification must have the four distinct eigenvalues

$$\{ie^{\phi/2}, -ie^{\phi/2}, ie^{-\phi/2}, -ie^{-\phi/2}\},$$

and there exists a complex basis of eigenvectors for T. The same vectors are also an eigenbasis for the complexification of $\tilde{T}$. However, an explicit calculation shows that $\tilde{T}$ has only two complex eigenvectors, and we have reached a contradiction. Thus $\tilde{T}$ belongs to no one-parameter group of rotations, and the exponential map $\exp : \underline{SO}(V) \to SO(V)$ is not surjective; the range does not even cover the connected component of $SO(V)$ containing I. It is straightforward to verify that $-I$ belongs to a one-parameter group of rotations. From this it also follows that the range of $\exp : \underline{Spin}(V) \to Spin(V)$ also does not cover the connected component of 1.

4.3 Euclidean Rotations

In this section, V denotes a Euclidean space. We begin by studying the double covering

$$p : Spin(V) \to SO(V)$$

from Proposition 4.1.9. In performing a rotation of vectors in V, this is typically done in a continuous way, that is, we do not just apply a rotation T discretely, but rather start at the identity I and rotate the object in a continuous way by a family of rotations that form a path in $SO(V)$ connecting I and T. As in Example 3.2.5, a fundamental observation is that even though we make a full rotation and return to the initial position at $T = I$, there is a kind of intrinsic memory of the path of rotations, which classifies the loop of rotations as being one of two possible kinds. What may happen is that even though the rotations make a loop from I to I, the rotors q representing T by $Tv = qvq^{-1}$ may only form a path of rotors from 1 to -1.

We recall the following concepts from topology.

Definition 4.3.1 (Fundamental group). Let M be a smooth connected manifold and fix $p, q \in M$. Define an equivalence relation on the set of *paths*, that is, continuous functions $\gamma : [0, 1] \to M$, with $\gamma(0) = p$ and $\gamma(1) = q$, where $\gamma_1 \sim \gamma_2$ if the two paths are *homotopic*, that is, if there is a continuous function

$$[0, 1] \times [0, 1] \to M : (s, t) \mapsto \Gamma(s, t)$$

such that $\Gamma(s, 0) = p$ and $\Gamma(s, 1) = q$ for all $s \in [0, 1]$ and $\Gamma(0, t) = \gamma_1(t)$ and $\Gamma(1, t) = \gamma_2(t)$ for all $t \in [0, 1]$. If $p = q$, we refer to the paths as *loops*, and if M is a Lie group, we choose $p = q = I$. The *fundamental group* of M is

$$\pi_1(M) = \{\text{equivalence classes } [\gamma] \text{ of loops } \gamma\}.$$

If there is only one equivalence class, so that all loops are homotopic, then we say that M is *simply connected*.

Exercise 4.3.2. Verify that the n-dimensional unit sphere S^n is simply connected if $n \geq 2$, for example by making use of stereographic projection.

Example 4.3.3. Consider the two-dimensional Euclidean plane V. Then both $\mathrm{SO}(V)$ and $\mathrm{Spin}(V)$ are diffeomorphic to the unit circle, and one shows that

$$\pi_1(\mathrm{SO}(V)) = \pi_1(\mathrm{Spin}(V)) = \mathbf{Z},$$

where the integer $k \in \mathbf{Z}$ represents the equivalence class of all loops winding k times counterclockwise around the origin, and negative k means clockwise winding.

Consider next the three-dimensional Euclidean space V. Then, as we have seen, $\mathrm{Spin}(V)$ is diffeomorphic to the three dimensional unit sphere S^3 in $\mathbf{H}$, and hence is simply connected. On the other hand, $\mathrm{SO}(V)$ is diffeomorphic to the three-dimensional real projective space, and as in Example 3.2.5 it follows that

$$\pi_1(\mathrm{SO}(V)) = \mathbf{Z}_2 = \{1, -1\},$$

that is, there are two distinct equivalence classes denoted by 1, for those loops that can be continuously deformed to the constant loop at p, and -1, for those which cannot. The loops in the class 1 are those for which the representing line in Example 3.2.5 has the same orientation at the start and end of the loop of rotation, whereas for those in the class -1, the line ends at the opposite orientation.

We now prove that in higher dimensions, these fundamental groups are the same as in three dimensions.

Theorem 4.3.4 (Universal cover of SO). *Let V be a Euclidean space. Then* $\mathrm{Spin}(V)$ *and* $\mathrm{SO}(V)$ *are connected and compact Lie groups. Moreover, if* $\dim V \geq 3$, *then* $\mathrm{Spin}(V)$ *is simply connected, while* $\pi_1(\mathrm{SO}(V)) = \mathbf{Z}_2$.

Proof. (i) Since $p : \mathrm{Spin}(V) \to \mathrm{SO}(V)$ is continuous and surjective, it suffices to prove that $\mathrm{Spin}(V)$ is compact and connected. The compactness is clear from Proposition 4.1.5, which shows that $\mathrm{Spin}(V)$ is a closed subset of the Euclidean unit sphere. To show connectedness, take any $q = a_1 \cdots a_{2k} \in \mathrm{Spin}(V)$. Turn a_{2j} continuously to a_{2j-1} on the unit sphere for each $1 \leq j \leq k$. This produces a path from q to $a_1 a_1 a_3 a_3 \cdots a_{2j-1} a_{2j-1} = 1$ in $\mathrm{Spin}(V)$. Thus $\mathrm{Spin}(V)$ is connected.

(ii) We prove that $\mathrm{Spin}(V)$ is simply connected by induction on the dimension $n = \dim V$, using the ideas from the proof of Theorem 4.1.3. For $n = 3$ we know that $\mathrm{Spin}(V) = S^3$ is simply connected. Let $n > 3$ and consider a loop $q(t)$ of rotors where $q(0) = q(1) = 1$. Fix a unit vector $a \in S^{n-1}$, the unit sphere in V, and let $V' := [a]^{\perp}$. Define $b(t) := q(t) a q(t)^{-1} \in S^{n-1}$. By a small perturbation of the loop $q(t)$, we can assume that $b(t) \neq a$ for $t \neq 0$, and that the one-sided derivatives $b'(0)$ and $b'(1)$ exist and satisfy $b'(0) = -b'(1) \in V \setminus \{0\}$. Let $b'(0)/|b'(0)| =: v$ and

$$v(t) := \frac{b(t) - a}{|b(t) - a|},$$

which is seen to be well defined and extends to a continuous loop $[0,1] \to S^{n-1}$, starting and ending at v. Consider the loop

$$\tilde{q}(t) := v(t) q(t)$$

in $\mathrm{Pin}(V) \cap \triangle^{\mathrm{od}} V$. We claim that $\lceil \tilde{q}(t) \rceil \subset V'$ for all $t \in [0,1]$. For the inhomogeneous multivector $\tilde{q}(t)$, by this we mean that this holds for all its homogeneous parts. Indeed, we note that

$$\tilde{q}(t) = |b(t) - a|^{-1}(q(t)a - aq(t)) = |b(t) - a|^{-1} q(t) \, \llcorner \, a.$$

Therefore $\tilde{q}(t) \, \llcorner \, a = 0$, or equivalently $a \in \lceil \tilde{q}(t) \rceil^{\perp} = \lfloor *\tilde{q}(t) \rfloor$ as in Corollary 2.6.10, which establishes the claim, since $\langle a, v \rangle = 0$ at $t = 0, 1$.

Since $\lceil \tilde{q}(t) \rceil \subset V'$, we have in fact a loop in $\mathrm{Pin}(V') \cap \triangle^{\mathrm{od}} V'$, which is diffeomorphic to the, by assumption, simply connected $\mathrm{Spin}(V')$. Therefore $\tilde{q}(t)$ is homotopic in $\mathrm{Pin}(V') \cap \triangle^{\mathrm{od}} V'$ to the constant loop v. Also, the loop $v(t)$ is homotopic in S^{n-1} to the constant loop v. Therefore

$$q(t) = v(t) \tilde{q}(t)$$

is homotopic in $\mathrm{Spin}(V)$ to the constant loop 1, by continuity of multiplication. This proves that $\mathrm{Spin}(V)$ is simply connected.

(iii) Consider finally the fundamental group for $\mathrm{SO}(V)$. Define the constant loop $q_0(t) := 1$ and the path

$$q_1(t) := \cos(\pi t) + e_{12} \sin(\pi t)$$

from 1 to -1 in $\mathrm{Spin}(V)$. Let $T(t)$ be any loop in $\mathrm{SO}(V)$. By Proposition 4.1.9 there exists a lifted path $q(t)$ starting at 1 and ending at ± 1 in $\mathrm{Spin}(V)$ such that $T(t) = pq(t)$. The simple connectedness of $\mathrm{Spin}(V)$ from (ii) shows that $q(t)$ is homotopic with fixed endpoints to either $q_0(t)$ or $q_1(t)$. Since p is continuous, this shows that $T(t)$ is homotopic to one of the two nonhomotopic loops $pq_0(t)$ and $pq_1(t)$ in $\mathrm{SO}(V)$. This completes the proof. $\square$

For the remainder of this section we study the exponential maps, aiming to prove that they parametrize all of $\mathrm{SO}(V)$ and $\mathrm{Spin}(V)$ respectively.

Example 4.3.5. Let V be a two-dimensional Euclidean plane, and let j be an orientation for V/an imaginary unit, that is, we assume that j is one of the two possible $j \in \triangle^2 V$ such that $|j| = 1$.

Any bivector $b \in \triangle^2 V$ can be written $b = \alpha j$, $\alpha \in \mathbf{R}$. By Proposition 4.2.3, all skew-symmetric maps of V are of the form $Av = \phi j \llcorner v$. By choosing j appropriately, we may assume that $\phi \geq 0$. If $\{e_1, e_2\}$ is an ON-basis such that $j = e_{12}$, then the matrix of A is

$$\begin{bmatrix} 0 & \phi \\ -\phi & 0 \end{bmatrix}.$$

Note that $\|A\| = |\phi|$. Recall from Section 3.2 that every rotation of V is of the form $v \mapsto \exp(\phi j/2) v \exp(-\phi j/2) = \exp(A)v$, where ϕ is the angle of rotation *clockwise* relative the orientation j. The matrix of this rotation is seen to be

$$\begin{bmatrix} \cos\phi & \sin\phi \\ -\sin\phi & \cos\phi \end{bmatrix}.$$

By choosing j appropriately, we may assume that $0 \leq \phi \leq \pi$.

Next we prove structure theorems for skew-symmetric maps and rotations in Euclidean space, which shows that in general, we simply have direct sums of the two-dimensional case.

Proposition 4.3.6 (Euclidean rotations).   (i) *Let T be a rotation of a Euclidean space V. Then there are two-dimensional subspaces $V_1, \ldots, V_k$ and a subspace V_0 such that V splits as a direct sum of orthogonal subspaces*

$$V = V_1 \oplus \cdots \oplus V_k \oplus V_0,$$

where all the subspaces are invariant under T, $T|_{V_j}$ is rotation in V_j, $1 \leq j \leq k$, and $T|_{V_0} = I$.

(ii) *Let A be a skew-symmetric map in a Euclidean space V. Then there are two-dimensional subspaces $V_1, \ldots, V_k$ and a subspace V_0 such that V splits as a direct orthogonal sum*

$$V = V_1 \oplus \cdots \oplus V_k \oplus V_0,$$

where all the subspaces are invariant under A, $A|_{V_j} \neq 0$ are skew-symmetric maps of the plane for $1 \leq j \leq k$, and $A|_{V_0} = 0$.

We refer to an orthogonal splitting $V = V_1 \oplus \cdots \oplus V_k \oplus V_0$ as above as an *orthogonal splitting for T and A respectively.*

Proof. (i) Define $V_0 := \mathsf{N}(T - I)$ and consider $\tilde{V} := V_0^\perp$. Lemma 4.1.2 shows that $T\tilde{V} = \tilde{V}$. By complexification as in Section 1.5, there exists an invariant one- or two-dimensional subspace $\tilde{V}_1 \subset \tilde{V}$, and Lemma 4.1.2 shows that $T\tilde{V}_1^\perp = \tilde{V}_1^\perp$ within $\tilde{V}$. Proceeding recursively like this, we find one- or two-dimensional invariant orthogonal subspaces such that $\tilde{V} = \tilde{V}_1 \oplus \cdots \oplus \tilde{V}_m$. If $\dim \tilde{V}_j = 1$, then $T|_{\tilde{V}_j} = -I$ since T is an isometry and $Tv \neq v$ on $\tilde{V}$. The two-dimensional subspaces V_j obtained from complex eigenvectors are all such that $\det T|_{V_j} = 1$. Thus $\dim \mathsf{N}(T+I)$ must be even, since $\det T|_{\tilde{V}} = 1$, and it can be split evenly into two-dimensional subspaces. Possibly combining subspaces $\tilde{V}_j$ to two-dimensional subspaces V_j gives the stated splitting.

(ii) Define $V_0 := \mathsf{N}(A)$ and consider $\tilde{V} := V_0^\perp$. Since $A^* = -A$, this is invariant under A. By complexification there exists an invariant one- or two-dimensional subspace $\tilde{V}_1 \subset \tilde{V}$, and $A\tilde{V}_1^\perp = \tilde{V}_1^\perp$ within $\tilde{V}$. Proceeding recursively like this, we find one- or two-dimensional invariant orthogonal subspaces such that $\tilde{V} = V_1 \oplus \cdots \oplus V_k$. Here we cannot have $\dim V_j = 1$, since A is skew-symmetric and injective on $\tilde{V}$. $\qquad\square$

Next we use Proposition 4.3.6 to show the surjectivity of the exponential map for $\mathrm{SO}(V)$.

Theorem 4.3.7 (SO surjectivity). *Let V be a Euclidean space. If $V = V_1 \oplus \cdots \oplus V_k \oplus V_0$ is an orthogonal splitting for $A \in \underline{\mathrm{SO}}(V)$, then this is also an orthogonal splitting for $\exp(A) \in \mathrm{SO}(V)$. If $\phi_j := \|A|_{V_j}\|$, then ϕ_j is the angle of rotation of $\exp(A)$ in V_j. The exponential map*

$$\exp : \underline{\mathrm{SO}}(V) \to \mathrm{SO}(V) : A \mapsto \exp(A)$$

is surjective.

Proof. If the splitting is orthogonal for A, write $A = A_1 + \cdots + A_k$, where

$$A_j(v_1 + \cdots + v_k + v_0) = A_j v_j$$

in the splitting. Then the maps A_j commute, so $\exp(A) = \exp(A_1) \cdots \exp(A_k)$, and we see that

$$\exp(A)(v_1 + \cdots + v_k + v_0) = \exp(A_1)v_1 + \cdots + \exp(A_k)v_k + v_0,$$

so this is also an orthogonal splitting for $\exp(A)$.

Let T be a rotation. Take an orthogonal splitting $V = V_1 \oplus \cdots \oplus V_k \oplus V_0$ for T, and consider $T_j := T|_{V_j}$. For this plane rotation, the matrix in an ON-basis is $\begin{bmatrix} \cos\phi_j & \sin\phi_j \\ -\sin\phi_j & \cos\phi_j \end{bmatrix}$ for some rotation angle ϕ. By choosing the orientation of the basis appropriately, we may assume that $0 \leq \phi_j \leq \pi$. Define A_j to be the skew-symmetric map with matrix $\begin{bmatrix} 0 & \phi_j \\ -\phi_j & 0 \end{bmatrix}$ in this basis. Define the skew-symmetric map A on V by

$$A(v_1 + \cdots + v_k + v_0) = A_1 v_1 + \cdots + A_k v_k.$$

Then $\exp(A) = T$.   □

We have seen that there is a one-to-one correspondence between bivectors and skew-symmetric maps, and a two-to-one correspondence between rotors and rotations. See Proposition 4.2.3 and Proposition 4.1.9. To prove analogues of the above results for Euclidean bivectors and rotors, we require the following lemma.

Lemma 4.3.8. *Let V be a Euclidean space and let $w_1, w_2 \in \hat{\Delta}^2 V \setminus \{0\}$ be two simple nonzero bivectors. Then w_1 and w_2 commute if and only if either $[w_1]$ and $[w_2]$ are orthogonal planes or $[w_1] = [w_2]$.*

Proof. By reducing to $[w_1] \oplus [w_2]$, we may assume that $\dim V = 4$ and that $w_1 = \alpha e_{12}$ and $w_2 = \sum_{1 \leq i < j \leq 4} \beta_{i,j} e_{i,j}$. Calculating the commutator, we see that this vanishes if and only if $w_2 = \beta_{12} e_{12} + \beta_{34} e_{34}$. However, we have seen that this can be simple only if $\beta_{12} = 0$ or $\beta_{34} = 0$. This proves the lemma.   □

The following is the analogue of Proposition 4.3.6 for bivectors and rotors.

Proposition 4.3.9 (Euclidean rotors).   (i) *Let $q \in \mathrm{Spin}(V)$ be a rotor in a Euclidean space V. Then there are simple linearly independent commuting $j_i \in \hat{\Delta}^2 V$ such that $|j_i| = 1$, and angles $0 < \phi_i \leq \pi$, $i = 1, \ldots, k$, and a sign $\epsilon = \pm 1$ such that*

$$q = \epsilon(\cos(\phi_1/2) + j_1 \sin(\phi_1/2)) \cdots (\cos(\phi_k/2) + j_k \sin(\phi_k/2)). \qquad (4.2)$$

(ii) *Let $b \in \Delta^2 V$ be a bivector in a Euclidean space V. Then there are simple nonzero linearly independent commuting bivectors $b_j \in \hat{\Delta}^2 V$ such that*

$$b = b_1 + \cdots + b_k. \qquad (4.3)$$

We refer to (4.3) and (4.2) as *orthogonal decompositions* of bivectors and rotors respectively.

Proof. (i) Consider the corresponding rotation $Tv = qvq^{-1}$ as in Proposition 4.1.9. Let $V = V_1 \oplus \cdots \oplus V_k \oplus V_0$ be the splitting from Proposition 4.3.6 for T. In the plane V_i, write the rotation T_i as $T_i v = \exp(\phi_i j_i / 2) v \exp(-\phi_i j_i / 2)$. By Lemma 4.3.8, the bivectors j_i commute, and we see that

$$Tv = \exp(\phi_1 j_1 / 2) \cdots \exp(\phi_k j_k / 2) v (\exp(\phi_1 j_1 / 2) \cdots \exp(\phi_k j_k / 2))^{-1}.$$

Proposition 4.1.9 now shows the existence of $\epsilon = \pm 1$ such that

$$q = \epsilon \exp(\phi_1 j_1 / 2) \cdots \exp(\phi_k j_k / 2).$$

(ii) Consider the corresponding skew-symmetric map $Av = b \llcorner v$ provided by Proposition 4.2.3. Let $V = V_1 \oplus \cdots \oplus V_k \oplus V_0$ be the splitting from Proposition 4.3.6 for A. Define $b_j \in \Delta^2 V_j$ to be the bivector corresponding to $A|_{V_j}$. Then $b = b_1 + \cdots + b_k$, since $b_j \llcorner v = 0$ if $v \perp V_j = [b_j]$, and the terms b_j commute by Lemma 4.3.8. $\qquad\square$

Example 4.3.10. Consider the bivector

$$b = 3e_{12} + 6e_{13} + 4e_{14} + 6e_{23} - 4e_{24} + 2e_{34}$$

in four-dimensional Euclidean space with ON-basis $\{e_1, e_2, e_3, e_4\}$. We want to write b as the sum of two simple commuting bivectors b_1 and b_2. The skew-symmetric map A corresponding to b by Proposition 4.2.3 has matrix

$$\begin{bmatrix} 0 & 3 & 6 & 4 \\ -3 & 0 & 6 & -4 \\ -6 & -6 & 0 & 2 \\ -4 & 4 & -2 & 0 \end{bmatrix}.$$

We compute the matrix of A^2 to be

$$\begin{bmatrix} 61 & -20 & 10 & 0 \\ -20 & -61 & -10 & 0 \\ 10 & -10 & -76 & 0 \\ 0 & 0 & 0 & -36 \end{bmatrix}.$$

We see that -36 is an eigenvalue of A^2. The corresponding eigenspace is $V_1 := [(2e_1 - 2e_2 + e_3) \wedge e_4]$. We calculate $V_2 = V_1^\perp = \text{span}\{e_1 - 2e_3, e_2 + 2e_3\}$, and we must have

$$b_1 = \lambda(2e_1 - 2e_2 + e_3) \wedge e_4 = \lambda(2e_{14} - 2e_{24} + e_{34}),$$
$$b_2 = \mu(e_1 - 2e_3) \wedge (e_2 + 2e_3) = \mu(e_{12} + 2e_{13} + 2e_{23}).$$

Checking the equation $b = b_1 + b_2$, we must have $\lambda = 2$ and $\mu = 3$, so

$$b = (4e_{14} - 4e_{24} + 2e_{34}) + (3e_{12} + 6e_{13} + 6e_{23}),$$

as desired. Note that if we want to factor $b_2 = 3e_{12} + 6e_{13} + 6e_{23} = 3(e_1 - 2e_3) \wedge (e_2 + 2e_3)$ into orthogonal vectors with the Clifford product, we calculate $(e_1 - 2e_3) \lrcorner b_2 = 3e_2 + 6e_3 + 12e_1 + 12e_2 = 12e_1 + 15e_2 + 6e_3$. This gives $b_2 = \frac{1}{5}(e_1 - 2e_3) \vartriangle (12e_1 + 15e_2 + 6e_3)$, so that

$$b = 2(2e_1 - 2e_2 + e_3) \vartriangle e_4 + \tfrac{3}{5}(e_1 - 2e_3) \vartriangle (4e_1 + 5e_2 + 2e_3).$$

Example 4.3.11. Let $\{e_1, e_2, e_3, e_4\}$ be an ON-basis in four-dimensional Euclidean space. Then for all $\alpha \in \mathbf{R}$ we have

$$e_1 e_2 + e_3 e_4 = (\cos \alpha e_1 + \sin \alpha e_3)(\cos \alpha e_2 + \sin \alpha e_4)$$
$$+ (\sin \alpha e_1 - \cos \alpha e_3)(\sin \alpha e_2 - \cos \alpha e_4),$$

showing that nonuniqueness in the decomposition $b = b_1 + b_2$ in Proposition 4.3.9 is possible when the simple bivectors b_1 and b_2 have different norms.

The following is the analogue of Proposition 4.3.7 for bivectors and rotors.

Theorem 4.3.12 (Spin surjectivity). *Let V be a Euclidean space. If $b \in \triangle^2 V$ is a bivector with orthogonal decomposition $b = \phi_1 j_1 + \cdots + \phi_k j_k$, then*

$$q = (\cos(\phi_1/2) + j_1 \sin(\phi_1/2)) \mathbin{\triangle} \cdots \mathbin{\triangle} (\cos(\phi_k/2) + j_k \sin(\phi_k/2))$$

is an orthogonal decomposition of $q := \exp(b/2) \in \mathrm{Spin}(V)$. The exponential map

$$\exp : \underline{\mathrm{Spin}}(V) = \triangle^2 V \to \mathrm{Spin}(V) : b \mapsto \exp(b/2)$$

is surjective.

Proof. If b has orthogonal decomposition $b = \phi_1 j_1 + \cdots + \phi_k j_k$, then $q = (\cos(\phi_1/2) + j_1 \sin(\phi_1/2)) \mathbin{\triangle} \cdots \mathbin{\triangle} (\cos(\phi_k/2) + j_k \sin(\phi_k/2))$, since the bivectors j_i commute. To prove surjectivity of the exponential map, let $q \in \mathrm{Spin}(V)$ and take an orthogonal decomposition

$$q = \epsilon(\cos(\phi_1/2) + j_1 \sin(\phi_1/2)) \mathbin{\triangle} \cdots \mathbin{\triangle} (\cos(\phi_k/2) + j_1 \sin(\phi_k/2))$$

for q as in Proposition 4.3.9(i). If $\epsilon = +1$, then $b := \phi_1 j_1 + \cdots + \phi_k j_k$ gives $\exp(b/2) = q$. If $\epsilon = -1$, then

$$b = (\phi_1 - 2\pi) j_1 + \phi_2 j_2 + \cdots + \phi_k j_k$$

gives $\exp(b/2) = q$.   □

The exponential maps

$$\underline{\mathrm{Spin}}(V) \to \mathrm{Spin}(V) : b \mapsto \exp(b/2)$$

and

$$\underline{\mathrm{SO}}(V) \to \mathrm{SO}(V) : A \mapsto \exp(A)$$

are both surjective maps. However, in contrast to $p : \mathrm{Spin}(V) \to \mathrm{SO}(V)$, they are in general not locally diffeomorphisms, as the following example shows.

Example 4.3.13. Let V be a three-dimensional Euclidean space, and let S^3 denote the unit sphere in $\mathbf{H}$. Then $\exp : \underline{\mathrm{Spin}}(V) \to S^3$ maps $\{b \in \underline{\mathrm{Spin}}(V) \; ; \; |b| < 2\pi\}$ diffeomorphically onto $S^3 \setminus \{-1\}$ and the boundary on the ball onto the south pole $q = -1$. Thus, if $|b| = 2\pi$, then the derivative of the exponential map $\underline{\exp}_b$ has a two-dimensional null space and a one-dimensional range.

4.4 Spacetime Rotations

In this section, W denotes a spacetime with $n \geq 1$ spatial dimensions. We write ON-bases as $\{e_0, e_1, \ldots, e_n\}$, where e_0 is the time-like basis vector. We refer to Section 1.3 for notation and the basic geometry of spacetimes. Our goal is to represent rotations of spacetimes by bivectors, generalizing the Euclidean theory from Section 4.3. The main novelty is that for rotors

$$q = \exp(b/2),$$

there are three types of bivectors in spacetime. The two-dimensional plane $[b] \subset W$ may be space-like, light-like, or time-like, corresponding to three fundamentally different types of spacetime rotations. *Elliptic rotations* are nothing but Euclidean rotations, and they arise from space-like bivectors. *Hyperbolic rotations* of spacetime arise from time-like bivectors. Such are of fundamental importance in special relativity theory, and in that context they are referred to as *Lorentz boosts*.

Example 4.4.1 (Lorentz boost). Let W be a two-dimensional spacetime with ON-basis $\{e_0, e_1\}$, and let T be a rotation. Introduce the basis $\{f_1, f_2\}$ of singular vectors $f_1 = (e_0 + e_1)/\sqrt{2}$ and $f_2 = (e_0 - e_1)/\sqrt{2}$. Since $\langle Tf_i \rangle^2 = \langle f_i \rangle^2 = 0$, we have $Tf_i \in W_l$. Since $\det T = 1$, we must have $Tf_1 = \lambda f_1$ and $Tf_2 = \mu f_2$ for some $\lambda, \mu > 0$ if we assume that T preserves direction of time. Since T is an isometry, we have

$$-1 = \langle f_1, f_2 \rangle = \langle Tf_1, Tf_2 \rangle = \lambda\mu\langle f_1, f_2 \rangle = -\lambda\mu.$$

Writing $\lambda = \exp(\phi)$ for some $\phi \in \mathbf{R}$, we have $\mu = \exp(-\phi)$. Switching back to the ON-basis $\{e_0, e_1\}$, we see that the matrix of T is

$$\begin{bmatrix} \cosh\phi & \sinh\phi \\ \sinh\phi & \cosh\phi \end{bmatrix} = \begin{bmatrix} 1/\sqrt{1-v^2} & v/\sqrt{1-v^2} \\ v/\sqrt{1-v^2} & 1/\sqrt{1-v^2} \end{bmatrix}.$$

In relativity theory, the hyperbolic rotation T is interpreted as a Lorentz boost, with velocity $v := \tanh\phi \in (-1, 1)$, which maps the inertial frame $\{e_0, e_1\}$ to another inertial frame $\{e_0', e_1'\} = \{Te_0, Te_1\}$. See Figure 4.2. If O is an inertial observer with world line $[e_0]$ in spacetime, then he will measure time t and distance x as the coordinates with respect to the ON-basis $\{e_0, e_1\}$. Another inertial observer O' with world line $[e_0']$ in spacetime will measure time t' and distance x' as the coordinates with respect to the ON-basis $\{e_0', e_1'\}$. The relation between coordinates is

$$\begin{bmatrix} t \\ x \end{bmatrix} = \begin{bmatrix} 1/\sqrt{1-v^2} & v/\sqrt{1-v^2} \\ v/\sqrt{1-v^2} & 1/\sqrt{1-v^2} \end{bmatrix} \begin{bmatrix} t' \\ x' \end{bmatrix},$$

so in particular, $x' = 0$ is equivalent to $x = vt$, showing that O' is traveling with speed v relative to O. We here use relativistic units, where the speed of light is $c = 1$, and no inertial observer O', related to O by a Lorentz boost, can travel faster than the speed of light.

Similarly to Euclidean rotations in the plane, the Lorentz boost T is represented by spacetime Clifford algebra and bivectors as

$$Tv = \exp(\phi j/2)v \exp(-\phi j/2) = \exp(\phi j)v = (\cosh \phi + j \sinh \phi)v = qvq^{-1},$$

where $j := e_0 e_1 \in \Delta^2 W$ and $q := \cosh(\phi/2) + j \sinh(\phi/2) \in \mathrm{Spin}_+(W)$. Note that $j^2 = +1$, and not -1 as in Euclidean space.

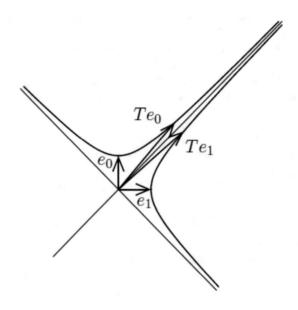

Figure 4.2: The Lorentz boost corresponding to an inertial observer travelling at 86% the speed of light, or equivalently a hyperbolic rotation through the angle $\phi = 1.3$.

Note that both elliptic and hyperbolic rotations are two-dimensional. We have seen in the Euclidean case that each such rotation is an orthogonal direct sum of two-dimensional rotations. In spacetime, surprisingly, there is a third type of rotation: *parabolic rotations*. In contrast, these are three-dimensional in the sense that although there is an invariant two-dimensional subspace, the space of the light-like bivector for the rotation, this has no invariant complementary line.

Example 4.4.2 (Parabolic rotation). Consider a three-dimensional spacetime W. Let $b \in \hat{\wedge}^2 W$ be a simple bivector with space $[b]$ a light-like plane. In this case we can choose the ON-basis $\{e_0, e_1, e_2\}$, so that $[b]$ intersects the light cone W_l along $[e_0 + e_1]$. We let

$$b = \phi(e_0 + e_1)e_2 = \phi j, \quad \phi \in \mathbf{R}, j := (e_0 + e_1)e_2.$$

Consider the spacetime rotation $T : v \mapsto qvq^{-1}$ with rotor $q := \exp(b/2)$. Note that $j^2 = (e_0 + e_1)e_2(e_0 + e_1)e_2 = -(e_0 + e_1)^2 e_2^2 = 0$, and therefore $q = \exp(b/2) = 1 + b/2$. To find the matrix of T in the basis $\{e_0, e_1, e_2\}$, we calculate, for example,

$$\exp(\phi j/2)e_0 \exp(-\phi j/2) = (1 + \tfrac{\phi}{2}e_{02} + \tfrac{\phi}{2}e_{12})e_0(1 - \tfrac{\phi}{2}e_{02} - \tfrac{\phi}{2}e_{12})$$

$$= (1 + \tfrac{\phi}{2}e_{02} + \tfrac{\phi}{2}e_{12})(e_0 - \tfrac{\phi}{2}e_2 - \tfrac{\phi}{2}e_{012})$$

$$= e_0 + \tfrac{\phi}{2}e_2 - \tfrac{\phi}{2}e_{012} + \tfrac{\phi}{2}e_2 + \tfrac{\phi^2}{4}e_0 + \tfrac{\phi^2}{4}e_1 + \tfrac{\phi}{2}e_{012} + \tfrac{\phi^2}{4}e_1 + \tfrac{\phi^2}{4}e_0$$

$$= (1 + \tfrac{\phi^2}{2})e_0 + \tfrac{\phi^2}{2}e_1 + \phi e_2.$$

A similar calculation of $T(e_1)$ and $T(e_2)$ gives us the matrix

$$\begin{bmatrix} 1 + \phi^2/2 & -\phi^2/2 & \phi \\ \phi^2/2 & 1 - \phi^2/2 & \phi \\ \phi & -\phi & 1 \end{bmatrix}.$$

To understand the action of T, though, it is more appropriate to change basis to $\{f_1, f_2, f_3\} = \{e_0 + e_1, e_2, e_0 - e_1\}$, in which the matrix becomes $\begin{bmatrix} 1 & \phi & \phi^2 \\ 0 & 1 & 2\phi \\ 0 & 0 & 1 \end{bmatrix}$.

Now it is straightforward to describe T. See Figure 4.3. Define the line $W_0 := [f_1] = [e_0 + e_1]$ and note that the orthogonal complement of this totally degenerate subspace is $W_0^\perp = [b] = [f_1 \wedge f_2]$, the degenerate tangent plane to the light cone along $W_0 \subset W_0^\perp$. Then T fixes W_0 and skews $W_0^\perp$ by translating f_2 along W_0. Furthermore, the last row in the matrix (for the basis $\{f_i\}$) shows that T maps the translated subspace $W_0^\perp + f_3$ onto itself. Note that since $W_0^\perp$ is tangent to W_l, the intersection $(W_0^\perp + f_3) \cap W_l$ is a parabola P. Since T is an isometry, T maps P onto itself, hence the terminology: as ϕ varies, f_3 moves along the parabola P. Note that the only eigenvector to T is f_1; thus there is no invariant line W_1 such that $W = [b] \oplus W_1$, as claimed.

After these examples, we are ready to study general spacetime rotations. In spacetime there are two types of rotations. Consider $T \in SO(W)$. Clearly T preserves the light cone W_l, as well as the space-like vectors W_s and the time-like vectors W_t. Since W_t consists of two connected components, we have two cases:

- The isometry T preserves the direction of time, so that $T : W_{t+} \to W_{t+}$. If $q \in \mathrm{Spin}(W)$ represents T as $Tv = qvq^{-1}$, then $\langle q, q \rangle = +1$ in this case.

- The isometry T reverses the direction of time, so that $T : W_{t+} \to W_{t-}$. If $q \in \mathrm{Spin}(W)$ represents T as $Tv = qvq^{-1}$, then $\langle q, q \rangle = -1$ in this case.

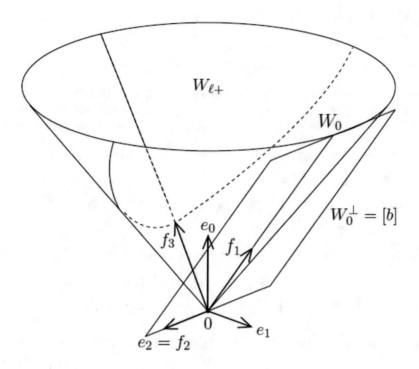

Figure 4.3: Parabolic spacetime rotation. The curve on $W_{\ell+}$ is the parabola parallel to the tangent plane $[b]$, along which f_3 is rotated.

The stated property of the rotor q follows from inspection of the spacetime proof of Theorem 4.1.3. This reveals that $\overline{q}q = 1$ if and only if the vector $Tv - v$, where v is time-like, used for the first reflection is space-like, since all remaining factors of q will be space-like. And it is indeed the case that the nonsingular vector $Tv - v$ is space-like if and only if Tv and v belong to the same connected component of W_t.

For Euclidean space, the space of rotations $O(V)$ has two connected components: $SO(V)$ and $O(V) \setminus SO(V)$. For spacetimes, each of these in turn splits into two connected components, since the set of time-like vectors is disconnected. Thus $O(W)$ has four connected components.

Definition 4.4.3 (Orthochronous rotation). Let W be a spacetime. Define the group of *orthochronous rotations* to be

$$SO_+(W) := \{T \in SO(W) \; ; \; T : W_{t+} \to W_{t+}\}.$$

Define the group of *orthochronous rotors* to be

$$\mathrm{Spin}_+(W) := \{q \in \mathrm{Spin}(W) \; ; \; \langle q, q \rangle = +1\}.$$

Orthochronous means that the direction of time is preserved. The condition on $q \in \mathrm{Spin}_+(W)$ is that q contains an even number of time-like vectors, as well as an even number of space-like vectors.

Proposition 4.4.4. *Let W be a spacetime. The map*

$$p : \mathrm{Spin}_+ \to SO_+(W) : q \mapsto Tv = qvq^{-1}$$

is a surjective homomorphism and has kernel ± 1.

The groups $SO_+(W)$ and $\mathrm{Spin}_+(W)$ are closed but not compact, and when $\dim W \geq 3$, then they are both connected.

Proof. What needs proof is the connectedness when $\dim W \geq 3$. That $q = a_1 \cdots a_{2k}$ $\in \mathrm{Spin}_+(W)$ means that the number of time-like vectors appearing is even, as well as the number of space-like vectors. To connect q to 1, it suffices to connect it to ± 1, since $-1 = v(-v)$ for any $v \in H(W_s)$, and connecting $-v$ to $+v$ within $H(W_s)$ yields a path from -1 to 1 in $\mathrm{Spin}_+(W)$. For a general q as above, perturb first all $a_j \in H(W_{t+})$ to the same vector $e_0 \in H(W_{t+})$, all vectors $a_j \in H(W_{t-})$ to the same vector $-e_0 \in H(W_{t-})$, and all vectors $a_j \in H(W_s)$ to the same vector $e_1 \in H(W_s)$ orthogonal to e_0. This connects q to ± 1. $\square$

Exercise 4.4.5. Let W be a two-dimensional spacetime. Show that $SO_+(W)$ is connected, but not $\mathrm{Spin}_+(W)$. The latter consists of the two hyperbolas $\alpha + j\beta$, $\alpha = \pm\sqrt{1 + \beta^2}$.

Proposition 4.4.6. *Let W be spacetime with $\dim W = 3$, and let $T \in SO_+(W)$. Then T is of exactly one of the following types if $T \neq I$:*

(i) *An elliptic rotation: T has no eigenvectors along the light cone.*

   In this case W splits orthogonally as $W_1 \oplus W_2$, where W_1 is a line spanned by a time-like vector fixed by T, and W_2 is a space-like two-dimensional plane in which T is a plane Euclidean rotation. In this case there is $b \in \widehat{\triangle}^2 W$ such that $b^2 < 0$, $[b] = W_2$, and $Tv = \exp(b/2)v\exp(-b/2)$ for all $v \in W$.

(ii) *A hyperbolic rotation: T has two eigenvectors along the light cone.*

   In this case W splits orthogonally as $W_1 \oplus W_2$, where W_1 is a two-dimensional time-like plane in which T is a plane Lorentz boost, and W_2 is a line spanned by a space-like vector fixed by T. In this case there is $b \in \widehat{\triangle}^2 W$ such that $b^2 > 0$, $[b] = W_1$, and $Tv = \exp(b/2)v\exp(-b/2)$ for all $v \in W$.

(iii) *A parabolic rotation: T has one eigenvector along the light cone.*

   In this case $W_0 \subset W_0^\perp \subset W$, where W_0 is a line spanned by a singular vector fixed by T, and $W_0^\perp$ is the plane tangent to the light cone along the line W_0, and $W_0^\perp$ is invariant under T. In this case there is $b \in \widehat{\triangle}^2 W$ such that $b^2 = 0$, $[b] = W_0^\perp$, and $Tv = \exp(b/2)v\exp(-b/2)$ for all $v \in W$.

Proof. Let $T \in \mathrm{SO}_+(W) \setminus \{I\}$. By the Cartan–Dieudonné theorem (Theorem 4.1.3), and since $T \neq I$ is a rotation, there are linearly independent vectors v_1, v_2 such that $Tx = v_1 v_2 x (v_1 v_2)^{-1}$. Note that since $v_1 v_2$ is orthochronous, we have $\langle v_1, v_2 \rangle^2 - (v_1 \wedge v_2)^2 = \langle v_1, v_2 \rangle^2 + \langle v_1 \wedge v_2 \rangle^2 > 0$. Three cases are possible:

(i) The space $[v_1 \wedge v_2]$ is a space-like plane. In this case, let $\{e_0, e_1, e_2\}$ be an ON-basis such that $\{e_1, e_2\}$ is an ON-basis for $[v_1 \wedge v_2]$. Then there exist $\alpha, \phi \in \mathbf{R}$ such that

$$\exp(\phi e_{12}) = \cos\phi + e_{12} \sin\phi = \alpha v_1 v_2.$$

This gives conclusion (i) with $b := \phi e_{12}$, $W_1 := [e_0]$, and $W_2 := [e_{12}]$.

(ii) The space $[v_1 \wedge v_2]$ is a time-like plane. In this case, let $\{e_0, e_1, e_2\}$ be an ON-basis such that $\{e_0, e_1\}$ is an ON-basis for $[v_1 \wedge v_2]$. Then there exist $\alpha, \phi \in \mathbf{R}$ such that

$$\exp(\phi e_{01}) = \cosh\phi + e_{01} \sinh\phi = \alpha v_1 v_2.$$

This gives conclusion (ii) with $b := \phi e_{01}$, $W_1 := [e_{01}]$, and $W_2 := [e_2]$.

(iii) The space $[v_1 \wedge v_2]$ is a light-like plane. In this case, let $\{e_0, e_1, e_2\}$ be an ON-basis such that $\{e_0 + e_1, e_2\}$ is a basis for $[v_1 \wedge v_2]$. Then there exist $\alpha, \phi \in \mathbf{R}$ such that

$$\exp(\phi(e_0 + e_1)e_2) = 1 + \phi(e_0 + e_1)e_2 = \alpha v_1 v_2.$$

This gives conclusion (iii) with $b := \phi(e_0 + e_1)e_2$, $W_0 := [e_0 + e_1]$, and $W_0 = [(e_0 + e_1) \wedge e_2]$. $\square$

Our next objective is to show that every orthochronous rotation decomposes into two-dimensional elliptic and hyperbolic rotations and three-dimensional parabolic rotations.

Proposition 4.4.7 (Spacetime rotations)**.** *Let* $T \in \mathrm{SO}_+(W)$ *be an orthochronous rotation in a spacetime* W. *Then there exist a time-like subspace* W_{-1} *of dimension two or three, two-dimensional space-like subspaces* $W_1, \ldots, W_k$, *and a space-like subspace* W_0 *such that* W *splits as a direct sum of orthogonal subspaces*

$$W = W_{-1} \oplus W_1 \oplus \cdots \oplus W_k \oplus W_0,$$

where all the subspaces are invariant under T, $T|_{W_{-1}}$ *is a Lorentz boost if* $\dim W_{-1} = 2$ *or a parabolic rotation if* $\dim W_{-1} = 3$, $T|_{W_j}$ *is a Euclidean rotation for* $j \geq 1$, *and* $T|_{W_0} = I$.

Proof. We first split off the subspace W_0. Define $\tilde{W} := \mathsf{N}(T - I)$. If $\tilde{W}$ is space- or time-like, let $W_0 := \tilde{W}$. If $\tilde{W}$ is light-like, take any space-like subspace $W_0 \subset \tilde{W}$ of one dimension less than $\tilde{W}$, which is possible by Proposition 1.3.7. Looking into $W_0^\perp$, replacing W by $W_0^\perp$, we assume from now on that $\mathsf{N}(T - I)$ is either $\{0\}$ or is a line along the light cone.

Complexification as in Section 1.5 shows that there exists a one- or two-dimensional subspace W' that is invariant, and we have the following possible cases:

1. A space-like plane W', where $\det T|_{W'} = 1$.

2. A space-like line W', where $T = -I$ due to the assumption on $\mathsf{N}(T - I)$.

3. A time-like plane W', where $\det T|_{W'} = 1$. A time-like line is not possible because $T|_{W'} \neq I$, by the assumption on $\mathsf{N}(T - I)$, and $T \neq -I$, since T is orthochronous.

4. A light-like plane or line. However, in a plane, any vector along the intersection with the light cone must be an eigenvector, since T is an isometry. Assume, therefore, that $Tv_1 = \lambda v_1$, where $v_1 \in W_l$. Since T is orthochronous, $\lambda > 0$, and we may assume that $\lambda = 1$. If this is not the case, since $0 = \det(T - \lambda) = \det(T^* - \lambda) = \det(T^{-1} - \lambda)$; we see that also λ^{-1} must be an eigenvalue to T, giving another eigenvector not parallel to v_1. But $\lambda^{-1} \neq 1$ can be an eigenvalue of an isometry only if the eigenvector belongs to W_l, which would give us two eigenvectors on W_l, and hence an invariant time-like plane, and we are in the previous case. To summarize, in this last case, we may assume that there exists a line along W_l spanned by v_1 such that $Tv_1 = v_1$.

In all but the last case, the invariant subspace is nondegenerate, and we can split W orthogonally into a direct sum

$$W = W' \oplus (W')^{\perp},$$

as in Proposition 1.3.4, with both subspaces invariant under T. In the last case, we need to find a larger invariant and nondegenerate subspace W' containing v_1. Let $E_1 := \mathrm{span}\{v_1\} = \mathsf{N}(T - I)$, and note as in Proposition 1.3.7 that $E_1^{\perp}$ is the tangent hyperplane to the light cone W_l along E_1. We have

$$\mathsf{R}(T - I) = \mathsf{N}(T^* - I)^{\perp} = \mathsf{N}(T^{-1} - I)^{\perp} = E_1^{\perp} \ni v_1.$$

Therefore we have v_2 such that $(T - I)v_2 = v_1$. Let $E_2 := \mathrm{span}\{v_1, v_2\}$, and note that v_2 is not parallel to v_1, since $Tv_1 = v_1$. We note that

$$0 = \langle v_1 \rangle^2 = \langle Tv_2 - v_2 \rangle^2 = 2\langle v_2 \rangle^2 - 2\langle Tv_2, v_2 \rangle = -2\langle v_1, v_2 \rangle,$$

so $v_2 \in E_1^{\perp}$. By Proposition 1.3.7, E_2 is a degenerate subspace, and we proceed by taking v_3 such that $(T-I)v_3 = v_2$, which is possible, since $v_2 \in E_1^{\perp} = \mathsf{R}(T-I)$. Let $E_3 := \mathrm{span}\{v_1, v_2, v_3\}$. We note that $T-I : E_2 \to E_1$, so that $T^*-I = (I-T)T^{-1} : E_1^{\perp} \to E_2^{\perp}$. Since $v_2 \notin E_2^{\perp}$ because $\langle v_2 \rangle^2 \neq 0$, it follows that $Tv_3 \notin E_1^{\perp}$. Since $T : E_1^{\perp} \to E_1^{\perp}$, we have $v_3 \notin E_1^{\perp}$. Thus we have constructed an invariant and nondegenerate subspace $W' = E_3$ by Proposition 1.3.7.

We proceed recursively and split the invariant complements $(W')^{\perp}$. Collecting all these orthogonal subspaces produced, we note that the only subspace for which $\det T_{W'} = -1$ is that of the space-like lines from 2. Since $\det T = 1$, the number of such lines must be even, and we can evenly collect them in invariant

2-planes. Note that these planes are nondegenerate and cannot be time-like, since T is orthochronous. Hence they are space-like, and T acts in them as Euclidean rotation through an angle π. This proves the orthogonal splitting of W.   □

As for Euclidean rotations, this orthogonal decomposition implies the following representation theorem for rotations of spacetime.

Theorem 4.4.8 (SO$_+$ surjectivity). *Let W be a spacetime. If $T \in \mathrm{SO}_+(W)$, then there exist $A \in \underline{\mathrm{SO}}(W)$ and $b \in \underline{\mathrm{Spin}}(W)$ such that*

$$Tv = \exp(A)v = \exp(b/2)v\exp(-b/2), \quad v \in W.$$

Thus each $T \in \mathrm{SO}_+(W)$ belongs to a one-parameter group of rotations $T(\phi)v :=$ $\exp(\phi A)v = \exp(\phi b/2)v\exp(-\phi b/2)$, and the exponential map $\exp : \underline{\mathrm{SO}}(W) \to \mathrm{SO}_+(W)$ is surjective.

Proof. Let $T \in \mathrm{SO}_+(W)$, and split spacetime $W = W_{-1} \oplus W_1 \oplus \cdots \oplus W_k \oplus W_0$ as in Proposition 4.4.7. By Proposition 4.4.6 there are $b_j \in \underline{\mathrm{Spin}}(W_j)$ such that $T|_{W_j}v = \exp(b_j/2)v\exp(-b_j/2)$, $v \in W_j$, $j = -1,1,2,\ldots,k$. The bivectors commute, so letting $b := b_{-1} + b_1 + \cdots + b_k$, we have

$$Tv = \exp(b_{-1}/2)\exp(b_1/2)\cdots\exp(b_k/2)v\exp(-b_k/2)\cdots\exp(-b_1/2)\exp(-b_{-1}/2)$$
$$= \exp(b/2)v\exp(-b/2).$$

Also, if $Av := b \,{\llcorner}\, v$, then $\exp(A) = T$.   □

We end this section with a number of exercises, which completes the extension of the Euclidean results to spacetime.

Exercise 4.4.9. Prove the following by modifying the proofs of Proposition 4.4.7, letting $\mathsf{N}(A)$ play the role of $\mathsf{N}(T - I)$.

Let $A \in \underline{\mathrm{SO}}(W)$ be a skew-symmetric map in a spacetime W. Then there exist a time-like subspace W_{-1} of dimension two or three, two-dimensional space-like subspaces $W_1,\ldots,W_k$, and a space-like subspace W_0 such that W splits as a direct sum of orthogonal subspaces

$$W = W_{-1} \oplus W_1 \oplus \cdots \oplus W_k \oplus W_0,$$

where all the subspaces are invariant under T. If $\dim W_{-1} = 2$, then the matrix of A in an ON-basis $\{e_0, e_1\}$ is $\begin{bmatrix} 0 & \phi \\ \phi & 0 \end{bmatrix}$, for some $\phi \in \mathbf{R}$, that is, $Av = \phi e_{0,1} \,{\llcorner}\, v$. If $\dim W_{-1} = 3$, then there is an ON-basis $\{e_0, e_1, e_2\}$ in which the matrix of A is

$$\begin{bmatrix} 0 & 0 & \phi \\ 0 & 0 & \phi \\ \phi & -\phi & 0 \end{bmatrix}$$

for some $\phi \in \mathbf{R}$, that is, $Av = \phi(e_0 + e_1)e_2 \,{\llcorner}\, v$. The restrictions $T|_{W_j}$, $j \geq 1$, are all Euclidean skew maps, and $T|_{W_0} = I$.

Exercise 4.4.10. Deduce the following from the above results for spacetime rotations and skew maps, similar to what was done in the Euclidean case in Proposition 4.3.9.

Let W be a spacetime.

(i) If $b \in \triangle^2 W$ is a bivector, then there exists an ON-basis $\{e_i\}$ in which b has the form

$$b = \phi_0 e_0 e_1 + \phi_1 e_2 e_3 + \phi_2 e_4 e_5 + \cdots + \phi_k e_{2k} e_{2k+1} \quad \text{or}$$
$$b = \phi_0 (e_0 + e_1) e_2 + \phi_1 e_3 e_4 + \phi_2 e_5 e_6 + \cdots \phi_k e_{2k+1} e_{2k+2}.$$

(ii) If $q \in \mathrm{Spin}_+(W)$ is an orthochronous rotor, then there exist an ON-basis $\{e_i\}$, a sign $\epsilon = \pm 1$, and angles $0 < \phi_j \le \pi$ such that q has the form

$$q = \epsilon\big(\cosh(\phi_0/2) + e_{0,1} \sinh(\phi_0/2)\big)\big(\cos(\phi_1/2) + e_{2,3} \sin(\phi_1/2)\big)$$
$$\cdots \big(\cos(\phi_k/2) + e_{2k,2k+1} \sin(\phi_k/2)\big), \quad \text{or}$$
$$q = \epsilon\big(1 + \tfrac{\phi_0}{2}(e_0 + e_1)e_2\big)\big(\cos(\phi_1/2) + e_{3,4} \sin(\phi_1/2)\big)$$
$$\cdots \big(\cos(\phi_k/2) + e_{2k+1,2k+2} \sin(\phi_k/2)\big).$$

In contrast to Theorem 4.4.8, the exponential map

$$\exp : \underline{\mathrm{Spin}}(W) \to \mathrm{Spin}_+(W)$$

is surjective in spacetime only when $\dim W \ge 5$. In dimensions two and three, exp is far from being surjective, whereas in dimension 4, it is only half of the orthochronous rotors representing parabolic rotations that cannot be represented by a bivector, as the following clarifies.

Exercise 4.4.11. Let W be a two-dimensional spacetime, and let $j \in \triangle^2 W$ be such that $j^2 = 1$. Show that

$$\mathrm{Spin}_+(W) = \left\{ \alpha + \beta j \; ; \alpha = \pm\sqrt{1 + \beta^2} \right\}$$
$$\supsetneq \exp(\underline{\mathrm{Spin}}(W)) = \left\{ \alpha + \beta j \; ; \alpha = \sqrt{1 + \beta^2} \right\}.$$

Exercise 4.4.12. Let W be a three-dimensional spacetime. Then

$$\mathrm{Spin}_+(W) = \{\alpha + \beta_1 j_1 + \beta_2 j_2 + \beta_3 j_3 \; ; \alpha^2 + \beta_3^2 = \beta_1^2 + \beta_2^2 + 1\}$$

is a connected Lie group by Proposition 4.4.4, where $\{e_0, e_1, e_2\}$ is an ON-basis with $e_0^2 = -1$ and $j_1 := e_{01}$, $j_2 := e_{02}$, and $j_3 := e_{12}$. Show that $\exp(\underline{\mathrm{Spin}}(W))$ contains all orthochronous rotors except those $\alpha + \beta_1 j_1 + \beta_2 j_2 + \beta_3 j_3 \in \mathrm{Spin}_+(W)$ for which $\beta_3^2 \le \beta_1^2 + \beta_2^2$ and $\alpha \le -1$.

Exercise 4.4.13. Let W be a four-dimensional spacetime as in relativity. Show that all $q \in \mathrm{Spin}_+(W)$ can be written $q = \exp(b/2)$ for some $b \in \underline{\mathrm{Spin}}(W)$, except those q that are of the form $q = -(1+b)$, where $b \in \triangle^2 W$ satisfies $b^2 = 0$.

Theorem 4.4.14 (Spin$_+$ surjectivity). *Let W be a spacetime with $\dim W \geq 5$. If $q \in \mathrm{Spin}_+(W)$, then there exists $b \in \underline{\mathrm{Spin}}(W)$ such that $q = \exp(b/2)$. Thus each $q \in \mathrm{Spin}_+(W)$ belongs to a one-parameter group of rotors $q(\phi) := \exp(\phi b/2)$, and the exponential map $\exp : \underline{\mathrm{Spin}}(W) \to \mathrm{Spin}_+(W)$ is surjective.*

Proof. If $q \in \mathrm{Spin}_+(W)$, consider the corresponding rotation $Tv := qvq^{-1}$. Theorem 4.4.8 shows that $Tv = \exp(b/2)v\exp(-b/2)$, where $b = b_{-1} + b_1 + \cdots + b_k$ for some commuting simple bivectors b_j. Thus $q = \pm\exp(b/2)$. In case of a minus sign, we can eliminate this as follows. If $k \geq 1$, then $b_k^2 < 0$ and we let $\phi := 2\pi/\sqrt{-b_k^2}$. This gives $\exp((b - \phi b_k)/2) = (-q)(-1) = q$, since b and b_k commute. If $b = b_{-1}$, then since $\dim W \geq 5$, there exists $b_1 \in \triangle^2 W$ that commutes with b and satisfies $b_1^2 = -\pi^2$. In this case $\exp((b - b_1)/2) = (-q)(-1) = q$ also.   □

4.5 Fractional Linear Maps

In this section we study a fundamental class of maps of a Euclidean space (X, V) that are closely related to isometries.

Definition 4.5.1 (Conformal map). Let (X, V) be a Euclidean space and let $D \subset X$ be an open set. A differentiable map $f : D \to X$ is called *conformal* in D if at each point $x \in D$, the derivative $\underline{f}_x : V \to V$ is a nonzero multiple of an isometry, that is, if there exists $\lambda : D \to \mathbf{R} \setminus \{0\}$ such that $\lambda(x)\underline{f}_x$ is an isometry for all $x \in D$.

Example 4.5.2. We have the following four classes of basic conformal maps in a Euclidean space X.

(i) Fix an origin in X, and identify X and V. Then the *isometry* $x \mapsto qxq^{-1}$, $x \in X$, is conformal for every $q \in \widehat{\triangle}V$.

(ii) For every $v \in V$, *translation* $x \mapsto x + v$, $x \in X$, by the vector v is a conformal map.

(iii) Fix an origin in X, and identify X and V. For every $c \in \mathbf{R} \setminus \{0\}$, *dilation* $x \mapsto cx$, $x \in X$, with scale factor c is a conformal map.

(iv) Fix an origin in X, and identify X and V. Then *inversion* in the unit sphere

$$x \mapsto \frac{1}{x} = \frac{x}{|x|^2}, \quad x \in X \setminus \{0\},$$

is a conformal map.

Exercise 4.5.3. Let $f(x) = x^{-1}$. Show that $|x|^2 f_{-x}$ is reflection in the hyperplane orthogonal to x, and in particular that f is conformal with derivative

$$\underline{f}_x(h) = -x^{-1}hx^{-1}, \quad h \in V, \ x \neq 0.$$

Clearly compositions of conformal maps are conformal, on appropriate domains of definition. Thus we have the following group consisting of conformal maps.

Definition 4.5.4 (Fractional linear map). Let (X, V) be a Euclidean space. A map $f : X \to X$ is said to be a *fractional linear map* if it is a finite composition of isometries, translations, dilations, and inversions.

In studying fractional linear maps it is convenient to extend the Euclidean space V by adding the point at infinity ∞. More precisely, we embed V in a Euclidean space V_∞ that is one dimension larger. A vector $e_\infty \in V_\infty$ with $|e_\infty| = 1$ is fixed and V is identified with the hyperplane orthogonal to e_∞. The *extended Euclidean space* $\overline{V}$ is the unit sphere in V_∞, where *the north pole e_∞ is identified with the point at infinity* ∞. The remaining points $\overline{V} \setminus \{e_\infty\}$ are identified with V through stereographic projection.

Definition 4.5.5 (Stereographic projection). *Stereographic projection* is the map

$$\overline{V} \setminus \{e_\infty\} \ni \overline{x} = x' + x_\infty e_\infty \mapsto x \in V,$$

for $x' \in V$ and $x_\infty \in [-1, 1)$, where the projection x is defined as the intersection of V and the straight line in V_∞ that intersects $\overline{V}$ at e_∞ and $\overline{x}$. See Figure 4.4. We may also refer to the inverse map $x \mapsto \overline{x}$ as stereographic projection.

Exercise 4.5.6. Prove the following explicit formulas for the stereographic projection and its inverse.

$$x = \frac{x'}{1 - x_\infty}, \quad \overline{x} = \frac{2}{|x|^2 + 1}x + \frac{|x|^2 - 1}{|x|^2 + 1}e_\infty.$$

In particular, the south pole $-e_\infty$ projects onto $0 \in V$ and the equator $x_\infty = 0$ projects onto the unit sphere $|x| = 1$.

After adding the point at infinity ∞ to V, the inversion map $x \mapsto x^{-1}$ becomes a smooth map $\overline{V} \to \overline{V}$. Indeed, it is seen to correspond to the equatorial reflection

$$\overline{x} = x' + x_\infty e_\infty \mapsto x' - x_\infty e_\infty = \widehat{e}_\infty \overline{x} e_\infty,$$

where $0 \leftrightarrow -e_\infty$ is mapped onto $\infty \leftrightarrow e_\infty$ and vice versa. We also extend isometries, translations, and dilation to continuous maps of the extended space $\overline{V}$, identified by V through stereographic projection, that fixes the point at infinity e_∞. Thus all fractional linear maps are homeomorphisms as maps $\overline{V} \to \overline{V}$.

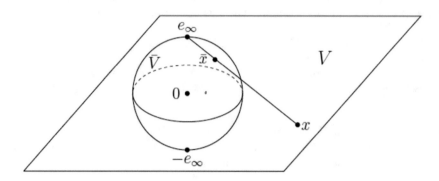

Figure 4.4: Stereographic projection.

Proposition 4.5.7 (Mapping of spheres). *Let $f : V \cup \{\infty\} \to V \cup \{\infty\}$ be a fractional linear map, and let $S \subset V \cup \{\infty\}$ be either a hypersphere, that is, of the form $S = \{x \in V \; ; \; |x - a|^2 = r^2\}$, or a hyperplane, that is, of the form $S = \{x \in V \; ; \; \langle x - a, b\rangle = 0\} \cup \{\infty\}$. Then the image $f(S)$ is a hypersphere if $f^{-1}(\infty) \notin S$, and $f(S)$ is a hyperplane if $f^{-1}(\infty) \in S$.*

Proof. It suffices to show that a hypersphere or hyperplane is mapped onto either a hypersphere or a hyperplane, since hyperspheres cannot be unbounded and hyperplanes cannot be bounded. It also suffices to prove this for isometries, translations, dilations, and inversions, and only the last case needs proof. Consider $x \mapsto x^{-1} = y$ and a hypersphere $|x - a|^2 = r^2$. This is mapped onto the points satisfying $|1 - ay|^2 = r^2|y|^2$, where we have used Lagrange's identity for the Clifford product. This yields

$$1 + |a|^2|y|^2 - 2\langle a, y\rangle = r^2|y|^2.$$

If $|a| = r$, that is, $\infty^{-1} = 0 \in S$, then this is a hyperplane $\langle a, y\rangle = 1/2$. If $|a| \neq r$, then this is the hypersphere $|y - a/(|a|^2 - r^2)|^2 = r^2/(|a|^2 - r^2)^2$.

On the other hand, consider a hyperplane $\langle a, x\rangle = b$. This is mapped onto the points satisfying $\langle a, y\rangle = b|y|^2$. If $b = 0$, this is a hyperplane, and if $b \neq 0$, this is the hypersphere $|y - a/(2b)|^2 = |a|^2/(4b^2)$. This proves the proposition.   □

This result can also be used in the larger space V_∞, where it turns out that the stereographic projection is a restriction of a fractional linear map.

Proposition 4.5.8. *The map*

$$V_\infty \setminus \{e_\infty\} \to V_\infty \setminus \{e_\infty\} : y \mapsto (e_\infty y + 1)(y - e_\infty)^{-1} \tag{4.4}$$

is a self-inverse fractional linear map. Its restriction to $\overline{V}$ coincides with the stereographic projection $\overline{V} \to V$, and its restriction to V coincides with the inverse $V \to \overline{V}$ of the stereographic projection.

Under the identification $V \cup \{\infty\} \leftrightarrow \overline{V}$, *hyperspheres in* V *correspond to hyperspheres in* $\overline{V}$ *not passing through* e_∞, *and hyperplanes in* V *correspond to hyperspheres in* $\overline{V}$ *passing through* e_∞. *By a hypersphere in* $\overline{V}$ *we mean a nontangential intersection of* $\overline{V}$ *and a hyperplane in* V_∞.

Proof. The map (4.4) can be written

$$y \mapsto (e_\infty(y - e_\infty) + 2)(y - e_\infty)^{-1} = e_\infty + 2(y - e_\infty)^{-1},$$

from which it is seen that it is a self-inverse fractional linear map.

If $y = \overline{x} \in \overline{V} \setminus \{e_\infty\}$, then $|\overline{x} - e_\infty|^2 = 2 - 2\langle \overline{x}, e_\infty \rangle = 2(1 - x_\infty)$, and thus

$$(e_\infty \overline{x} + 1)(\overline{x} - e_\infty)^{-1} = \frac{(e_\infty \overline{x} + 1)(\overline{x} - e_\infty)}{2(1 - x_\infty)} = \frac{e_\infty - e_\infty \overline{x} e_\infty + \overline{x} - e_\infty}{2(1 - x_\infty)}$$

$$= \frac{x'}{1 - x_\infty}.$$

On the other hand, if $y = x \in V$, then $|x - e_\infty|^2 = |x|^2 + 1$, and thus

$$(e_\infty x + 1)(x - e_\infty)^{-1} = \frac{(e_\infty x + 1)(x - e_\infty)}{|x|^2 + 1}$$

$$= \frac{e_\infty |x|^2 + x + x - e_\infty}{|x|^2 + 1}$$

$$= \frac{2}{|x|^2 + 1} x + \frac{|x|^2 - 1}{|x|^2 + 1} e_\infty.$$

To prove the mapping properties of the stereographic projection, we use Proposition 4.5.7 on the map (4.4). Given a hypersphere $S : |x - a|^2 = r^2$ in V, we view this as the intersection of V and the hypersphere $\overline{S} : |y - (a + te_\infty)|^2 = r^2 + t^2$ passing through e_∞, where $t := \frac{1}{2}(|a|^2 + 1 - r^2)$. Then (4.4) maps $\overline{S}$ onto a hyperplane not passing through e_∞. Similarly, given a hyperplane $S : \langle a, x \rangle = b$ in V, we view this as the intersection of V and the hyperplane $\overline{S} : \langle a + be_\infty, y \rangle = b$ passing through e_∞. Then (4.4) maps $\overline{S}$ onto a hyperplane passing through e_∞. This proves the proposition. $\qquad\square$

Exercise 4.5.9. Find a fractional linear map of three-dimensional space that maps the unit sphere onto itself and maps $(0, 0, 1/2)$ to the origin.

Definition 4.5.10 (Möbius map). A diffeomorphism $f : \overline{V} \to \overline{V}$ is said to be a *Möbius map* if it maps hyperspheres in $\overline{V}$ onto hyperspheres in $\overline{V}$. A map $f : V \to V$ is called a Möbius map if it extends to a Möbius map of $\overline{V}$. We denote the group of Möbius maps by $\mathrm{M\ddot{o}b}(\overline{V})$, or equivalently $\mathrm{M\ddot{o}b}(V)$.

Definition 4.5.11 (Global conformal map). A differentiable map $f : \overline{V} \to \overline{V}$, or the corresponding map $f : V \cup \{\infty\} \to V \cup \{\infty\}$, is said to be a *global conformal map* if f is conformal at each $p \in \overline{V}$. That $f : V \cup \{\infty\} \to V \cup \{\infty\}$ is globally conformal means that each point in V has a neighborhood, in which either $f(x)$ or $1/f(x)$ is a well-defined conformal map, and that $0 \in V$ has a neighborhood where either $f(1/x)$ or $1/f(1/x)$ is a well-defined conformal map. We here let $1/0 = \infty$ and $1/\infty = 0$, and well defined at x means that $f(x) \neq \infty$.

Theorem 4.5.12. *Assume* $\dim V \geq 2$. *For a diffeomorphism* $f : \overline{V} \to \overline{V}$, *the following are equivalent:*

(i) f *is a fractional linear map.*

(ii) f *is a Möbius map.*

(iii) f *is a global conformal map.*

Note that this is a somewhat remarkable result, in that (i) is an algebraic statement, (ii) is a geometric statement, and (iii) is an analytic statement.

Proof. We have seen that fractional linear maps are Möbius maps as well as global conformal maps. To show that all Möbius maps are conformal, assume that $f : \overline{V} \to \overline{V}$ maps hyperspheres to hyperspheres. Let $a \in \overline{V}$. Composing f by a suitable fractional linear map, we may assume that $a, f(a) \neq \infty$. We have

$$\frac{f(a + \epsilon h) - f(a)}{\epsilon} \to \underline{f}_a(h), \quad \epsilon \to 0,$$

uniformly for $|h| = 1$. Therefore the linear image $\underline{f}_a(S)$ of the unit sphere $S \subset V$ is the uniform limit of spheres, and is there itself a sphere. This is possible only if $\underline{f}_a$ is a multiple of an isometry, proving that f is conformal.

To show that all global conformal maps are fractional linear, assume that $f : \overline{V} \to \overline{V}$ is conformal. For $\dim V \geq 3$ it follows from Liouville's theorem (Theorem 11.4.2) on conformal maps that f is a fractional linear map. If $\dim V = 2$, by composing f with a suitable fractional linear map, we may assume that $f(\infty) = \infty$ and that f is orientation preserving. Identifying V and $\mathbf{C}$, we have in this case a bijective entire analytic function $f : \mathbf{C} \to \mathbf{C}$, and $1/f(1/z)$ is analytic around $z = 0$. It is well known from complex analysis that this implies that $f(z) = az + b$, $z \in \mathbf{C}$, for some $a, b \in \mathbf{C}$, $a \neq 0$. $\qquad\square$

Our next objective is to develop the algebra of fractional linear maps.

Exercise 4.5.13. Generalizing the algebra of fractional linear maps in complex analysis, we represent a map

$$\triangle V \ni w \mapsto (aw + b)(cw + d)^{-1} \in \triangle V, \quad w \in V,$$

where $a, b, c, d \in \triangle V$ are constants, by the matrix $\begin{bmatrix} a & b \\ c & d \end{bmatrix} \in \triangle V(2)$. Show that composition of matrices corresponds to composition of the corresponding maps. Be careful: the algebra is not commutative!

The following definition gives conditions on a, b, c, d such that $(aw + b)(cw + d)^{-1}$ preserves the vectors $V \subset \triangle V$, analogous to the Clifford cone for isometries. In Section 4.6 we will show that this rather ad hoc looking definition is very natural indeed and closely related to the Clifford group for spacetime.

Definition 4.5.14 (Vahlen matrices). Let V be a Euclidean space. A matrix $M = \begin{bmatrix} a & b \\ c & d \end{bmatrix} \in \triangle V(2)$ is called a *Vahlen matrix* if

(i) $a, b, c, d \in \widehat{\triangle} V \cup \{0\}$,

(ii) $a\bar{b}, c\bar{d} \in V = \triangle^1 V$,

(iii) $\Delta(M) := a\bar{d} - b\bar{c} \in \mathbf{R} \setminus \{0\} = \triangle^0 V \setminus \{0\}$.

Define the *Vahlen cone* $\widehat{\triangle} V(2)$ to be the set of Vahlen matrices in $\triangle V(2)$.

Lemma 4.5.15. *The Vahlen matrices* $\widehat{\triangle} V(2)$ *form a multiplicative group in* $\triangle V(2)$, *and the determinant satisfies* $\Delta(M_1 M_2) = \Delta(M_1)\Delta(M_2)$ *for* $M_1, M_2 \in \widehat{\triangle} V(2)$. *If* $M \in \widehat{\triangle} V(2)$ *is a Vahlen matrix, then* $M^{-1} = \frac{1}{\Delta(M)} \begin{bmatrix} \bar{d} & -\bar{b} \\ -\bar{c} & \bar{a} \end{bmatrix}$ *is a Vahlen matrix and*

$$a\bar{b} = b\bar{a} \in V, \quad \bar{a}b = \bar{b}a \in V, \quad c\bar{d} = d\bar{c} \in V, \quad \bar{c}d = \bar{d}c \in V,$$
$$a\bar{c} = c\bar{a} \in V, \quad \bar{a}c = \bar{c}a \in V, \quad b\bar{d} = d\bar{b} \in V, \quad \bar{b}d = \bar{d}b \in V,$$
$$a\bar{d} - b\bar{c} = d\bar{a} - c\bar{b} = \bar{d}a - \bar{b}c = \bar{a}d - \bar{c}b \in \mathbf{R} \setminus \{0\}.$$

Thus either $a, d \in \triangle^{\mathrm{ev}} V$ *and* $b, c \in \triangle^{\mathrm{od}} V$, *or* $a, d \in \triangle^{\mathrm{od}} V$ *and* $b, c \in \triangle^{\mathrm{ev}} V$.

Proof. (i) For the inverse of a Vahlen matrix M, the stated matrix is seen to be a right inverse, hence a left inverse also, which shows that

$$\begin{bmatrix} \Delta(M) & 0 \\ 0 & \Delta(M) \end{bmatrix} = \begin{bmatrix} \bar{d} & -\bar{b} \\ -\bar{c} & \bar{a} \end{bmatrix} \begin{bmatrix} a & b \\ c & d \end{bmatrix} = \begin{bmatrix} \bar{d}a - \bar{b}c & \bar{d}b - \bar{b}d \\ -\bar{c}a + \bar{a}c & -\bar{c}b + \bar{a}d \end{bmatrix}.$$

The diagonal entries show that $\Delta(M) = \bar{d}a - \bar{b}c$. Applying the reversion proves the remaining two determinant formulas.

Clearly $b\bar{a} = \overline{a\bar{b}} \in V$. To prove $\bar{a}b \in V$, we may assume that $a \neq 0$. In this case $V \ni a^{-1}(b\bar{a})a = a^{-1}b(\bar{a}a) = \bar{a}b$, since $a \in \widehat{\triangle} V$ and $a^{-1} = \bar{a}/|a|^2 = \bar{a}/(\bar{a}a)$. Repeating this argument, it suffices to prove that $a\bar{c} \in V$ and $b\bar{d} \in V$. For the first we may assume that $a, c \neq 0$. Consider the equation $a\bar{d} - b\bar{c} = \lambda \in \mathbf{R} \setminus \{0\}$. Multiplying by a^{-1} from the left and $\bar{c}^{-1}$ from the right, we get $\bar{c}^{-1}\bar{d} - a^{-1}b = \lambda(\bar{c}a)^{-1}$, from which $\bar{c}a \in V$ follows and thus $a\bar{c} \in V$. The proof of $b\bar{d} \in V$ is similar.

(ii) Next we show that $\widehat{\triangle} V(2)$ is closed under multiplication. Consider a product

$$M_1 M_2 = \begin{bmatrix} a_1 & b_1 \\ c_1 & d_1 \end{bmatrix} \begin{bmatrix} a_2 & b_2 \\ c_2 & d_2 \end{bmatrix} = \begin{bmatrix} a_1 a_2 + b_1 c_2 & a_1 b_2 + b_1 d_2 \\ c_1 a_2 + d_1 c_2 & c_1 b_2 + d_1 d_2 \end{bmatrix}.$$

To show that $a_1a_2 + b_1c_2 \in \widehat{\triangle}V \cup \{0\}$, we may assume that $a_1, c_2 \neq 0$. In this case

$$a_1a_2 + b_1c_2 = a_1(a_2c_2^{-1} + a_1^{-1}b_1)c_2,$$

where all factors belong to $\widehat{\triangle}V \cup \{0\}$. The proofs for the other three entries are similar.

To show that $(\bar{b}_2\bar{a}_1 + \bar{d}_2\bar{b}_1)(a_1a_2 + b_1c_2) \in V$, it suffices to show that $\bar{b}_2\bar{a}_1b_1c_2 + \bar{d}_2\bar{b}_1a_1a_2 \in V$, where we may assume that $a_2 \neq 0$, for otherwise c_2 is parallel to b_2, by the determinant condition on M_2, and the result follows. We need to show that

$$a_2(\bar{b}_2\bar{a}_1b_1c_2 + \bar{d}_2\bar{b}_1a_1a_2)\bar{a}_2 = (a_2\bar{b}_2)(\bar{a}_1b_1)(c_2\bar{a}_2) + (a_2\bar{d}_2)(\bar{b}_1a_1)(a_2\bar{a}_2)$$
$$= (a_2\bar{b}_2)(2\langle\bar{a}_1b_1, c_2\bar{a}_2\rangle - (c_2\bar{a}_2)(\bar{a}_1b_1)) + (a_2\bar{d}_2)(a_2\bar{a}_2)(\bar{b}_1a_1)$$
$$= 2\langle\bar{a}_1b_1, c_2\bar{a}_2\rangle(a_2\bar{b}_2) + a_2(-\bar{b}_2c_2 + \bar{d}_2a_2)\bar{a}_2(\bar{a}_1b_1)$$

is a vector, which is clear. A similar calculation shows that $\overline{(c_1b_2 + d_1d_2)}(c_1a_2 + d_1c_2) \in V$.

Finally, we calculate the determinant

$$\Delta(M_1M_2) = (a_1a_2 + b_1c_2)(\bar{b}_2\bar{c}_1 + \bar{d}_2\bar{d}_1) - (a_1b_2 + b_1d_2)(\bar{a}_2\bar{c}_1 + \bar{c}_2\bar{d}_1)$$
$$= a_1(a_2\bar{b}_2 - b_2\bar{a}_2)\bar{c}_1 + b_1(c_2\bar{d}_2 - d_2\bar{c}_2)\bar{d}_1 + a_1(a_2\bar{d}_2 - b_2\bar{c}_2)\bar{d}_1 + b_1(c_2\bar{b}_2 - d_2\bar{a}_2)\bar{c}_1$$
$$= a_1\Delta(M_2)\bar{d}_1 - b_1\Delta(M_2)\bar{c}_1 = \Delta(M_1)\Delta(M_2). \qquad \square$$

Theorem 4.5.16 (Representation by Vahlen matrices). *Let V be an n-dimensional Euclidean space. If $M = \begin{bmatrix} a & b \\ c & d \end{bmatrix}$ is a Vahlen matrix, then*

$$T : \overline{V} \to \overline{V} : x \mapsto (ax + b)(cx + d)^{-1}$$

is a well-defined fractional linear map. The map $M \mapsto T$ is a surjective homomorphism from the Vahlen cone $\widehat{\triangle}V(2)$ to the group of fractional linear maps. Its kernel consists of the Vahlen matrices

$$\alpha_0 \begin{bmatrix} 1 & 0 \\ 0 & 1 \end{bmatrix}, \quad \alpha_1 \begin{bmatrix} (-1)^{n+1}e_{\bar{n}} & 0 \\ 0 & e_{\bar{n}} \end{bmatrix},$$

where $0 \neq \alpha_0 \in \mathbf{R}$, $0 \neq \alpha_1 \in \mathbf{R}$.

Proof. To show that T is a fractional linear map, assume first that $c \neq 0$. Then

$$(ax + b)(cx + d)^{-1} = \big(ac^{-1}(cx + d) + (b - ac^{-1}d)\big)(cx + d)^{-1}$$
$$= ac^{-1} + (b - a\bar{c}d|c|^{-2})(x + c^{-1}d)^{-1}c^{-1}$$
$$= ac^{-1} + (b\bar{c} - a\bar{d})c|c|^{-2}(x + c^{-1}d)^{-1}c^{-1}$$
$$= ac^{-1} - (\Delta(M)/|c|^2)\,c(x + c^{-1}d)^{-1}c^{-1},$$

using that $\bar{c}d = \overline{c}\overline{d} = \bar{d}c$, since $\bar{c}d \in V$. Thus T is a composition of a translation, an inversion, an isometry, a dilation, and finally a translation. If $c = 0$, then $d^{-1} = |d|^{-2}\bar{d} = |d|^{-2}\Delta(M)a^{-1}$, and

$$(ax + b)(cx + d)^{-1} = (\Delta(M)/|d|^2)axa^{-1} + bd^{-1},$$

which is an isometry, a dilation, and a translation. In each case we have a fractional linear map.

It follows from Exercise 4.5.13 that $M \mapsto T$ is a homomorphism. Surjectivity is clear, since the Vahlen matrices

$$\begin{bmatrix} \hat{q} & 0 \\ 0 & q \end{bmatrix}, \quad \begin{bmatrix} 1 & v \\ 0 & 1 \end{bmatrix}, \quad \begin{bmatrix} \alpha & 0 \\ 0 & 1 \end{bmatrix}, \quad \begin{bmatrix} 0 & 1 \\ 1 & 0 \end{bmatrix},$$

represent isometries, translations, dilations, and inversions respectively.

To find the kernel, assume that $Tx = x$ for all $x \in V$, that is, $(ax+b) = x(cx+d)$. Letting $x = 0$ shows that $b = 0$, and $x = \infty$ shows that $c = 0$. The equation $ax = xd$ can hold for all vectors only if $a = \alpha_0 + \alpha_1 e_{\overline{n}}$ and $d = \alpha_0 + (-1)^{n-1}\alpha_1 e_{\overline{n}}$, as is seen by expressing a and d in an induced ON-basis and using the fact that if $s \neq \varnothing$ and $s \neq \overline{n}$, then there exist $i \in s$ and $j \notin s$ such that e_s has different commutation relations with e_i and e_j. Checking the conditions $a\overline{d} \in \mathbf{R} \setminus \{0\}$ and $a, d \in \widehat{\triangle}V$ in the four possible different dimensions modulo 4 shows that either $\alpha_0 = 0$ or $\alpha_1 = 0$. This proves the theorem. $\square$

Example 4.5.17. To find a fractional linear map that has certain desired mapping properties in an n-dimensional Euclidean space, one proceeds as in complex analysis, the only difference being that circles and lines are replaced by hyperspheres and hyperplanes. For example, to find the map

$$(e_\infty y + 1)(y - e_\infty)^{-1} = e_\infty + 2(y - e_\infty)^{-1}$$

from the properties that it should map the sphere $\overline{V}$ to the hyperplane V in V_∞, one composes the following maps. First the translation $x \mapsto x - e_\infty$, which maps e_∞ to 0, followed by inversion $x \mapsto 1/x$, which further maps this point to ∞, and also the translated sphere to the hyperplane $x_\infty = -1/2$. Finally, one dilates and translates by $x \mapsto 2x + e_\infty$ to obtain the image V of the sphere. In total, this gives

$$2(y - e_\infty)^{-1} + e_\infty = \left(2 + e_\infty(y - e_\infty)\right)(y - e_\infty)^{-1} = (e_\infty y + 1)(y - e_\infty)^{-1}.$$

Exercise 4.5.18. Extend Exercise 4.5.3 to cover general fractional linear maps. Use the factorization T in the proof of Theorem 4.5.16 to show that every fractional linear map T is conformal, with derivative

$$\underline{T}_x(h) = \frac{\Delta(M)}{|cx + d|^2}(cx + d)h(cx + d)^{-1}.$$

The general theory for fractional linear maps of Euclidean spaces can be expressed with complex numbers in the plane and with quaternions in three-dimensional space using the standard geometric representations of $\mathbf{C}$ and $\mathbf{H}$ from Section 3.2.

Example 4.5.19 (2D Möbius maps). Let $\dim V = 2$ and fix an ON-basis $\{e_1, e_2\}$. This gives an identification

$$V \in x = e_1 z \longleftrightarrow e_1 x = z \in \mathbf{C} = \triangle^{\mathrm{ev}} V$$

between vectors x and complex numbers z. On the one hand, if $(ax + b)(cx + d)^{-1}$ is a fractional linear map such that $a, d \in \mathbf{C}$ and $b, c \in V$, then in the complex representation of vectors this corresponds to

$$z \mapsto e_1(ae_1 z + b)(ce_1 z + d)^{-1} = (w_1 z + w_2)(w_3 z + w_4)^{-1},$$

where $w_1 := e_1 a e_1 = \bar{a}$, $w_2 := e_1 b$, $w_3 := ce_1$, and $w_4 := d$. The conditions on a, b, c, d translate to $w_i \in \mathbf{C}$ and $d\bar{a} - c\bar{b} = w_4 w_1 - w_3 e_1 \overline{w}_2 e_1 = w_1 w_4 - w_2 w_3 \in \mathbf{R} \setminus \{0\}$.

On the other hand, if $a, d \in V$ and $b, c \in \mathbf{C}$, then multiplying the two parentheses by e_1 from the right shows that the map is

$$z \mapsto e_1(a(e_1 z e_1) + be_1)(c(e_1 z e_1) + de_1)^{-1} = (w_1 \bar{z} + w_2)(w_3 \bar{z} + w_4)^{-1},$$

where $w_1 := e_1 a$, $w_2 := \bar{b}$, $w_3 := c$, and $w_4 := de_1$, and it follows that $w_i \in \mathbf{C}$ and $w_1 w_4 - w_2 w_3 \in \mathbf{R} \setminus \{0\}$. These two cases correspond to orientation-preserving and orientation-reversing Möbius maps respectively.

Example 4.5.20 (3D Möbius maps). Let $\dim V = 3$ and fix a volume element $J \in \triangle^3 V$ such that $J^2 = -1$. This gives an identification

$$V \ni x = -Jz = -zJ \longleftrightarrow z = Jx = xJ \in \mathbf{H} \cap \triangle^2 V$$

between vectors and pure quaternions, with the Hodge star map. Let $(ax + b)(cx + d)^{-1}$ be a fractional linear map. In this case we may assume that $a, d \in \triangle^{\mathrm{ev}} V$ and $b, c \in \triangle^{\mathrm{od}} \in V$, after possibly multiplying all coefficients by J. In quaternion representation, the fractional linear map is

$$z \mapsto J(-aJz + b)(-cJz + d)^{-1} = (q_1 z + q_2)(q_3 z + q_4)^{-1},$$

where $q_1 := a$, $q_2 := Jb$, $q_3 := -cJ$, and $q_4 := d$. The conditions on a, b, c, d translate to $q_i \in \mathbf{H}$ and $q_1 \bar{q}_4 + q_2 \bar{q}_3 \in \mathbf{R} \setminus \{0\}$. Orientation-preserving maps arise when $q_1 \bar{q}_4 + q_2 \bar{q}_3 > 0$, and orientation-reversing maps when $q_1 \bar{q}_4 + q_2 \bar{q}_3 < 0$.

4.6   Mappings of the Celestial Sphere

In this section we prove that there is a $2-1$ homomorphism taking spacetime isometries to Euclidean Möbius maps. To see the connection with Lorentz isometries, we identify the higher-dimensional Riemann sphere $\overline{V}$ with the celestial sphere in a spacetime W.

Definition 4.6.1 (Celestial sphere). Given a vector space W, the *projective space* $P(W)$ is

$$P(W) := \{[x] \; ; \; 0 \neq x \in W\},$$

so that an object in $P(W)$ is a one-dimensional line through the origin in W. Given a subspace $V_\infty \subset W$ of codimension one and a vector $e_0 \in W \setminus V_\infty$, we identify V_∞ and the subset $P(V_\infty - e_0) \subset P(W)$ with the injective map

$$V_\infty \to P(V_\infty - e_0) : v \mapsto [v - e_0].$$

The subset $P(V_\infty) := \{[v] \; ; \; 0 \neq v \in V_\infty\}$ is the complement of V_∞ in $P(W)$, and is referred to as the *hyperplane at infinity*.

If W is a spacetime, we assume that $e_0 \in W_{t+}$ and that $V_\infty = [e_0]^\perp$ is a space-like hyperplane. The *celestial sphere* is the image $P(W_l)$ of the light cone W_l, which coincides with the unit sphere $\overline{V}$ in V_∞ under the identification $P(V_\infty - e_0) \leftrightarrow V_\infty$. See Figure 4.5.

The relativity theory interpretation of this is as follows. Consider an inertial observer O with world line spanned by the future-pointing vector e_0. At the event in spacetime represented by the origin, what O can observe is all light sent out by past events, traveling exactly at the speed of light and reaching the eyes of O at the origin. For this to happen, the past events must lie on the past light cone W_{l-}. If we write the past event as $v - te_0 \in W_{l-}$, $v \in V_\infty$, then v represents the direction from which the light ray reaches O, and $t = |v|$ means that the time it took the light to reach O is proportional to the space distance to the past event. Thus the point $[v - te_0] \in P(W_l)$ represents all superimposed past events, the light from which O sees in direction v. In this way, the celestial sphere $P(W_l) \subset V_\infty$ represents what O observes of the universe W at the origin event.

Definition 4.6.2 (Induced map of the sphere). Let W be a spacetime. Fix a future-pointing vector $e_0 \in W_{t+}$ and the Euclidean subspace $V_\infty := [e_0]^\perp$, and identify the celestial sphere $P(W_l)$ and the unit sphere $\overline{V}$ in V_∞. Let $T \in O(W)$ be an isometry. Then T maps the light cone W_l onto itself, and thus induces a diffeomorphism $T_{\overline{V}} : \overline{V} \to \overline{V}$. We define this *induced map of the celestial sphere* by

$$[T_{\overline{V}}(v) - e_0] = [T(v - e_0)], \quad v \in \overline{V}.$$

As in Section 4.5, we fix a *zenith direction* $e_\infty \in \overline{V}$, write $V := V_\infty \cap [e_\infty]^\perp$, and identify $\overline{V}$ and $V \cup \{\infty\}$ under stereographic projection. Write T_V for the map in V corresponding to $T_{\overline{V}}$.

Proposition 4.6.3. *If $T : W \to W$ is an isometry, then the induced map $T_{\overline{V}} : \overline{V} \to \overline{V}$ of the celestial sphere is well defined and is a Möbius map, or equivalently, T_V is a fractional linear map.*

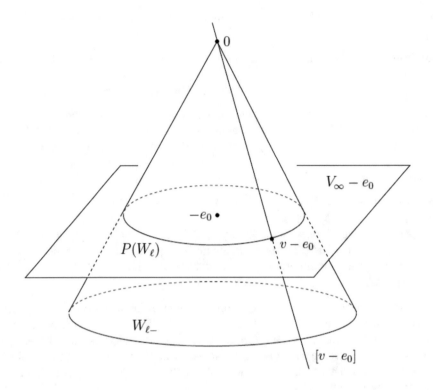

Figure 4.5: The celestial sphere and the past spacetime light cone.

Proof. Take any hyperplane $P \subset V_\infty$, and let P_W be the unique hyperplane in W passing through the origin in W and intersecting $V_\infty - e_0$ along P. Then the intersection of P_W and W_l corresponds to the hypersphere in $\overline{V}$, the intersection of P and $\overline{V}$. Since T is linear, it maps P_W onto another hyperplane $T(P_W)$ passing through the origin in W and intersecting $V_\infty - e_0$ along a hyperplane P'. Then by the definition of $T_{\overline{V}}$, the intersection $P' \cap \overline{V}$ is the image $T_{\overline{V}}(P \cap \overline{V})$. This proves that $T_{\overline{V}}$ maps hyperspheres to hyperspheres. $\square$

We wish to obtain an algebraic expression for T_V, given the spacetime rotor representing T.

Proposition 4.6.4 (Möbius reflection). *Fix an ON-basis $\{e_0, e_1, \ldots, e_n, e_\infty\}$ for spacetime W, and consider the reflection*

$$Tv = -(a + a_\infty e_\infty + a_0 e_0)v(a + a_\infty e_\infty + a_0 e_0)^{-1}, \quad v \in W$$

in the hyperplane orthogonal to $a + a_\infty e_\infty + a_0 e_0 \in W$, where $a \in V$, $V = [e_\infty \wedge$

$e_0]^\perp \subset V_\infty$, and $a_0^2 \neq |a|^2 + a_\infty^2$. Then

$$T_V(x) = (-ax + a_-)(a_+ x + a)^{-1}, \quad x \in V,$$

where $a_+ := a_\infty + a_0$ *and* $a_- := a_\infty - a_0$.

Proof. Consider first the induced map on $T_{\overline{V}}$ and write $a' := a + a_\infty e_\infty$. Let $x \in \overline{V}$ and consider the point $-e_0 + x \in W_{l-}$. This is mapped by T to

$$
\begin{aligned}
&- (a_0 e_0 + a')(-e_0 + x)(a_0 e_0 + a')^{-1} = \lambda(a_0 + a_0 e_0 x - a' e_0 + a' x)(a_0 e_0 + a') \\
&= \lambda(a_0^2 e_0 + a_0^2 x + a_0 a' + a_0 a' x e_0 + a_0 a' + a_0 e_0 x a' + |a'|^2 e_0 + a' x a') \\
&= \lambda((a_0^2 + 2a_0\langle a', x\rangle + |a'|^2)e_0 + (a_0^2 x + 2a_0 a' + a' x a')) \in W_l,
\end{aligned}
$$

where $\lambda \in \mathbf{R} \setminus \{0\}$. By normalizing the e_0 coordinate to -1, this means that

$$
\begin{aligned}
T_{\overline{V}}(x) &= -\frac{a_0^2 x + 2a_0 a' + a' x a'}{a_0^2 + 2a_0\langle a', x\rangle + |a'|^2} = -\frac{(a' x + a_0)(a_0 x + a')}{|a_0 x + a'|^2} \\
&= -(a' x + a_0)(a_0 x + a')^{-1}.
\end{aligned}
$$

Using matrix representation as in Exercise 4.5.13, through stereographic projection, in V this corresponds to the map

$$
\begin{aligned}
&\begin{bmatrix} e_\infty & 1 \\ 1 & -e_\infty \end{bmatrix} \begin{bmatrix} a' & a_0 \\ -a_0 & -a' \end{bmatrix} \begin{bmatrix} e_\infty & 1 \\ 1 & -e_\infty \end{bmatrix} \\
&= \begin{bmatrix} e_\infty a' e_\infty - a' & e_\infty a' + a' e_\infty - 2a_0 \\ a' e_\infty + e_\infty a' + 2a_0 & a' - e_\infty a' e_\infty \end{bmatrix} = \begin{bmatrix} -2a & 2(a_\infty - a_0) \\ 2(a_\infty + a_0) & 2a \end{bmatrix}.
\end{aligned}
$$

This proves that the map in V is $x \mapsto (-ax + a_-)(a_+ x + a)^{-1}$, which is a fractional linear map by Theorem 4.5.16. $\qquad\square$

We next consider some applications to special relativity theory.

Let $\{e_0, e_1, \ldots, e_n, e_\infty\}$ be an ON-basis for an inertial observer O. This means that the e_0 coordinate is the time of an event that O measures and that the e_∞ coordinate measures how far in the zenith direction in space the event lies as measured by O, and similarly for the other space coordinates. Assume that O' is another inertial observer passing by O at the origin event at relativistic speed, having an ON-basis $\{e_0', e_1', \ldots, e_n', e_\infty'\}$ relative to which he measures time and space. Denote by T the isometry that maps $T(e_i) = e_i'$, and let $A = (\alpha_{i,j})$ denote the matrix of T relative to the basis $\{e_i\}$, that is, $T(e_i) = \sum_j e_j \alpha_{j,i}$.

Consider an event with coordinates $X = \{x_i\}$ in the basis $\{e_i\}$ as observed by O. The same event observed by O' has coordinates $Y = \{y_i\}$ in the basis $\{e_i'\}$, where $Y = A^{-1}X$. If O and O' compare their observations by identifying $e_i = e_i'$, then passing from O's observation to O''s observation defines the *experienced map*

$$X \mapsto A^{-1}X,$$

in the basis $\{e_i\}$, that is, T^{-1}. In particular, the map taking O's observation of the celestial sphere to O''s observation of the celestial sphere is $T_{\overline{V}}^{-1}$.

Compare this result to how one experiences a usual rotation T in three-dimensional Euclidean space: if one does not realize that one has rotated by T, it looks as if space has been rotated by T^{-1}. Sometimes these two points of view are referred to as *active* and *passive* transformations. The above result is the analogue for the hyperbolic rotations of spacetime.

Example 4.6.5 (Watching a Lorentz boost). Consider four dimensional spacetime in relativity, and an inertial observer O with ON-basis $\{e_0, e_1, e_2, e_\infty\}$. Consider another observer O', viewed by O as travelling with speed $\tanh(\phi) > 0$ toward the north pole e_∞. As in Example 4.4.1, the Lorentz boost taking O:s ON-basis to O':s ON-basis is

$$Tv = \exp(\phi e_{0\infty}/2)v\exp(-\phi e_{0\infty}/2)$$
$$= (\cosh(\phi/2)e_\infty + \sinh(\phi/2)e_0)e_\infty ve_\infty^{-1}(\cosh(\phi/2)e_\infty + \sinh(\phi/2)e_0)^{-1}.$$

Computing the induced fractional linear map on $V = \operatorname{span}\{e_1, e_2\}$ by two applications of Proposition 4.6.4, we get

$$x \mapsto x^{-1} \mapsto (\cosh(\phi/2) - \sinh(\phi/2))/(\cosh(\phi/2) + \sinh(\phi/2))x = \exp(-\phi)x.$$

The experienced Möbius map of the celestial sphere $\overline{V}$, going from O's to O''s observation, is the inverse of this map corresponding to dilation $x \mapsto \exp(\phi)x$ on V. This means that the faster an observer travels in the direction e_∞, the more he will see the stars move in this direction! See Figure 4.6. Note that the two fixed points of the celestial sphere, $\pm e_\infty$, correspond to the two eigenvectors of the Lorentz boost on the light cone.

Example 4.6.6 (Relativistic sphere paradox). Consider a Lorentz boost from O to O' as in Example 4.6.5, and consider a spherical object at rest relative to O' described by the equation $(x'_\infty)^2 + (x'_1 - a_1)^2 + (x'_2 - a_2)^2 = 1$. As the observer O' passes O, O will see the object passing by at the same speed v as O' in the direction e_∞. However, according to O's measurements, it will not be a spherical object but rather an ellipsoid. Indeed, with the Lorentz boost in Example 4.4.1 in the $\{e_0, e_\infty\}$ plane, O will at time t describe the object by the equation

$$((x_\infty - vt)/\sqrt{1 - v^2})^2 + (x_1 - a_1)^2 + (x_2 + a_2)^2 = 1,$$

which is an ellipsoid that is shorter in the e_∞ direction. However, an amazing phenomenon occurs due to the finite speed of light. Even though O measures the object to be an ellipsoid after taking into account the finite propagation speed of light, the image he sees of the object is a circular shape, just like O', although any pattern on the surface of the sphere would be distorted. This is clear from Proposition 4.6.3, since the circular shape O' sees of the object is mapped by

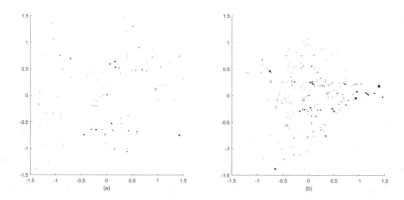

Figure 4.6: A selection of stars and constellations on the northern celestial sphere, with the Polar Star at $(0,0)$ and scale 1 corresponding to angle $\pi/4$. (a) Our view from Earth showing Cygnus, Cepheus, Cassiopeia, Perseus, Ursa Minor, Auriga, and Ursa Major. (b) The same view for an inertial observer passing Earth at $2/3$ the speed of light towards the Polar Star, showing also Aquila, Pegasus, Andromeda, Pisces, Taurus, Orion, Gemini, Leo, Virgo, Boötes, and Hercules. Note that since the constellations have changed by a conformal map, there is little distortion between the views. Note also that at speeds close enough to that of light, even the Southern Cross will move up from behind and be visible next to the Polar Star.

a Möbius map to another circular shape of the object on O's celestial sphere! This phenomenon occurs only for spherical objects. Differently shaped objects will become distorted by the Möbius map.

Example 4.6.7 (Watching a parabolic rotation)**.** Consider four-dimensional spacetime in relativity with ON-basis $\{e_0, e_1, e_2, e_\infty\}$ and the parabolic rotation

$$Tv = \exp(\phi(e_0 - e_\infty)a/2)v\exp(-\phi(e_0 - e_\infty)a/2),$$

where $a = a_1 e_1 + a_2 e_2$ is a unit vector in V. The rotor here is $q = \exp(\phi(e_0 - e_\infty)a/2) = (a + \frac{\phi}{2}(e_0 - e_\infty))a$. Proposition 4.6.4 applied twice shows that T induces a map of the celestial sphere that corresponds to the fractional linear map

$$x \mapsto -axa^{-1} \mapsto (-a(-axa^{-1}) - \phi)a^{-1} = x - \phi a$$

of V. Thus a parabolic spacetime rotation translates the celestial sphere, fixing only the north pole e_∞, which corresponds to the eigenvector of the rotation on the light cone.

We next turn to a more detailed study of the fractional linear maps appearing in Proposition 4.6.4. Collecting the four coefficients in a matrix, we have a map of spacetime vectors a:

$$C : a + a_\infty e_\infty + a_0 e_0 \mapsto \begin{bmatrix} -a & a_\infty - a_0 \\ a_\infty + a_0 & a \end{bmatrix}. \tag{4.5}$$

In fact, we already used such a map in the proof of Lemma 3.4.14, so the following should not come as a surprise.

Proposition 4.6.8 (Vahlen spacetime algebra). *Let W be a spacetime, and fix an orthogonal splitting $W = V \oplus [e_\infty] \oplus [e_0]$ as above. Let $\triangle V(2)$ denote the algebra of 2×2 matrices with coefficients in $\triangle V$. Then $(\triangle V(2), C)$, where C is defined in (4.5), is a Clifford algebra for W.*

Thus there is a unique algebra isomorphism $\triangle W \to \triangle V(2)$ that identifies $a + a_\infty e_\infty + a_0 e_0 \in W \subset \triangle W$ and $C(a + a_\infty e_\infty + a_0 e_0) \in \triangle V(2)$.

Proof. The basis vectors map to

$$C(e_\infty) = \begin{bmatrix} 0 & 1 \\ 1 & 0 \end{bmatrix}, \quad C(e_0) = \begin{bmatrix} 0 & -1 \\ 1 & 0 \end{bmatrix}, \quad C(e_j) = \begin{bmatrix} -e_j & 0 \\ 0 & e_j \end{bmatrix}, \quad j = 1, \dots, n.$$

The Clifford condition (C) is straightforward to verify. To verify (U), which is needed only when $n \equiv 3 \pmod 4$, it suffices by Proposition 3.3.3 to show that the volume element is not scalar. We have

$$C(e_0)C(e_\infty)C(e_1)\cdots C(e_n) = \begin{bmatrix} (-1)^{n+1} e_{\overline{n}} & 0 \\ 0 & e_{\overline{n}} \end{bmatrix},$$

which is not scalar.   □

To understand the connection between spacetime isometries and fractional linear operators, we need to identify the Clifford cone in $\triangle V(2)$. We use the following.

Lemma 4.6.9. *Let $M = \begin{bmatrix} a & b \\ c & d \end{bmatrix} \in \triangle V(2)$. Then the involution and reversion in the Clifford algebra $\triangle V(2) = \triangle W$ are*

$$\widehat{\begin{bmatrix} a & b \\ c & d \end{bmatrix}} = \begin{bmatrix} \widehat{a} & -\widehat{b} \\ -\widehat{c} & \widehat{d} \end{bmatrix}, \quad \overline{\begin{bmatrix} a & b \\ c & d \end{bmatrix}} = \begin{bmatrix} \widetilde{d} & \widetilde{b} \\ \widetilde{c} & \widetilde{a} \end{bmatrix}.$$

Proof. Recall that the involution and reversion act by $(-1)^k$ and $(-1)^{k(k-1)/2}$ on $\triangle^k W$. The corresponding subspace in $\triangle V(2)$ is spanned by

$$e_s = \begin{bmatrix} \widehat{e}_s & 0 \\ 0 & e_s \end{bmatrix}, \quad e_\infty e_t = \begin{bmatrix} 0 & e_t \\ \widehat{e}_t & 0 \end{bmatrix}, \quad e_0 e_{t'} = \begin{bmatrix} 0 & -e_{t'} \\ \widehat{e}_{t'} & 0 \end{bmatrix}, \quad e_0 e_\infty e_{t''} = \begin{bmatrix} -\widehat{e}_{t''} & 0 \\ 0 & e_{t''} \end{bmatrix},$$

where $|s| = k$, $|t| = |t'| = k - 1$, and $|t''| = k - 2$, and the result follows by inspection.   □

Proposition 4.6.10 (Vahlen = Clifford cone). *The Vahlen cone* $\widehat{\triangle V}(2)$ *is isomorphic to the Clifford cone* $\widehat{\triangle W}$ *under the isomorphism of Clifford algebras* $\triangle V(2) = \triangle W$ *determined by* (4.5). *In particular, every Vahlen matrix in* $\widehat{\triangle V}(2)$ *can be written as a product of at most* $\dim V + 2$ *matrices of the form* (4.5), *with* $a \in V$, $a_0, a_\infty \in \mathbf{R}$, *and* $a_0^2 \neq a_\infty^2 + |a|^2$.

Proof. To show that $\widehat{\triangle W} \subset \widehat{\triangle V}(2)$, note that spacetime vectors belong to $\widehat{\triangle V}(2)$ and that $\widehat{\triangle V}(2)$ is closed under multiplication by Lemma 4.5.15. Next consider the converse inclusion $\widehat{\triangle V}(2) \subset \widehat{\triangle W}$. Let $M = \begin{bmatrix} a & b \\ c & d \end{bmatrix}$ be a Vahlen matrix, and thus invertible. By Proposition 4.1.5 it suffices to prove, for all $v \in V$, $\alpha, \beta \in \mathbf{R}$, that

$$\begin{bmatrix} \widehat{a} & -\widehat{b} \\ -\widehat{c} & \widehat{d} \end{bmatrix} \begin{bmatrix} -v & \alpha \\ \beta & v \end{bmatrix} \begin{bmatrix} \overline{d} & -\overline{b} \\ -\overline{c} & \overline{a} \end{bmatrix}$$

is a spacetime vector, that is, the off-diagonal entries are scalars and the diagonals are $\pm$ a vector. By linearity, it suffices to check the cases in which only one of v, α, and β is nonzero. If $v = \beta = 0, \alpha = 1$, then the product is $\begin{bmatrix} -\widehat{a}\overline{c} & \widehat{a}\overline{a} \\ \widehat{c}\overline{c} & -\widehat{c}\overline{a} \end{bmatrix}$, which is a spacetime vector since $\widehat{a}\overline{c} = -\widehat{\overline{a}\overline{c}} = -\widehat{c}\overline{a}$, since $\widehat{a}\overline{c} \in V$. The case $v = \alpha = 0, \beta = 1$ is similar. For $\alpha = \beta = 0$, we get

$$\begin{bmatrix} -\widehat{a}v\overline{d} + \widehat{b}v\overline{c} & \widehat{a}v\overline{b} - \widehat{b}v\overline{a} \\ \widehat{c}v\overline{d} - \widehat{d}v\overline{c} & -\widehat{c}v\overline{b} + \widehat{d}v\overline{a} \end{bmatrix}.$$

To show that the lower left entry is a scalar, we may assume that $d \neq 0$. In this case, the question is whether

$$\overline{d}(\widehat{c}v\overline{d} - \widehat{d}v\overline{c})\widehat{d} = (\overline{d}\widehat{c})v(\overline{d}\widehat{d}) - (\overline{d}\widehat{d})v(\overline{c}\widehat{d}) = (\overline{d}\widehat{d})((\overline{d}\widehat{c})v - v(\overline{c}\widehat{d})) = (\overline{d}\widehat{d})((\overline{d}\widehat{c})v + v(\overline{d}\widehat{c}))$$

is scalar. This is clear, since $\overline{d}\widehat{c} \in V$ and $\overline{d}\widehat{d} \in \mathbf{R}$. Note that $\overline{c}\widehat{d} = -\widehat{\overline{c}\widehat{d}} = -\overline{d}\widehat{c}$, since $\overline{c}\widehat{d} \in V$. To show that the upper left entry is a vector, we may assume that $b \neq 0$, for otherwise, a and d are parallel, and the result follows. We need to show that

$$\overline{b}(-\widehat{a}v\overline{d} + \widehat{b}v\overline{c})\widehat{b} = -(\overline{b}\widehat{a})v(\widehat{d}\overline{b}) + (\overline{b}\widehat{b})v(\overline{c}\widehat{b}) = (v(\overline{b}\widehat{a}) - 2\langle v, \overline{b}\widehat{a}\rangle)(\widehat{d}\overline{b}) + v(\overline{b}\widehat{b})\overline{c}\widehat{b}$$
$$= v(\overline{b}\widehat{a})(\widehat{d}\overline{b}) + v(\overline{b}\widehat{b})\overline{c}\widehat{b} - 2\langle v, \overline{b}\widehat{a}\rangle(\widehat{d}\overline{b})$$
$$= -v\widehat{\overline{b}}(a\overline{d} - b\overline{c})\widehat{b} - 2\langle v, \overline{b}\widehat{a}\rangle(\widehat{d}\overline{b})$$

is a vector, which is clear. The right entries are shown to be scalars and vectors similarly. Finally,

$$-\widehat{a}v\overline{d} + \widehat{b}v\overline{c} = \widehat{\overline{a}v\overline{d}} - \widehat{\overline{b}v\overline{c}} = \widehat{c}v\overline{b} - \widehat{d}v\overline{a}.$$

This completes the proof of $\widehat{\triangle V}(2) = \widehat{\triangle W}$, from which the factorization result for Vahlen matrices follows by Proposition 4.1.5. $\square$

We end by summarizing the relation between spacetime isometries and Möbius maps. Let W be an n-dimensional spacetime, and let $V \leftrightarrow \overline{V} \subset V_\infty \subset W$ represent the celestial sphere as above. Then we have group homomorphisms

$$
\begin{array}{ccc}
q \in \widehat{\triangle}W & \xrightarrow{\;p_W\;} & O(W) \ni T \\[4pt]
\rho_0 \downarrow & & \downarrow \rho_1 \\[4pt]
M \in \widehat{\triangle}V(2) & \xrightarrow{\;p_V\;} & \mathrm{M\ddot{o}b}(V) \ni f
\end{array}
$$

such that $\rho_1 \circ p_W = p_V \circ \rho_0$.

- The map ρ_0 from Proposition 4.6.10 is an isomorphism.

- The map p_W as in Proposition 4.1.9 is surjective and has kernel $\triangle^0 W \setminus \{0\}$.

- The map p_V from Theorem 4.5.16 is surjective and has kernel $\rho_0(\triangle^0 W \setminus \{0\}) \cup \rho_0(\triangle^n W \setminus \{0\})$.

- The map ρ_1 from Proposition 4.6.3 is surjective and has kernel ± 1.

That $\rho_1 \circ p_W = p_V \circ \rho_0$ is straightforward to verify. We have seen the mapping properties for ρ_0, p_W, and p_V, and those for ρ_1 follow from this. Note that if we normalize $\widehat{\triangle}W$ to $\mathrm{Pin}(W)$, and also normalize $\widehat{\triangle}V(2)$ correspondingly, then p_W and ρ_1 are $2-1$ maps, p_V is a $4-1$ map, and ρ_0 is a $1-1$ map.

4.7   Comments and References

4.1 Our proof of the Cartan–Dieudonné theorem (Theorem 4.1.3) is from Grove [49], with a minor simplification using a continuity argument.

What we here call the Clifford cone $\widehat{\triangle}V$ is usually referred to as the *Clifford group* in the literature. Our terminology has been chosen to go together with the notion of the Grassmann cone $\widehat{\wedge}V$. The orthogonal and special orthogonal groups and their abbreviations are standard, and the name of the spin group is due to its connection to physics. J.-P. Serre introduced the name Pin by removing the S in Spin, in analogy with the orthogonal group.

The terminology *rotor* for an object in the spin group is not used in the literature.

4.2 Example 4.2.6, showing that the exponential map is not surjective for an ultrahyperbolic inner product space, is taken from [78].

4.4 The source of inspiration for the treatment of rotations in space and spacetime with Clifford algebra is M. Riesz [78], where most of the results are found.

4.5 Our treatment of Möbius maps with Clifford algebra follows L. V. Ahlfors [1]. Theorem 4.5.12 is from Hertrich–Jeromin [54].

4.6 I want to thank Malcolm Ludvigsen, whom I had as a teacher in a course in cosmology at Linköping University as an undergraduate student and who shared with us students many fascinating insights into relativity theory, including the sphere paradox described in Example 4.6.6.

Chapter 5

Spinors in Inner Product Spaces

Prerequisites:

This chapter builds on Chapters 3 and 4, and uses the material in Sections 1.4 and 1.5. Any knowledge of representation theory is helpful, but the presentation is self-contained and should be accessible to anyone with a solid background in linear algebra.

Road map:

In a certain sense, one can form a square root

$$\triangle\!\!\!\!/\, V = \sqrt{\triangle V}$$

of the Clifford algebra of a given inner product space V. Indeed, we have seen in Section 3.4 that $\triangle V$ is isomorphic to a matrix algebra. For example, the Clifford algebra of spacetime W with three space dimensions is isomorphic to

$$\mathcal{L}(\mathbf{R}^4) = \mathbf{R}^4 \otimes \mathbf{R}^4,$$

so in this sense, $\mathbf{R}^4 = \sqrt{\triangle W}$. Such spaces are referred to as spinor spaces, and have they deep applications in both physics and mathematics. Two problems with this construction need to be addressed, though. The first is that depending on the dimension and signature of the inner product space, by Theorem 3.4.13 the coefficients in the matrices may belong to $\mathbf{R}$, $\mathbf{R}^2$, $\mathbf{C}$, $\mathbf{H}$, or $\mathbf{H}^2$. However, the standard construction of spinor spaces is over the complex field, and indeed, complex Clifford algebras are always isomorphic to matrix algebras over $\mathbf{C}$ or $\mathbf{C}^2$ depending on the parity of the dimension, which simplifies matters.

The second and more fundamental problem is that isomorphisms $\triangle V_c \leftrightarrow \mathcal{L}(S)$ are a priori not unique. To show that a spinor space S has an invariant geometric meaning, we need to show that different choices of matrices used in setting up the isomorphism amount to only a renaming of the elements in S. This

© Springer Nature Switzerland AG 2019
A. Rosén, *Geometric Multivector Analysis*, Birkhäuser Advanced Texts Basler Lehrbücher,
https://doi.org/10.1007/978-3-030-31411-8_5

is the purpose of Section 5.2, which, through the principle of abstract algebra, completes the construction of the complex spinor space $S = \not{\triangle}V$ of an inner product space V.

In Section 5.3 we show how to map spinors between different spaces, something that becomes important, for example, in considering spinors on manifolds. The spinor space $\not{\triangle}V$ is a fundamental construction for the celebrated Atiyah–Singer index theorem for Dirac operators, which we look at in Chapter 12. In physics, spinors are famous for describing a certain intrinsic state for some elementary particles such as electrons in quantum mechanics, which we look at briefly in Section 9.2.

Spinor spaces generalize to higher dimensions, a topological feature of the complex square root

$$z \mapsto \sqrt{z}.$$

As we know, the complex square root has two possible values differing by sign. If we start at $\sqrt{1} = 1$ and move continuously around the unit circle, then we have $\sqrt{e^{2\pi i}} = -1$ and $\sqrt{e^{4\pi i}} = 1$. This means that it takes two full rotations for the square root to return to its original value. In higher dimensions, this is the characteristic behavior of spinor spaces: two full rotations of the physical, or vector, space are needed to return the spinors to their original positions. Beyond the standard linear representation $\not{\triangle}V$, there are countably infinitely many nonisomorphic spinor spaces possible for a given inner product space. This is the topic of Section 5.4.

Highlights:

- The standard representation of spinors: 5.1.5

- The main invariance theorem for spinors: 5.2.3

- Induced spinor maps: 5.3.5

- Finding all abstract spinor spaces in three and four dimensions: 5.4.2, 5.4.8

5.1   Complex Representations

We start with the following general idea of a spinor space.

Definition 5.1.1 (Group representation). Let V be a real inner product space, and consider the spin group $\mathrm{Spin}(V) \subset \triangle^{\mathrm{ev}}V$. A *complex* $\mathrm{Spin}(V)$ *representation* is a smooth group homomorphism

$$\rho : \mathrm{Spin}(V) \to \mathcal{L}(S)$$

into the space of linear operators on a complex linear space S. Consider $-1 \in \mathrm{Spin}(V)$. If $\rho(-1) = -I$, then S is called an *abstract spinor space* for V. If $\rho(-1) = I$, then S is called an *abstract tensor space* for V.

As spaces, abstract tensor and spinor spaces are nothing more than linear spaces in general. The point is that there is a coupled action of rotations in V and its various abstract tensor and spinor spaces. To explain the topological idea, let $T(t) \in \mathrm{SO}(V)$, $t \in [0,1]$, be a loop of rotations of V. Lift $T(t)$ to a path $q(t)$ in $\mathrm{Spin}(V)$ such that $p(q(t)) = T(t)$ with the covering map p from Proposition 4.1.9. Assuming that $T(0) = T(1) = I$ and that $q(0) = 1$, it may happen that $q(1) = 1$ or that $q(1) = -1$. See Theorem 4.3.4. Assume that $T(t)$ is such that $q(1) = -1$, that is, that $T(t)$ is a loop of rotations that is not homotopic to the constant loop.

If S is an abstract tensor space for V, then $\rho(-q) = \rho(q)$. This means that ρ is just a representation of $\mathrm{SO}(V)$, and as we rotate V by $T(t)$, the associated linear operators $\rho(q(t))$ rotate the abstract tensor space S a full loop.

If S is instead an abstract spinor space for V, then $\rho(-q) = -\rho(q)$. This means that as we rotate V by $T(t)$, the associated linear operators $\rho(q(t))$ rotate the abstract tensor space S. But when we have completed one full turn of V at $T(1) = I$, the space S is not in its initial position, since $\rho(-1) = -I$. See Figure 5.1. On performing the same rotation of V by $T(t)$ one more time, however, the associated linear operators $\rho(q(t))$ will return the abstract spinor space S to its original position.

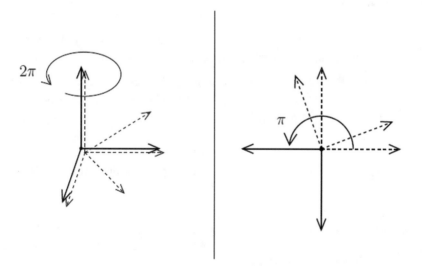

Figure 5.1: The coupled rotation of vector space V (left) and a spinor space S (right), where one full rotation of V corresponds to half a rotation of S. In this concrete example $V = \mathbf{R}^3$ and $S = \triangle\!\!\!\triangle\, \mathbf{R}^3 \leftrightarrow \mathbf{C}^2$, as in Definition 5.2.4, where only the real subspace $\mathbf{R}^2 \subset \mathbf{C}^2$ is shown to the right, and the spinor rotation is furnished by the two real Pauli matrices as in Examples 5.1.5 and 5.1.7.

Example 5.1.2. We list some abstract tensor and spinor spaces for a given Euclidean space V that we encounter elsewhere in this book.

(i) The exterior powers $\wedge^k V$ of V are (real) abstract tensor spaces for V, with the action of rotations T being the induced rotations $\rho(T)w = T(w)$ as in Definition 2.3.1. In terms of the rotor q representing T, we have $\rho(q)w = qwq^{-1}$ as in Proposition 4.1.10.

(ii) Similarly, the tensor product $V \otimes V$ from Definition 1.4.2 is a (real) abstract tensor space for V, with the action of a rotation T being the linear map induced by the bilinear map

$$(v_1, v_2) \mapsto T(v_1) \otimes T(v_k)$$

with the universal property of tensor algebras.

(iii) The Clifford algebra $\triangle V$, as well as the even subalgebra $\triangle^{ev} V$, is an abstract spinor space for V, with the action of a rotor q being $\rho(q)w = qw$.

(iv) The spaces $\mathcal{P}_k^{sh}$ of scalar-valued k-homogeneous harmonic polynomials from Definition 8.2.1 are (real) abstract tensor spaces for V with the action of a rotation T being

$$P(x) \mapsto P(T^{-1}(x)).$$

Note that the rotation invariance of the Laplace equation shows that the rotated polynomial $P \circ T^{-1}$ will be harmonic.

(v) The spaces $\mathcal{P}_k^{em}$ of k-homogeneous monogenic polynomials with values in the even subalgebra from Definition 8.2.1 are (real) abstract spinor spaces for V with the action of a rotor q being

$$P(x) \mapsto qP(q^{-1}xq).$$

Note that Proposition 8.1.14 shows that the so obtained polynomial will be monogenic, and that -1 acts by $-I$.

There are infinitely many nonisomorphic abstract tensor and spinor spaces for a given inner product space V, as we demonstrate in Section 5.4 in low dimension. We will mainly be concerned with one particular example of a spinor space, which builds on the following related notion.

Definition 5.1.3 (Vector representation). Let V be a real inner product space. A *complex V representation* is a real linear map $\rho : V \to \mathcal{L}(S)$ into the space of complex linear operators on a complex linear space S such that

$$\rho(v)^2 = \langle v \rangle^2 I, \quad v \in V.$$

We note that such V representations are nothing but algebra homomorphisms.

Lemma 5.1.4 (Vector=algebra representation). *Let V be a real inner product space, and consider its real Clifford algebra $\triangle V$ with complexification $\triangle V_c$. Then every complex V representation $\rho : V \to \mathcal{L}(S)$ extends in a unique way to a homomorphism*

$$\triangle V_c \to \mathcal{L}(S)$$

of complex algebras. Conversely, every such homomorphism restricts to a complex V representation.

Proof. Clearly, for any algebra homomorphism we have

$$\rho(v)^2 = \rho(v^2) = \rho(\langle v \rangle^2 1) = \langle v \rangle^2 \rho(1) = \langle v \rangle^2 I, \quad v \in V.$$

Given a complex V representation, it follows from the universal property of Clifford algebras as in Section 3.3 that it extends uniquely to a real algebra homomorphism

$$\triangle V \to \mathcal{L}(S).$$

As in Section 1.5, this complexifies in a unique way to a complex algebra homomorphism. □

A complex V representation is a special case of a complex $\mathrm{Spin}(V)$ representation. Indeed, if we restrict an algebra homomorphism $\triangle V_c \to \mathcal{L}(S)$ to the embedded Lie group $\mathrm{Spin}(V) \subset \triangle^{\mathrm{ev}} V \subset \triangle V_c$, we obtain a smooth group homomorphism. Moreover, since $\rho(-1) = -\rho(1) = -I$ for every algebra homomorphism, S will be an abstract spinor space for V.

We also note that in general, a complex $\mathrm{Spin}(V)$ representation does not arise in this way from a complex V representation, not even for abstract spinor spaces. Indeed, in general, a complex $\mathrm{Spin}(V)$ representation is not the restriction to $\mathrm{Spin}(V)$ of a linear map on $\triangle^{\mathrm{ev}} V$. In Example 5.1.2(v) above, this happens only for $k = 0$.

The spinor spaces that we mainly will use are the following.

Example 5.1.5 (The standard representation). We set out to construct a complex V representation, for a given Euclidean space V, which will be the most important example for us of a spinor space. The basic idea of this construction is Theorem 3.4.2.

(i) Consider first the case that $\dim V = n = 2m$ is even. Fix a complex structure J on V, as in Section 1.5, which is isometric. This turns V into a complex linear space $\mathcal{V} = (V, J)$. In fact, this is a Hermitian inner product space with complex inner product $(\cdot, \cdot)$ such that $\mathrm{Re}(\cdot, \cdot)$ is the original Euclidean inner product. Define the complex exterior algebra

$$S := \wedge \mathcal{V},$$

which has complex dimension 2^m. Generalizing the real theory from Chapter 2 to complex linear spaces, we obtain a complex bilinear exterior product $w_1 \wedge w_2$

and a dual complex sesquilinear product $w_1 \lrcorner w_2$ on $\wedge \mathcal{V}$. Define a real linear map $\rho : V \to \mathcal{L}(S)$ by

$$\rho(v)w := v \lrcorner w + v \wedge w, \quad w \in S.$$

Since

$$v \lrcorner (v \lrcorner w + v \wedge w) + v \wedge (v \lrcorner w + v \wedge w) = v \lrcorner (v \wedge w) + v \wedge (v \lrcorner w) = (v,v)w,$$

by the complex analogue of Theorem 2.8.1, where $(v,v) = \langle v, v \rangle = \langle v \rangle^2$, this ρ is a complex V representation.

To make the construction above more concrete, choose a complex ON-basis $\{e_k\}_{k=1}^m$ for $\mathcal{V}$. This means that this together with the vectors $e_{-k} := Je_k$, $k = 1, \dots, m$, forms a real ON-basis for V. Then

$$\rho(e_k)\psi = e_k \lrcorner \psi + e_k \wedge \psi,$$
$$\rho(e_{-k})\psi = i(-e_k \lrcorner \psi + e_k \wedge \psi), \quad \psi \in S.$$

(ii) Consider now the case that $\dim V = n = 2m + 1$ is odd. In this case, we fix a unit vector e_0 and consider $V' := [e_0]^\perp$. Proceeding as in (i) with the even-dimensional space V', fixing an isometric complex structure J on V', we obtain a complex V' representation on $S = \wedge \mathcal{V}'$. We extend this to a real linear map of V by defining

$$\rho(e_0)\psi := \widehat{\psi}, \quad \psi \in S,$$

where $\widehat{\psi}$ denotes the complex analogue of the involution from Definition 2.1.18. Note that we define this as a complex linear map. This yields a complex V representation. Indeed, $\rho(e_0)$ anticommutes with all $\rho(v)$, $v \in V'$, since these latter operators swap $\wedge^{\mathrm{ev}} \mathcal{V}'$ and $\wedge^{\mathrm{od}} \mathcal{V}'$.

Proposition 5.1.6 (Minimal vector representations). *Let V be a real inner product space of dimension $n = 2m$ or $n = 2m + 1$. Then $\dim_{\mathbf{C}} S \geq 2^m$ for every complex V representation*

$$\rho : V \to \mathcal{L}(S).$$

There exists such a complex V representation with $\dim_{\mathbf{C}} S = 2^m$. If n is even, such a minimal representation is an isomorphism of complex algebras $\triangle V_c \to \mathcal{L}(S)$. If n is odd, then $\triangle^{\mathrm{ev}} V_c \to \mathcal{L}(S)$ is an algebra isomorphism.

Proof. Consider a complex V representation ρ, and let $\{e_k\}$ be an ON-basis for V. As in the proof of Proposition 3.3.3, it follows that $\{\rho(e_s)\}_{s \subset \bar{n}}$ are linearly independent operators when n is even, and that $\{\rho(e_s)\}_{|s| \text{ is even}}$ are linearly independent operators when n is odd. This shows that $\dim_{\mathbf{C}} \mathcal{L}(S) \geq 2^{2m}$ and therefore that $\dim_{\mathbf{C}} S \geq 2^m$. It also proves the statements about algebra isomorphisms.

The existence of minimal V representations follows from Example 5.1.5 when V is a Euclidean space. For a non-Euclidean inner product space $(V, \langle \cdot, \cdot \rangle)$ we write $V = V_+ \oplus V_-$, where V_+ and V_- are orthogonal subspaces on which the inner product is positive and negative definite respectively. Write $\widetilde{V}$ for V made into a Euclidean space by changing the sign on the inner product on V_-, and let ρ be a standard $\widetilde{V}$ representation as in Example 5.1.5. Then it is straightforward to verify that

$$V = V_+ \oplus V_- \to \mathcal{L}(S) : v_+ + v_- \mapsto \rho(v_+) + i\rho(v_-)$$

is a complex V representation. □

We end this section by showing what these matrix representations look like in low-dimensional spaces.

Example 5.1.7 (2D and 3D matrices). (i) Let V be a two-dimensional Euclidean space. Fix an ON-basis $\{e_{-1}, e_1\}$ and a complex structure J such that $Je_1 = e_{-1}$. We take e_1 as an ON-basis for the one-dimensional Hermitian space $\mathcal{V}$, and for $S = \wedge \mathcal{V}$ we fix the ON-basis $\{1, e_1\}$. In this basis, the basis multivectors act as follows:

$$\rho(e_1) = \begin{bmatrix} 0 & 1 \\ 1 & 0 \end{bmatrix}, \quad \rho(e_{-1}) = \begin{bmatrix} 0 & -i \\ i & 0 \end{bmatrix},$$

and thus

$$\rho(1) = \begin{bmatrix} 1 & 0 \\ 0 & 1 \end{bmatrix}, \quad \rho(e_1 e_{-1}) = \begin{bmatrix} i & 0 \\ 0 & -i \end{bmatrix}.$$

(ii) Next add a basis vector e_0 and consider three-dimensional Euclidean space with ON-basis $\{e_{-1}, e_0, e_1\}$. With the same spinor space and basis as in two dimensions, we have the action

$$\rho(e_0) = \begin{bmatrix} 1 & 0 \\ 0 & -1 \end{bmatrix}.$$

Note that $\{\rho(e_1), \rho(e_{-1}), \rho(e_0)\}$ are the Pauli matrices from Example 3.4.19, and that $\rho(e_1 e_{-1} e_0) = \begin{bmatrix} i & 0 \\ 0 & i \end{bmatrix}$.

(iii) Let W be a spacetime with one space dimension. Fix an ON-basis $\{e_0, e_1\}$ and write $\{ie_0, e_1\}$ for the associated Euclidean ON-basis inside W_c. Identifying $\{ie_0, e_1\}$ and $\{e_{-1}, e_1\}$, we obtain from (i) the representation

$$\rho(e_0) = \begin{bmatrix} 0 & -1 \\ 1 & 0 \end{bmatrix}, \quad \rho(e_1) = \begin{bmatrix} 0 & 1 \\ 1 & 0 \end{bmatrix},$$

of W. This coincides with the representation we obtained in Example 3.4.18.

(iv) Consider next a spacetime with two space dimensions. Adding a Euclidean ON-basis vector e_2, with

$$\rho(e_2) = \begin{bmatrix} 1 & 0 \\ 0 & -1 \end{bmatrix},$$

we obtain a representation of this spacetime.

Example 5.1.8 (4D and 5D matrices). (i) Let V be a four-dimensional Euclidean space. Fix an ON-basis $\{e_{-2}, e_{-1}, e_1, e_2\}$ and a complex structure J such that $Je_1 = e_{-1}$ and $Je_2 = e_{-2}$. We take $\{e_1, e_2\}$ as ON-basis for the two-dimensional Hermitian space $\mathcal{V}$, and for $S = \wedge \mathcal{V}$ we fix the ON-basis $\{1, e_{12}, e_1, e_2\}$. In this basis, the basis vectors $\{e_1, e_2, e_{-1}, e_{-2}\}$ act by matrices

$$
\begin{bmatrix} 0 & 0 & 1 & 0 \\ 0 & 0 & 0 & 1 \\ 1 & 0 & 0 & 0 \\ 0 & 1 & 0 & 0 \end{bmatrix}, \quad
\begin{bmatrix} 0 & 0 & 0 & 1 \\ 0 & 0 & -1 & 0 \\ 0 & -1 & 0 & 0 \\ 1 & 0 & 0 & 0 \end{bmatrix}, \quad
\begin{bmatrix} 0 & 0 & -i & 0 \\ 0 & 0 & 0 & i \\ i & 0 & 0 & 0 \\ 0 & -i & 0 & 0 \end{bmatrix}, \quad
\begin{bmatrix} 0 & 0 & 0 & -i \\ 0 & 0 & -i & 0 \\ 0 & i & 0 & 0 \\ i & 0 & 0 & 0 \end{bmatrix}
$$

respectively.

(ii) Next add a basis vector e_0 and consider five-dimensional Euclidean space with ON-basis $\{e_{-2}, e_{-1}, e_0, e_1, e_2\}$. With the basis for S from (i), we have the action

$$
\rho(e_0) = \begin{bmatrix} 1 & 0 & 0 & 0 \\ 0 & 1 & 0 & 0 \\ 0 & 0 & -1 & 0 \\ 0 & 0 & 0 & -1 \end{bmatrix}.
$$

Example 5.1.9 (Dirac's γ-matrices). Let W be a spacetime with three space dimensions. Fix an ON-basis $\{e_0, e_1, e_2, e_3\}$ and write $\{ie_0, e_1, e_2, e_3\}$ for the associated Euclidean ON-basis inside W_c. Identifying $\{ie_0, e_1, e_2, e_3\}$ and $\{e_1, e_{-2}, -e_2, e_{-1}\}$, we obtain from Example 5.1.8(i) the representation

$$
\rho(e_0) = \begin{bmatrix} 0 & 0 & -i & 0 \\ 0 & 0 & 0 & -i \\ -i & 0 & 0 & 0 \\ 0 & -i & 0 & 0 \end{bmatrix}, \quad
\rho(e_1) = \begin{bmatrix} 0 & 0 & 0 & -i \\ 0 & 0 & -i & 0 \\ 0 & i & 0 & 0 \\ i & 0 & 0 & 0 \end{bmatrix},
$$

$$
\rho(e_2) = \begin{bmatrix} 0 & 0 & 0 & -1 \\ 0 & 0 & 1 & 0 \\ 0 & 1 & 0 & 0 \\ -1 & 0 & 0 & 0 \end{bmatrix}, \quad
\rho(e_3) = \begin{bmatrix} 0 & 0 & -i & 0 \\ 0 & 0 & 0 & i \\ i & 0 & 0 & 0 \\ 0 & -i & 0 & 0 \end{bmatrix}
$$

for W. The relation to standard representations used in quantum mechanics is the following. Consider instead spacetime $\widetilde{W}$ as an inner product space of signature $1 - 3$. Here the Dirac gamma matrices

$$
\gamma^0 = \begin{bmatrix} I & 0 \\ 0 & -I \end{bmatrix}, \quad
\gamma^1 = \begin{bmatrix} 0 & \sigma_1 \\ -\sigma_1 & 0 \end{bmatrix}, \quad
\gamma^2 = \begin{bmatrix} 0 & \sigma_2 \\ -\sigma_2 & 0 \end{bmatrix}, \quad
\gamma^3 = \begin{bmatrix} 0 & \sigma_3 \\ -\sigma_3 & 0 \end{bmatrix},
$$

represent an ON-basis for $\widetilde{W}$, where $\{\sigma_1, \sigma_2, \sigma_3\}$ are the Pauli matrices from Example 3.4.19. Another important matrix is the so-called fifth gamma matrix

$$
\gamma^5 := i\gamma^0 \gamma^1 \gamma^2 \gamma^3 = \begin{bmatrix} 0 & I \\ I & 0 \end{bmatrix}.
$$

In quantum mechanics this represents chirality, since $\gamma^5 \in \triangle^4 \widetilde{W}$.

However, matrix representations of the ON-basis vectors of W are quite arbitrary. Another representation is the Weyl representation of an ON-basis for $\widetilde{W}$, namely

$$\{\gamma^5, \gamma^1, \gamma^2, \gamma^3\}.$$

This gives another complex $\widetilde{W}$ representation, since $(\gamma^5)^2 = I$ and γ^5 anticommutes with $\gamma^1, \gamma^2, \gamma^3$. Comparing to our basis $\{e_0, e_1, e_2, e_3\}$ for spacetime with signature $3 - 1$, we see that

$$e_0 = -i\gamma^5, \quad e_1 = -i\gamma^1, \quad e_2 = -i\gamma^2, \quad e_3 = -i\gamma^3.$$

5.2 The Complex Spinor Space

To formulate the main theorem for spinor spaces, we need the following terminology.

Definition 5.2.1 (Main reflectors). Let V be a real inner product space of dimension n, and consider its complexified Clifford algebra $\triangle V_c$. The *main reflectors* in $\triangle V_c$ are the two n-vectors $w_n \in \triangle^n V_c$ satisfying $w_n^2 = 1$. For a given choice of main reflector w_n, define

$$w_n^+ := \tfrac{1}{2}(1 + w_n) \quad \text{and} \quad w_n^- := \tfrac{1}{2}(1 - w_n).$$

Note that if we fix $0 \neq w_0 \in \triangle^n V$, then any $w \in \triangle^n V_c$ can be written $w = \lambda w_0$ for some $\lambda \neq 0$. The equation $1 = \lambda^2 w_0^2$ has exactly two solutions over the complex field, which yield the two main reflectors.

Lemma 5.2.2. *Fix a main reflector $w_n \in \triangle^n V_c$ in a real inner product space V, and let $\rho : V \to \mathcal{L}(S)$ be a complex V representation. Then $\rho(w_n^+)$ and $\rho(w_n^-)$ are complementary projections in the sense that $\rho(w_n^\pm)^2 = \rho(w_n^\pm)$ and $\rho(w_n^+) + \rho(w_n^-) = I$.*

If $\dim V$ is even, then the dimensions of the ranges of the two projections $\rho(w_n^\pm)$ are equal. If $\dim V = 2m + 1$ is odd and if $\dim_{\mathbf{C}} S = 2^m$, then either $\rho(w_n^+) = 0$ or $\rho(w_n^-) = 0$. Equivalently, $\rho(w_n) = I$ or $\rho(w_n) = -I$.

Proof. That $\rho(w_n^\pm)$ are complementary projections is a consequence of ρ being an algebra homomorphism. In even dimension, we fix a unit vector $v \in V$ and note that $\rho(v)^2 = I$ and therefore $\rho(v)$ is self-inverse. Since $vw = -wv$, we deduce that $\rho(w_n^-) = \rho(v)\rho(w_n^+)\rho(v)^{-1}$, from which it follows in particular that $\rho(w_n^\pm)$ projects onto subspaces of S of equal dimension.

In odd dimension, it follows as in the proof of Proposition 3.3.3 for a minimal complex V representation that $\rho(w_n)$ and $\rho(1) = I$ must be linearly dependent. Since $\rho(w_n)^2 = I$, we have in fact $\rho(w_n) = \pm I$. $\qquad\square$

The main result needed for the geometric construction of spinors is the following uniqueness result. This is similar to the universal property for the exterior algebra from Proposition 2.1.4, and the universal property for Clifford algebras from Definition 3.3.1, in that it provides a useful way to construct mappings of spinors. See Section 5.3.

Theorem 5.2.3 (Uniqueness of minimal representations). *Let V be a real inner product space of dimension $n = 2m$ or $n = 2m + 1$. Assume that $\rho_j : V \to \mathcal{L}(S_j)$ are two complex V representations, both with dimension $\dim_{\mathbf{C}} S_j = 2^m$, $j = 1, 2$.*

(i) *If $n = 2m$, then there exists an invertible linear map $T : S_1 \to S_2$ such that*

$$\rho_2(w)T = T\rho_1(w), \quad w \in \triangle V_c.$$

If $\widetilde{T}$ is a second such map, then $\widetilde{T} = \lambda T$ for some $\lambda \in \mathbf{C} \setminus \{0\}$.

(ii) *If $n = 2m + 1$, then there exists an invertible linear map $T : S_1 \to S_2$ such that either*

$$\rho_2(w)T = T\rho_1(w), \quad w \in \triangle V_c, \quad or$$
$$\rho_2(w)T = T\rho_1(\widehat{w}), \quad w \in \triangle V_c.$$

The first case occurs when $\rho_1(w_n) = \rho_2(w_n)$ for the main reflectors w_n, and the second case occurs when $\rho_1(w_n) = -\rho_2(w_n)$.

In particular, $\rho_2(w)T = T\rho_1(w)$ for all $w \in \triangle^{\mathrm{ev}} V_c$. If $T \in \mathcal{L}(S_1; S_2)$ is a second map with this property, then $\widetilde{T} = \lambda T$ for some $\lambda \in \mathbf{C} \setminus \{0\}$.

This section is devoted to the proof of Theorem 5.2.3, but before embarking on this, we make use of this result to give a proper invariant geometric definition of spinors and the complex spinor space of V. To this end, recall the principle of abstract algebra, as discussed in the introduction of Chapter 1. At this stage the reader hopefully is so comfortable using multivectors that he or she has forgotten that we never actually defined a k-vector to be a specific object, but rather as an element in the range space of some arbitrarily chosen multilinear map with properties (A) and (U) in Definition 2.1.6. Proposition 5.1.6 and Theorem 5.2.3 now allow us to define spinors in a similar spirit, as elements in some arbitrarily chosen representation space S of minimal dimension 2^m. Indeed, Theorem 5.2.3 shows that any other choice of ρ and S amounts only to a renaming of the objects in the space S, and does not affect how $\triangle V_c$ acts on S.

Definition 5.2.4 (The complex spinor space). Let (X, V) be a real inner product space of dimension $n = 2m$ or $n = 2m + 1$. Fix one complex V representation $\rho : V \to \mathcal{L}(S)$ with $\dim_{\mathbf{C}} S = 2^m$. We denote this S by $\cancel{\triangle} V$, and refer to $(\cancel{\triangle} V, \rho)$ as *the complex spinor space* for V. We also use the shorthand notation

$$w.\psi := \rho(w)\psi$$

for the action of $w \in \triangle V_c$ on $\psi \in \mathbb{A}V$.

Fix a main reflector $w_n \in \triangle^n V_c$. When $\dim V$ is even, we denote by $\mathbb{A}^+V$ and $\mathbb{A}^-V$ the ranges of $\rho(w_n^{\pm})$, so that

$$\mathbb{A}V = \mathbb{A}^+V \oplus \mathbb{A}^-V.$$

When $\dim V$ is odd, we write $\rho_-(w)\psi := \hat{w}.\psi$ for the second nonisomorphic action of $\triangle V_c$ on $\mathbb{A}V$.

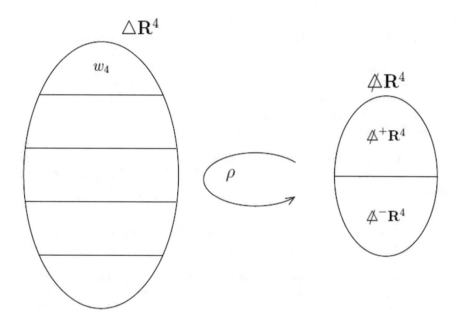

Figure 5.2: Multivector action on spinors in a four-dimensional inner product space.

We have constructed the complex spinor space $\mathbb{A}V$ for V in a way very similar in spirit to the construction of the exterior algebra $\wedge V$ for V. There is one big difference between the constructions of multivectors and spinors, though: with the map $\wedge^k$, a set of k vectors single out a certain k-vector, but vectors are not factors in spinors in this way. Rather vectors, and more generally multivectors, act as linear operators on spinors.

Exercise 5.2.5. Assume that $\dim V$ is even. Show that $\rho(v)$ maps $\mathbb{A}^+V \to \mathbb{A}^-V$ and $\mathbb{A}^-V \to \mathbb{A}^+V$, for every vector $v \in \triangle^1 V = V$. Using the standard representation from Example 5.1.5(i), show that $\mathbb{A}^+V$ and $\mathbb{A}^-V$ coincide with $\wedge^{\mathrm{ev}}V$ and $\wedge^{\mathrm{od}}V$.

We now turn to the proof of Theorem 5.2.3. The idea is roughly that two minimal complex V representations yield an isomorphism

$$\mathcal{L}(S_1) \to \mathcal{L}(S_2)$$

of complex algebras. Given an invertible map $T \in \mathcal{L}(S_1; S_2)$,

$$\mathcal{L}(S_1) \to \mathcal{L}(S_2) : X \to TXT^{-1}$$

is such an algebra isomorphism. We prove below that every isomorphism $\mathcal{L}(S_1) \to \mathcal{L}(S_2)$ arises in this way, and we deduce from this fact Theorem 5.2.3. For this proof, it does not matter whether we work over the real or complex field.

Definition 5.2.6 (Ideals). Let A be a complex associative algebra, as in Definition 1.1.4 but replacing $\mathbf{R}$ by $\mathbf{C}$. A linear subspace $I \subset A$ is called a *left ideal* if $xy \in I$ whenever $x \in A$ and $y \in I$. If $yx \in I$ whenever $x \in A$ and $y \in I$, then I is called a *right ideal*. A linear subspace I that is both a left and right ideal is called a *two-sided ideal*.

The notion of ideal is important in identifying the T representing a given algebra isomorphism as above. We also recall from Proposition 1.4.3 that there is a natural isomorphism

$$S \otimes S^* \leftrightarrow \mathcal{L}(S),$$

which identifies the simple tensor $v \otimes \theta \in S \otimes S^*$ and the rank-one linear operator $x \mapsto \langle \theta, x \rangle v$. We use both these views on linear operators below.

Proposition 5.2.7 (Matrix ideals). *Consider the algebra $\mathcal{L}(S)$ of all linear operators on a linear space S. There is a one-to-one correspondence between linear subspaces of S^* and left ideals in $\mathcal{L}(S)$ that identifies a subspace $U \subset S^*$ and the left ideal*

$$I_U^l := \{T \in \mathcal{L}(S) \; ; \; R(T^*) \subset U\} = \Big\{ \sum_k v_k \otimes \theta_k \; ; \; v_k \in S, \; \theta_k \in U \Big\}.$$

Similarly there is a one-to-one correspondence between linear subspaces of S and right ideals in $\mathcal{L}(S)$ that identifies a subspace $U \subset S$ and the right ideal

$$I_U^r := \{T \in \mathcal{L}(S) \; ; \; R(T) \subset U\} = \Big\{ \sum_k v_k \otimes \theta_k \; ; \; v_k \in U, \; \theta_k \in S^* \Big\}.$$

The only two-sided ideals in $\mathcal{L}(S)$ are the two trivial ones, namely $\{0\}$ and $\mathcal{L}(S)$ itself.

Note that in terms of matrices, a left ideal consists of all matrices with all row vectors in a given subspace, whereas a right ideal consists of all matrices with all column vectors in a given subspace.

Proof. Clearly I_U^l is a left ideal for any subspace $U \subset S^*$. Note that $\mathsf{R}(T^*) \subset U$ is equivalent to $U^\perp \subset \mathsf{N}(T)$. To see that all left ideals are of the form I_U^l, consider any left ideal $I \subset \mathcal{L}(S)$. Let $U \subset S^*$ be such that

$$U^\perp := \bigcap_{T \in I} \mathsf{N}(T).$$

Then $I \subset I_U^l$. For the converse we claim that there exists $T_0 \in I$ such that $\mathsf{N}(T_0) = U^\perp$. From this it follows that $I_U^l \subset I$. Indeed, if $\mathsf{N}(T_0) \subset \mathsf{N}(T)$, then $T = AT_0 \in I$ for some operator A. One way to prove the claim is to observe that

$$\mathsf{N}(T_1) \cap \mathsf{N}(T_2) = \mathsf{N}(T_1^*T_1 + T_2^*T_2),$$

and that $T_1^*T_1 + T_2^*T_2 \in I$ whenever $T_1, T_2 \in I$. Here the adjoints are with respect to any auxiliary Euclidean inner product. Using this observation a finite number of times, it follows that there exists $T_0 \in I$ with minimal null space.

The proof for right ideals is similar. Finally, if I is a two-sided ideal, then $I = I_U^l = I_V^r$ for some subspaces $U \subset S^*$ and $V \subset S$. This can happen only if $U = \{0\} = V$ or if $U = S^*$ and $V = S$. Thus $I = \{0\}$ or $I = \mathcal{L}(S)$. $\square$

The following is the key result in the proof of uniqueness of minimal spinor representations.

Proposition 5.2.8 (Completeness of matrix algebras). *The algebra $\mathcal{L}(S)$ of all linear operators on a linear space S is complete in the sense that every algebra automorphism $\phi : \mathcal{L}(S) \to \mathcal{L}(S)$ is inner, that is, there exists a linear invertible map $T \in \mathcal{L}(S)$ such that $\phi(X) = TXT^{-1}$ for all $X \in \mathcal{L}(S)$. Such T are unique up to scalar multiples.*

Proof. By Proposition 5.2.7, every minimal left ideal, that is, a left ideal I such that no left ideal J such that $\{0\} \subsetneq J \subsetneq I$ exists, must be of the form

$$L_\theta := \{v \otimes \theta \; ; \; v \in S\}, \quad \text{for some } \theta \in S^* \setminus \{0\}.$$

Similarly, all minimal right ideals are of the form

$$R_v := \{v \otimes \theta \; ; \; \theta \in S^*\}, \quad \text{for some } v \in S \setminus \{0\}.$$

Let $\phi : \mathcal{L}(S) \to \mathcal{L}(S)$ be an algebra automorphism, and fix $\theta' \in S^*$ and $v' \in S$ such that $\langle \theta', v' \rangle = 1$. Since ϕ only relabels the objects in $\mathcal{L}(S)$ without changing the algebraic structure, it is clear that $\phi(L_{\theta'}) = L_{\theta''}$ for some $\theta'' \in S^* \setminus \{0\}$. It follows that there is an invertible linear operator $T_1 \in \mathcal{L}(S)$ such that

$$\phi(v \otimes \theta') = T_1(v) \otimes \theta'', \quad \text{for all } v \in S. \tag{5.1}$$

Similarly, by considering the mapping of minimal right ideals, there exist an invertible linear operator $T_2 \in \mathcal{L}(S^*)$ and $v'' \in S$ such that

$$\phi(v' \otimes \theta) = v'' \otimes T_2(\theta), \quad \text{for all } \theta \in S^*.$$

Since ϕ is an automorphism, it follows that

$$\phi(v \otimes \theta) = \phi((v \otimes \theta')(v' \otimes \theta)) = (T_1(v) \otimes \theta'')(v'' \otimes T_2(\theta))$$
$$= \langle \theta'', v'' \rangle T_1(v) \otimes T_2(\theta), \quad \text{for all } v \in S, \theta \in S^*.$$

For example, by inversely rescaling T_1 and θ'', we may assume that $\langle \theta'', v'' \rangle = 1$. Furthermore, for all $v_1, v_2 \in S$ and $\theta_1, \theta_2 \in S^*$ we have

$$\langle \theta_1, v_2 \rangle \phi(v_1 \otimes \theta_2) = \phi((v_1 \otimes \theta_1)(v_2 \otimes \theta_2)) = (T_1(v_1) \otimes T_2(\theta_1))(T_1(v_2) \otimes T_2(\theta_2))$$
$$= \langle T_2(\theta_1), T_1(v_2) \rangle T_1(v_1) \otimes T_2(\theta_2),$$

so $(T_2)^{-1} = (T_1)^*$. Thus, with $T = T_1$, we get that $\phi(v \otimes \theta) = T(v) \otimes (T^*)^{-1}(\theta)$, and therefore $\phi(X) = TXT^{-1}$ by linearity.

The uniqueness result is a consequence of the fact that $\mathcal{Z}(\mathcal{L}(S)) = \text{span}\{I\}$. Indeed, if $\widetilde{T}X\widetilde{T}^{-1} = TXT^{-1}$ for all X, then $T^{-1}\widetilde{T} \in \mathcal{Z}(\mathcal{L}(S))$.   □

Note that formula (5.1) can be used to calculate T for a given automorphism ϕ. Fixing a basis for S and dual basis for S^*, we work with matrices and may assume $\theta' = \begin{bmatrix} 1 & 0 & \dots & 0 \end{bmatrix}$ and $\theta'' = \begin{bmatrix} a_1 & a_2 & \dots & a_k \end{bmatrix}$. Then we have for vectors $v \in S$ the matrix identity

$$\phi\left(\begin{bmatrix} v & 0 & \dots & 0 \end{bmatrix}\right) = \begin{bmatrix} a_1 T(v) & a_2 T(v) & \dots & a_k T(v) \end{bmatrix}.$$

Since at least one a_j is nonzero, we find that $T(v)$ can be defined as a nonzero column of the matrix $\phi(v \otimes \theta')$.

Proof of Theorem 5.2.3. (i) Assume $\dim V = 2m$. Consider two representations ρ_1 and ρ_2 with $\dim S_1 = 2^m = \dim S_2$, which by Proposition 5.1.6 are algebra isomorphisms $\triangle V_c \to \mathcal{L}(S_i)$. Then $\rho_2 \rho_1^{-1} : \mathcal{L}(S_1) \to \mathcal{L}(S_2)$ is an algebra isomorphism. Take any linear invertible map $T_0 : S_1 \to S_2$ and consider the induced algebra isomorphism $\rho_0 : \mathcal{L}(S_1) \to \mathcal{L}(S_2) : X \mapsto T_0 X T_0^{-1}$. Then $\rho_0^{-1} \rho_2 \rho_1^{-1}$ is an automorphism of $\mathcal{L}(S_1)$, and Proposition 5.2.8 shows the existence of $T_1 \in \mathcal{L}(S_1)$, unique up to scalar multiples, such that

$$\rho_0^{-1} \rho_2 \rho_1^{-1}(X) = T_1 X T_1^{-1}, \quad \text{for all } X \in \mathcal{L}(S_1).$$

Letting $T := T_0 T_1 : S_1 \to S_2$, this means that

$$\rho_2(w) = T\rho_1(w)T^{-1}, \quad \text{for all } w \in \triangle V_c.$$

(ii) Assume $\dim V = 2m + 1$. Consider two representations ρ_1 and ρ_2 with $\dim S_1 = 2^m = \dim S_2$. Then $\rho_i : \triangle^{\text{ev}} V \to \mathcal{L}(S)$ are both isomorphisms, so as in (i), we get from the algebra isomorphism $(\rho_2|_{\triangle^{\text{ev}} V})(\rho_1|_{\triangle^{\text{ev}} V})^{-1} : \mathcal{L}(S_1) \to \mathcal{L}(S_2)$ the existence of $T \in \mathcal{L}(S_1; S_2)$ such that

$$\rho_2(w) = T\rho_1(w)T^{-1}, \quad \text{for all } w \in \triangle^{\text{ev}} V_c,$$

unique up to multiples. Consider next an arbitrary multivector $w \in \triangle V_c$. This can be uniquely written as $w = w_1 + w_n w_2$, with $w_1, w_2 \in \triangle^{\mathrm{ev}} V_c$, if we fix a main reflection $w_n \in \triangle^n V_c$. If $\rho_1(w_n) = \rho_2(w_n)$, then

$$\rho_2(w)T = (\rho_2(w_1) + \rho_2(w_n)\rho_2(w_2))T = T(\rho_1(w_1) + \rho_1(w_n)\rho_1(w_2)) = T\rho_1(w),$$

since $w_1, w_2 \in \triangle^{\mathrm{ev}} V_c$ and $\rho_i(w_n) = \pm I$. If $\rho_1(w_n) = -\rho_2(w_n)$, then

$$\rho_2(w)T = (\rho_2(w_1) + \rho_2(w_n)\rho_2(w_2))T = T(\rho_1(w_1) - \rho_1(w_n)\rho_1(w_2)) = T\rho_1(\widehat{w}).$$

$\square$

5.3 Mapping Spinors

Consider two vector spaces V_1 and V_2 and a linear map $T : V_1 \to V_2$. In Section 2.3, we saw that this induces a unique linear map

$$T = T_\wedge : \wedge V_1 \to \wedge V_2$$

of multivectors, which is in fact a homomorphism with respect to the exterior product. When V_1 and V_2 are inner product spaces, we saw in Proposition 4.1.10 that $T_\wedge : \triangle V_1 \to \triangle V_2$ will be a homomorphism with respect to Clifford products if and only if T is an isometry.

In this section, we study in what sense a linear map $T : V_1 \to V_2$ induces a linear map of spinors

$$T_{\!\!/\!\!\backslash} : \!\!/\!\!\backslash V_1 \to \!\!/\!\!\backslash V_2.$$

To avoid extra technicalities, we consider only Euclidean spinors in this section. Consider first an invertible isometry T between Euclidean spaces. We have fixed complex V_i representations $\rho_i : V_i \to \mathcal{L}(\!\!/\!\!\backslash V_i)$, $i = 1, 2$. This means that on $V = V_1$, we have the two complex V representations ρ_1 and $\rho_2 T$. It follows from Theorem 5.2.3 that there exists an invertible map $T_{\!\!/\!\!\backslash} \in \mathcal{L}(\!\!/\!\!\backslash V_1; \!\!/\!\!\backslash V_2)$ such that

$$T_{\!\!/\!\!\backslash}(w.\psi) = (Tw).(T_{\!\!/\!\!\backslash}\psi), \quad w \in \triangle V_1, \psi \in \!\!/\!\!\backslash V_1.$$

In odd dimension, some care about how the main reflectors map is needed. See Proposition 5.3.5. This construction of induced maps $T_{\!\!/\!\!\backslash}$ of spinors leads to the following two questions, which we address in this section.

- How unique can we make the spinor map $T_{\!\!/\!\!\backslash}$ induced by the vector map T?

- Is there a natural way to define a spinor map $T_{\!\!/\!\!\backslash}$ for more general invertible vector maps T that are not isometries?

Concerning the first question, the problem is that $T_{\!\!/\!\!\backslash} \in \mathcal{L}(\!\!/\!\!\backslash V_1; \!\!/\!\!\backslash V_2)$ is unique only as a projective map

$$T_{\!\!/\!\!\backslash} : \!\!/\!\!\backslash V_1/\mathbf{C} \to \!\!/\!\!\backslash V_2/\mathbf{C},$$

that is, $T_{\!/\!\!\!\Delta}$ is unique only up to complex multiples $\lambda T_{\!/\!\!\!\Delta}$, $\lambda \in \mathbf{C} \setminus \{0\}$. When, for example, we are constructing and working with spinors over manifolds as we do in Chapter 12, this presents problems. To this end, we next define two additional natural structures, an inner product and a conjugation, on spinor spaces that allow us to obtain induced maps of spinors that are unique only up to sign $\pm T_{\!/\!\!\!\Delta}$. Such sign ambiguity will always be present, but this discrete nonuniqueness will not cause any problems.

Proposition 5.3.1 (Spinor inner product). *Let V be a Euclidean space, with complex spinor space $/\!\!\!\Delta V$. Then there exists a Hermitian complex inner product $(\cdot, \cdot)$ on $/\!\!\!\Delta V$ such that*

$$(\psi_1, v.\psi_2) = (v.\psi_1, \psi_2), \quad \psi_1, \psi_2 \in /\!\!\!\Delta V, \tag{5.2}$$

for all vectors $v \in V$. If $(\cdot, \cdot)'$ is another Hermitian inner product for which (5.2) holds, then there is a constant $\lambda > 0$ such that $(\psi_1, \psi_2)' = \lambda(\psi_1, \psi_2)$ for all $\psi_1, \psi_2 \in /\!\!\!\Delta V$.

Proof. Fix a basis for $/\!\!\!\Delta V$, view ψ_i as column vectors and $\rho(v)$ as matrices. Then a sesquilinear duality is uniquely represented by an invertible matrix M such that

$$(\psi_1, \psi_2) = \psi_1^* M \psi_2.$$

Condition (5.2) translated to M is that $M\rho(v) = \rho(v)^* M$ for all vectors $v \in V$. We note that $v \mapsto \rho(v)^*$ is a second complex V representation, which extends to the algebra homomorphism $w \mapsto \rho(\overline{w}^c)^*$. Therefore the existence of M, unique up to complex nonzero multiples, follows from Theorem 5.2.3. Note that in odd dimensions, these two representations coincide on $\triangle^n V_c$.

It remains to see that M can be chosen as a self-adjoint positive definite matrix. We note that when we are using the standard representation from Example 5.1.5, all matrices $\rho(v)$ are self-adjoint. Hence $M = I$ can be used in this basis. □

Proposition 5.3.2 (Spinor conjugation). *Let V be a Euclidean space of dimension n, with spinor space $/\!\!\!\Delta V$. Then there exists an antilinear map $/\!\!\!\Delta V \to /\!\!\!\Delta V : \psi \mapsto \psi^\dagger$ such that*

$$\begin{cases} (v.\psi)^\dagger = v.\psi^\dagger, & n \not\equiv 3 \mod 4, \\ (v.\psi)^\dagger = -v.\psi^\dagger, & n \equiv 3 \mod 4, \end{cases} \tag{5.3}$$

for all vectors $v \in V$ and spinors $\psi \in /\!\!\!\Delta V$, and satisfying

$$\begin{cases} (\psi^\dagger)^\dagger = \psi, & n \equiv 0, 1, 2, 7 \mod 8, \\ (\psi^\dagger)^\dagger = -\psi, & n \equiv 3, 4, 5, 6 \mod 8. \end{cases}$$

If $\psi \mapsto \psi^{\dagger'}$ is another such map for which this holds, then there is $\lambda \in \mathbf{C}$, $|\lambda| = 1$, such that $\psi^{\dagger'} = \lambda \psi^\dagger$ for all $\psi \in /\!\!\!\Delta V$.

Note that in even dimensions $n \equiv 0, 2 \mod 8$, this spinor conjugation provides a real structure on the complex spinor space, as in Section 1.5, in accordance with Theorem 3.4.13, which shows that in these dimensions the real-Euclidean Clifford algebras are isomorphic to real matrix algebras.

The technicalities about signs and dimension in the statement of Proposition 5.3.2 are best understood from its proof.

Proof. Fix a basis for $\not\!\!\triangle V$, view ψ as column vectors and $\rho(w)$ as matrices. Write N^c for the componentwise complex conjugation of a matrix N. Every antilinear map $\not\!\!\triangle V \to \not\!\!\triangle V : \psi \mapsto \psi^\dagger$ can be written

$$\psi^\dagger = (N\psi)^c$$

for some matrix N. The condition

$$(v.\psi)^\dagger = v.(\psi^\dagger) \tag{5.4}$$

is equivalent to $N\rho(v) = (\rho(v))^c N$, for all vectors v. We note that that $v \mapsto (\rho(v))^c$ is a second complex V representation, which extends to the algebra homomorphism $w \mapsto (\rho(w^c))^c$, where w^c denotes the real structure on $\triangle V_c$. Existence of antilinear maps satisfying (5.4), unique up to $\lambda \in \mathbf{C} \setminus \{0\}$, follows from Theorem 5.2.3, provided n is even, or if

$$\rho(e_1)^c \cdots \rho(e_n)^c = \rho(e_1) \cdots \rho(e_n)$$

when n is odd. Using a standard complex representation from Example 5.1.5, we see that this holds unless $n \equiv 3 \mod 4$, since all but m of the matrices representing the basis vectors are real. When $n \equiv 3 \mod 4$, by Theorem 5.2.3 the correct relation is $N\rho(\widehat{w}) = (\rho(w^c))^c N$.

To complete the proof, we claim that using a standard representation, we may choose

$$N := \begin{cases} \rho(e_{-1} \cdots e_{-m}), & n = 2m \equiv 0 \mod 4, \\ \rho(e_{-1} \cdots e_{-m}), & n = 2m + 1 \equiv 1 \mod 4, \\ \rho(e_1 \cdots e_m), & n = 2m \equiv 2 \mod 4, \\ \rho(e_1 \cdots e_m e_0), & n = 2m + 1 \equiv 3 \mod 4. \end{cases}$$

Indeed, we note that $\rho(e_k)^c = \rho(e_k)$, $k \geq 0$, and $\rho(e_k)^c = -\rho(e_k)$, $k < 0$, and we verify that $N^c N = I$ when $n \equiv 0, 1, 2, 7 \mod 8$ and $N^c N = -I$ when $n \equiv 3, 4, 5, 6 \mod 8$. This completes the proof, since $\lambda(\lambda\psi^\dagger)^\dagger = |\lambda|^2 (\psi^\dagger)^\dagger$. $\qquad\square$

Definition 5.3.3 (Normed spinor space). Let V be a Euclidean space, with spinor space $\not\!\!\triangle V$.

Fix a *spinor inner product*, by which we mean a Hermitian inner product on $\not\!\!\triangle V$ such that all vectors act as self-adjoint maps as in Proposition 5.3.1. This amounts to a choice of the parameter $\lambda > 0$.

Fix also a *spinor conjugation*, by which we mean an antilinear map on $\triangle V$ with properties as in Proposition 5.3.2. This amounts to a choice of the parameter $|\lambda| = 1$.

We refer to the triple $(\triangle V, (\cdot, \cdot), \cdot^\dagger)$ as a *normed spinor space*.

Lemma 5.3.4 (Compatibility). *A spinor inner product and a spinor conjugation are compatible in the sense that*

$$(\psi_1, \psi_2)^c = (\psi_1^\dagger, \psi_2^\dagger), \quad \psi_1, \psi_2 \in \triangle V.$$

Proof. Note that

$$(\psi_1, \psi_2)' := (\psi_1^\dagger, \psi_2^\dagger)^c$$

defines a second Hermitian spinor inner product. By uniqueness in Proposition 5.3.1, we have $(\psi_1, \psi_2)^c = \lambda (\psi_1^\dagger, \psi_2^\dagger)$ for some $\lambda > 0$. In particular,

$$(\psi, \psi) = \lambda(\psi^\dagger, \psi^\dagger) = \lambda^2((\psi^\dagger)^\dagger, (\psi^\dagger)^\dagger) = \lambda^2(\psi, \psi),$$

so $\lambda = 1$. $\qquad\qquad\qquad\qquad\qquad\qquad\qquad\qquad\qquad\qquad\qquad\qquad\square$

We can now answer the first question posed above, concerning uniqueness of induced spinor maps. For simplicity we write the inner product and conjugation in both spaces below with the same symbols.

Proposition 5.3.5 (Uniqueness of spinor maps). *Let V_1, V_2 be Euclidean spaces, with normed spinor spaces $\triangle V_1$ and $\triangle V_2$ respectively. Assume that $T : V_1 \to V_2$ is an invertible isometry. When $\dim V_1$ is odd, we assume that the main reflectors w_n and $T w_n$ in V_1 and V_2 respectively both act as $+I$ or as $-I$. Then there exists an isometric complex linear map $T_\triangle : \triangle V_1 \to \triangle V_2$ such that $(T_\triangle \psi)^\dagger = T_\triangle(\psi^\dagger)$ and*

$$T_\triangle(v.\psi) = (Tv).(T_\triangle \psi), \quad \psi \in \triangle V, v \in V. \tag{5.5}$$

If $T_\triangle'$ is another such map, then $T_\triangle' = \pm T_\triangle$.

Proof. We saw at the beginning of this section how Theorem 5.2.3 implies the existence of $T_\triangle$ satisfying (5.5), and every other such map is of the form $\lambda T_\triangle$ for some $\lambda \in \mathbf{C} \setminus \{0\}$.

To see that $T_\triangle$ can be chosen to be isometric, consider the Hermitian inner product $(\psi_1, \psi_2)' := (T_\triangle \psi_1, T_\triangle \psi_2)$ on $\triangle V_1$. We calculate

$$(\psi_1, v.\psi_2)' = (T_\triangle \psi_1, Tv.T_\triangle \psi_2) = (Tv.T_\triangle \psi_1, T_\triangle \psi_2) = (v.\psi_1, \psi_2)',$$

for $v \in V$. Proposition 5.3.1 shows that $(T_\triangle \psi_1, T_\triangle \psi_2) = \mu(\psi_1, \psi_2)$ for some $\mu > 0$, so $\lambda T_\triangle$ will be isometric if $|\lambda| = \mu^{-1/2}$.

To see that $T_\triangle$ can be chosen to be compatible with spinor conjugation, consider the antilinear map $\psi \mapsto T_\triangle^{-1}(T_\triangle \psi)^\dagger$ on $\triangle V_1$. We calculate

$$T_\triangle^{-1}(T_\triangle(v.\psi))^\dagger = T_\triangle^{-1}((Tv).(T_\triangle \psi))^\dagger = T_\triangle^{-1}(\epsilon(Tv).(T_\triangle \psi)^\dagger) = \epsilon v.T_\triangle^{-1}(T_\triangle \psi)^\dagger,$$

Fix also a *spinor conjugation*, by which we mean an antilinear map on $\not\!\!\!\Delta V$ with properties as in Proposition 5.3.2. This amounts to a choice of the parameter $|\lambda| = 1$.

We refer to the triple $(\not\!\!\!\Delta V, (\cdot, \cdot), \cdot^\dagger)$ as a *normed spinor space*.

Lemma 5.3.4 (Compatibility). *A spinor inner product and a spinor conjugation are compatible in the sense that*

$$(\psi_1, \psi_2)^c = (\psi_1{}^\dagger, \psi_2{}^\dagger), \quad \psi_1, \psi_2 \in \not\!\!\!\Delta V.$$

Proof. Note that

$$(\psi_1, \psi_2)' := (\psi_1{}^\dagger, \psi_2{}^\dagger)^c$$

defines a second Hermitian spinor inner product. By uniqueness in Proposition 5.3.1, we have $(\psi_1, \psi_2)^c = \lambda(\psi_1{}^\dagger, \psi_2{}^\dagger)$ for some $\lambda > 0$. In particular,

$$(\psi, \psi) = \lambda(\psi^\dagger, \psi^\dagger) = \lambda^2((\psi^\dagger)^\dagger, (\psi^\dagger)^\dagger) = \lambda^2(\psi, \psi),$$

so $\lambda = 1$. □

We can now answer the first question posed above, concerning uniqueness of induced spinor maps. For simplicity we write the inner product and conjugation in both spaces below with the same symbols.

Proposition 5.3.5 (Uniqueness of spinor maps). *Let V_1, V_2 be Euclidean spaces, with normed spinor spaces $\not\!\!\!\Delta V_1$ and $\not\!\!\!\Delta V_2$ respectively. Assume that $T : V_1 \to V_2$ is an invertible isometry. When $\dim V_1$ is odd, we assume that the main reflectors w_n and Tw_n in V_1 and V_2 respectively both act as $+I$ or as $-I$. Then there exists an isometric complex linear map $T_{\not\!\!\!\Delta} : \not\!\!\!\Delta V_1 \to \not\!\!\!\Delta V_2$ such that $(T_{\not\!\!\!\Delta}\psi)^\dagger = T_{\not\!\!\!\Delta}(\psi^\dagger)$ and*

$$T_{\not\!\!\!\Delta}(v.\psi) = (Tv).(T_{\not\!\!\!\Delta}\psi), \quad \psi \in \not\!\!\!\Delta V, v \in V. \tag{5.5}$$

If $T'_{\not\!\!\!\Delta}$ is another such map, then $T'_{\not\!\!\!\Delta} = \pm T_{\not\!\!\!\Delta}$.

Proof. We saw at the beginning of this section how Theorem 5.2.3 implies the existence of $T_{\not\!\!\!\Delta}$ satisfying (5.5), and every other such map is of the form $\lambda T_{\not\!\!\!\Delta}$ for some $\lambda \in \mathbf{C} \setminus \{0\}$.

To see that $T_{\not\!\!\!\Delta}$ can be chosen to be isometric, consider the Hermitian inner product $(\psi_1, \psi_2)' := (T_{\not\!\!\!\Delta}\psi_1, T_{\not\!\!\!\Delta}\psi_2)$ on $\not\!\!\!\Delta V_1$. We calculate

$$(\psi_1, v.\psi_2)' = (T_{\not\!\!\!\Delta}\psi_1, Tv.T_{\not\!\!\!\Delta}\psi_2) = (Tv.T_{\not\!\!\!\Delta}\psi_1, T_{\not\!\!\!\Delta}\psi_2) = (v.\psi_1, \psi_2)',$$

for $v \in V$. Proposition 5.3.1 shows that $(T_{\not\!\!\!\Delta}\psi_1, T_{\not\!\!\!\Delta}\psi_2) = \mu(\psi_1, \psi_2)$ for some $\mu > 0$, so $\lambda T_{\not\!\!\!\Delta}$ will be isometric if $|\lambda| = \mu^{-1/2}$.

To see that $T_{\not\!\!\!\Delta}$ can be chosen to be compatible with spinor conjugation, consider the antilinear map $\psi \mapsto T_{\not\!\!\!\Delta}^{-1}(T_{\not\!\!\!\Delta}\psi)^\dagger$ on $\not\!\!\!\Delta V_1$. We calculate

$$T_{\not\!\!\!\Delta}^{-1}(T_{\not\!\!\!\Delta}(v.\psi))^\dagger = T_{\not\!\!\!\Delta}^{-1}((Tv).(T_{\not\!\!\!\Delta}\psi))^\dagger = T_{\not\!\!\!\Delta}^{-1}(\epsilon(Tv).(T_{\not\!\!\!\Delta}\psi)^\dagger) = \epsilon v.T_{\not\!\!\!\Delta}^{-1}(T_{\not\!\!\!\Delta}\psi)^\dagger,$$

Note that in even dimensions $n \equiv 0, 2 \mod 8$, this spinor conjugation provides a real structure on the complex spinor space, as in Section 1.5, in accordance with Theorem 3.4.13, which shows that in these dimensions the real-Euclidean Clifford algebras are isomorphic to real matrix algebras.

The technicalities about signs and dimension in the statement of Proposition 5.3.2 are best understood from its proof.

Proof. Fix a basis for $\mathbb{A}V$, view ψ as column vectors and $\rho(w)$ as matrices. Write N^c for the componentwise complex conjugation of a matrix N. Every antilinear map $\mathbb{A}V \to \mathbb{A}V : \psi \mapsto \psi^\dagger$ can be written

$$\psi^\dagger = (N\psi)^c$$

for some matrix N. The condition

$$(v.\psi)^\dagger = v.(\psi^\dagger) \tag{5.4}$$

is equivalent to $N\rho(v) = (\rho(v))^c N$, for all vectors v. We note that that $v \mapsto (\rho(v))^c$ is a second complex V representation, which extends to the algebra homomorphism $w \mapsto (\rho(w^c))^c$, where w^c denotes the real structure on $\triangle V_c$. Existence of antilinear maps satisfying (5.4), unique up to $\lambda \in \mathbf{C} \setminus \{0\}$, follows from Theorem 5.2.3, provided n is even, or if

$$\rho(e_1)^c \cdots \rho(e_n)^c = \rho(e_1) \cdots \rho(e_n)$$

when n is odd. Using a standard complex representation from Example 5.1.5, we see that this holds unless $n \equiv 3 \mod 4$, since all but m of the matrices representing the basis vectors are real. When $n \equiv 3 \mod 4$, by Theorem 5.2.3 the correct relation is $N\rho(\widehat{w}) = (\rho(w^c))^c N$.

To complete the proof, we claim that using a standard representation, we may choose

$$N := \begin{cases} \rho(e_{-1} \cdots e_{-m}), & n = 2m \equiv 0 \mod 4, \\ \rho(e_{-1} \cdots e_{-m}), & n = 2m + 1 \equiv 1 \mod 4, \\ \rho(e_1 \cdots e_m), & n = 2m \equiv 2 \mod 4, \\ \rho(e_1 \cdots e_m e_0), & n = 2m + 1 \equiv 3 \mod 4. \end{cases}$$

Indeed, we note that $\rho(e_k)^c = \rho(e_k)$, $k \geq 0$, and $\rho(e_k)^c = -\rho(e_k)$, $k < 0$, and we verify that $N^c N = I$ when $n \equiv 0, 1, 2, 7 \mod 8$ and $N^c N = -I$ when $n \equiv 3, 4, 5, 6 \mod 8$. This completes the proof, since $\lambda(\lambda\psi^\dagger)^\dagger = |\lambda|^2 (\psi^\dagger)^\dagger$. $\square$

Definition 5.3.3 (Normed spinor space). Let V be a Euclidean space, with spinor space $\mathbb{A}V$.

Fix a *spinor inner product*, by which we mean a Hermitian inner product on $\mathbb{A}V$ such that all vectors act as self-adjoint maps as in Proposition 5.3.1. This amounts to a choice of the parameter $\lambda > 0$.

for $v \in V$, where $\epsilon = \pm 1$ depending on the dimension as in Proposition 5.3.2. Since also the square of this antilinear map coincides with the square of spinor conjugation on $\not{A}V_1$, we conclude from Proposition 5.3.2 that $T_{\not{A}}^{-1}(T_{\not{A}}\psi)^\dagger = \sigma\psi^\dagger$, for some $\sigma \in \mathbf{C}$, $|\sigma| = 1$. Therefore $e^{i\alpha}T_{\not{A}}$ will be compatible with spinor conjugation if $e^{-2i\alpha} = \sigma$. These two equations for the modulus and argument of λ have exactly two solutions differing by sign, which completes the proof. □

We next consider the second question posed above, concerning how to define a map of spinors $T_{\not{A}}$, when the map of vectors $T : V_1 \to V_2$ is not an isometry. Recall that when T is an isometry, writing $\rho_k : V_k \to \mathcal{L}(\not{A}V_k)$, $k = 1, 2$, for the complex V_k representations defining the spinor spaces, we used that $\rho_2 T$ was a second complex V_1 representation. Comparing this to ρ_1, the existence of $T_{\not{A}}$ followed from Theorem 5.2.3.

When T is not an isometry, this argument breaks down, since $\rho_2 T$ is not a complex V_1 representation. Indeed,

$$\rho_2(Tv)^2 = \langle Tv \rangle^2 I \neq \langle v \rangle^2 I.$$

What we do in this case is to produce an isometry $U : V_1 \to V_2$ from T by polar factorization. Proposition 1.4.4 shows that there is a unique isometry $U : V_1 \to V_2$ such that

$$T = US_1 = S_2 U,$$

for some positive symmetric maps $S_1 \in \mathcal{L}(V_1)$ and $S_2 \in \mathcal{L}(V_2)$. The formula for this U, which we refer to as the *polar isometric factor* of T, is

$$U := T(T^*T)^{-1/2} = (TT^*)^{-1/2}T.$$

Definition 5.3.6 (Induced spinor map). Let V_1, V_2 be Euclidean spaces, with normed spinor spaces $\not{A}V_1$ and $\not{A}V_2$ respectively. Assume that $T : V_1 \to V_2$ is an invertible linear map. Denote by $U : V_1 \to V_2$ the polar isometric factor of T. If $\dim V_1$ is odd, we assume that the main reflectors w_n and Uw_n in V_1 and V_2 respectively both act as $+I$ or as $-I$. Then we refer to the two maps

$$T_{\not{A}} := U_{\not{A}} : \not{A}V_1 \to \not{A}V_2$$

constructed from U as in Proposition 5.3.5 as the *spinor maps induced* by T.

Exercise 5.3.7. Consider the Euclidean plane V, with ON-basis $\{e_1, e_{-1}\}$, and consider the standard representation of the complex spinor space $\not{A}V$ from Example 5.1.5(i), equipped with the spinor duality and conjugation from the proofs of Propositions 5.3.1 and 5.3.2. Calculate the two spinor maps $T_{\not{A}} : \not{A}V \to \not{A}V$ induced by the linear map $T : V \to V$ with matrix

$$T = \begin{bmatrix} 1 & 0 \\ 3 & 2 \end{bmatrix}$$

in the basis $\{e_1, e_{-1}\}$.

Recall from Section 2.3 that if vector maps T induce multivector maps $T_\wedge$, then

$$(T^{-1})_\wedge = (T_\wedge)^{-1} \quad \text{and} \quad (T_2 \circ T_1)_\wedge = (T_2)_\wedge \circ (T_1)_\wedge,$$

but $(\lambda T)_\wedge \neq \lambda T_\wedge$ and $(T_1 + T_2)_\wedge \neq (T_1)_\wedge + (T_2)_\wedge$. The corresponding result for induced spinor maps holds for isometries, but in general the composition rule fails for non-isometries.

Exercise 5.3.8 (Failure of transitivity). (i) Let $T : V_1 \to V_2$ be an invertible linear map with polar isometric factor $U : V_1 \to V_2$. Show that T^{-1} has polar isometric factor U^{-1}. Conclude that

$$(T_{\cancel{\triangle}})^{-1}$$

are the spinor maps induced by T^{-1}.

(ii) Let $T_1 = T$ be as in (i), and let $T_2 : V_2 \to V_3$ be an invertible linear map with polar isometric factor $U_2 : V_2 \to V_3$. Construct T_1 and T_2 such that $U_2 U_1$ is not the polar isometric factor of $T_2 T_1$. Conclude that

$$(T_2)_{\cancel{\triangle}} \circ (T_1)_{\cancel{\triangle}}$$

are not in general the spinor maps induced by $T_2 \circ T_1$. Show, however, that this is the case when at least one of the maps T_1 and T_2 is an isometry.

For the action of general multivectors on spinors, we note the following somewhat surprising result.

Proposition 5.3.9 (Polar factorization of induced maps). *Let $T : V_1 \to V_2$ be an invertible linear map between Euclidean spaces, and define its polar isometric factor $U : V_1 \to V_2$ as above. Let $T_\wedge : \wedge V_1 \to \wedge V_2$ be the $\wedge$ homomorphism induced by T from Proposition 2.3.2, and let $U_\wedge = U_\triangle : \triangle V_1 \to \triangle V_2$ be the induced $\triangle$ (as well as $\wedge$) homomorphism induced by U. Then the polar isometric factor of $T_\wedge$ equals $U_\wedge$.*

Proof. We have

$$T_\wedge((T_\wedge)^* T_\wedge)^{-1/2} = T_\wedge((T^*)_\wedge T_\wedge)^{-1/2} = T_\wedge((T^*T)_\wedge)^{-1/2}$$
$$= T_\wedge((T^*T)^{-1/2})_\wedge = (T(T^*T)^{-1/2})_\wedge = U_\wedge.$$

For the first equality, see Section 2.7. The third equality uses $(A^2)_\wedge = ((A^{-1})_\wedge)^{-2}$ for $A = (T^*T)^{1/2}$, which is true.   □

5.4   Abstract Spinor Spaces

In Section 3.3 we introduced the notion of abstract Clifford algebras, among which we treat $\triangle V = (\wedge V, \triangle)$ as the standard Clifford algebra. Similarly, we introduced the concept of abstract spinor spaces in Definition 5.1.1, for which it takes two full

V-rotations two complete one full rotation of the spinor space. Among these we treat the spinor space $\not\triangle V$ from Section 5.2 as the standard spinor space. However, there is one important difference: all abstract Clifford algebras are isomorphic to $\triangle V$, but there are infinitely many nonisomorphic abstract spinor spaces, as we shall see. The goal of the present section is to identify all possible abstract complex spinor and tensor spaces of three and four-dimensional Euclidean space, up to isomorphism.

We start by collecting the basic tools from the theory of representations of compact Lie groups, which we need. The only groups we use are $G = \mathrm{SO}(V)$ and $G = \mathrm{Spin}(V)$, where V is a given Euclidean space.

- Let G denote a compact Lie group. A complex *representation* (S, ρ) of G is a complex linear space S together with a smooth homomorphism $\rho : G \to \mathcal{L}(S)$. If (S, ρ) is a representation of G, and if $S' \subset S$ is a subspace such that

$$\rho(g)w \in S', \quad \text{for all } w \in S', g \in G,$$

then (S', ρ') is said to be a *subrepresentation* of (S, ρ), where $\rho'(g)$ denotes the restriction of $\rho(g)$ to the invariant subspace S'. If (S, ρ) has no nontrivial subrepresentations, that is none besides $S' = \{0\}$ and $S' = S$, then we say that (S, ρ) is an *irreducible representation* of G.

- Let (S_1, ρ_1) and (S_2, ρ_2) be two representations of G. We write

$$\mathcal{L}_G(S_1, S_2) := \{T \in \mathcal{L}(S_1, S_2) \; ; \; \rho_2(g)T = T\rho_1(g), \, g \in G\}$$

and call $T \in \mathcal{L}_G(S_1, S_2)$ a *G-intertwining map*. The representations (S_1, ρ_1) and (S_2, ρ_2) are *isomorphic* if there exists a bijective map $T \in \mathcal{L}_G(S_1, S_2)$. Let $\widehat{G}$ be the set of equivalence classes of mutually isomorphic irreducible representations of G.

  Schur's lemma shows that if (S_1, ρ_1), (S_2, ρ_2) are irreducible representations of G, then

  (i) $\mathcal{L}_G(S_1, S_2) = \{0\}$ if they are nonisomorphic, and

  (ii) $\mathcal{L}_G(S_1, S_2) = \{\lambda T \; ; \; \lambda \in \mathbf{C} \setminus \{0\}\}$ if $T \in \mathcal{L}_G(S_1, S_2)$ is an isomorphism.

  The proof follows from the observation that if $T \in \mathcal{L}_G(S_1, S_2)$, then $\mathrm{N}(T)$ is a subrepresentation of S_1 and $\mathrm{R}(T)$ is a subrepresentation of S_2. Irreducibility allows us to conclude.

- Given a representation (S, ρ) of G, we can write

$$S = S_1 \oplus S_2 \oplus \cdots \oplus S_k, \tag{5.6}$$

where each S_j is an invariant subspace of $\rho(g)$ for all $g \in G$, and each S_j is an irreducible representation of G. For the proof, we construct an auxiliary Hermitian inner product $(\cdot, \cdot)$ such that G acts isometrically on S, that is,

$$(\rho(g)w)^2 = (w)^2, \quad \text{for all } w \in S, g \in G.$$

Such invariant inner products are not unique. The existence follows from the well-known result in measure theory that there exists a Haar measure on G, that is, a Borel measure $d\mu$ with total measure $\mu(G) = 1$ that is left and right invariant in the sense that

$$\int_G f(gx)d\mu(x) = \int_G f(x)d\mu(x) = \int_G f(xg)d\mu(x),$$

for all Borel measurable functions $f : G \to \mathbf{C}$ and $g \in G$. Starting from any Hermitian inner product $(\cdot, \cdot)'$ on S, we see that

$$(w_1, w_2) := \int_G (\rho(g)w_1, \rho(g)w_2)'d\mu(g)$$

defines a Hermitian inner product that is invariant under G.

To obtain a decomposition of the form (5.6), we simply note that if S_1 is a subrepresentation, then so is $S_1^\perp$, using a complex analogue of Lemma 4.1.2. We continue to split S_1 and $S_1^\perp$ further until irreducible subrepresentations have been obtained.

- Given a representation (S, ρ) of G, the number of irreducible subrepresentations from each equivalence class $\alpha \in \widehat{G}$ that it contains is well defined. To see this, assume that

$$S = S_1 \oplus S_2 \oplus \cdots \oplus S_k = S_1' \oplus S_2' \oplus \cdots \oplus S_l'$$

are two decompositions of S into irreducible representations. Fix $\alpha \in \widehat{G}$ and consider any $S_i \in \alpha$ appearing in the first decomposition, and

$$p_j' : S_i \to S_j' : w \mapsto p_j'(w),$$

where p_j' denotes orthogonal projection onto S_j'. Schur's lemma implies that $S_i \subset \oplus_{S_j' \in \alpha} S_j'$. With this and the reverse result obtained by swapping the roles of S_i and S_j', we get $k = l$ and

$$\oplus_{S_i \in \alpha} S_i = \oplus_{S_j' \in \alpha} S_j', \quad \alpha \in \widehat{G}.$$

- A tool for identifying which irreducible representations $\alpha \in \widehat{G}$ are present in a given representation (S, ρ), and their multiplicities, is the *character* χ_S of (S, ρ). This is the function

$$\chi_S : G \to \mathbf{C} : g \mapsto \chi_S(g) := \mathrm{Tr}(\rho_S(g)),$$

where Tr denotes the trace functional as in Section 1.4. Since $\mathrm{Tr}(T\rho(g)T^{-1}) = \mathrm{Tr}(\rho(g))$, isomorphic representations have the same character. The *Peter–Weyl theorem* in representation theory shows that the characters χ_α, $\alpha \in \widehat{G}$, of the irreducible representations form an ON-basis for

$$L_2^{cl}(G) := \{f \in L_2(G) \; ; \; f(gxg^{-1}) = f(x), \text{ for all } x, g \in G\}.$$

Such functions f are referred to as *class functions* on G.

We shall not use the general fact that the class functions span $L_2^{\mathrm{cl}}(G)$. To see the orthogonality of χ_{S_1} and χ_{S_2} for two nonisomorphic irreducible representations, we consider the auxiliary representation of G on $S := \mathcal{L}(S_1; S_2)$ given by

$$\rho(g)T := \rho_{S_2}(g) \circ T \circ \rho_{S_1}(g^{-1}), \quad g \in G.$$

This is seen to have character $\chi(g) = \chi_{S_1}(g)^c \chi_{S_2}(g)$. Define the linear operator $P := \int_G \rho(g) d\mu(g) \in \mathcal{L}(S)$, using componentwise integration, and note that

$$P^2 = \left(\int_G \rho(g) d\mu(g) \right) \left(\int_G \rho(h) d\mu(h) \right) = \int_G \int_G \rho(gh) d\mu(g) d\mu(h)$$
$$= \int_G \left(\int_G \rho(g) d\mu(g) \right) d\mu(h) = P.$$

Thus P is a projection, and we check that its range is $\mathcal{L}_G(S_1; S_2)$, which gives

$$\int_G \chi_1(g)^c \chi_{S_2}(g) d\mu(g) = \mathrm{Tr} P = \dim \mathcal{L}_G(S_1; S_2).$$

The orthonormality of characters therefore follows from Schur's lemma.

- As a corollary of the Peter–Weyl theorem, it follows that there are at most countably many nonisomorphic irreducible representations of G. Moreover, the number of irreducible subrepresentations that a given representation contains equals $\int_G |\chi_S|^2$.

We now apply these tools from Lie group representation theory, starting with the three-dimensional spin group. Fixing an ON-basis for V, we write

$$\mathrm{Spin}(n) := \mathrm{Spin}(\mathbf{R}^n), \quad \text{and} \quad \mathrm{SO}(n) := \mathrm{SO}(\mathbf{R}^n).$$

Recall from Theorem 3.4.13 that $\triangle \mathbf{R}^3$ is isomorphic to $\mathbf{C}(2)$. By restricting such an isomorphism to $\mathrm{Spin}(3) \subset \triangle \mathbf{R}^3$, this Lie group is seen to be isomorphic to the complex isometries on $\mathbf{C}^2$.

Definition 5.4.1. Let the *special unitary group* SU(2) in two complex dimensions be

$$\mathrm{SU}(2) := \left\{ \begin{bmatrix} a & b \\ -b^c & a^c \end{bmatrix} ; a, b \in \mathbf{C}, |a|^2 + |b|^2 = 1 \right\}.$$

By the *standard isomorphism* $\mathrm{SU}(2) \leftrightarrow \mathrm{Spin}(3)$, we mean

$$\mathrm{SU}(2) \ni \begin{bmatrix} a & b \\ -b^c & a^c \end{bmatrix} \leftrightarrow a_1 - a_2 j_1 - b_1 j_2 - b_2 j_3 \in \mathrm{Spin}(3) = S^3 \subset \mathbf{H},$$

where $j_1 = e_{23}$, $j_2 = e_{31}$, $j_3 = e_{12}$, and $a = a_1 + ia_2$ and $b = b_1 + ib_2$. Here $i \in \mathbf{C}$ is the algebraic imaginary unit, not related to $\mathbf{H}$.

We set out to find the characters of all irreducible representations of $SU(2) =$ $Spin(3) = S^3$. Note that if f is a class function on this group, then it is uniquely determined by its values at points

$$SU(2) \ni \begin{bmatrix} \exp(-it) & 0 \\ 0 & \exp(it) \end{bmatrix} \leftrightarrow \cos t + j_1 \sin t = \exp(tj_1) \in Spin(3),$$

for $t \in [0, \pi]$. Indeed, as noted at the end of Section 3.2, all rotations of pure quaternions $\wedge^2 V$ can be represented $x \mapsto qxq^{-1}$, $q \in Spin(3)$. In particular, we can rotate any pure quaternion to the line $[j_1]$ in this way.

From

$$\langle qx \rangle^2 = \langle x \rangle^2 = \langle xq \rangle^2, \qquad \text{for all } q \in S^3, \ x \in \mathbf{H},$$

it is clear that Lebesgue surface measure on $Spin(V)$ is invariant under $Spin(3)$, and therefore equals the Haar measure, modulo the normalizing factor $\int_{S^3} |d\hat{x}| = 2\pi^2$.

Consider representations (V_k, ρ_k) of $SU(2)$, where

$$V_k := \{\text{polynomials } P : \mathbf{C}^2 \to \mathbf{C} \ ; \ P(\lambda z, \lambda w) = \lambda^k P(z, w), \ \lambda \in \mathbf{R}, \ z, w \in \mathbf{C}\},$$

and $\rho_k(T)P := P \circ T^{-1}$. Concretely, V_k is spanned by $\{z^{k-j}w^j \ ; \ 0 \leq j \leq k\}$ and

$$(\rho_k(T)P)(z, w) = P(a^c z - bw, b^c z + aw) \quad \text{if } T = \begin{bmatrix} a & b \\ -b^c & a^c \end{bmatrix}.$$

Proposition 5.4.2 (Finding all 3D spinor spaces). *The irreducible representations of the Lie group* $SU(2) = Spin(3)$ *are indexed by the natural numbers* $\widehat{Spin}(3) =$ $\{0, 1, 2, 3, \dots\}$. *The representations in the equivalence class* $k \in \widehat{Spin}(3)$ *are* $(k+1)$-*dimensional and are uniquely determined by the character values*

$$\chi_k(\exp(tj_1)) = \begin{cases} 1 + 2 \displaystyle\sum_{m=1}^{k/2} \cos(2mt), & k \text{ even}, \\ 2 \displaystyle\sum_{m=0}^{(k-1)/2} \cos((2m+1)t), & k \text{ odd}. \end{cases}$$

With the terminology of Definition 5.1.1, these irreducible representations are abstract spinor spaces if k *is odd and abstract tensor spaces if* k *is even.*

Proof. Note that $z^{k-m}w^m$ is an eigenvector to $\rho_k(\exp(tj_1))$ with eigenvalue $e^{it(k-2m)}$. Summing these gives the stated characters. We calculate

$$\int_{S^3} |\chi_k(g)|^2 d\mu(g) = \frac{1}{2\pi^2} \int_0^\pi |\chi_k(\exp(tj_1))|^2 4\pi \sin^2 t \, dt$$

$$= \frac{2}{\pi} \int_0^\pi \left| \frac{1}{2i} \sum_{m=0}^k e^{it(k-2m)}(e^{it} - e^{-it}) \right|^2 dt = \frac{2}{\pi} \int_0^\pi |\sin((k+1)t)|^2 dt = 1,$$

so these representations are irreducible. Moreover, it is known from Fourier theory that $\{\cos(jx)\}_{j=0}^{\infty}$ is dense L_2, so these are all the irreducible representations. For the last statement, we note that $-1 \in \mathrm{Spin}(V)$ acts by multiplication with $(-1)^k$ on V_k. □

Example 5.4.3 ($\triangle \mathbf{R}^3$). The standard representation of $\mathrm{Spin}(V)$ on the three-dimensional vector space V_c itself is given by $\rho = p$ from Proposition 4.1.9. The rotor $q = \exp(tj_1)$ acts by rotation the angle $2t$ and with plane of rotation $[j_1]$. Summing the diagonal elements of the matrix for this rotation in the standard basis yields the character $\chi(\exp(tj_1)) = 1 + 2\cos(2t)$. Thus the tensor space $2 \in \widehat{\mathrm{Spin}}(V)$ is the standard representation $V_c = \wedge^1 V_c$. Through the Hodge star map and Proposition 4.1.10, $\wedge^2 V_c$ is an isomorphic representation. Even more trivial: the representations on $\wedge^0 V_c$ and $\wedge^3 V_c$ are both representatives of $0 \in \widehat{\mathrm{Spin}}(V)$, where all rotors act as the identity on a one-dimensional space.

Example 5.4.4 ($\triangle \mathbf{R}^3$). Consider the spinor space $\triangle V$ for three-dimensional Euclidean space V from Definition 5.2.4, and restrict this to a complex $\mathrm{Spin}(V)$ representation. Using the matrices from Example 5.1.7(ii), we have

$$\rho(j_1) = \rho(e_1 e_{-1}) = \begin{bmatrix} 0 & 1 \\ 1 & 0 \end{bmatrix} \begin{bmatrix} 0 & -i \\ i & 0 \end{bmatrix} = \begin{bmatrix} i & 0 \\ 0 & -i \end{bmatrix},$$

and therefore the character for this representation is $\chi(e^{tj_1}) = \mathrm{Tr} \begin{bmatrix} e^{it} & 0 \\ 0 & e^{-it} \end{bmatrix} = 2\cos t$. We conclude that the smallest abstract V-spinor space $1 \in \widehat{\mathrm{Spin}}(V)$ is the standard spinor space $\triangle V$.

The following two examples build on Section 8.2, but are not used elsewhere and may be omitted and returned to after Section 8.2 has been read.

Example 5.4.5 (Harmonic polynomials). Consider the abstract V-tensor space of scalar (complexified) k-homogeneous harmonic polynomials $\mathcal{P}_k^{sh}$ from Example 5.1.2(iii). To avoid dealing with bases for these spaces, we recall Proposition 8.2.3, which amounts to a statement about decomposition of the representation on all k-homogeneous polynomials into subrepresentations. We have the representation $\rho(q)P(x) = P(q^{-1}xq)$ on

$$\mathcal{P}_k^s = \mathcal{P}_k^{sh} \oplus |x|^2 \mathcal{P}_{k-2}^s,$$

where both the terms are subrepresentations, and $|x|^2 \in \mathcal{L}_{\mathrm{Spin}(3)}(\mathcal{P}_{k-2}^s, |x|^2 \mathcal{P}_{k-2}^s)$ is an isomorphism. It follows that the character for the representation $\mathcal{P}_k^{sh}$ is the difference between the characters of the two representations $\mathcal{P}_k^s$ and $\mathcal{P}_{k-2}^s$. To avoid unnecessarily technical trace computations, we choose the convenient basis

$$\{x_1^\alpha z^\beta \bar{z}^\gamma\}_{\alpha+\beta+\gamma=k} = \{x_1^\alpha (x_2 + ix_3)^\beta (x_2 - ix_3)^\gamma\}_{\alpha+\beta+\gamma=k}$$

for $\mathcal{P}_k^s$. Here $\{x_1, x_2, x_3\}$ are the coordinates in the ON-basis $\{e_1, e_2, e_3\}$ for V, that is, the dual basis for $V^* = \mathcal{P}_1^s$. From $\rho(e^{tj_1})P(x) = P(e^{-tj_1}xe^{tj_1})$ we see that

$\{x_1^\alpha z^\beta \bar{z}^\gamma\}$ is an eigenbasis of the operator $\rho(e^{tj_1})$ with eigenvalues $e^{2it(\beta-\gamma)}$. This gives the character

$$\chi(e^{tj_1}) = \sum_{\alpha+\beta+\gamma=k} e^{2it(\beta-\gamma)} = \sum_{\beta+\gamma\leq k} e^{2it(\beta-\gamma)}$$

for the representation $\mathcal{P}_k^s$, and therefore the character

$$\chi(e^{tj_1}) = \sum_{\beta+\gamma=k-1,k} e^{2it(\beta-\gamma)} = 1 + \sum_{m=0}^{k} \cos(2mt)$$

for the representation $\mathcal{P}_k^{sh}$. We conclude that $\mathcal{P}_k^{sh}$ is the irreducible V-tensor space $2k \in \widehat{\mathrm{Spin}}(V)$.

Example 5.4.6 (Monogenic polynomials). Consider the abstract V-spinor space of k-homogeneous monogenic polynomials $\mathcal{P}_k^m$ from Example 5.1.2(iv), for a three-dimensional Euclidean space V. On this real linear space of dimension $8(k+1)$, there is a rather natural complex structure, namely multiplication by the orientation $e_{123} \in \triangle^3 V$, which belongs to the center of the algebra. In this way, we consider $\mathcal{P}_k^m$ as a $4(k+1)$-dimensional complex spinor space with rotors acting as $\rho(q)P(x) = qP(q^{-1}xq)$.

Similar to Example 5.4.5, having Proposition 8.2.3 in mind, we first compute the character of the spinor space $\mathcal{P}_k$ (also considered as a complex linear space) with same same action. To handle the multivector-valued polynomials, we note that we have an isomorphism of representations

$$\mathcal{P}_k \leftrightarrow \mathbf{H} \otimes \mathcal{P}_k^s,$$

where $q \in \mathrm{Spin}(V)$ acts on $\mathbf{H}$ as $x \mapsto qx$, and on $\mathcal{P}_k^s$ as $P(x) \mapsto P(q^{-1}xq)$. It follows that the character is

$$\chi_{\mathcal{P}_k}(e^{tj_1}) = \chi_{\mathbf{H}}(e^{tj_1})\chi_{\mathcal{P}_k^s}(e^{tj_1}) = 4\cos t \sum_{\beta+\gamma\leq k} e^{2it(\beta-\gamma)},$$

from which we get

$$\chi_{\mathcal{P}_k^m}(e^{tj_1}) = \chi_{\mathcal{P}_k}(e^{tj_1}) - \chi_{\mathcal{P}_{k-1}}(e^{tj_1}) = 4\cos t \sum_{j=0}^{k} e^{2it(k-2j)} = 4\sum_{j=0}^{k} \cos((2j+1)t).$$

Proposition 5.4.2 now shows that the V-spinor space $\mathcal{P}_k^m$ is not irreducible but contains two copies of the irreducible spinor space $2k+1 \in \widehat{\mathrm{Spin}}(3)$. We can find such a subrepresentation as a minimal left ideal $S \subset \triangle V$. Then the subspace of monogenic polynomials $\mathcal{P}_k^m$ with values in S will be such an irreducible spinor subrepresentation. Note that such subrepresentations are not unique. And indeed, there are infinitely many left ideals S that can be used.

We next study representations of the group Spin(4) of rotors in four-dimensional Euclidean space V. It is a fortunate fact that we have an isomorphism

$$\text{Spin}(4) \leftrightarrow \text{Spin}(3) \times \text{Spin}(3),$$

as we shall see. This should not come as a surprise, since $\triangle^{\text{ev}}V$, by Proposition 3.3.5, is isomorphic to the Clifford algebra for $\mathbf{R}^3$ with negative definite inner product, which in turn is isomorphic to $\mathbf{H} \oplus \mathbf{H}$ according to Theorem 3.4.13. To make these isomorphisms explicit, fix an ON-basis $\{e_1, e_2, e_3, e_4\}$ and an orientation e_{1234} for V, and define $p_\pm := \frac{1}{2}(1 \pm e_{1234})$. We have

$$p_\pm^2 = p_\pm, \quad p_+ + p_- = 1,$$

and $p_\pm \in \mathcal{Z}(\triangle^{\text{ev}}V)$. Therefore $p_\pm$ split the even subalgebra

$$\triangle^{\text{ev}}V = \triangle_+^{\text{ev}}V \oplus \triangle_-^{\text{ev}}V,$$

where $\triangle_\pm^{\text{ev}}V := p_\pm \triangle^{\text{ev}}V$ are the two two-sided ideals in the even subalgebra. The subspaces $\triangle_\pm^{\text{ev}}V$ are real algebras in themselves, each being isomorphic to $\mathbf{H}$, but should not be considered subalgebras of $\triangle^{\text{ev}}V$, since $p_\pm$, and not 1, is the identity element in $\triangle_\pm^{\text{ev}}V$. To make the isomorphism $\triangle^{\text{ev}}V = \mathbf{H} \oplus \mathbf{H}$ explicit, we identify basis elements as

$$\triangle_+^{\text{ev}}V \ni p_+ = p_+ e_{1234} = \tfrac{1}{2}(1 + e_{1234}) \leftrightarrow e \in \mathbf{H},$$
$$\triangle_+^{\text{ev}}V \ni p_+ e_{41} = p_+ e_{23} = \tfrac{1}{2}(e_{41} + e_{23}) \leftrightarrow j_1 = -i \in \mathbf{H},$$
$$\triangle_+^{\text{ev}}V \ni p_+ e_{42} = -p_+ e_{13} = \tfrac{1}{2}(e_{42} - e_{13}) \leftrightarrow j_2 = -j \in \mathbf{H},$$
$$\triangle_+^{\text{ev}}V \ni p_+ e_{43} = p_+ e_{12} = \tfrac{1}{2}(e_{43} + e_{12}) \leftrightarrow j_3 = -k \in \mathbf{H},$$

with notation as in Section 3.2, but writing $e \in \mathbf{H}$ for the identity in $\mathbf{H}$ to avoid confusion with $1 \in \triangle^{\text{ev}}V$. We then use the self-inverse automorphism

$$\triangle^{\text{ev}}V \to \triangle^{\text{ev}}V : w \mapsto e_4 w e_4$$

to identify $\triangle_-^{\text{ev}}V = \triangle_+^{\text{ev}}V$, and write $e' := e_4 e e_4 = p_-$ and $j_k' := e_4 j_k e_4$, $k = 1, 2, 3$.

Proposition 5.4.7 (Spin(4) = Spin(3) × Spin(3)). *Let V be a four-dimensional Euclidean space. We have an algebra isomorphism $\triangle^{\text{ev}}V \to \mathbf{H} \oplus \mathbf{H} : w \mapsto (p_+ w, e_4 p_- w e_4)$, where we identify $\mathbf{H} = \triangle_+^{\text{ev}}V$ as above. Letting $\langle \cdot, \cdot \rangle_{\mathbf{H}} = 2\langle \cdot, \cdot \rangle_{\triangle_+^{\text{ev}}V}$, this isomorphism is an isometry. This algebra isomorphism restricts to a group isomorphism $\text{Spin}(V) \to S^3 \times S^3$, where $S^3 = \text{Spin}(3)$ is the unit quaternion 3-sphere, as in the following diagram:*

$$
\begin{array}{ccc}
\triangle^{\text{ev}}V & \longrightarrow & \mathbf{H} \oplus \mathbf{H} \\
\uparrow & & \uparrow \\
\text{Spin}(V) & \longrightarrow & S^3 \times S^3 \\
\downarrow{\scriptstyle p} & & \downarrow{\scriptstyle \tilde{p}} \\
\text{SO}(V) & \longrightarrow & \text{SO}(\mathbf{H})
\end{array}
\qquad (5.7)
$$

The standard covering map $p : \mathrm{Spin}(V) \to \mathrm{SO}(V) : q \mapsto q(\cdot)q^{-1}$ *corresponds to the covering map*

$$\tilde{p} : S^3 \times S^3 \to \mathrm{SO}(\mathbf{H}) : (q_1, q_2) \mapsto q_1(\cdot)q_2^{-1}$$

if we use the linear identification $V \to \mathbf{H} : v \mapsto p_+ v e_4$.

Proof. From the above discussion it is clear that

$$\triangle^{\mathrm{ev}} V \to \mathbf{H} \oplus \mathbf{H} : w \mapsto (p_+ w, e_4 p_- w e_4)$$

is an algebra isomorphism. To check the stated mapping of the spin group, write $w \in \triangle^{\mathrm{ev}} V$ as

$$\begin{aligned}
w &= (w_0 1 + w_{1234} e_{1234}) + (w_{12} e_{12} + w_{34} e_{34}) + (w_{13} e_{13} + w_{24} e_{24}) \\
&\quad + (w_{14} e_{14} + w_{23} e_{23}) \\
&= (w_0 + w_{1234})e + (w_0 - w_{1234})e' + (-w_{14} + w_{23})j_1 + (w_{14} + w_{23})j_1' \\
&\quad + (-w_{13} - w_{24})j_2 + (-w_{13} + w_{24})j_2' + (w_{12} - w_{34})j_3 + (w_{12} + w_{34})j_3' \\
&= (x_0 e + x_1 j_1 + x_2 j_2 + x_3 j_3) + (y_0 e' + y_1 j_1' + y_2 j_2' + y_3 j_3').
\end{aligned}$$

It follows that $x_0^2 + x_1^2 + x_2^2 + x_3^2 = 1 = y_0^2 + y_1^2 + y_2^2 + y_3^2$ if and only if $|w| = 1$ and $w_0 w_{1234} - w_{14} w_{23} + w_{13} w_{24} - w_{12} w_{34}$. By Example 4.1.8 this is equivalent to $w \in \mathrm{Spin}(V)$.

Next consider the action of $\mathrm{Spin}(V)$ on V with p. We have

$$qvq^{-1} = qv\bar{q} = (p_+ q + p_- q)v(p_+ \bar{q} + p_- \bar{q}) = q_1 v e_4 \bar{q}_2 e_4 + e_4 q_2 e_4 v \bar{q}_1,$$

where $q_1 := p_+ q,\, q_2 := e_4 p_- q e_4 \in \mathbf{H} = \triangle_+^{\mathrm{ev}} V$. Note that two terms vanish, since $v p_+ = p_- v$ and $p_+ p_- = 0$. To write this action entirely in terms of quaternions, we need to identify $V = \mathbf{H}$. Multiplying $v \in V$ by e_4, we have $v e_4 \in \triangle^{\mathrm{ev}} V$. Projecting onto the subspace $\triangle_+^{\mathrm{ev}} V$, we verify that

$$p_+ v e_4 = v_1 i + v_2 j + v_3 k + v_4 e \in \mathbf{H} = \triangle_+^{\mathrm{ev}} V$$

if $v = v_1 e_1 + v_2 e_2 + v_3 e_3 + v_4 e_4 \in V$. We obtain

$$p_+(qvq^{-1})e_4 = q_1(p_+ v e_4)\bar{q}_2 + (p_+ e_4 q_2 e_4)v\bar{q}_1 e_4 = q_1(p_+ v e_4)q_2^{-1},$$

since $p_+ e_4 q_2 = 0$, which completes the proof. $\square$

We now construct all irreducible representations of $\mathrm{Spin}(V)$ for a four-dimensional Euclidean space V. Fix an ON-basis $\{e_1, e_2, e_3, e_4\}$ and consider the subgroup

$$T := \{e^{\theta_1 e_{12} + \theta_2 e_{34}} \ ; \ \theta_1, \theta_2 \in \mathbf{R}\} \subset \mathrm{Spin}(V).$$

In representation theory, such a subgroup is referred to as a *maximal torus*. This subgroup plays the same role for $\mathrm{Spin}(4)$ as did the circle, the one-dimensional torus, $\{e^{tj_1} ; t \in \mathbf{R}\}$ for Spin(3). Using the isomorphism from Proposition 5.4.7, we have $e_{12} \leftrightarrow (j_3, j_3)$ and $e_{34} \leftrightarrow (-j_3, j_3)$, and thus

$$e^{\theta_1 e_{12} + \theta_2 e_{34}} \leftrightarrow (e^{t_1 j_3}, e^{t_2 j_3}), \quad \text{where } t_1 = \theta_1 - \theta_2, t_2 = \theta_1 + \theta_2.$$

The importance of the maximal torus $T \subset \text{Spin}(V)$ is that a class function $f : \text{Spin}(V) \to \mathbf{C}$ is uniquely determined by its values on T, in the sense that for each element $q \in \text{Spin}(V)$ there is $q_1 \in \text{Spin}(V)$ such that $q_1 q q_1^{-1} \in T$. In fact, only the values on part of T are needed. Using the $S^3 \times S^3$ characterization, this result carries over from three dimensions, but we can also prove it directly for $\text{Spin}(V)$. According to Proposition 4.3.9(i), which we note is a result on maximal tori in the Lie groups $\text{Spin}(V)$, each $q \in \text{Spin}(V)$ belongs to one such maximal torus (with $e_1' e_2', e_3' e_4'$ instead of e_{12}, e_{34}). We can then find $q_1 \in \text{Spin}(V)$ such that the induced rotation $w \mapsto q_1 w q_1^{-1}$ of $\triangle V$ maps $e_1' e_2', e_3' e_4'$ to e_{12}, e_{34}, possibly after first having adjusted the angles so that the two bases have the same orientation.

Proposition 5.4.8 (Finding all 4D spinor spaces). *For the Lie group* $\text{Spin}(4) = \text{Spin}(3) \times \text{Spin}(3)$, *the irreducible representations are indexed by the pairs of natural numbers* $\widehat{\text{Spin}(4)} = \mathbf{N}^2 = \{(k,l) ; k,l = 0,1,2,3,\ldots\}$. *The representations in the equivalence class* $(k,l) \in \widehat{\text{Spin}(4)}$ *are* $(k+1)(l+1)$-*dimensional and are uniquely determined by the character values*

$$\chi_{(k,l)}(e^{\theta_1 e_{12} + \theta_2 e_{34}}) = \chi_k(e^{(\theta_1 - \theta_2)j_3}) \chi_l(e^{(\theta_1 + \theta_2)j_3}),$$

where χ_k *denotes the character for the irreducible representation* $k \in \widehat{\text{Spin}(3)}$ *of* $\text{Spin}(3)$ *from Proposition 5.4.2. The irreducible abstract V-tensor spaces correspond to those pairs for which $k + l$ is even, and the irreducible abstract V-spinor spaces correspond to those for which $k + l$ is odd.*

The integral $\int_{\text{Spin}(4)} f(x) d\mu(x)$ *of a class function f on $\text{Spin}(4)$ with respect to Haar measure equals*

$$\frac{4}{\pi^2} \int_0^\pi \int_0^\pi f(\exp(\tfrac{1}{2}(t_1 + t_2)e_{12} + \tfrac{1}{2}(-t_1 + t_2)e_{34})) \sin^2 t_1 \sin^2 t_2 dt_1 dt_2.$$

Proof. It is clear that the Haar measure on $\text{Spin}(3) \times \text{Spin}(3)$ is the product measure $d\mu(x_1)d\mu(x_2)$. We see that a class function is determined by its values on the quarter $\{(e^{t_1 j_3}, e^{t_2 j_3}) ; 0 < t_1, t_2 < \pi\}$ of the maximal torus, and the stated integral formula follows by translating back to $\text{Spin}(V)$.

Now let (V_k, ρ_k) be the irreducible representations of $\text{SU}(2) = \text{Spin}(3)$ used in the proof of Proposition 5.4.2. For $(k,l) \in \mathbf{N}^2$, define the representation $\rho_{(k,l)}$ of $\text{Spin}(3) \times \text{Spin}(3)$ on the tensor product space $V_k \otimes V_l$ by applying the universal property to the bilinear map

$$(P_1, P_2) \mapsto (\rho_k(q_1)P_1) \otimes (\rho_l(q_2)P_2)$$

to obtain a linear map $\rho_{(k,l)}(q_1, q_2) \in \mathcal{L}(V_k \otimes V_l)$ for each $(q_1, q_2) \in \text{Spin}(3) \times \text{Spin}(3)$. The character of this representation is seen to be $\chi_{(k,l)}(q_1, q_2) =$

$\chi_k(q_1)\chi_l(q_2)$, and Fubini's theorem shows that $\int_{\mathrm{Spin}(4)} |\chi_{(k,l)}(q_1,q_2)|^2 = 1$, so this is an irreducible representation of $\mathrm{Spin}(3) \times \mathrm{Spin}(3)$. In particular, for $q_1 = e^{(\theta_1-\theta_2)j_3}$ and $q_2 = e^{(\theta_1+\theta_2)j_3}$ we obtain the stated character values.

Since $\{\chi_k(e^{tj_3})\}_{k=0}^\infty$ is an ON-basis for $L_2((0,\pi); \frac{2}{\pi}\sin^2 t dt)$, the functions $\{\chi_k(e^{t_1 j_3})\chi_l(e^{t_2 j_3})\}_{k,l=0}^\infty$ form an ON-basis for

$$L_2((0,\pi) \times (0,\pi); \tfrac{4}{\pi^2}\sin^2 t_1 \sin^2 t_2 dt_1 dt_2).$$

Hence the representations $\rho_{(k,l)}$ constitute all the possible irreducible representations of $\mathrm{Spin}(3) \times \mathrm{Spin}(3) = \mathrm{Spin}(4)$.   □

Example 5.4.9 ($\triangle \mathbf{R}^4$). The standard representation of $\mathrm{Spin}(V)$ on the four-dimensional space V_c itself is given by $\rho = p$ from Proposition 4.1.9. The rotor $q = e^{\theta_1 e_{12}+\theta_2 e_{34}}$ acts by rotation in the plane $[e_{12}]$ through an angle $2\theta_1$ and rotation in the plane $[e_{34}]$ through an angle $2\theta_2$. This gives the character

$$\chi(e^{\theta_1 e_{12}+\theta_2 e_{34}}) = 2\cos(2\theta_1) + 2\cos(2\theta_2) = (2\cos(\theta_1-\theta_2))(2\cos(\theta_1+\theta_2)),$$

so $(1,1) \in \widehat{\mathrm{Spin}}(4)$ is the standard V-tensor space.

Turning to the induced representation of $\mathrm{Spin}(V)$ on $\wedge^2 V$, this is not irreducible for a four-dimensional space V. Indeed, it splits into two subrepresentations on $\wedge_\pm^2 V := \triangle_\pm^{\mathrm{ev}} V \cap \wedge^2 V$, where $\triangle_\pm^{\mathrm{ev}} V$ are the two two-sided ideals of the even subalgebra as above. Using the basis $\{e_{41}+e_{23}, e_{42}-e_{13}, e_{43}+e_{12}\}$ for $\wedge_+^2 V$, we obtain the character

$$\chi_{\wedge_+^2 V}(e^{\theta_1 e_{12}+\theta_2 e_{34}}) = 1 + 2\cos(2(\theta_1-\theta_2)),$$

so this is the V-tensor space $(2,0) \in \widehat{\mathrm{Spin}}(4)$. On the other hand, a trace calculation with the basis $\{-e_{41}+e_{23}, e_{42}+e_{13}, -e_{43}+e_{12}\}$ for $\wedge_-^2 V$ gives $\chi_{\wedge_-^2 V}(e^{\theta_1 e_{12}+\theta_2 e_{34}}) = 1 + 2\cos(2(\theta_1+\theta_2))$, and we conclude that $\wedge_-^2 V$ is the V-tensor space $(0,2) \in \widehat{\mathrm{Spin}}(4)$.

Note the somewhat surprising result that the standard four-dimensional V-tensor space $(1,1)$ is not the smallest: there are two nonisomorphic three-dimensional V-tensor spaces $(2,0)$ and $(0,2)$!

Example 5.4.10 ($\mathbb{A} \mathbf{R}^4$). Consider the spinor space $\mathbb{A} V$ from Definition 5.2.4 for a four-dimensional space V and the restriction of ρ to $\mathrm{Spin}(V) \subset \triangle^{\mathrm{ev}} V$. We pick bases $\{1, e_{12}, e_1, e_2\}$ for $\mathbb{A} V$ and $\{e_1, e_{-1}, e_2, e_{-2}\}$ for V. Using the matrices from Example 5.1.8(i), we obtain

$$\rho(e^{\theta_1 e_{12}+\theta_2 e_{34}}) = \begin{bmatrix} \cos(\theta_1-\theta_2) & -\sin(\theta_1-\theta_2) & 0 & 0 \\ \sin(\theta_1-\theta_2) & \cos(\theta_1-\theta_2) & 0 & 0 \\ 0 & 0 & \cos(\theta_1+\theta_2) & \sin(\theta_1+\theta_2) \\ 0 & 0 & -\sin(\theta_1+\theta_2) & \cos(\theta_1+\theta_2) \end{bmatrix}.$$

The block structure of this matrix is due to the two subrepresentations $\triangle^+ V$ and $\triangle^- V$. For these we obtain the characters $\chi_{S+} = 2\cos(\theta_1 - \theta_2)$ and $\chi_{S-} = 2\cos(\theta_1 + \theta_2)$ respectively, so these two representations are the two irreducible V-spinor spaces $(1,0)$ and $(0,1) \in \widehat{\mathrm{Spin}}(4)$ respectively.

Exercise 5.4.11. Extend Examples 5.4.5 and 5.4.6 to four-dimensional Euclidean space, and find the irreducible subrepresentations contained in $(\mathcal{P}_k^{sh})_c$ and $(\mathcal{P}_k^{em})_c$. Note that in this case, $(\mathcal{P}_k^{em})_c \neq \mathcal{P}_k^m$.

5.5 Comments and References

5.1 Spinors in general were discovered by Élie Cartan. Here is a quotation from his book [26]: "Spinors were first used under that name, by physicists, in the field of Quantum Mechanics. In their most general form, spinors were discovered in 1913 by the author of this work, in his investigations on the linear representations of simple groups."

There exist many variations of the construction of spinors. The complex spinors that we construct here are usually referred to as *Dirac spinors* in physics. The method of imaginary rotations used in the proof of Proposition 5.1.6 to reduce to the case of Euclidean space is, in the case of spacetime, referred to as *Wick rotation*.

An inspirational book on Clifford algebras and spinors is Lounesto [64].

5.2 In Chapter 3, we recycled Grassmann's exterior algebra $\wedge V$, and identified the Clifford algebra $\triangle V$ and the exterior algebra $\wedge V$. This space of multivectors is the same as a linear space, but we have two different associative products $\triangle$ and $\wedge$ on it, leading to two different algebras. In view of Example 5.1.5, it is tempting to try to realize the spinor space $\triangle V$ as some suitable subspace of a Clifford algebra. One obvious way is to view $\triangle V$ as a minimal left ideal in $\triangle V$, as first proposed by M. Riesz. However, there is no canonical choice of minimal left ideal, and there is no geometric reason for such an identification. The ideal point of view is also problematic in considering spinor bundles as in Section 11.6. The approach in this book is that the spinor space $\triangle V$ is a new independent universe, without any relation to other spaces other than that multivectors act as linear operators on spinors, and that it is unnatural to try to set up any identification between the spinor space and any other space. As before we follow the principle of abstract algebra, as explained in the introduction of Chapter 1, to construct $\triangle V$. However, to calculate with spinors one may choose any favorite representation of them, which may consist in using some ad hoc identification. In this book, we choose the identification with a certain exterior algebra in Example 5.1.5.

The slash notation $I\!\!\!/$ was introduced by Feynman in physics to denote the representation of a vector by gamma matrices. In this book we use the notation differently. We have chosen to use the slash as a symbol for spinor objects, to distiguish them from the related Clifford algebra objects. For example $\not\!\!\triangle V$ for spinor space as compared to $\triangle V$ for Clifford algebra, and $I\!\!\!/$ will denote Dirac operators acting on spinor fields as compared to $\mathbf{D}$ for Dirac operators acting on multivector fields.

We mainly denote objects in spinor spaces by ψ, following the tradition from quantum mechanics, where they represent the wave functions of particles.

5.3 The notion of spinor conjugation is related to charge conjugation in physics, which is a transformation that switches particles and antiparticles. See Hladik [59]. A reference for spinor inner products is Harvey [50]. Spinor inner products and conjugation are well-defined natural structures on spinor spaces also for general inner product spaces. In Section 9.2 we study the case of space-time with three space dimensions, and explain how spinors are used in the Dirac theory in quantum mechanics.

The construction of induced spinor maps using polar decomposition of non-isometries is due to Bourgignon [22]. A motivating application is to construct a map of spinor fields between nonisometric manifolds, and this was used in studying perturbations of $\not\!\!\triangle$-Dirac operators in Bandara, McIntosh, and Rosén [17].

5.4 The representation theory for compact Lie groups used in this section can be found in many textbooks on the subject, including results for higher-dimensional spin groups beyond the three- and four-dimensional examples that we limit ourselves to in this book. Our discussion of small spin groups gathers inspiration from Bröcker and tom Dieck [24]. See also Fulton and Harris [41] and Gilbert and Murray [42]. The spin representation from Example 5.4.6 was introduced by Sommen [86, 87, 88].

Chapter 6

Interlude: Analysis

Road map:

This chapter is **not** where to start reading the second part of this book on multivector analysis, which rather is Chapter 7. The material in the present chapter is meant to be used as a reference for some background material from analysis, which we use in the remaining chapters. A main idea in this second part is that of *splittings of function spaces*:

$$\mathcal{H} = \mathcal{H}_1 \oplus \mathcal{H}_2. \tag{6.1}$$

When $\mathcal{H}_1$ is a k-dimensional subspace of a linear space $\mathcal{H}$ of finite dimension n, then every subspace $\mathcal{H}_2$ of dimension $n - k$ that intersects $\mathcal{H}_1$ only at 0 is a subspace complementary to $\mathcal{H}_1$.

When $\mathcal{H}$ is an infinite-dimensional linear space, which is typically the case for the linear spaces of functions that we use in analysis, it is seldom of any use to have only an algebraic splitting (6.1), meaning that every $x \in \mathcal{H}$ can be written $x = x_1 + x_2$, for unique $x_1 \in \mathcal{H}_1$ and $x_2 \in \mathcal{H}_2$. Using the axiom of choice, one can show that every subspace $\mathcal{H}_1$ has an algebraic complement $\mathcal{H}_2$. Instead, we restrict attention to topological splittings of Hilbert and Banach spaces $\mathcal{H}$, meaning that we assume that $\mathcal{H}_1$ and $\mathcal{H}_2$ are closed subspaces and that we have an estimate

$$\|x_1\| + \|x_2\| \leq C\|x\|$$

for some $C < \infty$. See Definition 6.4.5. The Hilbert space interpretation of this latter reverse of the triangle inequality is that the angle between $\mathcal{H}_1$ and $\mathcal{H}_2$ is positive. We consider splittings of Banach spaces also, and it should be noted that closed subspaces $\mathcal{H}_1$ of a Banach space $\mathcal{H}$ in general do not have any topological complement $\mathcal{H}_2$. Such a complementary subspace exists precisely when there exists a bounded projection onto $\mathcal{H}_1$. A well-known example is the subspace c_0 of the sequence space ℓ_∞, which does not have any topological complement at all. In

© Springer Nature Switzerland AG 2019
A. Rosén, *Geometric Multivector Analysis*, Birkhäuser Advanced Texts Basler Lehrbücher,
https://doi.org/10.1007/978-3-030-31411-8_6

Hilbert space, though, we can always use the orthogonal complement $\mathcal{H}_2 = \mathcal{H}_1^\perp$. In this case we have $\|x_1\|^2 + \|x_2\|^2 = \|x\|^2$, and $C = \sqrt{2}$ suffices.

When doing analysis, not only with scalar valued functions but with more general multivector fields, we can replace the Laplace operator Δ by more fundamental first-order partial differential operators d, δ, and $\mathbf{D}$. When working with these first-order operators on domains D, splittings of function spaces occur frequently. In Section 7.6, Chapter 10, and Section 9.6 we encounter the *Hodge splittings* associated with the exterior and interior derivatives d and δ. These are orthogonal splittings of $L_2(D)$, involving also a finite-dimensional third subspace. In Section 8.3 we encounter the *Hardy splittings* associated to the Dirac operator $\mathbf{D}$. These are in general nonorthogonal, but topological, splittings of $L_2(\partial D)$. In Section 9.3 we consider generalizations of the Hardy splittings for time-harmonic waves, and in Sections 9.4 and 9.5 we show that the fundamental structure behind elliptic boundary value problems consists not only of one splitting, but two independent splittings: one related to the differential equation and one related to the boundary conditions. For Dirac's original equation for the time evolution of spin-1/2 particles in relativistic quantum mechanics, we also see in Section 9.2 how splittings appear, for example, in the description of antiparticles and chirality. Splittings also appear in index theory for Dirac operators in Chapter 12, where the Dirac operators are considered in the splittings $L_2(M; \triangle M) = L_2(M; \triangle^{\mathrm{ev}} M) \oplus L_2(M; \triangle^{\mathrm{od}} M)$ and $L_2(M; \not\triangle M) = L_2(M; \not\triangle^+ M) \oplus L_2(M; \not\triangle^- M)$, respectively, on our manifold M.

In the present chapter, the material in Section 6.4 should not be needed before Section 9.4, with the exception of splittings of function spaces.

6.1   Domains and Manifolds

An extremely useful notation for estimates in analysis is the following, which we use in the remainder of this book.

Definition 6.1.1 (The analyst's (in)equality). By the notation $X \lesssim Y$, we mean that there exists $C < \infty$ such that $X \leq CY$ for all relevant values of the quantities X and Y. By $X \gtrsim Y$ we mean $Y \lesssim X$, and $X \approx Y$ means $X \lesssim Y$ and $X \gtrsim Y$.

We use the following standard terminology and notation concerning the regularity of functions. Consider a function $f : D \to L$, with D an open subset of some affine space X, and L a linear space, or possibly affine. Denote by $B(x, r) := \{y \in X \; ; \; |y - x| < r\}$ the ball with center x and radius $r > 0$ when X is Euclidean space.

- The function f is C^k-*regular* in D, $k = 0, 1, 2, \ldots$, if all directional/partial derivatives $\partial_{v_1} \cdots \partial_{v_m} f(x)$ of order $m \leq k$ exist as continuous functions of $x \in D$, for all directions $v_i \in V$. Here

$$\partial_v f(x) := \lim_{h \to 0} (f(x + hv) - f(x))/h.$$

Given a basis $\{e_i\}$, we write $\partial_i := \partial_{e_i}$. We say that f is C^∞-*regular* if it is C^k-regular for all $k < \infty$.

- The function f is *Hölder regular* of order $0 < \alpha < 1$ in D if

$$|f(x) - f(y)| \lesssim |x - y|^\alpha, \quad \text{for all } x, y \in D,$$

and we write $f \in C^\alpha(D; L)$. For $\alpha = 0$, $f \in C^0(D; L) = C(D; L)$ means that f is continuous on D. When $\alpha = 1$, we say that f is *Lipschitz regular* and write $f \in C^{0,1}(D; L)$. Note that the precise value of the implicit constant C as in Definition 6.1.1, but not the Hölder or Lipschitz property of f, depends on the choice of Euclidean norm $|\cdot|$ on X.

- A bijective function $f : D \to D'$, with an open set $D' \subset L$, is a *homeomorphism* if $f \in C^0(D; D')$ and $f^{-1} \in C^0(D'; D)$. Lipschitz diffeomorphisms and C^k-*diffeomorphisms* are defined similarly. A *diffeomorphism* refers to a C^∞-diffeomorphism.

- The *support* of a function defined in X is the closed set

$$\operatorname{supp} f := \overline{\{x \in X \ ; \ f(x) \neq 0\}}.$$

If $f \in C^\infty(D)$, D an open set, then we write $f \in C_0^\infty(D)$ if $\operatorname{supp} f$ is a compact subset of D.

- Write $C^k(\overline{D}) := \{F|_D \ ; \ F \in C^k(X)\}$, and similarly for C^α and C^∞. When the range L of the function is clear from the context, we suppress L in the notation and abbreviate $C^k(D; L)$ to $C^k(D)$.

Definition 6.1.2 (Total derivative). Let $D \subset X$ be an open set in an affine space X with vectors V, and let (X', V') be a second affine space. If $\rho : D \to X'$ is differentiable at $x \in D$, then we define its *total derivative* at x to be the unique linear map $\underline{\rho}_x : V \to V'$ such that

$$\rho(x + v) - \rho(x) = \underline{\rho}_x(v) + o(v),$$

where $o(v)$ denotes a function $\lambda(v)$ such that $\lambda(v)/|v| \to 0$ when $v \to 0$. With respect to bases $\{e_i\}$ and $\{e_i'\}$, $\underline{\rho}_x$ has matrix

$$\begin{bmatrix} \partial_1 \rho_1(x) & \cdots & \partial_k \rho_1(x) \\ \vdots & \ddots & \vdots \\ \partial_1 \rho_n(x) & \cdots & \partial_k \rho_n(x) \end{bmatrix},$$

where $\rho = \sum \rho_i e_i'$ and partial derivatives $\partial_i = \partial_{e_i}$ are with respect to e_i. Equivalently, the total derivative is $\underline{\rho}_x(v) = \partial_v \rho(x)$.

Note that when ρ maps between affine spaces, then the total derivative $\underline{\rho}_x$ maps between the vector spaces since differences of points are vectors. To simplify notation, we shall often drop subscripts x and write $\underline{\rho}$.

The total derivative of a differential map between affine spaces extends from a map of vectors to a map of multivectors as in Section 2.3. With our notation, for example, the chain rule takes the form

$$\underline{\rho_2 \circ \rho_1}_x(w) = \underline{\rho_2}_{\rho_1(x)}\underline{\rho_1}_x(w), \quad w \in \wedge V_1,$$

for the composition of maps $\rho_1 : X_1 \to X_2$ and $\rho_2 : X_2 \to X_3$.

Definition 6.1.3 (Jacobian). Let $\rho : D \to X'$ be as in Definition 6.1.2, with total derivative $\underline{\rho}_x : V \to V'$. Denote by

$$\underline{\rho}_x : \wedge V \to \wedge V'$$

the induced linear map. Assume that X, X' are oriented n-dimensional affine spaces with orientations $e_{\overline{n}}$ and $e'_{\overline{n}}$ respectively. Then its *Jacobian* $J_\rho(x)$ is the scalar function representing $\underline{\rho}_x|_{\wedge^n V}$, that is, the determinant

$$J_\rho(x) := \langle e'^*_{\overline{n}}, \underline{\rho}_x(e_{\overline{n}}) \rangle$$

of $\underline{\rho}_x$.

The main use of Jacobians is in the change of variables formula

$$\int_{\rho(D)} f(y)dy = \int_D f(\rho(x))J_\rho(x)dx \tag{6.2}$$

for integrals. For Lipschitz change of variables ρ, this continues to hold. Note that in this case J_ρ is well defined almost everywhere, since Lipschitz maps ρ are differentiable almost everywhere by Rademacher's theorem.

We use the following standard terminology for domains $D \subset X$.

Definition 6.1.4 (Domains). Let D be a domain, that is, an open subset, in an n-dimensional affine space (X, V). We say that D is a C^k-*domain*, $k = 1, 2, \ldots$, if its boundary $\overline{D}$ is C^k-smooth in the following sense. At each $p \in \partial D$, we assume that there exists a C^k diffeomorphism $\rho : \Omega_p \to D_p$ between a neighborhood $\Omega_p \subset \mathbf{R}^n$ of 0 and a neighborhood $D_p \subset X$ such that

$$\rho(\{x \in \Omega_p \ ; \ x_n > 0\}) = D_p \cap D,$$
$$\rho(\{x \in \Omega_p \ ; \ x_n = 0\}) = D_p \cap \partial D, \text{ and}$$
$$\rho(\{x \in \Omega_p \ ; \ x_n < 0\}) = D_p \setminus \overline{D}.$$

Lipschitz domains are defined similarly, by requiring that the local parametrizations ρ be $C^{0,1}$ diffeomorpisms.

In a Euclidean space X, we denote by ν the outward-pointing unit normal vector field on ∂D. For a C^k-domain, ν is a C^{k-1}-regular vector field defined on all ∂D. For a Lipschitz domain, by Rademacher's theorem, ν is well defined at almost every point $p \in \partial D$.

In many cases it is important to consider domains beyond C^1, such as Lipschitz domains. For example, the intersection and union of two C^1 domains is much more likely to be Lipschitz than C^1. However, as the following example indicates, Lipschitz domains constitute a far wider class than domains with a finite number of corners, edges, etc.

Example 6.1.5 (Lipschitz scale invariance). We consider how a function $\phi : \mathbf{R} \to \mathbf{R}$ scales. Assume that $\phi(0) = 0$ and let $\phi_n(x) := n\phi(x/n)$. Thus the graph of ϕ_n represents what ϕ looks like around 0 through a magnifying glass that magnifies n times. If ϕ is C^1 regular, then $|\phi_n(x) - \phi'(0)x| \leq \epsilon_n |x|$, $|x| < 1$, where $\epsilon_n \to 0$ when $n \to \infty$. This means that ϕ "looks flat" on small enough scales, since it is well approximated by the straight line $y = \phi'(0)x$. On the other hand, if ϕ is a Lipschitz function, then ϕ_n is another Lipschitz function with the same Lipschitz constant C. In contrast to the C^1 case, ϕ_n will not converge to a linear function, as is seen, for example, from $\phi(x) = |x|$, for which $\phi_n(x) = |x|$ for all n. However, this example is very atypical for Lipschitz functions. In general, each ϕ_n will give an entirely new function. This means that a Lipschitz function is nontrivial, that is, nonflat, on each scale, but still nondegenerate, that is, still a Lipschitz function.

By the implicit function theorem, the boundary of a C^k domain, $k = 1, 2, \ldots$, is locally the graph of a C^k function, in the sense that the local parametrization ρ can be written

$$\rho(x', x_n) = (x', x_n + \phi(x')), \quad x' \in \mathbf{R}^{n-1}, \tag{6.3}$$

in a suitable basis for $X = V$, where $\phi : \mathbf{R}^{n-1} \to \mathbf{R}$ is a C^k-regular function. In stark contrast, this is not true for Lipschitz domains.

Example 6.1.6 (Bricks and spirals). (i) In $\mathbf{R}^3$, let

$$D_1 := \{(x, y, z) \; ; \; -1 < x < 0, \; -1 < y < 1, \; -1 < z < 0\} \quad \text{and}$$
$$D_2 := \{(x, y, z) \; ; \; -1 < x < 1, \; -1 < y < 0, \; 0 < z < 1\}.$$

Placing the "brick" D_2 on top of D_1, consider the *two-brick domain* D with $\overline{D} = \overline{D}_1 \cup \overline{D}_2$. Then D is a Lipschitz domain, but at the origin ∂D is not the graph of a Lipschitz function.

(ii) In polar coordinates (r, θ) in $\mathbf{R}^2$, consider the *logarithmic spiral*

$$D := \{(r \cos \theta, r \sin \theta) \; ; \; e^{-(\theta+a)} < r < e^{-(\theta+b)}, \; \theta > 0\},$$

where $b < a < b + 2\pi$ are two constants. Then D is a Lipschitz domain, but at the origin ∂D is not the graph of a Lipschitz function.

If D is a Lipschitz domain in which all local parametrizations ρ of ∂D are of the form (6.3) with $C^{0,1}$ functions ϕ, then we say that D is a *strongly Lipschitz domain*.

Exercise 6.1.7 (Star-shaped domains). We say that a domain D is *star-shaped* with respect to some point $p \in D$ if for each $x \in D$, the line $\{p+t(x-p) \; ; \; t \in [0,1]\}$ in contained in D. Show that every bounded domain in a Euclidean space that is star-shaped with respect to each point in some ball $B(p; \epsilon) \subset X$, $\epsilon > 0$, is a strongly Lipschitz domain. Conversely, show that every bounded strongly Lipschitz domain is a finite union of such domains that are star-shaped with respect to some balls.

Exercise 6.1.8 (Rellich fields). Let D be a bounded strongly Lipschitz domain in a Euclidean space (X, V). Show that there exists a vector field $\theta \in C_0^\infty(X; V)$ such that

$$\inf_{x \in \partial D} \langle \nu(x), \theta(x) \rangle > 0.$$

A partition of unity, see below, may be useful.

Besides open subsets, that is, domains in affine space, we also make use of lower-dimensional curved surfaces. More generally, we require the notion of a manifold from differential geometry, which we now fix notation. We consider only compact manifolds, but both with and without boundary, and in many cases embedded in an affine space. For simplicity, we consider only regularity $k \geq 1$. Our notation is the following. Let $H_+^n := \{(x', x_n) \; ; \; x' \in \mathbf{R}^{n-1}, \; x_n \geq 0\}$ and $\mathbf{R}_+^n := \{(x', x_n) \; ; \; x' \in \mathbf{R}^{n-1}, \; x_n > 0\}$ denote the closed and open upper half-spaces, and identify $\mathbf{R}^{n-1}$ and $\mathbf{R}^{n-1} \times \{0\}$. In general, let M be a compact (second countable Hausdorff) topological space, for example a compact subset of an affine space X.

- We assume that M is locally homeomorphic to H_+^n, in the sense that we are given a collection of *charts*, that is, homeomorphisms $\{\mu_\alpha : D_\alpha \to M_\alpha\}_{\alpha \in \mathcal{I}}$, the *atlas* for M, between open sets $D_\alpha \subset H_+^n$ and $M_\alpha \subset M$ such that $M = \bigcup_{\alpha \in \mathcal{I}} M_\alpha$. By compactness, we may assume that the index set $\mathcal{I}$ is finite.

- Define open sets $D_{\beta\alpha} := \mu_\alpha^{-1}(M_\beta) \subset D_\alpha$, and *transition maps*

$$\mu_{\beta\alpha} : D_{\beta\alpha} \to D_{\alpha\beta} : x \mapsto \mu_{\beta\alpha}(x) := \mu_\beta^{-1}(\mu_\alpha(x))$$

for $\alpha, \beta \in \mathcal{I}$. We say that M is a (compact) C^k-*manifold* if $\mu_{\beta\alpha} \in C^k(\overline{D_{\beta\alpha}})$ for all $\alpha, \beta \in \mathcal{I}$. In this case, these transition maps are C^k diffeomorphisms, since $\mu_{\beta\alpha}^{-1} = \mu_{\alpha\beta}$. A *manifold* refers to a C^∞-manifold. If all these transition maps are orientation-preserving, then we say that M is *oriented*. When it is possible to find another atlas with all transition maps between its charts orientation-preserving, then we say that M is *orientable*.

More generally, a *chart* for M refers to any homeomorphism $\mu' : D' \to M'$ between open sets $D' \subset H_+^n$ and $M' \subset M$ such that $\mu'^{-1} \circ \mu_\alpha \in C^k(\overline{\mu_\alpha^{-1}(M')})$ for all $\alpha \in \mathcal{I}$.

- If $D_\alpha \subset \mathbf{R}_+^n$ for all $\alpha \in \mathcal{I}$, then we say that M is a *closed manifold*. This means that M is a compact manifold without boundary. If $D_\alpha \cap \mathbf{R}^{n-1} \neq \emptyset$ for some $\alpha \in \mathcal{I}$, then we say that M is a *manifold with boundary*. In this case, the *boundary* of M, denoted by ∂M, is the closed manifold defined as follows. Let $D'_\alpha := A_\alpha \cap \mathbf{R}^{n-1}$,

$$\mu'_\alpha := \mu_\alpha|_{\mathbf{R}^{n-1}},$$

and $M'_\alpha := \mu'_\alpha(D'_\alpha)$. It suffices to consider α such that $D'_\alpha \neq \emptyset$, and we may assume that $D'_\alpha \subset \mathbf{R}_+^{n-1}$. Then ∂M is the closed manifold $\bigcup_{\alpha \in \mathcal{I}} M'_\alpha$ with atlas $\{\mu'_\alpha : D'_\alpha \to M'_\alpha\}_{\alpha \in \mathcal{I}}$.

- When M is a compact n-dimensional C^k-manifold that is also a subset of an affine space X, with the topology inherited from X, then we say that M is an *n-surface* in X if the derivative $\underline{\mu}_\alpha$ of $\mu_\alpha : D_\alpha \to M_\alpha \subset X$ is injective for all $x \in D_\alpha$ and all $\alpha \in \mathcal{I}$. If $\mu_\alpha \in C^k(D_\alpha; X)$, then we say that M is a C^k-*regular n-surface* in X. By the inverse function theorem, an n-surface is locally the graph of a C^k-regular function in n variables, in a suitably rotated coordinate system for X. As above, n-surfaces may be closed or may have a boundary. If $D \subset X$ is a bounded C^k-domain in an affine space X as in Definition 6.1.4, then we see that $M = \overline{D}$ is a compact C^k-regular n-surface with boundary.

   More generally but similarly, we can consider n-surfaces M embedded in some, in general higher-dimensional, manifold N.

- For a function $f : M \to L$ on a C^k manifold M, with values in a linear space L, we define $f \in C^j(M; L)$ to mean that $f \circ \mu_\alpha \in C^j(\overline{D}_\alpha; L)$ for all $\alpha \in \mathcal{I}$, when $j \leq k$.

   A *partition of unity* for a C^k-manifold M, subordinate to a finite covering $M = \bigcup_{\alpha \in \mathcal{I}} M_\alpha$ by open sets $M_\alpha \subset M$, is a collection $\{\mu_\alpha\}_{\alpha \in \mathcal{I}}$ of functions such that $\mathrm{supp}\, \eta_\alpha \subset M_\alpha$ and $\sum_{\alpha \in \mathcal{I}} \eta_\alpha(x) = 1$ for all $x \in M$. There exists such a partition of unity with $\eta_\alpha \in C^k(M; [0, 1])$ on every C^k-manifold M.

The standard use of a partition of unity is to localize problems: Given a function f on M, we write

$$f = \sum_\alpha \eta_\alpha f.$$

Here $\mathrm{supp}\, \eta_\alpha f \subset M_\alpha$, and by working locally in this chart, we can obtain results for $\eta_\alpha f$, which then we can sum to a global result for f.

6.2 Fourier Transforms

This section collects computations of certain Fourier transforms that are fundamental to the theory of partial differential equations. Fix a point of origin in an oriented affine space X and identify it with its vector space V. In particular, V is an abelian group under addition, and as such it comes with a Fourier transform. This is the linear operator

$$\mathcal{F}(f)(\xi) = \hat{f}(\xi) := \int_V f(x)e^{-i\langle \xi, x\rangle} dx, \quad \xi \in V^*.$$

This Fourier transform maps a complex-valued function f on V to another complex-valued function on V^*. If instead f takes values in some complex linear space L, we let $\mathcal{F}$ act componentwise on f. Assuming that V is a Euclidean space and $V^* = V$, the fundamental theorem of Fourier analysis is *Plancherel's theorem*, which states that $\mathcal{F}$ defines, modulo a constant, an L_2 isometry:

$$\int_V |f(x)|^2 dx = \frac{1}{(2\pi)^n} \int_V |\hat{f}(\xi)|^2 d\xi.$$

We recall that the inverse Fourier transform is given by

$$\mathcal{F}^{-1}(\hat{f})(x) = f(x) = \int_{V^*} \hat{f}(\xi)e^{i\langle \xi, x\rangle} d\xi, \quad x \in V,$$

and basic formulas

$$\mathcal{F}(\partial_k f(x)) = i\xi_k \hat{f}(\xi),$$

$$\mathcal{F}(f(x) * g(x)) = \hat{f}(\xi) \cdot \hat{g}(\xi),$$

where the convolution of $f(x)$ and $g(x)$ is the function

$$(f * g)(x) := \int_V f(x - y)g(y)dy.$$

The most fundamental of Fourier transforms is

$$\mathcal{F}\{e^{-|x|^2/2}\} = (2\pi)^{n/2}e^{-|\xi|^2/2},$$

that is, the Gauss function $e^{-|x|^2/2}$ is an eigenfunction to $\mathcal{F}$.

Proposition 6.2.1 (Gaussians and homogeneous functions). *Let $f(x)$ be a homogeneous polynomial of degree j that is harmonic on an n-dimensional Euclidean space V. Then for every constant $s > 0$, we have Fourier transforms*

$$\mathcal{F}\{f(x)e^{-s|x|^2}\} = 2^{-j}cs^{-(n/2+j)}f(\xi)e^{-|\xi|^2/(4s)},$$

where $c = \pi^{n/2}(-i)^j$. For every constant $0 < \alpha < n$, we have Fourier transforms

$$\mathcal{F}\{f(x)/|x|^{n-\alpha+j}\} = 2^\alpha c \frac{\Gamma((\alpha + j)/2)}{\Gamma((n - \alpha + j)/2)} f(\xi)/|\xi|^{\alpha+j},$$

where $\Gamma(z) := \int_0^\infty e^{-t}t^{z-1}dt$ is the gamma function, with $\Gamma(k) = (k - 1)!$.

Proof. (i) Calculating the Fourier integral, we have

$$\int_V f(x)e^{-s|x|^2}e^{-i\langle x,\xi\rangle}\,dx = e^{-|\xi|^2/(4s)}\int_V f(x)e^{-s(x+i\xi/(2s))^2}\,dx$$

$$= e^{-|\xi|^2/(4s)}\int_V f(x-i\xi/(2s))e^{-s|x|^2}\,dx$$

$$= e^{-|\xi|^2/(4s)}\int_0^\infty e^{-sr^2}r^{n-1}\left(\int_{|\omega|=1} f(r\omega - i\xi/(2s))d\omega\right)dr,$$

where we have extended f to a polynomial of n complex variables. According to the mean value theorem for harmonic functions,

$$\int_{|\omega|=1} f(r\omega + y)d\omega = \sigma_{n-1}f(y),$$

for every $y \in V$, where σ_{n-1} is the area of the unit sphere in V. By analytic continuation, this formula remains valid for all complex $y \in V_c$. Since

$$\int_0^\infty e^{-sr^2}r^{n-1}dr = \frac{1}{2s^{n/2}}\int_0^\infty e^{-u}u^{n/2-1} = \frac{1}{2s^{n/2}}\Gamma(n/2)$$

and $\sigma_{n-1} = 2\pi^{n/2}/\Gamma(n/2)$, the stated identity follows.

(ii) To establish the second Fourier transform identity, we use the identity

$$\int_0^\infty s^{(n-\alpha+j)/2-1}e^{-sr^2}\,ds = \frac{1}{r^{n-\alpha+j}}\int_0^\infty x^{(n-\alpha+j)/2-1}e^{-x}\,dx = \frac{\Gamma((n-\alpha+j)/2)}{r^{n-\alpha+j}}.$$

Writing $r^{-(n-\alpha+j)}$ as a continuous linear combination of functions e^{-sr^2} in this way, we deduce that

$$\mathcal{F}\{f(x)/|x|^{n-\alpha+j}\} = \frac{1}{\Gamma((n-\alpha+j)/2)}\int_0^\infty s^{(n-\alpha+j)/2-1}\mathcal{F}\{f(x)e^{-s|x|^2}\}ds$$

$$= \frac{1}{\Gamma((n-\alpha+j)/2)}\int_0^\infty s^{(n-\alpha+j)/2-1}2^{-j}cs^{-(n/2+j)}f(\xi)e^{-|\xi|^2/(4s)}ds$$

$$= \frac{2^{-j}cf(\xi)}{\Gamma((n-\alpha+j)/2)}\int_0^\infty s^{-(\alpha+j)/2-1}e^{-(1/s)(|\xi|/2)^2}ds$$

$$= 2^\alpha c\frac{\Gamma((\alpha+j)/2)}{\Gamma((n-\alpha+j)/2)}\frac{f(\xi)}{|\xi|^{\alpha+j}}. \qquad \square$$

The following functions, or more precisely distributions in dimension ≥ 3, appear in solving the wave equation.

Proposition 6.2.2 (Riemann functions). *Let R_t, for $t > 0$, be the Fourier multiplier*

$$\mathcal{F}(R_t f)(\xi) = \frac{\sin(t|\xi|)}{|\xi|}\mathcal{F}(f)(\xi).$$

In low dimensions, the Riemann function R_t has the following expression for $t > 0$:

$$R_t f(x) = \frac{1}{2} \int_{|y|<t} f(x-y)dy, \quad \dim V = 1,$$

$$R_t f(x) = \frac{1}{2\pi} \iint_{|y|<t} \frac{f(x-y)}{\sqrt{t^2-|y|^2}}dy, \quad \dim V = 2,$$

$$R_t f(x) = \frac{1}{4\pi t} \iint_{|y|=t} f(x-y)dy, \quad \dim V = 3.$$

In higher dimensions, we have the following expressions:

$$R_t f(x) = \frac{1}{(n-2)!!\,\sigma_{n-1}} \left(\frac{1}{t}\frac{\partial}{\partial t}\right)^{n/2-1} \int_{|y|<t} \frac{f(x-y)}{\sqrt{t^2-|y|^2}}dy, \quad \dim V = n \geq 4 \text{ even,}$$

$$R_t f(x) = \frac{1}{(n-2)!!\,\sigma_{n-1}} \left(\frac{1}{t}\frac{\partial}{\partial t}\right)^{(n-3)/2} \frac{1}{t}\int_{|y|=t} f(x-y)dy, \quad \dim V = n \geq 5 \text{ odd.}$$

Here $\sigma_{n-1} = 2\pi^{n/2}/\Gamma(n/2)$ is the measure of the unit sphere. In the formulas for odd dimension, dy denotes $n-1$ measure on the sphere $|y| = t$.

Proof. We are looking for $R_t(x)$ such that $\hat{R}_t(\xi) = \sin(t|\xi|)/|\xi|$. Since $\hat{R}_t$ is rotation-invariant, so is R_t, and it suffices to find a rotation-invariant function R_t such that $\hat{R}_t(ae_1) = \sin(ta)/a$ for $a \geq 0$. The low-dimensional formulas follow from the computations

$$\int_{-t}^{t} e^{-ix\xi}dx = 2\sin(t\xi)/\xi,$$

$$\iint_{x_1^2+x_2^2<t^2} (t^2-x_1^2-x_2^2)^{-1/2} e^{-ix_1 a}dx_1 dx_2 = \int_{-t}^{t}\left(\int_{-1}^{1}(1-u^2)^{-1/2}du\right)e^{-ix_1 a}dx_1$$

$$= \cdots = \iint_{x_1^2+x_2^2+x_3^2=t^2} e^{-ix_1 r} = t^2 \int_0^{2\pi}\int_0^{\pi} e^{-irt\cos\theta}\sin\theta\,d\theta\,d\varphi = \cdots .$$

Similarly, when $\dim V = n = 2k+1$, $k = 2, 3, \ldots$, we have

$$\int_{|x|=t} e^{-ix_1 a}dx = \int_{-t}^{t}\sqrt{1+\left(\frac{\partial}{\partial x_1}\sqrt{t^2-x_1^2}\right)^2}\,\sigma_{n-2}\sqrt{t^2-x_1^2}^{\,n-2} e^{-ix_1 a}dx_1$$

$$= t\sigma_{n-2}\int_{-t}^{t}(t^2-x_1^2)^{k-1}e^{-ix_1 a}dx_1.$$

Using that $\frac{1}{2t}\frac{\partial}{\partial t}(t^2-x_1^2)^m = (t^2-x_1^2)^{m-1}$, we obtain after $k-1$ iterations that

$$\frac{1}{2\sigma_{n-2}(k-1)!}\left(\frac{1}{2t}\frac{\partial}{\partial t}\right)^{k-1}\frac{1}{t}\int_{|x|=t}e^{-ix_1 a}dx = \sin(ta)/a$$

from the one-dimensional result.

Finally, consider $\dim V = n = 2k$, $k = 2, 3, \ldots$ By iterated integration, we have

$$\int_{|x|<t} \frac{e^{-ix_1 a}}{\sqrt{t^2 - |x|^2}} dx = \int_{-t}^{t} \left(\int_{|u|<1} (t^2 - x_1^2)^{k-1} \frac{du}{\sqrt{1 - |u|^2}} \right) e^{-ix_1 a} dx_1$$

$$= \sigma_{n-2} \int_{|u|<1} \frac{du}{\sqrt{1 - |u|^2}} \int_{-t}^{t} (t^2 - x_1^2)^{k-1} e^{-ix_1 a} dx_1$$

$$= \sigma_{n-2} \frac{\pi(n-3)!!}{2(n-2)!!} \int_{-t}^{t} (t^2 - x_1^2)^{k-1} e^{-ix_1 a} dx_1.$$

Applying $\frac{1}{2t} \frac{\partial}{\partial t}$ as above, we obtain after $k - 1$ iterations that

$$\frac{1}{\pi \sigma_{n-2}(k-1)!} \frac{(n-2)!!}{(n-3)!!} \left(\frac{1}{2t} \frac{\partial}{\partial t} \right)^{k-1} \int_{|x|<t} \frac{e^{-ix_1 a}}{\sqrt{t^2 - |x|^2}} dx = \sin(ta)/a.$$

The proof is now completed by noting the iteration formula

$$n!! \, \sigma_{n+1} = \begin{cases} \pi(n-1)!! \, \sigma_n, & n \text{ even}, \\ 2(n-1)!! \, \sigma_n, & n \text{ odd}, \end{cases}$$

for the measure of the unit spheres. $\qquad \Box$

Corollary 6.2.3 (Klein–Gordon Riemann functions). *Let R_t^m, $t, m > 0$, be the Fourier multiplier*

$$\mathcal{F}(R_t^m f)(\xi) = \frac{\sin\left(t\sqrt{|\xi|^2 + m^2}\right)}{\sqrt{|\xi|^2 + m^2}} \mathcal{F}(f)(\xi).$$

If $\dim V = 3$, then

$$R_t^m f(x) = \frac{1}{4\pi t} \iint_{|y|=t} f(x-y)dy - \frac{m}{4\pi} \iint_{|y|<t} \frac{J_1\left(m\sqrt{t^2 - |y|^2}\right)}{\sqrt{t^2 - |y|^2}} f(x-y)dy,$$

where J_n, $n \in \mathbf{Z}$, denote the Bessel functions

$$J_n(r) = \frac{1}{\pi} \int_0^{\pi} \cos(n\theta - r\sin\theta)d\theta.$$

Proof. Proceeding by the method of descent, we use the Riemann function

$$R_t(x, x_4) = \frac{1}{4\pi^2 t} \partial_t \left(\frac{\chi_t(x, x_4)}{\sqrt{t^2 - |x|^2 - x_4^2}} \right), \quad x \in V,$$

in four dimensions from Proposition 6.2.2, where χ_t denotes the characteristic function of the ball $B(0,t)$. We have

$$R_t^m(x) = \int_{\mathbf{R}} R_t(x, x_4) e^{-imx_4} dx_4 = \frac{1}{4\pi^2 t} \partial_t \left(\int_{|x_4| < t^2 - |x|^2} \frac{e^{-imx_4} dx_4}{\sqrt{t^2 - |x|^2 - x_4^2}} \chi_t(x) \right)$$

$$= \frac{1}{4\pi t} \partial_t \left(J_0 \big(m\sqrt{t^2 - |x|^2} \big) \chi_t(x) \right).$$

Calculating this distributional derivative leads to the stated identity.   □

Corollary 6.2.4 (Helmholtz fundamental solution). *Let $\Phi_k(x) = \Phi_k(r)$ be the radial function with*

$$\mathcal{F}(\Phi_k)(\xi) = (k^2 - |\xi|^2)^{-1},$$

where $k \in \mathbf{C}$ with $\operatorname{Im} k > 0$. In low dimensions, we have the following formulas:

$$\Phi_k(r) = \frac{1}{2ik} e^{ikr}, \quad \dim V = 1,$$

$$\Phi_k(r) = -\frac{1}{2\pi} \int_1^\infty \frac{e^{ikrt} dt}{\sqrt{t^2 - 1}}, \quad \dim V = 2,$$

$$\Phi_k(r) = -\frac{1}{4\pi r} e^{ikr}, \quad \dim V = 3.$$

In higher dimensions, we have the following:

$$\Phi_k(r) = \frac{-1}{(n-2)!! \, \sigma_{n-1}} \left(\frac{\partial}{\partial r} \frac{-1}{r} \right)^{n/2-1} \int_1^\infty \frac{e^{ikrt} dt}{t^{n-2}\sqrt{t^2-1}}, \quad \dim V = n \geq 4 \text{ even,}$$

$$\Phi_k(r) = \frac{-1}{(n-2)!! \, \sigma_{n-1}} \left(\frac{-1}{r} \frac{\partial}{\partial r} \right)^{(n-3)/2} \left(\frac{e^{ikr}}{r} \right), \quad \dim V = n \geq 5 \text{ odd.}$$

In terms of Hankel functions,

$$H_\nu^{(1)}(z) = \frac{2(-i)^{2\nu}}{i\sqrt{\pi}\,\Gamma(\nu + 1/2)} \left(\frac{z}{2} \right)^\nu \int_1^\infty e^{izt}(t^2 - 1)^{\nu - 1/2} dt,$$

$$\operatorname{Im} z > 0, \; \nu = 0, \tfrac{1}{2}, 1, \tfrac{3}{2}, 2, \ldots,$$

we have

$$\Phi_k(r) = -\frac{i}{4} \left(\frac{k}{2\pi r} \right)^{n/2-1} H_{n/2-1}^{(1)}(kr), \quad n = 1, 2, 3, 4, \ldots.$$

Proof. The proof uses the identity

$$\int_0^\infty e^{ikt} R_t f(x) dt = -\Phi_k * f(x),$$

which is proved by Fourier transformation. The formulas for R_t from Proposition 6.2.2 yield the stated formulas for Φ_k. To express these in terms of the Hankel funktions, we use for odd n that

$$H_{1/2}^{(1)}(z) = -i \left(\frac{2}{\pi z} \right)^{1/2} e^{iz},$$

$$H_{\nu+1}^{(1)}(z) = \left(\frac{\nu}{z} - \frac{d}{dz} \right) H_\nu^{(1)}(z) = -z^\nu \frac{d}{dz} (z^{-\nu} H_\nu^{(1)}(z)).$$

For even n, we use that

$$\left(\frac{d}{dz} + \frac{\nu}{z} \right) H_\nu^{(1)}(z) = H_{\nu-1}^{(1)}(z),$$

and note that the same recurrence formula holds for $f_\nu(z) := 4i(2\pi r/k)^{n/2-1} \Phi_k(r)$, with $z = kr$, $\nu := n/2 - 1$. $\qquad \square$

6.3 Partial Differential Equations

In classical analysis of partial differential equations, abbreviated as PDEs, the focus is on scalar-valued functions. This leads to a theory based on second-order equations. In this section we recall the basic second-order partial differential equations for scalar-valued functions.

Given an inner product space (X, V), we have a canonical second-order partial differential operator

$$\langle \nabla, \nabla \rangle u = \sum_{i=1}^n \sum_{j=1}^n \langle e_i^*, e_j^* \rangle \partial_i \partial_j u,$$

where ∂_i denotes the partial derivatives along basis vectors e_i, with dual basis vectors e_i^*. The chain rule and the bilinearity of the inner product ensure that this operator is basis-independent. If V is a Euclidean space, then this is the *Laplace operator*, given by

$$\Delta u = \langle \nabla, \nabla \rangle u = \sum_{i=1}^n \partial_i^2 u$$

in an ON-basis. If V is a Minkowski spacetime, then this is the *d'Alembertian*, given by

$$\Box u = \langle \nabla, \nabla \rangle u = -\partial_0^2 u + \sum_{i=1}^n \partial_i^2 u$$

in an ON-basis, with e_0 the time-like vector.

Example 6.3.1 (The Laplace equation). Consider the Laplace operator Δ on a Euclidean space V. Given a function $f : V \to \mathbf{R}$, we consider the Poisson equation $\Delta u = f$ for u. Applying the Fourier transform, this is equivalent to $-|\xi|^2 \hat{u} = \hat{f}$, and we obtain the solution

$$\hat{u}(\xi) = -\frac{1}{|\xi|^2} \hat{f}(\xi).$$

It follows from Proposition 6.2.1 that $u = \Phi * f$ under suitable conditions on f, where Φ is the fundamental solution to the Laplace operator given by

$$\Phi(x) := \begin{cases} -1/((n-2)\sigma_{n-1}|x|^{n-2}), & n = \dim V \geq 3, \\ (2\pi)^{-1} \ln|x|, & n = 2. \end{cases}$$

Here σ_{n-1} is the $(n-1)$-volume of the unit sphere in V. The Laplace equation $\Delta u = 0$ is the fundamental example of an *elliptic* partial differential equation, which typically models a state of equilibrium.

Example 6.3.2 (The wave equation). Consider next the wave equation $\Box u = 0$ for a function $u : W \to \mathbf{R}$ on a spacetime W. In a fixed ON-basis, this reads

$$\partial_t^2 u = \Delta u,$$

where we write $\partial_t = \partial_0$ for the time derivative and Δ denotes the Laplace operator on the space-like orthogonal complement V to e_0. Applying the partial Fourier transform in the spatial variable x only, we obtain the ordinary differential equation

$$\partial_t^2 \hat{u} = -|\xi|^2 \hat{u},$$

for $\hat{u} = \hat{u}(t, \xi) = \int_V u(t, x) e^{-i\langle \xi, x \rangle} dx$. For the Cauchy inital value problem, where $u(0, \cdot) = f$ and $\partial_t u(0, \cdot) = g$ are given, we obtain the solution $\hat{u} = \cos(t|\xi|)\hat{f} + \frac{1}{|\xi|} \sin(t|\xi|)\hat{g}$. It follows from Proposition 6.2.2 that

$$u(t, \cdot) = \partial_t R_t f + R_t g, \qquad t > 0,$$

in terms of the Riemann operators.

The wave equation $\Box u = 0$ is the fundamental example of a *hyperbolic* partial differential equation, which typically models a time evolution of scalar waves, such as three-dimensional acoustic waves. This wave evolution is symmetric and reversible in time. We note from Proposition 6.2.2 that this wave equation has finite propagation speed ≤ 1, since the value $u(t, x)$ depends only on the inital data $f(y)$ and $g(y)$ for $|y - x| \leq |t|$. In odd dimension we have a Huygens principle and the propagation speed $= 1$.

Example 6.3.3 (The heat equation). With a fixed time direction, we can also consider time evolution with the Laplace operator on an n-dimensional Euclidean space V. The heat equation for a function $u : \mathbf{R} \times V \to \mathbf{R}$ is

$$\partial_t u = \Delta u.$$

Applying the partial Fourier transform in the spatial variable x, we have

$$\partial_t \hat{u} = -|\xi|^2 \hat{u}.$$

Solving this ordinary differential equation, we get $\hat{u} = e^{-t|\xi|^2} \hat{f}$, if $f = u(0, \cdot)$ denotes the given inital data. It follows from Proposition 6.2.1, with $j = 0$ and $s = 1/(4t)$, that

$$u(t, x) = \frac{1}{(4\pi t)^{n/2}} e^{-|x|^2/(4t)} * f(x), \quad t > 0.$$

The heat equation $\partial_t u = \Delta u$ is the fundamental example of a *parabolic* partial differential equation, which typically models an irreversible and smoothing process, for example heat flow or diffusion. Unlike the wave equation, the heat equation cannot be solved stably backward in time. It also has an infinite propagation speed in that $u(t, x)$ depends on $f(y)$ for all $y \in \mathbf{R}$.

Some important variations of these three main equations are the following.

Example 6.3.4 (The Helmholtz equation). If we instead apply the partial Fourier transform of the time variable in the wave equation, writing $k = \xi$ for a fixed frequency and keeping the notation $u : V \to \mathbf{C}$, we obtain the Helmholtz equation

$$\Delta u + k^2 u = 0.$$

This describes stationary waves at a fixed frequency, the Laplace equation being the static special case. Starting instead with a dampened wave equation $\partial_t^2 u = \Delta u - a\partial_t u$, with parameter $a > 0$, leads to a Helmholtz equation with $0 < \operatorname{Im} k < \operatorname{Re} k$. Note that even for $k \in \mathbf{R}$, it is important to allow for complex-valued solutions u, since $\arg u$ encodes the phase of the oscillation.

Like the Laplace equation, the Helmholtz equation is an elliptic equation. For given sources g, the solution to $\Delta u + k^2 u = g$ is given by

$$u = \Phi_k * g,$$

where Φ_k is the Helmholtz fundamental solution from Corollary 6.2.4.

For $\operatorname{Im} k = 0$, the decay properties of solutions to the Helmholtz equation as $x \to \infty$ are more subtle. A fundamental result that we require is the following.

Lemma 6.3.5 (Rellich). *Let $k \in \mathbf{R} \backslash \{0\}$ and assume that $u : D \to \mathbf{R}$ solves $\Delta u + k^2 u$ in a connected complement D of a compact set. If*

$$\lim_{R \to \infty} \int_{|x|=R} |u(x)|^2 dx = 0,$$

then $u = 0$ identically.

Example 6.3.6 (The Schrödinger equation). An imaginary analogue of the heat equation is the free time-dependent Schrödinger equation

$$\partial_t u(t,x) = i\Delta u$$

for a function $u : \mathbf{R} \times V \to \mathbf{C}$. Although it resembles the heat equation, this is a hyperbolic equation more closely related to the wave equation. Formally setting $s = 1/(4it)$ in Proposition 6.2.1, we obtain the intriguing solution formula

$$u(t,x) = \frac{1}{(4\pi it)^{n/2}} e^{-|x|^2/(4it)} * f(x), \quad t \in \mathbf{R},$$

given $u(0,\cdot) = f$. The free Schrödinger equation describes the time evolution of the wave equation for particles in quantum mechanics, and $|u(x)|^2$ represents a probability density for the position of the particle. With physical units and an external energy potential $V(x)$ in which the particle moves, the Schrödinger equation reads

$$i\hbar\partial_t u = -\frac{\hbar^2}{2m}\Delta u + Vu,$$

where m is the mass of the particle and $\hbar$ is Planck's constant. The standard derivation is to start from the classical expression for the total energy E as the sum of kinetic energy $p^2/(2m)$ and potential energy V, and to replace E and p by the differential operators $i\hbar\partial_t$ and $-i\hbar\nabla$ respectively.

Example 6.3.7 (Harmonic oscillator). An important potential in quantum mechanics is $V(x)$ proportional to x^2, referred to as the harmonic oscillator. We consider, for a parameter $\omega > 0$ and $x \in \mathbf{R}$, the corresponding parabolic equation

$$\partial_t f(t,x) = \partial_x^2 f(t,x) - \omega^2 x^2 f(t,x),$$

which generalizes the heat equation when $\omega = 0$. The solution to the initial value problem is given by *Mehler's formula*

$$f(t,x) = \int_{\mathbf{R}} \sqrt{\frac{\omega}{2\pi\sinh(2\omega t)}}$$
$$\cdot \exp\left(\frac{\omega}{\sinh(2\omega t)}\left(-\cosh(2\omega t)(x^2+y^2)/2 + xy\right)\right) f(0,y)dy.$$

6.4   Operator Theory

When our linear space is infinite-dimensional, which is the case for the spaces of functions that we use in analysis, we consider only Banach spaces beyond the Hilbert spaces, the infinite-dimensional analogue of a Euclidean space. Recall from functional analysis what this means more precisely: that these are the spaces with a unit sphere without any holes, with Hilbert spaces being the case of a round sphere. In general, we use $|\cdot|$ for norms in finite-dimensional spaces and $\|\cdot\|$ for norms in infinite-dimensional spaces, where $\|f\| = \sqrt{\langle f, f \rangle}$ for Hilbert spaces. When using the complex field, we use a Hermitian inner product $\langle f, f \rangle$. According to the Riesz representation theorem, there is for Hilbert spaces an identification $g : \mathcal{H} \to \mathcal{H}^*$, the analysis twin of Proposition 1.2.3. For Banach spaces, we recall that in general there is no invertible linear identification of $\mathcal{H}$ and its dual space $\mathcal{H}^*$ of bounded linear functionals on $\mathcal{H}$. Two norms $\|\cdot\|$ and $\|\cdot\|'$ on a Banach space are said to be equivalent if $\|f\|' \approx \|f\|$ in the sense of Definition 6.1.1 and define the same topology on the space. Infinite-dimensional linear spaces with an indefinite inner product are called Krein spaces. Although we will not use any theory of Krein spaces, our main idea of splittings of function spaces as explained in the introduction of this chapter is closely related to the Krein space structure.

Example 6.4.1 (Classical Banach spaces). (i) The Lebesgue spaces $L_p(D)$, $1 \leq p \leq \infty$, on a measure space D are defined by the norm $\|f\|_p := \left(\int_D |f(x)|^p dx \right)^{1/p}$, with the endpoint case $p = \infty$ being the supremum norm $\|f\|_\infty := \sup_D |f|$.

(ii) The Hölder spaces $C^\alpha(D)$, $0 < \alpha < 1$, on a metric space D are defined as

$$C^\alpha(D) := \{f : D \to \mathbf{R} \; ; \; |f(x)| \leq C, |f(x) - f(y)| \leq C|x - y|^\alpha, \text{ for all } x, y \in D\},$$

with norm $\|f\|_\alpha$ defined to be the smallest such constant $C < \infty$. The endpoint space $\alpha = 1$, the Lipschitz continuous functions, denoted by $C^{0,1}(D)$. At the endpoint $\alpha = 0$ we require f to be continuous and use the supremum norm on $C(D) = C^0(D)$.

(iii) Among the above spaces, only $L_2(D)$ is a Hilbert space. Via the Fourier transform and Plancherel's theorem, we define the L_2-based Sobolev spaces $H^s(X)$, $s \in \mathbf{R}$, in general for $s < 0$ containing distributions, on a Euclidean space X via the norm

$$\|f\|_{H^s(X)}^2 := \int_V |\hat{f}(\xi)|^2 (1 + |\xi|^2)^s d\xi < \infty.$$

For example, if $s = 1$, then $\|f\|_{H^1(X)} = (\|f\|_{L_2(X)}^2 + \|\nabla f\|_{L_2(X)}^2)^{1/2}$. For a bounded domain $D \subset X$, two ways to define Sobolev spaces are

$$H^s(\overline{D}) := \{f|_D \; ; \; f \in H^s(X)\},$$
$$H^s_{\overline{D}}(X) := \{f \in H^s(X) \; ; \; \operatorname{supp} f \subset \overline{\Omega}\},$$

where the latter smaller space typically entails some boundary conditions on f. For example, for Lipschitz domains and $s = 1$, one shows that $H^1(\overline{D}) = H^1(D) := \{f \in L_2(D) \; ; \; \nabla f \in L_2(D)\}$, whereas $H^1_{\overline{D}}(X) = H^1_0(D)$, defined as the closure of $C_0^\infty(D)$ in $H^1(D)$.

We can also define Sobolev spaces on manifolds M, with or without boundary, using a partition of unity and an atlas, in the obvious way.

For a linear operator $T : \mathcal{H}_1 \to \mathcal{H}_2$, or map, between Banach spaces $\mathcal{H}_i$, not necessarily bounded or defined on all $\mathcal{H}_1$, we define

- its *domain* $\mathsf{D}(T) \subset \mathcal{H}_1$, the linear subspace, not necessarily closed, on which T is defined,

- its *null space* $\mathsf{N}(T) := \{f \in \mathsf{D}(T) \ ; \ Tf = 0\}$,

- its *range* $\mathsf{R}(T) = T\mathcal{H}_1 := \{g \in \mathcal{H}_2 \ ; \ g = Tf \text{ for some } f \in \mathsf{D}(T)\}$, and

- its *graph* $\mathsf{G}(T) := \{(f, Tf) \in \mathcal{H}_1 \oplus \mathcal{H}_2 \ ; \ f \in \mathsf{D}(T)\}$.

The operator T is said to be *closed* if $\mathsf{G}(T)$ is a closed subspace of $\mathcal{H}_1 \oplus \mathcal{H}_2$, and is said to be *densely defined* if $\mathsf{D}(T)$ is a dense subspace of $\mathcal{H}_1$. By the closed graph theorem, a closed operator defined on a closed subspace $\mathsf{D}(T)$ of $\mathcal{H}_1$ is bounded. Closed operators appear naturally as inverses of bounded operators and as differential operators with domains defined by suitable boundary conditions. Note that as an operator $T : \mathsf{G}(T) \to \mathcal{H}_2$, identifying $\mathsf{D}(T) \leftrightarrow \mathsf{G}(T)$, every closed operator may be viewed as a bounded operator.

Exercise 6.4.2 (Sums and compositions). Let T be a bounded operator and A a densely defined closed operator. Prove the following. The operator $A + T$, with domain $\mathsf{D}(A+T) := \mathsf{D}(A)$, is closed and densely defined, with $(A+T)^* = A^* + T^*$. The operator TA, with domain $\mathsf{D}(TA) := \mathsf{D}(A)$, is densely defined with $(TA)^* = A^*T^*$, where $\mathsf{D}(A^*T^*) = \{f \ ; \ T^*f \in \mathsf{D}(A^*)\}$. If furthermore $\|Tf\| \gtrsim \|f\|$ for all $f \in \mathsf{R}(A)$, then both TA and A^*T^* are closed and densely defined, and $(A^*T^*)^* = TA$.

We say that T is an *integral operator* on some measure space D if there a kernel function $k(x, y)$, $x, y \in D$, such that

$$Tf(x) = \int_D k(x, y)f(y)d\mu(y), \quad x \in D.$$

If D is an open subset of an n-dimensional Euclidean space, and if $|k(x, y)| \lesssim 1/|x - y|^\alpha$ for some $\alpha < n$, then T is called a *weakly singular* integral operator.

Exercise 6.4.3 (Schur estimates). Let T be an integral operator on D and $1 < p < \infty$ and $1/p + 1/q = 1$. Using Hölder's inequality, give a short proof of the interpolation estimate

$$\|T\|_{L_p(D) \to L_p(D)} \leq M_1^{1/p} M_\infty^{1/q},$$

where

$$M_1 := \|T\|_{L_1(D) \to L_1(D)} = \sup_{y \in D} \int_D |k(x, y)| dx \quad \text{and}$$

$$M_\infty := \|T\|_{L_\infty(D) \to L_\infty(D)} = \sup_{x \in D} \int_D |k(x, y)| dy.$$

Deduce that a weakly singular integral operator on a bounded set is bounded on $L_p(D)$, $1 \le p \le \infty$.

We recall that for a bounded operator $T : \mathcal{H}_1 \to \mathcal{H}_2$, its *adjoint*, or *dual*, operator is the bounded operator $T^* : \mathcal{H}_2^* \to \mathcal{H}_1^*$ uniquely determined by

$$\langle g, Tf \rangle = \langle T^* g, f \rangle, \quad f \in \mathcal{H}_1, g \in \mathcal{H}_2^*.$$

The naive extension of this to unbounded operators is to require $\langle g, Tf \rangle = \langle Sg, f \rangle$ for all $f \in \mathsf{D}(T)$ and $g \in \mathsf{D}(S)$, in which case we say that T and S are *formally adjoint operators*. For the stronger adjoint notion, we identify T and its graph $\mathsf{G}(T)$, and note that the adjoint T^* is closely related to the orthogonal complement of the subspace of $\mathsf{G}(T)$. Using the "complex structure" $J(f_1, f_2) := (-f_2, f_1)$ on $\mathcal{H}_1^* \oplus \mathcal{H}_2^*$, we see for bounded operators that

$$\mathsf{G}(T)^{\perp} = J\mathsf{G}(T^*). \tag{6.4}$$

For an unbounded operator T, $J\mathsf{G}(T)^{\perp}$ is the graph of a linear operator T^* if and only if T is densely defined. In this case the *adjoint* $T^* : \mathcal{H}_2^* \to \mathcal{H}_1^*$ *in the sense of unbounded operators*, and its domain are defined by the identity (6.4). Concretely, $g \in \mathsf{D}(T^*)$ means that

$$|\langle g, Tf \rangle| \lesssim \|f\|, \quad f \in \mathsf{D}(T).$$

By construction the adjoint T^* is always a closed operator.

The best-known result in operator theory is the following, which tells when an unbounded operator can be diagonalized, in possibly a continuous way.

Theorem 6.4.4 (Spectral theorem). *Let T be a self-adjoint operator on a Hilbert space $\mathcal{H}$, that is, $T^* = T$ in the sense of unbounded operators. Then there exist a measure space (Ω, μ), a measurable function $m : \Omega \to \mathbf{R}$, and a unitary operator $U : \mathcal{H} \to L_2(\Omega, \mu)$ such that*

$$UTx = MUx, \quad x \in \mathcal{H},$$

where M is the multiplication operator $f \mapsto mf$ on $L_2(\Omega, \mu)$ and $U(D(T)) = \{f \in L_2(\Omega, \mu) \; ; \; mf \in L_2(\Omega; \mu)\}$.

The following notion is central to this book.

Definition 6.4.5 (Splittings of Banach spaces). Two closed subspaces $\mathcal{H}_1, \mathcal{H}_2 \subset \mathcal{H}$ are *complementary* if they give a direct sum

$$\mathcal{H} = \mathcal{H}_1 \oplus \mathcal{H}_2.$$

We refer to such direct sums of closed subspaces as *splittings*.

By the bounded inverse theorem, we have estimates

$$\|f_1 + f_2\| \approx \|f_1\| + \|f_2\|, \quad f_1 \in \mathcal{H}_1, f_2 \in \mathcal{H}_2,$$

for any splitting $\mathcal{H} = \mathcal{H}_1 \oplus \mathcal{H}_2$. For Hilbert space splittings we can write this as

$$|\langle f_1, f_2 \rangle| \leq \cos\theta \|f_1\| \|f_2\|, \quad f_1 \in \mathcal{H}_1, \ f_2 \in \mathcal{H}_2,$$

where $0 < \theta \leq \pi/2$ is the angle between the subspaces. Then with $\theta = \pi/2$ we have the optimal case of an orthogonal splitting.

In terms of operators, a splitting is described by projections. By a projection P, we mean a bounded linear operator such that $P^2 = P$. A splitting $\mathcal{H} = \mathcal{H}_1 \oplus \mathcal{H}_2$ corresponds to two projections P_1 and P_2 such that

$$P_1 + P_2 = I,$$

that is, two complementary projections. Here the ranges are $\mathcal{H}_1 = \mathsf{R}(P_1)$ and $\mathcal{H}_2 = \mathsf{R}(P_2)$, and the null spaces are $\mathsf{N}(P_1) = \mathcal{H}_2$ and $\mathsf{N}(P_2) = \mathcal{H}_1$ respectively. Orthogonal Hilbert space splittings correspond to self-adjoint, or equivalently orthogonal, projections. In a Banach space, it is important to note that in general, a given subspace $\mathcal{H}_1$ does not have any complement $\mathcal{H}_2$. This happens precisely when there exists a projection onto $\mathcal{H}_1$, which is not always the case.

Definition 6.4.6 (Compact operators). A bounded operator $T : \mathcal{H}_1 \to \mathcal{H}_2$ between Banach spaces is *compact* if the closure of the image $\overline{\{Tf \ ; \ \|f\| \leq 1\}}$ of the unit ball is a compact set in $\mathcal{H}_2$.

If T_k are compact operators and if $\lim_{k\to\infty} \|T_k - T\| = 0$, then T is a compact operator. A Hilbert space operator is compact if and only if it is the norm limit of operators with finite-dimensional ranges. However, this is not true in general for Banach spaces.

Example 6.4.7. (i) An integral operator T with kernel $k(x, y)$ is called a *Hilbert–Schmidt operator* if

$$\int_D \int_D |k(x, y)|^2 dxdy < \infty.$$

Every such operator is compact on $L_2(D)$.

(ii) If D is a compact metric space and $0 < \alpha < 1$, then the inclusion map

$$C^\alpha(D) \to C(D)$$

is a compact map by the Arzelà–Ascoli theorem.

(iii) By the Rellich–Kondrachov theorem, the inclusion map

$$I : H^1(\overline{D}) \to L_2(D)$$

is compact for every bounded Lipschitz domain D.

Example 6.4.8 (Operator ideals). Among the classical Banach spaces ℓ_p consisting of p-summable sequences $(x_k)_{k=1}^{\infty}$, the three spaces $\ell_1 \subset \ell_2 \subset \ell_\infty$ are particularly important. We recall that ℓ_2 is a Hilbert space and that the Cauchy–Schwarz inequality shows that $(x_k y_k) \in \ell_1$ if $(x_k) \in \ell_2$ and $(y_k) \in \ell_2$.

There is an important analogue of this for operators on a given Hilbert space $\mathcal{H}$. In this case, the space of bounded operators $\mathcal{L}(\mathcal{H})$ plays the role of ℓ_∞, the space $\mathcal{L}_2(\mathcal{H})$ of Hilbert–Schmidt operators plays the role of ℓ_2, and the space $\mathcal{L}_1(\mathcal{H})$ of trace class operators plays the role of ℓ_1. If $T \in \mathcal{L}(\mathcal{H})$, let m be the multiplier for the bounded self-adjoint operator T^*T given by the spectral theorem (Theorem 6.4.4). Then the norms on these spaces are $\|T\|_{\mathcal{L}_2(\mathcal{H})} := \int_\Omega m \, d\mu$ and $\|T\|_{\mathcal{L}_1(\mathcal{H})} := \int_\Omega \sqrt{m} \, d\mu$ respectively. We have

$$\mathcal{L}_1(\mathcal{H}) \subset \mathcal{L}_2(\mathcal{H}) \subset \mathcal{L}(\mathcal{H}),$$

and an analogue of the Cauchy–Schwarz inequality shows that $T_1 T_2 \in \mathcal{L}_1(\mathcal{H})$ if $T_1 \in \mathcal{L}_2(\mathcal{H})$ and $T_2 \in \mathcal{L}_2(\mathcal{H})$.

By the Hilbert–Schmidt kernel theorem, every $T \in \mathcal{L}_2(\mathcal{H})$ is unitarily equivalent to a Hilbert–Schmidt integral operator with kernel $k \in L_2(\Omega \times \Omega)$ on a measure space (Ω, μ), as in Example 6.4.7. Furthermore, for trace class operators $T \in \mathcal{L}_1(\mathcal{H})$, the trace functional

$$\mathrm{Tr}(T) = \int_\Omega k(x, x) d\mu(x)$$

is continuous in the trace norm. All trace class and Hilbert–Schmidt operators are compact.

If splittings concern halves of invertible operators and if compact operators concern almost finite-dimensional operators, the following important class of operators comprises the almost invertible operators.

Definition 6.4.9 (Fredholm operators). A bounded operator $T : \mathcal{H}_1 \to \mathcal{H}_2$ between Banach spaces is called a *Fredholm operator* if $\mathsf{R}(T)$ is a closed subspace of finite codimension in $\mathcal{H}_2$ and if $\mathsf{N}(T)$ is finite-dimensional. If $\mathsf{R}(T)$ is closed and if at least one of $\mathsf{N}(T)$ and $\mathcal{H}_2/\mathsf{R}(T)$ is finite-dimensional, then T is called a *semi-Fredholm operator*. The *index* of T is $\mathrm{Ind}(T) := \dim \mathsf{N}(T) - \dim(\mathcal{H}_2/\mathsf{R}(T))$.

Fredholm operators are the invertible operators modulo finite-dimensional spaces. Indeed, T is a Fredholm operator if and only if we have splittings $\mathcal{H}_1 = \mathsf{N}(T) \oplus \mathcal{H}_1'$ and $\mathcal{H}_2 = \mathsf{R}(T) \oplus \mathcal{H}_2'$, with $\mathsf{N}(T)$ and $\mathcal{H}_2'$ finite-dimensional and

$$T : \mathcal{H}_1' \to \mathsf{R}(T)$$

an invertible operator. Such operators appear frequently in solving linear equations

$$Tf = g, \tag{6.5}$$

where $g \in \mathcal{H}_2$ is given, $T : \mathcal{H}_1 \to \mathcal{H}_2$ is bounded, and we are looking for $f \in \mathcal{H}_1$ solving the equation. With Fredholm theory we investigate solvability of this equation as follows.

- The first and main task is to prove a *lower bound* modulo compact operators, that is, an estimate of the form

$$\|f\|_{\mathcal{H}_1} \lesssim \|Tf\|_{\mathcal{H}_2} + \|Kf\|_{\mathcal{H}_3}, \quad f \in \mathcal{H}_1,$$

  with the error term K being some auxiliary compact operator $K : \mathcal{H}_1 \to \mathcal{H}_3$ into some Banach space $\mathcal{H}_3$. Then T is a semi-Fredholm operator with finite-dimensional null space, which gives a uniqueness and stability result for solutions to (6.5).

  A more general perturbation result is that if T is a semi-Fredholm operator and K is a compact operator, then $T + K$ is a semi-Fredholm operator with the same index $\mathrm{Ind}(T + K) = \mathrm{Ind}(T)$.

- To show existence, the most flexible tool is the *method of continuity*, which is the perturbation result that $\mathrm{Ind}(T_0) = \mathrm{Ind}(T_1)$ whenever T_s is a family of semi-Fredholm operators depending continuously on $s \in [0, 1]$. More generally, the Fredholm property is stable, and the index is constant, not only under compact perturbations, but also under small perturbations in the norm topology.

  The standard way to apply this to obtain an existence result for a given semi-Fredholm operator T is to continuously perturb this, usually through a spectral parameter or the geometry, to an operator that is known to be invertible.

- A third main technique for Fredholm operators is *duality*. The closed range theorem, which applies more generally to any closed and densely defined operator $T : \mathcal{H}_1 \to \mathcal{H}_2$ between Banach spaces, states that $\mathsf{R}(T^*)$ is closed if and only if $\mathsf{R}(T)$ is closed. In this case, $\mathsf{R}(T) = \mathsf{N}(T^*)^\perp$. Note that for every closed and densely defined operator T, the null space $\mathsf{N}(T)$ is closed and $\mathsf{R}(T)^\perp = \mathsf{N}(T^*)$.

  The standard way to apply duality to obtain existence results for solutions to (6.5) is to show that $\mathsf{N}(T^*)$ is finite-dimensional. If T, or equivalently T^*, has been shown to be a semi-Fredholm operator, then this implies that $\mathsf{R}(T)$ is a closed subspace of finite codimension in $\mathcal{H}_2$.

6.5   Comments and References

6.1 For the linear approximation of a nonlinear map ρ, we use the terminology total derivative and notation $\dot{\rho}$, or $\nabla \otimes \rho$ when viewing this linear map as a tensor as in Section 1.4. Standard for this object in the literature is $d\rho$, the differential of ρ, a notation that we avoid, since d also denotes other objects such as the exterior derivative and infinitesimals.

A reference for Rademacher's theorem and on Lipschitz change of variables formulas is Evans and Gariepy [36].

A reference for nonsmooth domains is Grisvard [48]. A reference for manifolds and differential geometry is Taubes [90].

6.2–6.3 Our main reference for partial differential equations, which gives a comprehensive and modern treatment and makes good use of differential forms, is Taylor [91, 92]. In particular, a proof of Mehler's formula stated in Example 6.3.7 and further Fourier calculations are found here. Another excellent general reference for PDEs is Evans [35]. A classical reference for Bessel functions is Watson [95]. A proof of Rellich's Lemma 6.3.5 is found in, for example, [67].

6.4 Our main reference for operator theory is Kato [61]. In particular, perturbation results for Fredholm operators are found in [61, Sec. 4.5]. A very readable account of the functional analysis that we need is found in [91, App. A]. Unpublished handwritten lecture notes on Fredholm theory, singular integrals and Tb theorems from 2011 are Rosén [81]. The first part of these notes contains full proofs of all the main results in Fredholm theory. The author plans to publish a book on singular integral equations based on these notes in the near future.

Chapter 7

Multivector Calculus

Prerequisites:

This chapter is where the second part of this book on analysis starts. Chapter 6 is meant to be used as a reference in reading this and later chapters. Otherwise, a solid background in multivariable and vector calculus should suffice for this chapter. In particular, the reader is assumed to have a solid understanding of gradient, divergence, and curl, as well as the classical Gauss, Stokes, and Green integral theorems. Some basic understanding of the Hilbert spaces $L_2(D)$ and $H^1(D)$ is helpful for Section 7.6.

Road map:

Roughly speaking, this chapter is the analysis twin of Chapter 2. The main players are the exterior derivative

$$d : F(x) \mapsto \nabla \wedge F(x),$$

generalizing in particular gradient and curl, and the dual interior derivative

$$\delta : F(x) \mapsto \nabla \lrcorner F(x),$$

generalizing in particular divergence, acting on multivector fields $F(x)$ in Euclidean spaces. Without an inner product, the exterior derivative acts on multi-covector fields. We know from Section 2.3 that induced linear multivector maps T are $\wedge$-homomorphisms, and as in Section 2.7 we have dual transformations for the interior product. Amazingly, there is an extension of this to nonlinear vector maps $T = \rho$: a pullback of multicovector fields ρ^* that commutes with exterior differentiation

$$\rho^*(\nabla \wedge F) = \nabla \wedge (\rho^* F).$$

This, as well as the dual result for the interior derivative and pushforwards, is the topic of Section 7.2.

© Springer Nature Switzerland AG 2019
A. Rosén, *Geometric Multivector Analysis*, Birkhäuser Advanced Texts Basler Lehrbücher,
https://doi.org/10.1007/978-3-030-31411-8_7

We all know the fundamental theorem of calculus $\int_a^b f'(x)dx = f(b) - f(a)$, as well as its generalizations to dimensions two and three: the Green, Gauss, and Stokes theorems. In Section 7.3, we prove the general Stokes theorem, from which all integral theorems of this kind in affine space follow. In the case that we will use the most in this book, namely the case of domains D, the formula reads

$$\int_D \theta(\dot{x}, \nabla)dx = \int_{\partial D} \theta(y, \nu(y))dy.$$

Here $\theta(y, \nu)$ is an expression depending differentiably on position y and linearly on the unit normal vector field ν on ∂D, and the Stokes formula allows you to pass to a solid integral by simply replacing $\nu = \sum_i \nu_i e_i$ with the nabla symbol $\nabla = \sum_i \partial_i e_i$ and acting with the partial derivatives on all x in the expression. We also prove the Stokes theorem when D is replaced by the lower-dimensional surface M in affine space, in which case νdy is replaced by the oriented measure $\underline{dy} \in \widehat{\wedge}^k V$ for a k-dimensional surface as in Section 2.4.

We also study fundamental splittings of multivector fields on Euclidean domains, referred to as Hodge decompositions. In the classical case of vector fields, this is the Helmholtz decomposition in three dimensions, known as the fundamental theorem of vector calculus, and saying that on some domain D, every vector field $F(x)$ can be split into essentially two parts,

$$F(x) = \nabla U(x) + G(x) + \nabla \times V(x),$$

a gradient vector field ∇U and a curl vector field $\nabla \times V$. In the presence of a boundary ∂D, we need to require either Dirichlet boundary conditions on U or normal boundary conditions on V, but not both. The remaining piece G lives in a finite-dimensional subspace, the dimension of which is determined by the topology of D only. In Section 7.6 we show how Hodge decompositions work and how they can be used to solve boundary value problems. Chapter 10 is devoted to a more thorough study of these Hodge decompositions.

Highlights:

- Commutation of pullback and exterior derivative: 7.2.9

- The general Stokes theorem: 7.3.9

- Cartan's magic formula: 7.4.7

- The Poincaré theorem: 7.5.2

- Hodge decomposition: 7.6.6

7.1 Exterior and Interior Derivatives

Fix an an affine space (X, V). We aim to develop the basic differential and integral calculus for *multivector fields*

$$F : D \to \wedge V : x \mapsto F(x)$$

and *multicovector fields*

$$\Theta : D \to \wedge V^* : x \mapsto \Theta(x)$$

defined on $D \subset X$. Recall that when we have fixed a duality on V, then it suffices to consider multivector fields, since a multicovector field $\Theta(x)$ is pointwise identified with the corresponding multivector field $F(x)$ as in Proposition 1.2.3.

We make use of the *nabla symbol* ∇, which is the formal expression

$$\nabla := \sum_{i=1}^{n} e_i^* \partial_i,$$

where as usual ∂_i is the partial derivative along the basis vector e_i, and $\{e_i^*\}$ is the dual basis to $\{e_i\}$. Thus we view ∇ as a covector, but with the usual scalar coordinates replaced by the partial derivatives. We now show that whenever there is a bilinear product

$$V^* \times L \to L' : (\theta, v) \mapsto \theta \bullet v,$$

then the nabla symbol induces a differential operator that sends L-valued functions to L'-valued functions. The most general such operator is the following.

Example 7.1.1 (Total derivative). Let $D \subset X$ be an open subset and L a linear space, and let $F : D \to L$ be differentiable at $x \in D$. Form its total derivative $\underline{F}_x \in \mathcal{L}(V, L)$ at $x \in D$ as in Definition 6.1.2. As in Section 1.4, we may view this linear map $V \to L$ as a tensor in $V^* \otimes L$. In terms of the nabla symbol, we have

$$\underline{F}_x = \nabla \otimes F(x).$$

Concretely, if $\{e_i\}_{i=1}^n \subset V$ and $\{e_i^*\}_{i=1}^n \subset V^*$ are dual bases and $\{v_k\}_{k=1}^m \subset L$ is a basis for L, then the matrix element on row k, column i in the matrix for $\underline{F}_x$ is $\partial_i F_k$, with F_k the kth coordinate function of F. The corresponding tensor is $\partial_i F_k \, e_i^* \otimes v_k$, and thus

$$\nabla \otimes F(x) = \sum_{i=1}^{n} \sum_{k=1}^{m} \partial_i F_k(x) \, e_i^* \otimes v_k = \sum_{i=1}^{n} e_i^* \otimes \partial_i F(x).$$

Thus the total derivative $\nabla \otimes F$ is nothing but the collection of all partial derivatives of all coordinate functions, collected in a matrix or a tensor depending on the formulation.

Definition 7.1.2 (Nabla operators). Let $D \subset X$ be an open subset and L a linear space, and let $F : D \to L$ be differentiable at $x \in D$. Assume that $V^* \times L \to L'$: $(\theta, v) \mapsto \theta \bullet v$ is a bilinear map. Then we define the derivative

$$\nabla \bullet F(x) := T(\nabla \otimes F(x)) \in L',$$

where $T : V^* \otimes L \to L'$ is the lifted linear map such that $\theta \bullet v = T(\theta \otimes v)$, given by the universal property for tensor products.

This means that we start from all partial derivatives $\nabla \otimes F$ and map by T, leaving only some combinations $\nabla \bullet F$ of the partial derivatives, as compared to the total derivative.

Proposition 7.1.3 (Coordinate expression for nabla). *If $\{e_i\}$ is a basis for V, ∂_i denotes the partial derivative along e_i, and $\{e_i^*\}$ denotes the dual basis for V^*, then*

$$\nabla \bullet F(x) = \sum_{i=1}^{n} e_i^* \bullet \partial_i F(x).$$

Proof. Apply the linear lift T to the identity $\nabla \otimes F(x) = \sum_{i=1}^{n} e_i^* \otimes \partial_i F(x)$ from Example 7.1.1. $\square$

Note that when V is an inner product space, then we identify V^* and V, viewing in particular e_i^* as a vector. If $\{e_i\}$ is an ON-basis, then $\nabla \bullet F(x) = \sum_{i=1}^{n} \langle e_i \rangle^2 e_i \bullet \partial_i F(x)$. In particular, if $\{e_i\}$ is a Euclidean ON-basis, then

$$\nabla \bullet F(x) = \sum_{i=1}^{n} e_i \bullet \partial_i F(x).$$

Exercise 7.1.4. From the above definition of $\nabla \bullet F(x) = \sum_{i=1}^{n} e_i^* \bullet \partial_i F(x)$, it is clear that $\nabla \bullet F$ is independent of the choice of basis $\{e_i\}$. Give a direct proof of this, using the chain rule and the dual change of basis.

In this chapter, we shall study the nabla operators induced by the exterior and interior products. If X is also an inner product space, then the nabla operators induced by the Clifford product and the action on the spinor space yield Dirac operators, which are topics for later chapters.

Definition 7.1.5 (Exterior and interior derivatives). Let $D \subset X$ be an open set. If $\Theta : D \to \wedge V^*$ is a multicovector field in D that is differentiable at $x \in D$, then its *exterior derivative* at x is

$$d\Theta(x) := \nabla \wedge \Theta(x) = \sum_{i=1}^{n} e_i^* \wedge \partial_i \Theta(x).$$

If $F : D \to \wedge V$ is a multivector field in D that is differentiable at $x \in D$, then its *interior derivative* at x is

$$\delta F(x) := \nabla \lrcorner F(x) = \sum_{i=1}^{n} e_i^* \lrcorner \partial_i F(x).$$

These two fundamental differential operators d and δ are usually pronounced *dee* and *dell*. A main property is their nilpotence.

Proposition 7.1.6 (Nilpotence). *For every C^2 regular multicovector field Θ, we have*

$$\nabla \wedge (\nabla \wedge \Theta) = 0.$$

Similarly, for every C^2 multivector field F, we have $\nabla \lrcorner (\nabla \lrcorner F) = 0$.

Proof. This is a differential operator version of the fact that $v \wedge v = 0$ for every vector v. Indeed, we have

$$\nabla \wedge (\nabla \wedge \Theta) = \sum_{i,j} e_i^* \wedge e_j^* \wedge \partial_i \partial_j \Theta = 0,$$

since $\partial_i \partial_j \Theta = \partial_j \partial_i \Theta$ and $e_i^* \wedge e_j^* = -e_j^* \wedge e_i^*$. A similar calculation applies to the interior derivative, using Proposition 2.6.3. $\qquad\square$

Algebraic identities for the exterior and interior products induce relations between the exterior and interior derivatives.

Proposition 7.1.7 (Dualities for d and δ). *Let (X, V) be an oriented affine space.*

(i) *For all differentiable k-covector fields Θ, we have the Hodge star intertwining*

$$\nabla \lrcorner (\Theta*) = (-1)^k (\nabla \wedge \Theta)*.$$

(ii) *The exterior derivative and the negative of the interior derivative are formally adjoint in the sense that*

$$\int_X \langle \nabla \wedge \Theta(x), F(x) \rangle \, dx = - \int_X \langle \Theta(x), \nabla \lrcorner F(x) \rangle \, dx,$$

for all C^1 regular and compactly supported $\Theta : X \to \wedge V^*$ and $F : X \to \wedge V$.

Proof. (i) According to Proposition 2.6.8(ii), we have

$$\theta \lrcorner (\Theta*) = (\Theta \wedge \theta)* = (-1)^k (\theta \wedge \Theta)*, \quad \theta \in V^*.$$

Therefore the differential operators corresponding to these bilinear products are equal.

(ii) It follows from the product rule and Proposition 7.1.3 that

$$\sum_{i=1}^n \partial_i \langle e_i^* \wedge \Theta(x), F(x) \rangle = \sum_{i=1}^n \langle e_i^* \wedge \partial_i \Theta(x), F(x) \rangle + \langle e_i^* \wedge \Theta(x), \partial_i F(x) \rangle$$

$$= \langle \nabla \wedge \Theta(x), F(x) \rangle + \langle \Theta(x), \nabla \lrcorner F(x) \rangle.$$

Thus integration yields $0 = \int_X \langle \nabla \wedge \Theta, F \rangle \, dx + \int_X \langle \Theta, \nabla \lrcorner F \rangle \, dx$, since the integral of a partial derivative of a compactly supported function vanishes. $\qquad\square$

Considering how homogeneous multi(co)vector fields are mapped, we have the following diagrams:

$$\wedge^0 V^* \xrightarrow{\;d_0\;} \wedge^1 V^* \xrightarrow{\;d_1\;} \wedge^2 V^* \xrightarrow{\;d_2\;} \cdots \xrightarrow{\;d_{n-3}\;} \wedge^{n-2} V^* \xrightarrow{\;d_{n-2}\;} \wedge^{n-1} V^* \xrightarrow{\;d_{n-1}\;} \wedge^n V^*$$

$$\wedge^0 V \xleftarrow{\;\delta_1\;} \wedge^1 V \xleftarrow{\;\delta_2\;} \wedge^2 V \xleftarrow{\;\delta_3\;} \cdots \xleftarrow{\;\delta_{n-2}\;} \wedge^{n-2} V \xleftarrow{\;\delta_{n-1}\;} \wedge^{n-1} V \xleftarrow{\;\delta_n\;} \wedge^n V$$

The exterior and interior derivatives are generalizations of the classical vector calculus derivatives gradient, divergence, and curl as follows.

d_0 If $f : D \to \mathbf{R} = \wedge^0 V^*$ is a scalar function, then its *gradient* grad $f = \nabla f$ equals its exterior derivative, as well as its total derivative, which is the covector field

$$\operatorname{grad} f(x) = \nabla f(x) = \nabla \wedge f(x) = \nabla \otimes f(x) = \underline{f}_x = \sum_{i=1}^{n} \partial_i f(x)\, e_i^*.$$

By Proposition 7.1.7(i), the derivative δ_n also corresponds to the gradient. If X is an inner product space, then grad f can be viewed as a vector field.

δ_1 If $F : D \to V = \wedge^1 V$ is a vector field, then its *divergence* div $F = \langle \nabla, F \rangle$ equals its interior derivative, which is the scalar field

$$\operatorname{div} F(x) = \langle \nabla, F(x) \rangle = \nabla \lrcorner F(x) = \sum_{i=1}^{n} e_i^* \lrcorner \partial_i F(x) = \sum_{i=1}^{n} \langle e_i^*, \partial_i F(x) \rangle.$$

By Proposition 7.1.7(i), the derivative $(-1)^{n-1} d_{n-1}$ also corresponds to the divergence.

d_1 If $\Theta : D \to V^* = \wedge^1 V^*$ is a covector field, then its *curl* curl $\Theta = \nabla \wedge \Theta$ equals its exterior derivative, which is the bi-covector field

$$\operatorname{curl} \Theta(x) = \nabla \wedge \Theta(x) = \sum_{i=1}^{n} e_i^* \wedge \partial_i \Theta(x).$$

By Proposition 7.1.7(i), the derivative $-\delta_{n-1}$ also corresponds to the curl. If X is an inner product space, viewing $\Theta = F$ as a vector field, the curl $\nabla \wedge F$ can be viewed as a bivector field. Moreover, if X is a three-dimensional oriented Euclidean space, then by Hodge duality one can identify the bivector field $\nabla \wedge F$ and the vector field

$$\nabla \times F(x) = *(\nabla \wedge F(x)) = \sum_{i=1}^{3} e_i^* \times \partial_i F(x).$$

Classically, this vector field $\nabla \times F$ is referred to as the curl of F.

d_2 In dimension $n \geq 4$, the exterior and interior derivatives also contain nabla derivatives beyond these three classical vector calculus derivatives. The simplest such derivative is d_2 in four dimensions. We will see in Section 9.2 that two of Maxwell's equations, the Faraday and magnetic Gauss laws, can be written $d_2 F = 0$ for the electromagnetic field F in spacetime.

Exercise 7.1.8. Investigate what Proposition 7.1.6 means for the gradient, divergence, and curl in three dimensions, and recover two well-known results from vector calculus: $\nabla \times (\nabla u) = 0$ and $\langle \nabla, \nabla \times F \rangle = 0$.

When applying the product rule to nabla derivatives, one encounters additional problems: Noncommutativity and nonassociativity. Starting from a general formulation of the problem, we consider two C^1 regular functions $F : D \to L_1$ and $G : D \to L_2$ defined on some open set $D \subset X$. Consider some bilinear maps $L_1 \times L_2 \to L_3 : (w_1, w_2) \mapsto w_1 * w_2$ and $V^* \times L_3 \to L_4 : (\theta, w) \mapsto \theta \bullet w$. As in Definition 7.1.2, we can apply the nabla operator $\nabla\bullet$ to the product of functions $F * G : D \to L_3$. Let $\{e_i\}$ be a basis for X's vector space V with dual basis $\{e_i^*\}$, and let $\{e_\alpha'\}$ and $\{e_\beta''\}$ be bases for L_1 and L_2 respectively. Then if $F(x) = \sum_\alpha F_\alpha(x) e_\alpha'$ and $G(x) = \sum_\beta G_\beta(x) e_\beta''$, the product rule applied to the coordinate functions shows that

$$\nabla \bullet (F(x) * G(x)) = \left(\sum_i e_i^* \partial_i \right) \bullet \left(\left(\sum_\alpha F_\alpha e_\alpha' \right) * \left(\sum_\beta G_\beta(x) e_\beta'' \right) \right)$$

$$= \sum_{i,\alpha,\beta} \partial_i \left(F_\alpha(x) G_\beta(x) \right) e_i^* \bullet (e_\alpha' * e_\beta'')$$

$$= \sum_{i,\alpha,\beta} \left((\partial_i F_\alpha(x)) G_\beta(x) + F_\alpha(x)(\partial_i G_\beta(x)) \right) e_i^* \bullet (e_\alpha' * e_\beta'')$$

$$= \sum_i e_i^* \bullet \left(\left(\sum_\alpha \partial_i F_\alpha e_\alpha' \right) * \left(\sum_\beta G_\beta e_\beta'' \right) \right) + \sum_i e_i^* \bullet \left(\left(\sum_\alpha F_\alpha e_\alpha' \right) * \left(\sum_\beta \partial_i G_\beta e_\beta'' \right) \right)$$

$$= \nabla \bullet (F(\dot{x}) * G(x)) + \nabla \bullet (F(x) * G(\dot{x})).$$

Note that after having applied the partial derivative ∂_i to the products $F_\alpha G_\beta$ of component functions, we have moved ∂_i to the factor that ∂_i differentiates. However, in general we may have neither associativity or commutativity properties for products $\bullet$ and $*$, so the basis covector e_i^* corresponding to ∂_i cannot be moved along to the factor it differentiates. To handle such expressions resulting from the product rule, it is convenient to use the *dot notation* $\dot{x}$ to indicate the factor on which the partial derivatives fall. Thus $\nabla \bullet (F(\dot{x}) * G(x))$ means that the variable in G is treated as constant in performing the differentiation, and $\nabla\bullet (F(x)*G(\dot{x}))$ means that the variable in F is treated as constant.

Example 7.1.9 (Product rules for d and δ). Let X be an inner product space, and consider exterior and interior derivatives of multivector fields.

First, let $\bullet$ and $*$ both be the exterior product, and consider $\nabla \wedge (F(x) \wedge G(x)) = \nabla \wedge (F(\dot{x}) \wedge G(x)) + \nabla \wedge (F(x) \wedge G(\dot{x}))$. In this case, we can use that the exterior product is associative for the first term and that we have the commutation relation $e_i^* \wedge F = \widehat{F} \wedge e_i^*$ for the second term. Performing the algebra, we get

$$d(F \wedge G) = \nabla \wedge (F(\dot{x}) \wedge G(x)) + \nabla \wedge (F(x) \wedge G(\dot{x}))$$
$$= (\nabla \wedge F(x)) \wedge G(x) + \widehat{F}(x) \wedge (\nabla \wedge G(x)) = (dF) \wedge G + \widehat{F} \wedge (dG).$$

Next let $\bullet = \lrcorner$ and $* = \wedge$, and consider the product rule application $\nabla \lrcorner (F(x) \wedge G(x)) = \nabla \lrcorner (F(\dot{x}) \wedge G(x)) + \nabla \lrcorner (F(x) \wedge G(\dot{x}))$. In this case, it is harder to move along ∇ to respective factor algebraically. The best we can do is to use the expansion rule from Corollary 2.8.4 for the interior product, which allows this expression to be rewritten as

$$\delta(F \wedge G) = \nabla \lrcorner (F(\dot{x}) \wedge G(x)) + \nabla \lrcorner (F(x) \wedge G(\dot{x}))$$
$$= \big((\delta F) \wedge G + \widehat{F}(\dot{x}) \wedge (\nabla \lrcorner G(x))\big) + \big((\nabla \lrcorner F(x)) \wedge G(\dot{x}) + \widehat{F} \wedge (\delta G)\big).$$

However, two of the terms still cannot be expressed in terms of d and δ. In the case of two vector fields $F, G : D \to V$, the two remaining terms can be written as directional derivatives. We get $-F(\dot{x}) \wedge (\nabla \lrcorner G(x)) = -\langle G(x), \nabla \rangle F(\dot{x}) = -\partial_G F(x)$, and similarly that $(\nabla \lrcorner F(x)) \wedge G(\dot{x})$ is the directional derivative of G along F. Using the Lie bracket from Lemma 7.4.3, we get the final result

$$\delta(F \wedge G) = (\operatorname{div} F)G - F(\operatorname{div} G) + [F, G].$$

7.2 Pullbacks and Pushforwards

When we have a multivector field $F(x)$ or a multicovector field $\Theta(x)$ in an affine space X, and have a diffeomorphism $\rho : X' \to X$, then we can of course make a simple change of variables and obtain functions $F \circ \rho$ and $\Theta \circ \rho$ defined on the affine space X'. However, we do not obtain multi(co)vector fields in X', since their ranges are still in the exterior algebras $\wedge V$ and $\wedge^* V$ for X respectively. It turns out that the natural ways to map back their ranges to the corresponding exterior algebras for X', namely by $\underline{\rho}^{-1}$ and $\underline{\rho}^*$ respectively, yield very fundamental and useful operations.

Definition 7.2.1 (Pullback and pushforward). Let $\rho : D_1 \to D_2$ be a C^1 map between open sets $D_1 \subset X_1$ and $D_2 \subset X_2$, with total derivative $\underline{\rho}_y : V_1 \to V_2$, $y \in D_1$, as in Definition 6.1.2. Consider the induced linear map $\underline{\rho}_y : \wedge V_1 \to \wedge V_2$, and the dual linear map $\underline{\rho}_y^* : \wedge V_2^* \to \wedge V_1^*$, as in Sections 2.3 and 2.7.

(i) If $\Theta : D_2 \to \wedge V_2^*$ is a multicovector field in D_2, then its *pullback* by ρ is the multicovector field $\rho^* \Theta$ in D_1 defined by

$$(\rho^* \Theta)(y) := \underline{\rho}_y^* (\Theta(x)), \quad y \in D_1, x = \rho(y).$$

(ii) Further assume that ρ is a C^1 diffeomorphism. If $F : D_1 \to \wedge V_1$ is a multi-vector field in D_1, then its *pushforward* by ρ is the multivector field $\rho_* F$ in D_2 defined by

$$(\rho_* F)(x) := \underline{\rho}_y(F(y)), \quad x = \rho(y) \in D_2.$$

As usual, if V is an inner product space, then we identify $\wedge V^*$ and $\wedge V$, and in this case both ρ^* and ρ_* act on multivector fields. Note that both ρ and its inverse ρ^{-1} are needed to define the pushforward as a function on D_2. The definition of the pullback, however, does not require ρ to be invertible.

Exercise 7.2.2. Let $V_0 \subset V$ be a hyperplane, with unit normal n, in a Euclidean space. If $i : V_0 \to V$ is the inclusion map, show that the pullback of a multivector field F in V is

$$i^*(F) = n \lrcorner (n \wedge F|_{V_0}).$$

By Proposition 2.8.7, this means that $i^*(F)$ is the restriction of the tangential part of F to V_0.

Exercise 7.2.3 (Tangential projection). Let X_1 and X_2 be Euclidean spaces. Then the pullback ρ^* and the inverse pushforward $(\rho^{-1})_* = (\rho_*)^{-1}$ map between the same domains. Show that $\rho^* = (\rho^{-1})_*$ if and only if ρ is an isometry.

Clearly there is a dual relation between the pullback and pushforward operations. In fact, it is important to note that in fact, there are two types of dualities: the pointwise duality and the L_2 duality.

Definition 7.2.4 (Normalized pushforward). Let $\rho : D_1 \to D_2$ be a C^1 diffeomorphism between open sets $D_1 \subset X_1$ and $D_2 \subset X_2$ in oriented affine spaces. If $F : D_1 \to \wedge V_1$ is a multivector field in D_1, then its *normalized pushforward* is the multivector field $\tilde{\rho}_* F$ in D_2 defined by

$$\tilde{\rho}_* F(x) := |J_\rho(y)|^{-1} \underline{\rho}_y F(y), \quad x = \rho(y) \in D_2, y \in D_1,$$

where $J_\rho(y)$ denotes the Jacobian of ρ at $y \in D_1$.

The following proposition makes it clear why we define this rescaled version $\tilde{\rho}_* = |J_\rho|^{-1} \rho_*$ of the pushforward operation.

Proposition 7.2.5 (Push and pull dualities). *Let $\rho : D_1 \to D_2$ be a C^1 diffeomorphism between open sets $D_1 \subset X_1$ and $D_2 \subset X_2$ in affine spaces.*

(i) *At a fixed point $y \in D_1$, $x = \rho(y) \in D_2$, the pullback ρ^* and pushforward ρ_* are pointwise dual in the sense that*

$$\langle (\rho^* \Theta)(y), F(y) \rangle = \langle \Theta(x), (\rho_* F)(x) \rangle,$$

for all $\Theta(x) \in \wedge V_2^*$ and $F(y) \in \wedge V_1$.

(ii) *Assume that X_1 and X_2 are oriented. The pullback ρ^* and the normalized pushforward $\tilde{\rho}_*$ are L_2 dual in the sense that*

$$\int_{D_1} \langle \rho^*\Theta(y), F(y) \rangle \, dy = \int_{D_2} \langle \Theta(x), \tilde{\rho}_*F(x) \rangle \, dx, \qquad (7.1)$$

for $\Theta \in L_2(D_2; \wedge V_2^*)$ and $F \in L_2(D_1; \wedge V_1)$.

Proof. (i) is immediate from Definition 7.2.1. To obtain (ii), we change variables in the integral:

$$\int_{D_2} \langle \Theta(x), \tilde{\rho}_*F(x) \rangle \, dx = \int_{D_1} \langle \Theta(\rho(y)), |J_\rho(y)|^{-1}\underline{\rho}_y F(y) \rangle \, |J_\rho(y)| dy$$

$$= \int_{D_1} \langle \underline{\rho}_y^*\Theta(\rho(y)), F(y) \rangle \, dy = \int_{D_1} \langle \rho^*\Theta(y), F(y) \rangle \, dy. \quad \square$$

Exercise 7.2.6 (Transitivity). Show under suitable assumptions that

$$(\rho_2 \circ \rho_1)^* = \rho_1^* \rho_2^*,$$
$$(\rho_2 \circ \rho_1)_* = (\rho_2)_*(\rho_1)_*, \text{ and}$$
$$\widetilde{(\rho_2 \circ \rho_1)}_* = (\tilde{\rho}_2)_*(\tilde{\rho}_1)_*.$$

Proposition 7.2.7 (Homomorphism). *Let $\rho : D_1 \to D_2$ be a C^1 diffeomorphism between open sets $D_1 \subset X_1$ and $D_2 \subset X_2$ in oriented affine spaces. For multicovector fields Θ_i in D_2 and multivector fields F_i in D_1, we have the identities*

$$\rho^*(\Theta_1 \wedge \Theta_2) = (\rho^*\Theta_1) \wedge (\rho^*\Theta_2), \qquad \rho_*(F_1 \wedge F_2) = (\rho_*F_1) \wedge (\rho_*F_2),$$
$$\rho^*(\rho_*(F_1) \lrcorner \Theta_2) = F_1 \lrcorner \rho^*(\Theta_2), \qquad \rho_*(\rho^*(\Theta_1) \lrcorner F_2) = \Theta_1 \lrcorner \rho_*(F_2).$$

In particular, when the spaces are oriented, we have for a multivector field F in D_1 that

$$*(\tilde{\rho}_*F) = \pm(\rho^*)^{-1}(*F),$$

where $\pm$ is the sign of the Jacobian J_ρ.

Note that in general, the normalized pushforward $\tilde{\rho}_*$ is not a homomorphism. For example, if f is a scalar function with range in $\mathbf{R} = \wedge^0 V = \wedge^0 V^*$, then

$$\rho^*f = f \circ \rho, \quad \rho_*f = f \circ \rho^{-1}, \quad \tilde{\rho}_*f = \frac{1}{|J_{\rho \circ \rho^{-1}}|} f \circ \rho^{-1}.$$

At the opposite end, ρ^* and ρ_* act by multiplication by the Jacobian J_ρ on n-(co)vectors, but $\tilde{\rho}_*$ preserves n-vectors if ρ is orientation-preserving.

Proof. The homomorphism formulas follow pointwise from Propositions 2.3.2 and 2.7.1.

To see the Hodge star identity, we use a version of the interior homomorphisms for the right interior product and obtain

$$\rho^*(*(\rho_*F)) = J_\rho(*F),$$

since ρ^* acts by J_ρ on $\wedge^n V_2^*$. Dividing by $|J_\rho|$ yields the stated formula. $\quad \square$

A characteristic geometric property of pullbacks and pushforwards is the following. Consider a diffeomorphism $\rho : X_1 \to X_2$ between Euclidean spaces mapping a hypersurface $M_1 \subset X_1$ onto a hypersurface $M_2 = \rho(M_1)$, with derivative $\underline{\rho}_y$ mapping the tangent hyperplane V_1' to M_1 at y onto the tangent hyperplane V_2' to M_2 at $x = \rho(y)$. Then at these points,

- if a multivector $w \in \wedge V_1$ is tangential to V_1', then $\rho_* w$ (and $\tilde{\rho}_* w$) is tangential to V_2', and

- if a multivector $w \in \wedge V_2$ is normal to V_2', then $\rho^* w$ is normal to V_1'.

To see this, note that if n_2 is a vector normal to V_2', then $n_1 := \rho^* n_2$ is normal to V_1'. Indeed, for all $v \in V_1'$,

$$\langle \underline{\rho}_y^* n_2, v \rangle = \langle n_2, \underline{\rho}_y v \rangle = 0,$$

since $\underline{\rho}_y v \in V_2'$. From this, the above mapping properties follow from Definition 2.8.6, using that

$$n_2 \lrcorner (\rho_* w) = \rho_* (n_1 \lrcorner w) \quad \text{and}$$
$$n_1 \wedge (\rho^* w) = \rho^* (n_2 \wedge w),$$

by Proposition 7.2.7.

Example 7.2.8. Consider the elliptic–polar change of variables ρ given by

$$\begin{cases} x_1 = ay_1 \cos(y_2), \\ x_2 = by_1 \sin(y_2), \end{cases}$$

for some constants $a, b > 0$ in the Euclidean plane $\mathbf{R}^2$. If, say, $D_1 := \{(y_1, y_2) ; y_1 > 0, 0 < y_2 < \pi/2\}$ and $D_2 := \{(x_1, x_2) ; x_1 > 0, x_2 > 0\}$, then $\rho : D_1 \to D_2$ is a diffeomorphism. Let F be the constant vector field $F(y) = e_1$ parallel to the y_1-axis. To push forward and pull back F to the $x_1 x_2$-plane, we calculate the derivative

$$\underline{\rho}_y = \begin{bmatrix} a \cos y_2 & -ay_1 \sin y_2 \\ b \sin y_2 & by_1 \cos y_2 \end{bmatrix}.$$

This gives the pushed forward vector field

$$\rho_* F = \begin{bmatrix} a \cos y_2 & -ay_1 \sin y_2 \\ b \sin y_2 & by_1 \cos y_2 \end{bmatrix} \begin{bmatrix} 1 \\ 0 \end{bmatrix} = \begin{bmatrix} a \cos y_2 \\ b \sin y_2 \end{bmatrix} = \frac{1}{\sqrt{(x_1/a)^2 + (x_2/b)^2}} \begin{bmatrix} x_1 \\ x_2 \end{bmatrix}.$$

On the other hand, pulling back F by ρ^{-1} gives

$$(\rho^{-1})^* F = \begin{bmatrix} a^{-1} \cos y_2 & -a^{-1} y_1^{-1} \sin y_2 \\ b^{-1} \sin y_2 & b^{-1} y_1^{-1} \cos y_2 \end{bmatrix} \begin{bmatrix} 1 \\ 0 \end{bmatrix} = \begin{bmatrix} a^{-1} \cos y_2 \\ b^{-1} \sin y_2 \end{bmatrix}$$

$$= \frac{1}{\sqrt{(x_1/a)^2 + (x_2/b)^2}} \begin{bmatrix} a^{-2} x_1 \\ b^{-2} x_2 \end{bmatrix}.$$

Note that $\rho_* F$ is tangent to the radial lines $\rho(\{y_2 = \text{constant}\})$, and that $(\rho^{-1})^* F$ is normal to the ellipses $\rho(\{y_1 = \text{constant}\})$, in accordance with the discussion above. See Figure 7.1(b)–(c).

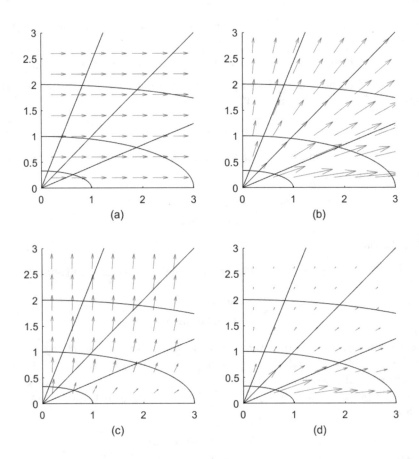

Figure 7.1: (a) Change of variables $F \circ \rho^{-1}$. (b) Pushforward $\rho_* F$. (c) Inverse pullback $(\rho^{-1})^* F$. (d) Normalized pushforward $\tilde{\rho}_* F$. The field has been scaled by a factor 0.3, that is, the plots are for $F = 0.3e_1$.

Since F is constant, it is of course divergence- and curl-free: $\nabla\lrcorner F = 0 = \nabla\wedge F$. By direct calculation, we find that $(\rho^{-1})^* F$ is curl-free. For the pushforward, we note that

$$\text{div}(\rho_* F) = 1/\sqrt{(x_1/a)^2 + (x_2/b)^2} \neq 0.$$

However, the normalized pushforward

$$\tilde{\rho}_* F = \frac{1}{ab((x_1/a)^2 + (x_2/b)^2)} \begin{bmatrix} x_1 \\ x_2 \end{bmatrix}$$

is seen to be divergence-free. See Figure 7.1(d). This is in accordance with Theorem 7.2.9 below.

We now show that in general, pullbacks commute with the exterior derivative, and dually that normalized pushforwards commute with the interior derivative. At first it seems that taking the exterior derivative of a pulled back multicovector field would give two terms, a first-order term when the derivatives hit $\Theta(\rho(x))$ according to the chain rule and a zero-order term when the derivatives hit $\underline{\rho}_x$ according to the product rule. However, it turns out that the zero-order term vanishes miraculously due to the alternating property of the exterior product and the equality of mixed derivatives.

Theorem 7.2.9 (The commutation theorem). *Let $\rho : D_1 \to D_2$ be a C^2 map between open sets $D_1 \subset X_1$ and $D_2 \subset X_2$.*

(i) *If $\Theta : D_2 \to \wedge V_2^*$ is a C^1 multicovector field in D_2, then the pullback $\rho^* \Theta : D_1 \to \wedge V_1^*$ is C^1 and $\nabla \wedge (\rho^* \Theta)(y) = \rho^*(\nabla \wedge \Theta)(y)$ for $y \in D_1$, that is,*

$$d(\rho^* \Theta) = \rho^*(d\Theta).$$

(ii) *Further assume that ρ is a C^2 diffeomorphism. If $F : D_1 \to \wedge V_1$ is a C^1 multivector field in D_1, then the normalized pushforward $\tilde{\rho}_* F : D_2 \to \wedge V_2$ is C^1 and $\nabla \lrcorner (\tilde{\rho}_* F)(x) = \tilde{\rho}_*(\nabla \lrcorner F)(x)$ for $x \in D_2$, that is,*

$$\delta(\tilde{\rho}_* F) = \tilde{\rho}_*(\delta F).$$

The proof uses the following lemma, the proof of which we leave as an exercise.

Lemma 7.2.10. *Let $\{e_i\}$ and $\{e'_i\}$ be bases for V_1 and V_2, with dual bases $\{e_i^*\}$ and $\{e_i'^*\}$ respectively. Then the pullback of a covector field $\theta(x) = \sum_i \theta_i(x) e_i'^*$ is*

$$\rho^* \theta(y) = \sum_{i,j} \theta_i(x) \, \partial_j \rho_i(y) e_j^*, \quad x = \rho(y) \in D_2, y \in D_1,$$

and the pushforward of a vector field $v(y) = \sum_i v_i(y) e_i$ is

$$\rho_* v(x) = \sum_{i,j} v_i(y) \, \partial_i \rho_j(y) e'_j, \quad x = \rho(y) \in D_2, \ y \in D_1.$$

Proof of Theorem 7.2.9. Since both $y \mapsto \underline{\rho}_y^*$ and $y \mapsto \Theta(\rho(y))$ are C^1, so is $\rho^* \Theta$.

(i) When $\Theta = f$ is a scalar field, then the formula is the chain rule. Indeed, changing variables $x = \rho(y)$ in the scalar function $f(x)$, for $\rho^* f = f \circ \rho$ we have

$$\nabla_y(f(\rho(y))) = \sum_{i,k} e_i^*(\partial_i \rho_k(y))(\partial_{x_k} f)(x) = \underline{\rho}_y^*(\nabla f)(x),$$

using Lemma 7.2.10.

(ii) Next consider a vector field $\Theta = \theta = \sum_i \theta_i e_i : D_2 \to \wedge V_2^*$. Fix bases $\{e_i\}$ and $\{e_i'\}$ for V_1 and V_2 respectively and write $\{e_i^*\}$ and $\{e_i'^*\}$ for the dual bases and ∂_i and ∂_i' for the partial derivatives. From Lemma 7.2.10 we have

$$\nabla \wedge (\rho^* \theta) = \nabla_y \wedge \left(\sum_{i,j} \theta_i(\rho(y)) \partial_j \rho_i(y) e_j^* \right)$$

$$= \sum_{i,j,k} (\partial_k \theta_i \, \partial_j \rho_i + \theta_i \, \partial_k \partial_j \rho_i) e_k^* \wedge e_j^* = \sum_{i,j,k} \partial_k \theta_i \, \partial_j \rho_i \, e_k^* \wedge e_j^*,$$

since $\partial_k \partial_j = \partial_j \partial_k$ and $e_k \wedge e_l = -e_l \wedge e_k$. This is the key point of the proof. On the other hand, we have

$$\rho^*(\nabla \wedge \theta) = \rho^* \left(\sum_{i,j} \partial_j' \theta_i \, e_j'^* \wedge e_i'^* \right) = \sum_{i,j} \partial_j' \theta_i \, \rho^*(e_j'^* \wedge e_i'^*) = \sum_{i,j,k,l} \partial_j' \theta_i \, \partial_k \rho_j \, \partial_l \rho_i \, e_k^* \wedge e_l^*.$$

Note that $\sum_j \partial_j' \theta_i \, \partial_k \rho_j = \partial_k \theta_i$ by the chain rule. Thus changing the dummy index l to j proves the formula for covector fields.

(iii) Next consider a general multicovector field Θ. By linearity, we may assume that $\Theta(x) = \theta^1(x) \wedge \cdots \wedge \theta^k(x)$ for C^1 covector fields θ^j. We need to prove

$$\sum_{i,j} e_i^* \wedge \rho^* \theta^1 \wedge \cdots \wedge \partial_i(\rho^* \theta^j) \wedge \cdots \wedge \rho^* \theta^k = \sum_{i,j} \rho^* e_i'^* \wedge \rho^* \theta^1 \wedge \cdots \wedge \rho^*(\partial_i' \theta^j) \wedge \cdots \wedge \rho^* \theta^k.$$

For this, it suffices to show that

$$\sum_i e_i^* \wedge \partial_i(\rho^* \theta) = \sum_j \rho^* e_j'^* \wedge \rho^*(\partial_j' \theta)$$

for all C^1 covector fields θ in D_2. But this follows from step (ii) of the proof.

(iv) From the hypothesis it follows that $x \mapsto \rho_{\rho^{-1}(x)}$, $x \mapsto |J_\rho(\rho^{-1}(x))|$, and $x \mapsto F(\rho^{-1}(x))$ are C^1. Therefore the product rule shows that $\tilde{\rho}_* F$ is C^1. Let $\Theta : D_2 \to \wedge V_2^*$ be any compactly supported smooth multicovector field. Then Propositions 7.1.7(ii) and 7.2.5(ii) and step (iii) above show that

$$\int_{D_2} \langle \Theta, \nabla \lrcorner \tilde{\rho}_* F \rangle dx = -\int_{D_2} \langle \nabla \wedge \Theta, \tilde{\rho}_* F \rangle dx = -\int_{D_1} \langle \rho^*(\nabla \wedge \Theta), F \rangle dy$$

$$= -\int_{D_1} \langle \nabla \wedge \rho^* \Theta, F \rangle dy = \int_{D_1} \langle \rho^* \Theta, \nabla \lrcorner F \rangle dy = \int_{D_2} \langle \Theta, \tilde{\rho}_*(\nabla \lrcorner F) \rangle dx.$$

Since Θ is arbitrary, we must have $\nabla \lrcorner (\tilde{\rho}_* F) = \tilde{\rho}_*(\nabla \lrcorner F)$. $\qquad\square$

Example 7.2.11 (Orthogonal curvilinear coordinates). Let $\rho : \mathbf{R}^3 \to X$ be curvilinear coordinates in three-dimensional Euclidean space X. Important examples treated in the standard vector calculus curriculum are spherical and cylindrical coordinates. The pushforward of the standard basis vector fields are

$$\tilde{e}_i := \rho_* e_i = \partial_{y_i} \rho(y), \quad i = 1, 2, 3,$$

where $\{e_i\}$ denotes the standard basis in $\mathbf{R}^3$. The frame $\{\tilde{e}_i\}$ is in general not an ON-frame in X, but in important examples such as the two mentioned above, these frame vector fields are orthogonal at each point. Assuming that we have such orthogonal curvilinear coordinates, we define

$$h_i(y) := |\rho_* e_i|$$

and $e_i := \tilde{e}_i / h_i$, for $y \in \mathbf{R}^3$. This gives us an ON-frame $\{e_i(y)\}$ in X.

We now show how the well-known formulas for gradient, divergence, and curl follow from Theorem 7.2.9. Note that

$$\underline{\rho} = \underline{\rho}^* = \begin{bmatrix} h_1 & 0 & 0 \\ 0 & h_2 & 0 \\ 0 & 0 & h_3 \end{bmatrix}$$

with respect to the ON-bases $\{e_i\}$ and $\{e_i\}$. For the gradient, we have

$$\nabla u = (\rho^*)^{-1} \operatorname{grad}(\rho^* u) = (\rho^*)^{-1} \sum_i (\partial_i u) e_i = \sum_i h_i^{-1} (\partial_i u) e_i.$$

Note that ρ^* acts on scalar functions just by changing variables, whereas ρ^* acts on vectors by the above matrix.

For the curl of a vector field $F = \sum_i F_i e_i$ in X, we similarly obtain

$$\nabla \wedge F = (\rho^*)^{-1} \operatorname{curl}(\rho^* F) = (\rho^*)^{-1} \nabla \wedge \sum_i h_i F_i e_i$$

$$= (\rho^*)^{-1} \sum_j \sum_i \partial_j (h_i F_i) e_j \wedge e_i = \sum_j \sum_i (h_i h_j)^{-1} \partial_j (h_i F_i) e_j \wedge e_i$$

$$= \frac{1}{h_1 h_2 h_3} \begin{vmatrix} h_1 e_1 & \partial_1 & h_1 F_1 \\ h_2 e_2 & \partial_2 & h_2 F_2 \\ h_3 e_3 & \partial_3 & h_3 F_3 \end{vmatrix}.$$

Note that ρ^* acts on $\wedge^2 V$ by two-by-two subdeterminants of the above matrix as in Example 2.3.4.

For the divergence of a vector field $F = \sum_i F_i e_i$ in X, we use instead the normalized pushforward to obtain

$$\nabla \lrcorner F = \tilde{\rho}_* \operatorname{div}(\tilde{\rho}_*)^{-1} F = \tilde{\rho}_* \nabla \lrcorner \left(h_1 h_2 h_3 \sum_i h_i^{-1} F_i e_i \right) = \tilde{\rho}_* \sum_i \partial_i (h_1 h_2 h_3 h_i^{-1} F_i)$$

$$= (h_1 h_2 h_3)^{-1} \big(\partial_1 (h_2 h_3 F_1) + \partial_2 (h_1 h_3 F_2) + \partial_3 (h_1 h_2 F_3) \big).$$

Note that $\tilde{\rho}_* = \rho_* / (h_1 h_2 h_3)$ and that ρ_* acts on vectors by the above matrix and simply by change of variables on scalars.

Example 7.2.12 (Pullback of Laplace equation). To see how the Laplace operator Δ transforms under a change of variables, let $\rho : D_1 \to D_2$ be a C^2-diffeomorphism of Euclidean domains and let $u : D_2 \to \mathbf{R}$ be harmonic, that is, $\Delta u = 0$. Changing variables, u corresponds to a function $v(y) = u(\rho(y)) = \rho^* u(y)$ in D_1. According to the commutation theorem (Theorem 7.2.9), we have

$$0 = \nabla \lrcorner \left(\nabla \wedge ((\rho^*)^{-1} v) \right) = \nabla \lrcorner \left((\rho^*)^{-1}(\nabla v) \right) = \tilde{\rho}_* \nabla \lrcorner \left((\tilde{\rho}_*)^{-1}(\rho^*)^{-1}(\nabla v) \right)$$

$$= \tilde{\rho}_* \nabla \lrcorner \left((\rho^* \tilde{\rho}_*)^{-1}(\nabla v) \right) = \tilde{\rho}_* \operatorname{div}(A^{-1} \operatorname{grad} v).$$

Since $\tilde{\rho}_*$ is invertible, we see that the Laplace equation transforms into the divergence-form equation $\operatorname{div}(A^{-1} \operatorname{grad} v) = 0$. Here the linear map $A(y)$, for fixed $y \in D_1$, in an ON-basis $\{e_i\}$ has matrix elements

$$A_{i,j}(y) = \langle e_i, \rho^* \tilde{\rho}_* e_j \rangle = |J_\rho(y)|^{-1} \langle \rho_y(e_i), \rho_y(e_j) \rangle = g(y)^{-1/2} g_{i,j}(y),$$

where $g_{i,j}$ is the metric in D_1 representing the Euclidean metric in D_2 and $g(x) := \det(g_{i,j}(x)) = \det(\rho^* \rho_*) = |J_\rho(x)|^2$ is the determinant of the metric matrix. Thus the Laplace equation $\sum_i \partial_i^2 u = 0$ transforms to the divergence-form equation

$$\sum_{i,j} \partial_i(\sqrt{g}\, g^{i,j} \partial_j v) = 0,$$

where $g^{i,j}$ denotes the inverse of the metric matrix.

Example 7.2.12 is a special case of the following pullback formulas for exterior and interior derivatives.

Proposition 7.2.13 (Pullback of interior derivatives). *Let X_1, X_2 be oriented Euclidean spaces, let $\rho : D_1 \to D_2$ be a C^2-diffeomorphism, and denote by $G(y) : \wedge V_1 \to \wedge V_1$, $y \in D_1$, the metric of D_2 pulled back to D_1, that is,*

$$\langle Gw_1, w_2 \rangle = \langle \rho_* w_1, \rho_* w_2 \rangle$$

for multivector fields w_1, w_2 in D_1. Write $g(y) = \det G|_{\wedge^1 V_1}(y) = |J_\rho(y)|^2$. Then we have pullback formulas

$$\rho^*(dF) = d(\rho^* F),$$
$$\rho^*(\delta F) = (g^{-1/2}G)\delta(g^{1/2}G^{-1})(\rho^* F).$$

7.3 Integration of Forms

In this section, we develop integration theory for forms over k-surfaces. To avoid technicalities concerning bundles, at this stage we limit ourselves to k-surfaces in affine spaces, but the integration theory we develop generalizes with minor changes to general manifolds.

Definition 7.3.1 (k-Form). A *k-form* defined on a subset D of an affine space (X, V) is a function

$$\Theta : D \times \widehat{\wedge}^k V \to L,$$

with range in a finite-dimensional linear space L, such that

$$\Theta(x, \lambda w) = \lambda \Theta(x, w), \quad x \in M, \ w \in \widehat{\wedge}^k V, \ \lambda > 0.$$

We say that Θ is a *linear k-form* if for each $x \in D$, $w \mapsto \Theta(x, w)$ extends to a linear function of $w \in \wedge^k V$.

The idea with k-forms is that in integrating at a point $x \in D$, the integrand also depends on the orientation at $x \in M$ of the k-surface M that we integrate over.

Definition 7.3.2 (Integral of form). Let M be a compact oriented C^1-regular k-surface in an affine space (X, V), and let $\Theta : M \times \widehat{\wedge}^k V \to L$ be a continuous k-form. We define the *integral* of Θ over M to be

$$\int_M \Theta(x, \underline{dx}) := \sum_{\alpha \in \mathcal{I}} \int_{D_\alpha} \eta_\alpha(\mu_\alpha(y)) \Theta(\mu_\alpha(y), \underline{\mu_\alpha}_y(e_1 \wedge \cdots \wedge e_k)) dy,$$

where $\{e_i\}$ is the standard basis in $\mathbf{R}^k$ and dy is standard Lebesgue measure in $\mathbf{R}^k \supset D_\alpha$. Here $\{\mu_\alpha\}_{\alpha \in \mathcal{I}}$ is the atlas of M, and $\{\eta_\alpha\}_{\alpha \in \mathcal{I}}$ denotes a partition of unity for M.

Note as in Section 2.4 how the induced action of the derivative $\underline{\mu_\alpha}_y : \wedge^k \mathbf{R}^k \to \wedge^k V$ maps the oriented volume element $e_1 \wedge \cdots \wedge e_k \, dy$ in $\mathbf{R}^k$ to the oriented volume element

$$\underline{dx} = \underline{\mu_\alpha}_y(e_1 \wedge \cdots \wedge e_k \, dy) = \underline{\mu_\alpha}_y(e_1 \wedge \cdots \wedge e_k) dy$$

on M. Note that this infinitesimal k-vector is simple, and hence Θ in general, needs to be defined only on the Grassmann cone. However, it is only to linear forms that Stokes's theorem applies, as we shall see.

The following proposition shows that such integrals do not depend on the precise choice of atlas and partition of unity for M, but only on the orientation of M.

Proposition 7.3.3 (Independence of atlas). *Consider a compact oriented C^1-regular k-surface M, with atlas $\{\mu_\alpha : D_\alpha \to M_\alpha\}_{\alpha \in \mathcal{I}}$, in an affine space (X, V). Let $\{\mu_\beta : D_\beta \to M_\beta\}_{\beta \in \mathcal{I}'}$, $\mathcal{I}' \cap \mathcal{I} = \emptyset$, be a second atlas for M such that all transition maps between D_α and D_β, $\alpha \in \mathcal{I}$, $\beta \in \mathcal{I}'$, are C^1-regular and orientation-preserving. Further assume that $\{\eta_\alpha\}_{\alpha \in \mathcal{I}}$ is a partition of unity for $\{\mu_\alpha\}_{\alpha \in \mathcal{I}}$ and that $\{\eta_\beta\}_{\beta \in \mathcal{I}'}$*

is a partition of unity for $\{\mu_\beta\}_{\beta \in \mathcal{I}'}$. *Then for every continuous k-form* Θ, *we have*

$$\sum_{\alpha \in \mathcal{I}} \int_{D_\alpha} \eta_\alpha(\mu_\alpha(y))\Theta(\mu_\alpha(y), \underline{\mu_\alpha}_y(e_1 \wedge \cdots \wedge e_k))dy$$
$$= \sum_{\beta \in \mathcal{I}'} \int_{D_\beta} \eta_\beta(\mu_\beta(z))\Theta(\mu_\beta(z), \underline{\mu_\beta}_z(e_1 \wedge \cdots \wedge e_k))dz.$$

Proof. Inserting $1 = \sum_{\beta \in \mathcal{I}'} \eta_\beta$ in the integral on the left-hand side and $1 = \sum_{\alpha \in \mathcal{I}} \eta_\alpha$ in the integral on the right-hand side, it suffices to show that

$$\int_{D_{\beta\alpha}} \Theta_{\alpha\beta}(\mu_\alpha(y), \underline{\mu_\alpha}_y(e_1 \wedge \cdots \wedge e_k))dy = \int_{D_{\alpha\beta}} \Theta_{\alpha\beta}(\mu_\beta(z), \underline{\mu_\beta}_z(e_1 \wedge \cdots \wedge e_k))dz,$$

where $\Theta_{\alpha\beta}(x, w) := \eta_\alpha(x)\eta_\beta(x)\Theta(x, w)$, since $\operatorname{supp} \eta_\alpha\eta_\beta \subset M_\alpha \cap M_\beta$. Changing variables $z = \mu_{\beta\alpha}(y)$ in the integral on the right-hand side, we get

$$\int_{D_{\beta\alpha}} \Theta_{\alpha\beta}(\mu_\beta(\mu_{\beta\alpha}(y)), \underline{\mu_\beta}_{\mu_{\beta\alpha}(y)}(e_1 \wedge \cdots \wedge e_k))J_{\mu_{\beta\alpha}}(y)dy.$$

Since $\mu_\beta \circ \mu_{\beta\alpha} = \mu_\alpha$, and therefore $\underline{\mu_\beta}(e_1 \wedge \cdots \wedge e_k)J_{\mu_{\beta\alpha}} = \underline{\mu_\alpha}(e_1 \wedge \cdots \wedge e_k)$, the stated formula follows from the homogeneity of $w \mapsto \Theta_{\alpha\beta}(x, w)$. $\square$

Example 7.3.4 (Oriented and scalar measure). The simplest linear k-form in an affine space is

$$\Theta(x, w) = w \in L := \wedge^k V.$$

In this case, $\int_M \Theta(x, \underline{dx}) = \int_M \underline{dx} = \wedge^k(M)$ is the oriented measure of M discussed in Section 2.4.

In a Euclidean space, given a continuous function $f : M \to L$, the integral of the k-form

$$\Theta(x, w) := f(x)|w|$$

is seen to be the standard surface integral $\int_M f(x)dx$, where $dx = |\underline{dx}|$. Note that these are not linear k-forms. In particular, $f = 1$ yields the usual (scalar) measure $|M| := \int_M dx$ of M.

Using that $dx = |\underline{dx}|$ and the usual triangle inequality for integrals, we obtain from the definitions that oriented and scalar measure satisfy the triangle inequality

$$\left| \int_M \underline{dx} \right| \leq \int_M dx.$$

We continue with Example 2.4.4, where we calculated the oriented area element

$$\underline{dx} = (e_{12} + 2y_2e_{13} - 2y_1e_{23})dy_1dy_2.$$

Hence

$$dx = |\underline{dx}| = \sqrt{1 + 4y_1^2 + 4y_2^2}\,dy_1dy_2,$$

giving an area of the paraboloid equal to

$$\int_{|y|<1} \sqrt{1 + 4y_1^2 + 4y_2^2}\, dy_1 dy_2 = 2\pi \int_0^1 r\sqrt{1 + 4r^2}\, dr = \frac{\pi}{6}(5\sqrt{5} - 1).$$

The triangle inequality above shows that the size of the oriented area of the paraboloid is less than its area, and indeed we verify that $\pi < \pi(5\sqrt{5} - 1)/6$.

Example 7.3.5 (Linear k-forms and k-covectors). A k-covector field $D \to \wedge^k V^*$: $x \mapsto \Theta_1(x)$ in an affine space (X, V) gives, by Proposition 2.5.3, a linear k-form

$$\Theta(x, w) = \langle \Theta_1(x), w \rangle, \quad x \in D, w \in \wedge^k V. \tag{7.2}$$

In fact, a general linear k-form $(x, w) \mapsto \Theta(x, w) \in L$ is just a vector-valued generalization of this. Indeed, if we fix a basis $\{v_i\}$ for L, with dual basis $\{v_i^*\}$, then by Proposition 2.5.3 we can write

$$\Theta(x, w) = \langle \Theta_1(x), w \rangle v_1 + \cdots + \langle \Theta_N(x), w \rangle v_N,$$

where Θ_i is the k-covector field such that $\langle \Theta_i(x), w \rangle = \langle v_i^*, \Theta(x, w) \rangle$ for $w \in \wedge^k V$. Equivalently, a linear k-form can be viewed as a function

$$D \to L \otimes \wedge^k V^*.$$

Exercise 7.3.6 (Pullback of k-forms). Consider compact oriented C^1 regular k-surfaces M and M' in affine spaces (X, V) and (X', V') respectively. Let $\rho : D \to D'$ be a C^1-diffeomorphism between open neighborhoods of M and M' such that $\rho(M) = M'$. Given a continuous k-form $\Theta : D' \times \widehat{\wedge}^k V' \to L$, we define the pulled back k-form

$$\rho^*\Theta(y, w) := \Theta(\rho(y), \rho_y(w)), \quad y \in D, \ w \in \wedge^k V.$$

Show that for scalar-valued linear k-forms this pullback corresponds to the usual pullback of k-covector fields from Definition 7.2.1, under the correspondence in Example 7.3.5. Prove that

$$\int_{\rho(M)} \Theta(x, \underline{dx}) = \int_M \rho^*\Theta(y, \underline{dy}).$$

For the main result in this section, the general Stokes theorem, we need to discuss two further concepts: the exterior derivative of a linear k-form and the orientation of the boundary ∂M. We know from Section 7.1 the exterior derivative $\nabla \wedge \Theta_1$ of a k-covector field Θ_1. If we view Θ_1 as a scalar-valued linear k-form Θ as in (7.2), then $\nabla \wedge \Theta_1$ corresponds to the scalar-valued linear $(k+1)$-form

$$(x, w) \mapsto \langle \nabla \wedge \Theta_1(x), w \rangle = \sum_i \partial_{x_j} \langle \Theta_1(x), e_i^* \lrcorner w \rangle = \sum_i \partial_{x_j} \Theta(x, e_i^* \lrcorner w),$$

$x \in D, w \in \wedge^{k+1} V$, using a basis $\{e_i\}$ for V, with dual basis $\{e_i^*\}$.

Definition 7.3.7 (Exterior derivative). Let Θ be a C^1-regular linear k-form defined on an open set D in an affine space (X, V), with basis $\{e_i\}$. Then its *exterior derivative* is the linear $(k+1)$-form

$$(x, w) \mapsto \Theta(\dot{x}, \nabla \lrcorner w) := \sum_i \partial_{x_i}\Theta(x, e_i^* \lrcorner w), \quad w \in \wedge^{k+1}V, \ x \in D.$$

As for the boundary ∂M of a k-surface M in an affine space (X, V), we point out that when $k < \dim X$, this is of course not the same as the boundary of M as a subset of X, since this would be the whole compact set M. Rather, we define ∂M as a closed $(k-1)$-surface, by restricting the charts for M to the relevant hyperplanes in $\mathbf{R}^k$. The standard relation between the orientations of M and ∂M, which is compatible with the (left) exterior derivative as above, is the following.

Definition 7.3.8 (Boundary orientation). Let M be a compact C^1-regular oriented k-surface with boundary ∂M in an affine space (X, V). By an *orientation* of M at a point $x \in M$, we mean an orientation for the tangent space $T_xM \subset V$ in the sense of Definition 2.4.1. If $\rho = \eta_\alpha : D_\alpha \to M_\alpha$ is a chart, $x = \rho(y)$, $y \in D_\alpha$, then the orientation of M at x is given by the k-vector $\underline{\rho}_y(e_1 \wedge \cdots \wedge e_k)$.

Consider $x \in \partial M$ and $y \in D_\alpha \cap \mathbf{R}^{k-1}$. The orientation $e_1 \wedge \cdots \wedge e_k$ of $\mathbf{R}^k$ determines an orientation

$$\nu_\alpha \lrcorner (e_1 \wedge \cdots \wedge e_k) = \nu_\alpha* \in \wedge^{k-1}\mathbf{R}^k$$

of ∂D_α at y as in Definition 2.4.1, where ν_α is the unit normal vector to ∂D_α, pointing out from D_α. Mapping by $\underline{\rho}_y$ and using Proposition 2.7.1, we have

$$\underline{\rho}_y(\nu_\alpha*) = \nu \lrcorner \underline{\rho}_y(e_1 \wedge \cdots \wedge e_k),$$

with $\nu \in V^*$ being such that $\underline{\rho}_y^*\nu = \nu_\alpha$. The orientation of ∂M determined by $\underline{\rho}_y(\nu_\alpha*)$ is called the *boundary orientation induced by M* for ∂M.

We note that by continuity, it suffices to specify the orientation of ∂M at one point in each connected component of ∂M. Concerning the outward-pointing covector field ν in a Euclidean space, we can make this more explicit as follows. Consider the invertible linear map

$$\underline{\rho}_y : \mathbf{R}^k \to T_xM.$$

Then $\nu = (\underline{\rho}_y^*)^{-1}\nu_\alpha$, which is seen to be an outward-pointing normal vector to ∂M along T_xM. It is, however, not necessarily a unit vector.

Theorem 7.3.9 (Stokes's theorem for forms). *Let $M \subset X$ be a C^1-regular oriented k-surface, $k = 1, 2\ldots, n$, in an affine space (X, V) of dimension n and having boundary ∂M with the orientation induced by M as in Definition 7.3.8. Let Θ :*

$\Omega \times \wedge^{k-1} V \to L$ be a C^1-regular linear $(k-1)$-form, defined on a neighborhood of M, with values in a finite-dimensional linear space L. Then

$$\int_M \Theta(\dot{x}, \nabla \lrcorner \underline{dx}) = \int_{\partial M} \Theta(y, \underline{dy}), \tag{7.3}$$

where the integrand on the left-hand side is the exterior derivative of Θ as in Definition 7.3.7. When $k = 1$, the right-hand side should be interpreted as $\sum_i \Theta(p_i, \epsilon_i)$, where $\partial M = \{p_i\}$ is a point set with orientation $\epsilon_i = \pm 1$.

When $k = n$, we have the following Hodge dual reformulation in a Euclidean space X. Let M be the closure of a bounded C^1-regular domain D in X, with boundary $\partial D = \partial M$ and outward-pointing unit normal vector field ν. Let $\theta :$ $M \times V \to L$ be a C^1-regular linear 1-form. Then

$$\int_D \theta(\dot{x}, \nabla) dx = \int_{\partial D} \theta(y, \nu(y)) dy, \tag{7.4}$$

where $dx = |\underline{dx}|$ and $dy = |\underline{dy}|$.

We have chosen to state the special case of domains $k = n$ explicitly in the theorem, since by far, this is the most useful instance of Stokes' theorem in this book.

Note that for scalar-valued linear forms, we can rewrite the Stokes formula (7.3) as

$$\int_M \langle \nabla \wedge \Theta(x), \underline{dx} \rangle = \int_{\partial M} \langle \Theta(y), \underline{dy} \rangle, \tag{7.5}$$

for a k-covector field Θ.

Proof. (i) We may assume that Θ is scalar-valued, that is, a $k - 1$ covector field, since the general case follows by componentwise integration in L. Let $\{\mu_\alpha\}_{\alpha \in \mathcal{I}}$ be the atlas of M, and $\{\eta_\alpha\}_{\alpha \in \mathcal{I}}$ a partition of unity for M. We can assume that $\eta_\alpha \in C^1(\Omega)$ and $1 = \sum_{\alpha \in \mathcal{I}} \eta_\alpha(x)$ for all $x \in \Omega$. It suffices to show that

$$\int_M \langle \nabla \wedge \Theta_\alpha(x), \underline{dx} \rangle = \int_{\partial M} \langle \Theta_\alpha(y), \underline{dy} \rangle,$$

for each $\alpha \in \mathcal{I}$, where $\Theta_\alpha(x) := \eta_\alpha(x) \Theta(x)$. Since $\mathrm{supp}\, \Theta_\alpha \cap M = M_\alpha$, we can pull back this equation to D_α by the chart μ_α as in Exercise 7.3.6. By the commutation theorem (Theorem 7.2.9), it suffices to prove

$$\int_{D_\alpha} \langle \nabla \wedge \tilde{\Theta}_\alpha(x), e_1 \wedge \cdots \wedge e_k \rangle dx = \int_{D_\alpha \cap \mathbf{R}^{k-1}} \langle \tilde{\Theta}_\alpha(y), -e_k \lrcorner (e_1 \wedge \cdots \wedge e_k) \rangle dy,$$

where $\tilde{\Theta}_\alpha := \mu_\alpha^* \Theta_\alpha$. And indeed, using the fundamental theorem of calculus k

times, we obtain

$$\int_{D_\alpha} \langle \nabla \wedge \tilde{\Theta}_\alpha(x), e_1 \wedge \cdots \wedge e_k \rangle dx = \sum_{j=1}^{k} \int_{D_\alpha} \partial_j \langle \tilde{\Theta}_\alpha(x), e_j \lrcorner (e_1 \wedge \cdots \wedge e_k) \rangle dx$$

$$= \sum_{j=1}^{k-1} 0 + 0 - \int_{D_\alpha \cap \mathbf{R}^{k-1}} \langle \tilde{\Theta}_\alpha(y), e_k \lrcorner (e_1 \wedge \cdots \wedge e_k) \rangle dy,$$

by performing the x_j integral first in term j and using the support properties of $\tilde{\Theta}_\alpha$. This proves the general Stokes formula.

(ii) To prove the Hodge dual formulation for domains, we fix an auxiliary orientation $e_{\overline{n}}$ for X. Given the linear 1-form $\theta : M \times V \to \mathbf{R}$, we define the Hodge dual linear $(n-1)$-form

$$\Theta(x, w) := \theta(x, *w), \quad w \in \wedge^{n-1} V.$$

Its exterior derivative is

$$\Theta(\dot{x}, \nabla \lrcorner e_{\overline{n}}) = \sum_{i=1}^{n} \partial_i \theta(x, *(e_i \lrcorner e_{\overline{n}})) = \sum_{i=1}^{n} \partial_i \theta(x, e_i) = \theta(\dot{x}, \nabla),$$

$x \in D$. From (i) we get

$$\int_{\partial D} \theta(y, \nu(y)) dy = \int_{\partial M} \Theta(y, \underline{dy}) = \int_M \Theta(\dot{x}, \nabla \lrcorner \underline{dx}) = \int_D \theta(\dot{x}, \nabla) dx,$$

using that the oriented volume elements are $\underline{dy} = (\nu*)dy$ on ∂M and $\underline{dx} = e_{\overline{n}} dx$ in M.   □

We end this section by demonstrating through examples how to apply this general Stokes's theorem.

Example 7.3.10 (Classical Stokes's theorem). The classical Stokes's theorem is the special case of the Stokes formula (7.3), when X is an oriented three-dimensional Euclidean space, and M is a C^1-regular compact oriented 2-surface. In this case, the covector field Θ can be identified with a vector field F, and its curl with the Hodge dual vector field

$$\nabla \times F := *(\nabla \wedge F).$$

We get

$$\langle \nabla \wedge F, \underline{dx} \rangle = \langle (\nabla \times F)*, \underline{dx} \rangle = \langle e_{123}, (\nabla \times F) \wedge \underline{dx} \rangle = \langle *\underline{dx}, \nabla \times F \rangle.$$

Thus Stokes's theorem can be rewritten as

$$\int_M \langle \nabla \times F(x), *\underline{dx} \rangle = \int_{\partial M} \langle F(y), \underline{dy} \rangle.$$

In classical notation, we write $*\underline{dx}$ as an infinitesimal vector ndS normal to the surface M, with dS denoting scalar surface measure, and the right-hand side is the usual path integral along the boundary curve ∂M as in vector calculus. By Definition 7.3.8, the orientations of M and ∂M must satisfy $\underline{dx}/dx = \nu \wedge \underline{dy}/dy$ at ∂M, or equivalently $n = \nu \times \underline{dy}/dy$. See Figure 7.2.

Green's theorem is the special case in which the surface M is flat, in which case we have

$$\int_D \langle \nabla \wedge F(x), \underline{dx} \rangle = \int_{\partial D} \langle F(y), \underline{dy} \rangle,$$

where F is a vector field in a bounded C^1-domain in the oriented Euclidean plane.

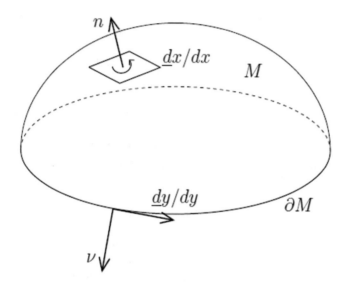

Figure 7.2: Orientations for the Stokes theorem.

Example 7.3.11 (Gauss's theorem). Let D be a bounded C^1-domain in a Euclidean space X, and consider a C^1 vector field F on $\overline{D}$. Applying the Stokes formula (7.4) to the linear 1-form

$$\theta(x, v) = \langle F(x), v \rangle,$$

we obtain Gauss's theorem/divergence theorem

$$\int_D \operatorname{div} F(x)\, dx = \int_{\partial D} \langle \nu(y), F(y) \rangle dy.$$

A special case of this that frequently is used in the study of partial differential equations is the Green formulas, which we obtain by applying Gauss's theorem to a vector field of the form $F(x) = u(x)\nabla v(x)$ for scalar C^1 functions u and C^2 functions v. Using the product rule, we obtain *Green's first identity*

$$\int_D \left(\langle \nabla u(x), \nabla v(x) \rangle + u(x)\Delta v(x) \right) dx = \int_{\partial D} u(y)\partial_\nu v(y)\, dy,$$

where $\partial_\nu v = \langle \nu, \nabla v \rangle$ is the *normal derivative* of v on ∂D. Switching the roles of u and v and subtracting, we have *Green's second identity*

$$\int_D \left(u(x)\Delta v(x) - v(x)\Delta u(x) \right) dx = \int_{\partial D} \left(u(y)\partial_\nu v(y) - v(y)\partial_\nu u(y) \right) dy.$$

Example 7.3.12 (Cauchy's integral theorem). Let D be a bounded C^1-domain in $\mathbf{R}^2 = \mathbf{C}$, where we as usual identify the vector $xe_1 + ye_2 \in \mathbf{R}^2$ and the complex number $x + iy \in \mathbf{C}$. Consider a C^1-regular function $f : D \to \mathbf{C}$. Define the linear 1-form

$$\theta(z, v) := f(z)v, \quad z, v \in \mathbf{C} = \mathbf{R}^2,$$

using the complex product. Its exterior derivative is

$$\theta(\dot{z}, \nabla \lrcorner\, e_{12}) = \partial_1 \theta(z, e_2) + \partial_2 \theta(z, -e_1) = \partial_1 f(z)i + \partial_2 f(z)(-1) = i\overline{\partial} f(z),$$

where the d-bar operator is $\overline{\partial} := \partial_1 + i\partial_2$. This is the nabla operator induced by the complex product as in Definition 7.1.2. The Stokes formula (7.3) therefore reduces to the Cauchy integral theorem

$$i \int_D \overline{\partial} f(z) dz = \int_{\partial D} f(w) \underline{dw},$$

where dz denotes scalar area measure and $\underline{dw}$ denotes oriented measure along the curve ∂D. The classical notation is to write dw for the oriented measure $\underline{dw}$ along the boundary. In the best-known case, that f is an analytic function, it satisfies the Cauchy–Riemann equation $\overline{\partial} f(z) = 0$, and in this case the Cauchy integral theorem reduces to $\int_{\partial D} f(w) \underline{dw} = 0$.

Example 7.3.13 (Higher-dimensional difference quotients). The one-dimensional derivative is defined as the limit of difference quotients

$$f'(x) = \lim_{h \to 0} \frac{f(x+h) - f(x)}{h}.$$

Stokes's theorem can be used to give a similar characterization of the nabla derivative $\nabla \bullet F(x)$ as in Definition 7.1.2. Indeed, applying the Stokes formula (7.4) to the linear 1-form

$$\theta(x, v) := v \bullet F(x)$$

shows that

$$\int_D \nabla \bullet F(x) dx = \int_{\partial D} \nu(y) \bullet F(y) dy.$$

Applying this to balls $B(x, \epsilon)$ centered at x shows that for C^1-regular fields F we have

$$\nabla \bullet F(x) = \lim_{\epsilon \to 0} \frac{1}{|B(x, \epsilon)|} \int_{|y-x|=\epsilon} \nu(y) \bullet F(y)dy,$$

which can be seen as a limit of higher-dimensional difference quotients. Note that the difference at the zero-dimensional boundary of the interval $[x, x+h]$ has been replaced by a boundary integral.

Example 7.3.14 (Graph of an analytic function). Consider the 2-surface M from Example 2.4.5 in Euclidean 4-dimensional space $\mathbf{R}^4$. Let us verify Stokes's theorem for the vector field

$$F(x_1, x_2, x_3, x_4) = x_2 e_1 + 2x_4 e_3 - 3x_3 e_2 + 4x_1 e_4.$$

Its exterior derivative is $\nabla \wedge F = -e_{12} + 4e_{14} + 3e_{23} - 2e_{34}$. Parametrizing by $(y_1, y_2) = (x_1, x_2)$, using the oriented area element calculated in Example 2.4.5 gives

$$\int_M \langle \nabla \wedge F, \underline{dx} \rangle$$

$$= \iint \langle -e_{12} + 4e_{14} + 3e_{23} - 2e_{34},$$

$$e_{12} - e^{y_1} \sin y_2 e_{13} + e^{y_1} \cos y_2 e_{14} - e^{y_1} \cos y_2 e_{23} - e^{y_1} \sin y_2 e_{24} + e^{2y_1} e_{34} \rangle dy_1 dy_2$$

$$= \int_0^{2\pi} \int_0^1 (-1 + e^{y_1} \cos y_2 - 2e^{2y_1})dy_1 dy_2 = -2\pi e^2.$$

The boundary consists of four smooth curves N^1, N^2, N^3, N^4. Let N^1 be parametrized by $x_1 \in [0, 1]$, where $x_2 = 0$. Since its outside orientation is $-e_2 \lrcorner e_{12} = e_1$ and $\rho(e_1) = e_1 + e^{x_1}e_3$, we get

$$\int_{N^1} \langle F, \underline{dy} \rangle = \int \langle x_2 e_1 + 2x_4 e_3 - 3x_3 e_2 + 4x_1 e_4, e_1 + e^{x_1}e_3 \rangle dx_1 = \int_0^1 0 dx_1 = 0.$$

Let N^2 be parametrized by $x_2 \in [0, 2\pi]$, where $x_1 = 1$. Since its outside orientation is $e_1 \lrcorner e_{12} = e_2$ and $\rho(e_2) = e_2 + e(-\sin x_2 e_3 + \cos x_2 e_4)$, we get

$$\int_{N^2} \langle F, \underline{dy} \rangle = \int \langle x_2 e_1 + 2x_4 e_3 - 3x_3 e_2 + 4x_1 e_4, e_2 + e(-\sin x_2 e_3 + \cos x_2 e_4) \rangle dx_2$$

$$= \int_0^{2\pi} (e \cos x_2 - 2e^2 \sin^2 x_2)dx_2 = -2\pi e^2.$$

Let N^3 be parametrized by $x_1 \in [0, 1]$, where $x_2 = 2\pi$. Since its outside orientation is $e_2 \lrcorner e_{12} = -e_1$ and $\rho(-e_1) = -e_1 - e^{x_1}e_3$, we get

$$\int_{N^3} \langle F, \underline{dy} \rangle = \int \langle x_2 e_1 + 2x_4 e_3 - 3x_3 e_2 + 4x_1 e_4, -e_1 - e^{x_1}e_3 \rangle dx_1 = \int_0^1 (-2\pi)dx_1$$

$$= -2\pi.$$

Finally, let N^4 be parametrized by $x_2 \in [0, 2\pi]$, where $x_1 = 0$. Since its outside orientation is $-e_1 \lrcorner e_{12} = -e_2$ and $\underline{\rho}(-e_2) = -e_2 - (-\sin x_2 e_3 + \cos x_2 e_4)$, we get

$$\int_{N^4} \langle F, \underline{dy} \rangle = \int \langle x_2 e_1 + 2x_4 e_3 - 3x_3 e_2 + 4x_1 e_4, -e_2 - (-\sin x_2 e_3 + \cos x_2 e_4) \rangle dx_2$$

$$= \int_0^{2\pi} (3\cos x_2 + 2\sin^2 x_2) dx_2 = 2\pi.$$

Thus

$$\int_{\partial M} \langle F, \underline{dy} \rangle = 0 - 2\pi e^2 - 2\pi + 2\pi = -2\pi e^2 = \int_M \langle \nabla \wedge \Theta, \underline{dx} \rangle,$$

verifying Stokes's theorem in this case.

Proposition 7.3.15 (Oriented measure and boundary). *If M is a C^1-regular compact oriented k-surface in X, and $p \in X$, then*

$$\int_M \underline{dx} = \frac{1}{k} \int_{\partial M} (y - p) \wedge \underline{dy}.$$

In particular, if two k-surfaces have the same boundary, with the same orientation, then they have the same oriented measure. The oriented measure of any closed k-surface is zero.

Proof. Consider the linear $(k-1)$-form $\Theta(x, w) = (x - p) \wedge w$, $x \in X, w \in \wedge^{k-1}V$, where $p \in X$ is any fixed point. Its exterior derivative is

$$\Theta(\dot{x}, \nabla \lrcorner w) = \sum_{i=1}^n (\partial_i x) \wedge (e_i^* \lrcorner w) = \sum_{i=1}^n e_i \wedge (e_i^* \lrcorner w) = kw \in \wedge^k V.$$

The stated formula follows from the Stokes formula (7.3) applied to Θ. $\square$

Example 7.3.16 (The area Pythagorean theorem). Let $\{e_1, e_2, e_3\}$ be an ON-basis for three-dimensional Euclidean space. Consider the simplex

$$M := \{x_1 e_1 + x_2 e_2 + x_3 e_3 \; ; \; x_i \geq 0, a_1 x_1 + a_2 x_2 + a_3 x_3 \leq b\},$$

where a_1, a_2, a_3, and b are positive constants. We have $\partial M = S \cup S_1 \cup S_2 \cup S_3$, where S_i are the boundary surfaces in the coordinate planes and S is the boundary surface in the positive octant. By Proposition 7.3.15 we have

$$w + w_1 + w_2 + w_3 = 0,$$

where $w := \wedge^2(S)$ and $w_i := \wedge^2(S_i)$ are the oriented measures of the parts of the boundary. It is clear that w_i are orthogonal bivectors, so we have

$$|w|^2 = |w_1|^2 + |w_2|^2 + |w_3|^2.$$

This Pythagorean theorem for three-dimensional simplices states that the square of the area of the "hypotenuse triangle" S is the sum of the squares of the areas of the three "leg triangles" S_i.

7.4 Vector Fields and Cartan's Formula

We have seen that algebraically, vectors are the building blocks for multivectors. For multivector calculus, vector fields and the flows that they generate are fundamental. Recall that pointwise, a vector $v \in V$ in X is by definition a translation $x \mapsto x + v$ of X. We write

$$\mathcal{F}_v^t(x) := x + tv, \quad x \in X, \ t \in \mathbf{R}, \ v \in V.$$

Conversely, v is uniquely determined by $\mathcal{F}_v$, since $v = \partial_t \mathcal{F}_v^t(x)$. This generalizes to nonconstant vector fields: if v is a C^1 vector field in a neighborhood of a point x_0, then we can solve the ordinary differential equation

$$y'(t) = v(y(t)), \quad y(0) = x.$$

Picard's existence theorem shows that there exists a unique solution $y : \mathbf{R} \to X$ defined in a neighborhood of $t = 0$. Write $\mathcal{F}_v^t(x) := y(t)$. This defines locally a family $\{x \mapsto \mathcal{F}_v^t(x)\}$ of C^k maps if v is C^k-regular.

Definition 7.4.1 (Flow). The family $\mathcal{F}_v^t$ is called the *flow* of the vector field v. It is the unique family of functions for which

$$\partial_t \mathcal{F}_v^t(x) = v(\mathcal{F}_v^t(x)), \quad \mathcal{F}_v^0(x) = x,$$

locally around $(t, x) = (0, x_0)$. Note that $\mathcal{F}_v^t \circ \mathcal{F}_v^s = \mathcal{F}_v^{t+s}$ and $\mathcal{F}_v^t = \mathcal{F}_{tv}^1$.

In this section we develop the theory for the exterior and interior derivatives and pullbacks and pushforwards further, based on the anticommutation formula from Theorem 2.8.1. This yields the two algebraic identities

$$\langle \theta, v \rangle w = \theta \,\lrcorner\, (v \wedge w) + v \wedge (\theta \,\lrcorner\, w), \quad \theta \in V^*, v \in V, w \in \wedge V,$$
$$\langle \theta, v \rangle \Theta = \theta \wedge (v \,\lrcorner\, \Theta) + v \,\lrcorner\, (\theta \wedge \Theta), \quad \theta \in V^*, v \in V, \Theta \in \wedge V^*,$$

where we have switched the roles of V and V^* to obtain the second identity. If v is a constant vector field in an open set D and $w = F : D \to \wedge V$ and $\Theta : D \to \wedge V^*$ are C^1 regular, then plugging in $\theta = \nabla$ gives

$$\partial_v F = \nabla \,\lrcorner\, (v \wedge F) + v \wedge (\nabla \,\lrcorner\, F),$$
$$\partial_v \Theta = \nabla \wedge (v \,\lrcorner\, \Theta) + v \,\lrcorner\, (\nabla \wedge \Theta),$$

since $\langle \nabla, v \rangle = \sum_i \langle e_i^*, v \rangle \partial_i$. These identities are not true for general vector fields v, since the partial derivatives do not satisfy associativity as scalars do, but rather follow the product rule for derivatives. Compare Example 7.1.9 and the discussion preceding it. However, if the directional derivatives ∂_v are replaced by Lie derivatives, which we now define, the identities continue to hold.

Definition 7.4.2 (Lie derivatives). Let v be a C^1 vector field in an open set D, and write $\mathcal{F}^t = \mathcal{F}_v^t$ for its flow. For multicovector fields Θ and multivector fields F in D, define the *Lie derivatives*

$$\mathcal{L}_v^*\Theta := \partial_t(\mathcal{F}^t)^*\Theta|_{t=0}, \quad \mathcal{L}_v F := \partial_t(\mathcal{F}^{-t})_* F|_{t=0}, \quad \tilde{\mathcal{L}}_v F := \partial_t(\tilde{\mathcal{F}}^{-t})_* F|_{t=0}.$$

Note that a Lie derivative of a k-(co)vector field is again a k-(co)vector field. When v is a constant vector field, then the derivative $\underline{\mathcal{F}}_x^t$ of the translation $\mathcal{F}^t = \mathcal{F}_v^t$ at any point x is the identity map. Thus the pushforwards $(\mathcal{F}^{-t})_* F(x) = F(x+tv)$ and pullbacks $(\mathcal{F}^t)^*\Theta(x) = \Theta(x + tv)$ simply translate the fields, so that $\mathcal{L}_v F = \partial_v F$ and $\mathcal{L}_v^*\Theta = \partial_v\Theta$. For general flows, the pullbacks and pushforwards also change the orientation of the fields, giving rise to zero-order terms in the Lie derivatives.

Lemma 7.4.3. *Let v be a C^1 vector field in an open set D.*

(i) *If $\Theta : D \to \wedge V^*$ is a C^1 multicovector field, then*

$$\mathcal{L}_v^*\Theta = \partial_v\Theta + \sum_i e_i^* \wedge ((\partial_i v) \lrcorner \Theta).$$

(ii) *If $F : D \to \wedge V$ is a C^1 multivector field, then*

$$\mathcal{L}_v F = \partial_v F - \sum_i \partial_i v \wedge (e_i^* \lrcorner F).$$

In particular, for vector fields $F = v'$ the Lie derivative equals $\mathcal{L}_v v' = \partial_v v' - \partial_{v'} v$, which by definition is the *Lie bracket* $[v, v']$.

Proof. Let $\mathcal{F}^t = \mathcal{F}_v^t$ be the flow of v. By definition of the flow, we have $\partial_t \mathcal{F}^t(x) = v(x)$ at $t = 0$. To prove (i), note that by the chain and product rules, at $t = 0$, we have

$$\partial_t(\underline{\mathcal{F}}_x^t)^*\left(\Theta(\mathcal{F}^t(x))\right) = \partial_{v(x)}\Theta(x) + (\partial_t\underline{\mathcal{F}}_x^t)^*(\Theta(x)).$$

To find the derivative $\partial_t \underline{\mathcal{F}}_x^t$ at $t = 0$, we fix a vector $v_0 \in V$ and a covector $\theta_0 \in V^*$ and calculate at $t = 0$ that

$$\partial_t\langle\theta_0, \underline{\mathcal{F}}_x^t v_0\rangle = \sum_i \partial_t\langle\theta_0, \partial_i\mathcal{F}^t(x)\langle e_i^*, v_0\rangle\rangle = \sum_i\langle\theta_0, \partial_i v(x)\rangle\langle e_i^*, v_0\rangle. \tag{7.6}$$

If $\Theta = \theta_j$ is a covector field, then (7.6) shows that the zero order term is

$$\partial_t(\underline{\mathcal{F}}_x^t)^*\theta_j = \sum_i e_i^*\langle\theta_j, \partial_i v(x)\rangle.$$

In general, by linearity, we may assume that $\Theta = \theta_1 \wedge \cdots \wedge \theta_k$, and in this case the product rule for derivatives and the expansion rule from Corollary 2.8.3 yield

$$\partial_t(\underline{F}_x^t)^*\Theta = \sum_j (\underline{F}_x^t)^*\theta_1 \wedge \cdots \wedge \partial_t((\underline{F}_x^t)^*\theta_j) \wedge \cdots \wedge (\underline{F}_x^t)^*\theta_k$$

$$= \sum_{i,j} e_i^* \wedge (-1)^{j-1}\langle \theta_j, \partial_i v\rangle\, \theta_1 \wedge \cdots \breve{\theta}_j \cdots \wedge \theta_k = \sum_i e_i^* \wedge (\partial_i v \lrcorner \Theta).$$

To prove (ii), we use the chain and product rules to calculate

$$\partial_t(\underline{F}_{F^t(x)}^{-t})\big(F(\mathcal{F}^t(x))\big) = \partial_{v(x)}F(x) + (\partial_t\underline{F}_x^{-t})(F(x)) + \partial_t\underline{F}_{F^t(x)}^0(F(x)).$$

However, the last extra term is zero, since $\underline{F}_y^0 = I$ for all y. For the second term, if $F = v_j$ is a vector field, then (7.6) gives

$$\partial_t(\underline{F}_x^{-t})v_j = -\sum_i \partial_i v(x)\langle e_i^*, v_j\rangle.$$

As in the proof of (i), this generalizes to general multivector fields F using the expansion rule. $\qquad\square$

The corresponding result for the normalized Lie derivative of vector fields, is the following.

Lemma 7.4.4 (Divergence and volume expansion). *Let v be a C^1 vector field in an open set D, and write $\mathcal{F}^t = \mathcal{F}_v^t$ for its flow. Then the n-volume expansion that the flow gives equals the divergence of the vector field, i.e.,*

$$\partial_t J_{\mathcal{F}^t}(x)|_{t=0} = \mathrm{div}(v)(x).$$

In particular, if $F : D \to \wedge V$ is a C^1 multivector field, then

$$\tilde{\mathcal{L}}_v F = \mathcal{L}_v F + \mathrm{div}(v)\, F = \partial_v F + \sum_i e_i^* \lrcorner (\partial_i v \wedge F).$$

Proof. As in the proof of Lemma 7.4.3, we calculate that $\partial_t \underline{F}_x^t e_i|_{t=0} = \partial_i v(x)$. This gives

$$(\partial_t J_{\mathcal{F}^t}(x)) \cdot (e_1 \wedge \cdots \wedge e_n) = \sum_i (\underline{F}_x^t e_1) \wedge \cdots \wedge (\partial_t \underline{F}_x^t e_i) \wedge \cdots \wedge (\underline{F}_x^t e_n)$$

$$= \sum_i e_1 \wedge \cdots \wedge \partial_i v \wedge \cdots \wedge e_n,$$

at $t = 0$, and thus

$$\partial_t J_{\mathcal{F}_x^t} = \sum_i \langle e_1^* \wedge \cdots \wedge e_n^*, e_1 \wedge \cdots \wedge \partial_i v \wedge \cdots \wedge e_n\rangle = \sum_i \langle e_i^*, \partial_i v\rangle = \mathrm{div}(v)$$

through iterated use of the expansion rule. From the definition of $(\tilde{\mathcal{F}}^{-t})_*$ it is now clear that $\partial_t(\tilde{\mathcal{F}}^{-t})_*F|_{t=0} = \partial_t(\mathcal{F}^{-t})_*F|_{t=0} + \operatorname{div}(v)\,F$. The second identity follows from the anticommutation rule

$$\operatorname{div}(v)F = \sum_i \langle e_i^*, \partial_i v\rangle F = \sum_i \partial_i v \wedge (e_i^* \lrcorner F) + \sum_i e_i^* \lrcorner (\partial_i v \wedge F). \qquad \square$$

Proposition 7.4.5 (Duality for Lie derivatives). *Let $v : X \to V$ be a C^1 vector field and let $\Theta : X \to \wedge V^*$ and $F : X \to \wedge V$ be C^1 regular and compactly supported fields. Then*

$$\int_X \langle \mathcal{L}_v^* \Theta, F\rangle\, dx = - \int_X \langle \Theta, \tilde{\mathcal{L}}_v F\rangle\, dx.$$

Proof. Integration by parts yields

$$\int_X \langle \mathcal{L}_v^* \Theta, F\rangle\, dx = \sum_i \int_X \langle v_i \partial_i \Theta + e_i^* \wedge (\partial_i v \lrcorner \Theta), F\rangle\, dx$$

$$= \sum_i \int_X \langle \Theta, -\partial_i(v_i F) + \partial_i v \wedge (e_i^* \lrcorner F)\rangle\, dx$$

$$= - \int_X \langle \Theta, \operatorname{div}(v)F\rangle dx - \sum_i \int_X \langle \Theta, v_i \partial_i F - \partial_i v \wedge (e_i^* \lrcorner F)\rangle\, dx$$

$$= - \int_X \langle \Theta, \tilde{\mathcal{L}}_v F\rangle\, dx. \qquad \square$$

Exercise 7.4.6. Prove the duality directly from Proposition 7.2.5(ii), using that $\mathcal{L}_v^* \Theta = \partial_t(\mathcal{F}^t)^* \Theta|_{t=0}$ and $\tilde{\mathcal{L}}_v F = \partial_t(\tilde{\mathcal{F}}^{-t})_*F|_{t=0}$.

We can finally prove the following *Cartan formulas* for the Lie derivatives.

Theorem 7.4.7 (Cartan's formula). *Let v be a C^1 vector field in an open set D, and consider C^1 multicovector fields Θ and multivector fields F in D. Then we have*

$$\tilde{\mathcal{L}}_v F = \nabla \lrcorner (v \wedge F) + v \wedge (\nabla \lrcorner F),$$
$$\mathcal{L}_v^* \Theta = \nabla \wedge (v \lrcorner \Theta) + v \lrcorner (\nabla \wedge \Theta).$$

Proof. The derivative of F is

$$\tilde{\mathcal{L}}_v F = \sum_i \langle e_i^*, v\rangle \partial_i F + e_i^* \lrcorner (\partial_i v \wedge F)$$

$$= \sum_i v \wedge (e_i^* \lrcorner \partial_i F) + e_i^* \lrcorner (v \wedge \partial_i F) + e_i^* \lrcorner (\partial_i v \wedge F)$$

$$= \sum_i v \wedge (e_i^* \lrcorner \partial_i F) + e_i^* \lrcorner \partial_i(v \wedge F) = v \wedge (\nabla \lrcorner F) + \nabla \lrcorner (v \wedge F).$$

The derivative of Θ is

$$
\begin{aligned}
\mathcal{L}_v^* \Theta &= \sum_i \langle e_i^*, v \rangle \partial_i \Theta + e_i^* \wedge (\partial_i v \lrcorner \Theta) \\
&= \sum_i v \lrcorner (e_i^* \wedge \partial_i \Theta) + e_i^* \wedge (v \lrcorner \partial_i \Theta) + e_i^* \wedge (\partial_i v \lrcorner \Theta) \\
&= \sum_i v \lrcorner (e_i^* \wedge \partial_i \Theta) + e_i^* \wedge \partial_i (v \lrcorner \Theta) = v \lrcorner (\nabla \wedge \Theta) + \nabla \wedge (v \lrcorner \Theta). \qquad \square
\end{aligned}
$$

Exercise 7.4.8. Let F be a vector field and G a multivector field in an inner product space. Use Cartan's formula to prove the product rule

$$
\delta(F \wedge G) = (\operatorname{div} F)G - F \wedge (\delta G) + \mathcal{L}_F G.
$$

Deduce as a corollary the last formula in Example 7.1.9.

The Cartan formulas have good invariance properties, which can be used to extend the theory from affine spaces to general manifolds. If $\rho : D_1 \to D_2$ is a C^1 diffeomorphism and v is a vector field in D_2, then the pushed forward vector field $v_1 := \rho_*^{-1} v$ has flow $\mathcal{F}_{\rho_*^{-1} v}^t$ that satisfies $\rho \circ \mathcal{F}_{v_1}^t = \mathcal{F}_v^t \circ \rho$, so that $\rho^* (\mathcal{F}_v^t)^* = (\mathcal{F}_{v_1}^t)^* \rho^*$. Thus, if Θ is a multicovector field in D_2, then pulling back the formula for $\mathcal{L}_v^* \Theta$ to D_1 gives

$$
\mathcal{L}_{v_1}^* \rho^* \Theta = \nabla \wedge (v_1 \lrcorner \rho^* \Theta) + v_1 \lrcorner (\nabla \wedge \rho^* \Theta),
$$

according to Theorem 7.2.9 and Proposition 7.2.7.

Similarly, if v is a vector field in D_1, then the pushed forward vector field $v_2 := \rho_* v$ has flow $\mathcal{F}_{v_2}^{-t}$ that satisfies $\rho \circ \mathcal{F}_v^{-t} = \mathcal{F}_{v_2}^{-t} \circ \rho$, so that $\tilde{\rho}_* (\tilde{\mathcal{F}}_v^{-t})_* = (\tilde{\mathcal{F}}_{v_2}^{-t})_* \tilde{\rho}_*$. Thus, if F is a multivector field in D_1, then pushing forward the formula for $\tilde{\mathcal{L}}_v F$ to D_2 gives

$$
\tilde{\mathcal{L}}_{v_2} \tilde{\rho}_* F = \nabla \lrcorner (v_2 \wedge \tilde{\rho}_* F) + v_2 \wedge (\nabla \lrcorner \tilde{\rho}_* F).
$$

Exercise 7.4.9 (Expansion rule for d). Let $v_0, v_1, \ldots, v_k$ be C^1 vector fields and let Θ be a C^1 k-covector field. Show that

$$
\begin{aligned}
\langle d\Theta, v_0 \wedge \cdots \wedge v_k \rangle = {} & \sum_{i=0}^k (-1)^i \partial_{v_i} \langle \Theta, v_0 \wedge \cdots \check{v}_i \cdots \wedge v_k \rangle \\
& + \sum_{0 \le i < j \le k} (-1)^{i+j} \langle \Theta, [v_i, v_j] \wedge v_0 \wedge \cdots \check{v}_i \cdots \check{v}_j \cdots \wedge v_k \rangle.
\end{aligned}
$$

7.5 Poincaré's Theorem

Given a multicovector field Θ in an open set D in an affine space X, we ask whether there exists a second multicovector field U in D such that

$$
\Theta = \nabla \wedge U.
$$

In an oriented three-dimensional Euclidean space, for a vector field Θ this amounts to finding a scalar potential U for the vector field with gradient equal to Θ, and for a bivector field Θ this amounts to finding a vector potential with curl equal to Θ. In general, we note from Proposition 7.1.6 that a necessary condition for a potential U to exist is that

$$\nabla \wedge \Theta = 0.$$

This nilpotence of the exterior derivative also shows that in the case that a potential U does exist, it is in general highly nonunique, since for every multicovector field V,

$$\nabla \wedge (U + \nabla \wedge V) = \Theta + 0 = \Theta.$$

We also consider the Hodge dual question of when, for a given multivector field F in D, there exists a second multivector field U such that

$$F = \nabla \lrcorner U.$$

We note that a necessary condition is that $\nabla \lrcorner F = 0$, and that in general, such U are also highly nonunique.

Definition 7.5.1 (Closed vs. exact). Let D be an open subset of an affine space (X, V). Then $\Theta \in C^1(D; \wedge V^*)$ is said to be *closed* if $dF = 0$ in D, and it is said to be *exact* if there exists a *potential* $U \in C^1(D; \wedge V^*)$ such that $F = dU$ in D.

Similarly, $F \in C^1(D; \wedge V)$ is said to be *coclosed* if $\delta F = 0$ in D, and is said to be *coexact* if there exists a *potential* $U \in C^1(D; \wedge V)$ such that $F = \delta U$ in D.

In general, the necessary conditions are almost sufficient. We shall see in Chapter 10 that if D is a bounded domain, under suitable regularity assumptions on the fields and the domain, if Θ is closed and satisfies a finite number of linear constraints depending on the topology of D, then Θ is exact. In this section, we prove the following classical result in this direction. For topologically trivial domains, the star-shaped ones, this shows that all closed fields, except the scalar constants, are exact.

Theorem 7.5.2 (Poincaré's theorem). *Let D be an open set in an affine space (X, V) that is star-shaped with respect to a point $p \in D \subset X$, which we identify with $0 \in V$. If $\Theta \in C^1(D; \wedge^k V^*)$ is closed, and $1 \leq k \leq n = \dim X$, then $\Theta = \nabla \wedge U$ if*

$$U(x) := x \lrcorner \int_0^1 \Theta(tx)\, t^{k-1} dt, \quad x \in D.$$

Similarly, if $F \in C^1(D; \wedge^k V)$ is coclosed, and $0 \leq k \leq n-1$, then $F = \nabla \lrcorner U$ if

$$U(x) := x \wedge \int_0^1 F(tx)\, t^{n-k-1} dt, \quad x \in D.$$

Note that closed scalar functions $\Theta \in C^1(D; \wedge^0 V^*)$ are constant. However, only $\Theta = 0$ among these is exact, since $\wedge^{-1} V^* = \{0\}$. Similarly, all coclosed $F \in C^1(D; \wedge^n V)$ are constant. However, only $F = 0$ among these is exact, since $\wedge^{n+1} V = \{0\}$.

Proof. If $\nabla \wedge \Theta = 0$, then Cartan's formula (Theorem 7.4.7) reduces to

$$\mathcal{L}_v^* \Theta(x) = \nabla \wedge (v(x) \lrcorner \Theta(x)).$$

If v is a vector field in D such that $\mathcal{F}_v^t(D) \subset D$ for all $t \geq 0$, then

$$\partial_t (\mathcal{F}_v^t)^* \Theta|_{t=s} = \partial_t (\mathcal{F}_v^{t-s})^* ((\mathcal{F}_v^s)^* \Theta)|_{t=s} = \mathcal{L}_v^* (\mathcal{F}_v^s)^* \Theta = \nabla \wedge (v \lrcorner (\mathcal{F}_v^s)^* \Theta)$$

for all $s \geq 0$, so

$$(\mathcal{F}_v^t)^* \Theta(x) - \Theta(x) = \nabla \wedge \left(v(x) \lrcorner \int_0^t (\mathcal{F}_v^s)^* \Theta(x) ds \right).$$

We now would like to choose v such that $\lim_{t \to \infty} (\mathcal{F}_v^t)^* \Theta(x) = 0$. From the definition of the pullback, we see that this happens if all the orbits $\{\mathcal{F}_v^t(x)\}_{t \geq 0}$ converge to 0. Choose, for example, the vector field $v(x) = -x$. We calculate the corresponding flow

$$\mathcal{F}_v^t(x) = e^{-t}(x),$$

which has the constant derivative $\underline{\mathcal{F}}_v^t = e^{-t} I : V \to V$. The induced derivative acting on k-vectors is $\underline{\mathcal{F}}_v^t = e^{-kt} I : \wedge^k V \to \wedge^k V$, and therefore

$$(\mathcal{F}_v^t)^* \Theta(x) = \underline{\mathcal{F}}_v^t \Theta(\mathcal{F}_v^t(x)) = e^{-kt} \Theta(e^{-t} x).$$

This yields the identity

$$e^{-kt} \Theta(e^{-t} x) - \Theta(x) = \nabla \wedge \left(-x \lrcorner \int_0^t e^{-ks} \Theta(e^{-s} x) ds \right)$$

$$= -\nabla \wedge \left(x \lrcorner \int_{e^{-t}}^1 \Theta(sx) s^{k-1} ds \right),$$

after a change of variables in the integral. Letting $t \to \infty$ proves Poincaré's formula for the potential to Θ, since the integral is convergent if $k \geq 1$.

Next consider a coclosed k-vector field F. Then Proposition 7.1.7(i) shows that $*F$ is a closed $(n-k)$-vector field, and thus

$$*F(x) = \nabla \wedge \left(x \lrcorner \int_0^1 *F(tx) \, t^{n-k-1} dt \right), \quad x \in D.$$

Taking the Hodge star of this equation, using Proposition 7.1.7(i) and Proposition 2.6.8(ii) proves the second Poincaré formula. $\qquad \square$

Example 7.5.3 (Gradient potentials). If $F : D \to V$ is a curl-free vector field, that is, a closed 1-vector field, on a star-shaped set D in a Euclidean space, then Poincaré's formula reads

$$F(x) = \nabla \int_0^1 \langle F(x), x \rangle dt.$$

That is, the scalar potential is obtained as a path integral of Θ along the straight line from p to x. This is the usual path integral formula for a scalar potential to a curl-free vector field. Stokes's theorem shows that the straight line can be replaced by any curve from p to x.

Example 7.5.4 (Divergence potentials). Every scalar function $u : D \to \mathbf{R} = \wedge^0 V$ is coclosed. Thus if D is star-shaped with respect to $p = 0$, then Poincaré's formula for the interior derivative shows that u is the divergence of some vector field F. One such field is the radial vector field F given by the formula

$$F(x) = \left(\int_0^1 u(tx) \, t^{n-1} dt \right) x.$$

Example 7.5.5 (Curl potentials). Consider a star-shaped set D in an oriented three-dimensional Euclidean space X. A bivector field F in D is closed if and only if $*F$ is a divergence-free vector field. For such F, Poincaré's formula

$$U(x) = x \lrcorner \int_0^1 F(tx) t \, dt$$

yields an angular vector field U such that $\operatorname{curl} U = F$.

Exercise 7.5.6 (Homotopy relations). Generalize Theorem 7.5.2 to general, not necessarily (co)closed, fields on a domain D star-shaped with respect to 0. Prove the following. If $\Theta \in C^1(D; \wedge^k V^*)$ and $1 \le k \le n$, then

$$\Theta(x) = \nabla \wedge \left(x \lrcorner \int_0^1 \Theta(tx) \, t^{k-1} dt \right) + x \lrcorner \int_0^1 (\nabla \wedge \Theta)(tx) \, t^k dt, \quad x \in D,$$

and if $F \in C^1(D; \wedge^k V)$ and $0 \le k \le n - 1$, then

$$F(x) = \nabla \lrcorner \left(x \wedge \int_0^1 F(tx) \, t^{n-k-1} dt \right) + x \wedge \int_0^1 (\nabla \lrcorner F)(tx) \, t^{n-k} dt, \quad x \in D.$$

7.6 Hodge Decompositions

In this section, we assume that (X, V) is a Euclidean space. This enables us to compare the exterior and interior derivatives, since they now both act on multivector fields. We address the problem of finding potentials for these derivatives as in Section 7.5, but in the Hilbert space $L_2(D) = L_2(D; \wedge V)$ of square integrable multivector fields

$$\|F\| = \left(\int_D |F(x)|^2 dx \right)^{1/2} < \infty,$$

on a bounded Lipschitz domain $D \subset X$ as in Definition 6.1.4. Allowing for a nonempty and nonregular boundary is important for applications to many classical boundary value problems. Some of these are discussed in this section, and the proofs of the main results on Hodge decompositions stated here are deferred to Chapter 10. To avoid technicalities about vector bundles, we stay in Euclidean affine space in this section. The extension of the theory of Hodge decompositions to more general manifolds, with or without boundaries, is, however, rather straightforward given the commutation theorem (Theorem 7.2.9). We treat the case of closed manifolds in Chapter 11.

Our point of departure is the following application of Stokes's theorem (Theorem 7.3.9):

$$\int_D \left(\langle \nabla \wedge F(x), G(x) \rangle + \langle F(x), \nabla \lrcorner G(x) \rangle \right) dx$$

$$= \int_{\partial D} \langle \nu(y) \wedge F(y), G(y) \rangle dy = \int_{\partial D} \langle F(y), \nu(y) \lrcorner G(y) \rangle dy, \qquad (7.7)$$

which holds for smooth enough multivector fields $F, G : \overline{D} \to \wedge V$ and domains D. Indeed, apply (7.4) to the scalar linear 1-form $\theta(x, v) = \langle v \wedge F(x), G(x) \rangle = \langle F(x), v \lrcorner G(x) \rangle$. This motivates the following weak definitions.

Definition 7.6.1 (Weak boundary conditions). Let D be a bounded Lipschitz domain in Euclidean space. If $F, F' \in L_2(D)$, then we say that $\nabla \wedge F = F'$ if $\int_D \langle F(x), \nabla \lrcorner G(x) \rangle dx = - \int_D \langle F'(x), G(x) \rangle dx$ for all test functions $G \in C_0^\infty(D)$. If $G, G' \in L_2(D)$, then we say that $\nabla \lrcorner G = G'$ if $\int_D \langle \nabla \wedge F(x), G(x) \rangle dx = - \int_D \langle F(x), G'(x) \rangle dx$ for all test functions $F \in C_0^\infty(D)$.

If $F, \nabla \wedge F \in L_2(D)$, then we say that F is *normal*, or equivalently $\nu \wedge F = 0$, on ∂D if

$$\int_D \left(\langle \nabla \wedge F(x), G(x) \rangle + \langle F(x), \nabla \lrcorner G(x) \rangle \right) dx = 0 \qquad (7.8)$$

for all test functions $G \in C^\infty(\overline{D})$. If $G, \nabla \lrcorner G \in L_2(D)$, then we say that G is *tangential*, or equivalently $\nu \lrcorner G = 0$, on ∂D if (7.8) holds for all test functions $F \in C^\infty(\overline{D})$.

For the exterior derivative it is important to note that $\nabla \wedge F \in L_2(D)$ does not entail that all partial derivatives of F are in $L_2(D)$. Indeed, the exterior product by a vector is not injective, and in Definition 7.1.2 we throw away half, in a certain sense, of the derivatives in $\nabla \otimes F$ when forming $\nabla \wedge F$. The same remark applies to the interior derivative.

Exercise 7.6.2 (Interior nonregularity). Let P be a hyperplane dividing D into two parts D_1 and D_2. For $F_1, F_2 \in C_0^\infty(D)$, let

$$F(x) := \begin{cases} F_1(x), & x \in D_1, \\ F_2(x), & x \in D_2. \end{cases}$$

Show that $\nabla \wedge F \in L_2(D)$ if and only if the tangential part of F is continuous across P, and that $\nabla \lrcorner F \in L_2(D)$ if and only if the normal part of F is continuous across P.

Numerical examples of such half-regularity for divergence-free fields and curl-free fields are shown in Figures 7.3, 7.4, 10.2 and 10.3.

Definition 7.6.3 (Hodge subspaces). On a bounded Euclidean Lipschitz domain D, we define the following linear subspaces of $L_2(D)$:

$\mathsf{R}(d) := \{F \in L_2(D) \ ; \ F = \nabla \wedge U \text{ for some } U \in L_2(D)\},$

$\mathsf{N}(d) := \{F \in L_2(D) \ ; \ \nabla \wedge F = 0\},$

$\mathsf{R}(\underline{\delta}) := \{F \in L_2(D) \ ; \ F = \nabla \lrcorner U \text{ for some } U \in L_2(D) \text{ with } \nu \lrcorner U = 0 \text{ on } \partial D\},$

$\mathsf{N}(\underline{\delta}) := \{F \in L_2(D) \ ; \ \nabla \lrcorner F = 0 \text{ and } \nu \lrcorner F = 0 \text{ on } \partial D\},$

$\mathcal{C}_{\|}(D) := \mathsf{N}(d) \cap \mathsf{N}(\underline{\delta}),$

$\mathsf{R}(\underline{d}) := \{F \in L_2(D) \ ; \ F = \nabla \wedge U \text{ for some } U \in L_2(D) \text{ with } \nu \wedge U = 0 \text{ on } \partial D\},$

$\mathsf{N}(\underline{d}) := \{F \in L_2(D) \ ; \ \nabla \wedge F = 0 \text{ and } \nu \wedge F = 0 \text{ on } \partial D\},$

$\mathsf{R}(\delta) := \{F \in L_2(D) \ ; \ F = \nabla \lrcorner U \text{ for some } U \in L_2(D)\},$

$\mathsf{N}(\delta) := \{F \in L_2(D) \ ; \ \nabla \lrcorner F = 0\},$

$\mathcal{C}_{\perp}(D) := \mathsf{N}(\underline{d}) \cap \mathsf{N}(\delta).$

These are the subspaces of *exact, closed, tangential coexact, tangential coclosed, tangential cohomology, normal exact, normal closed, coexact, coclosed,* and *normal cohomology* fields respectively. When decomposing multivector fields into their homogeneous k-vector parts

$$L_2(D) = L_2(D; \wedge V) = L_2(D; \wedge^0 V) \oplus L_2(D; \wedge^1 V) \oplus \cdots \oplus L_2(D; \wedge^n V),$$

each of these ten subspaces splits into homogeneous parts, and we write, for example, $\mathsf{N}(d; \wedge^k) := \mathsf{N}(d) \cap L_2(D; \wedge^k V)$. Define the *Betti numbers*

$$b_k(D) := \dim \mathcal{C}_{\|}(D; \wedge^k), \quad k = 0, 1, 2, \ldots, n.$$

Note that normal boundary conditions are natural for the exterior derivative, and tangential boundary conditions go with the interior derivative. The other way around does not work.

Lemma 7.6.4 (Nilpotence and boundary conditions). *Let D be a bounded Lipschitz domain in Euclidean space. If $F, \nabla \wedge F \in L_2(D)$ and F is normal on ∂D, then*

$\nabla \wedge F$ is also normal on ∂D. If $G, \nabla \lrcorner G \in L_2(D)$ and G is tangential on ∂D, then $\nabla \lrcorner G$ is also tangential on ∂D. We have inclusions

$$R(d) \subset N(d), \quad R(\underline{\delta}) \subset N(\underline{\delta}), \quad R(\underline{d}) \subset N(\underline{d}), \quad and \quad R(\delta) \subset N(\delta).$$

Proof. If F is normal, we note for its exterior derivative $F' = \nabla \wedge F$ that

$$\int_D \left(\langle \nabla \wedge F', G \rangle + \langle F', \nabla \lrcorner G \rangle \right) dx = \int_D (0 + \langle \nabla \wedge F, \nabla \lrcorner G \rangle) dx$$

$$= - \int_D \langle F, \nabla \lrcorner (\nabla \lrcorner G) \rangle dx = 0$$

for all test functions G, since $\nabla \wedge (\nabla \wedge F) = 0$ and $\nabla \lrcorner (\nabla \lrcorner G) = 0$. This means that F' is normal. The invariance of tangential boundary conditions under the interior derivative is proved similarly, and the stated inclusions are straightforward to verify. Recall Proposition 7.1.6. □

Exercise 7.6.5. Clearly, if $F \in N(d)$ and $\nu \wedge F = 0$, then $F \in N(\underline{d})$. The situation for exact forms is more subtle. Give an example of a domain D and a gradient vector field F that is normal at ∂D, but which has no scalar potential U with Dirichlet boundary conditions $U = 0$ at ∂D. Conclude that in general, it is not sufficient that $F \in R(d)$ and $\nu \wedge F = 0$ for $F \in R(\underline{d})$, and it is not sufficient that $F \in R(\delta)$ and $\nu \lrcorner F = 0$ for $F \in R(\underline{\delta})$.

The following main results in this section show that under appropriate boundary conditions the exterior and interior derivatives split each multivector field into an exact and a coexact field, modulo a small cohomology space. For the five subspaces associated to tangential boundary conditions, we have the following splitting.

Theorem 7.6.6 (Tangential Hodge decomposition). *Let D be a bounded Lipschitz domain in Euclidean space. Then the spaces $R(d)$ and $R(\underline{\delta})$ of exact and tangential coexact fields are closed subspaces of $L_2(D)$, and the tangential cohomology space $C_{\parallel}(D)$ is finite-dimensional. Furthermore, we have an orthogonal splitting*

$$L_2(D) = R(d) \oplus C_{\parallel}(D) \oplus R(\underline{\delta}), \tag{7.9}$$

where $R(d) \oplus C_{\parallel}(D) = N(d)$, $C_{\parallel}(D) \oplus R(\underline{\delta}) = N(\underline{\delta})$, and $C_{\parallel}(D) = N(d) \cap N(\underline{\delta})$. At the level of k-vector fields, each $L_2(D; \wedge^k V)$ splits in this way.

Road map to the proof. The orthogonal splitting (7.9) is proved in Propositions 10.1.2 and 10.2.3. That $R(d)$ and $R(\underline{\delta})$ are closed subspaces of $L_2(D)$ and that the space $C_{\parallel}(D)$ is finite-dimensional is proved in Theorem 10.3.1.

At a formal level, the proof of the orthogonal splitting is as follows. Note that

$$\int_D \langle \nabla \wedge F(x), G(x) \rangle dx = \int_{\partial D} \langle F(y), \nu(y) \lrcorner G(y) \rangle dy - \int_D \langle F(x), \nabla \lrcorner G(x) \rangle dx$$

for all fields F and G. For fixed G, the left-hand side vanishes for all F precisely when G is orthogonal to $\mathsf{R}(d)$, whereas the right-hand side vanishes for all F precisely when $G \in \mathsf{N}(\underline{\delta})$. This indicates that we have an orthogonal splitting $L_2(D) = \mathsf{R}(d) \oplus \mathsf{N}(\underline{\delta})$.

Next note that

$$\int_D \langle F(x), \nabla \lrcorner\, G(x)\rangle dx = -\int_D \langle \nabla \wedge F(x), G(x)\rangle dx$$

for all fields G that are tangential on ∂D and all fields F. For fixed F, the left-hand side vanishes for all tangential G precisely when F is orthogonal to $\mathsf{R}(\delta_T)$, whereas the right-hand side vanishes for all tangential G precisely when $F \in \mathsf{N}(d)$. This indicates that we have an orthogonal splitting $L_2(D) = \mathsf{N}(d) \oplus \mathsf{R}(\underline{\delta})$.

Since $\mathsf{R}(d) \subset \mathsf{N}(d)$ and $\mathsf{R}(\underline{\delta}) \subset \mathsf{N}(\underline{\delta})$, taking the intersection of the splittings from (i) and (ii) yields the Hodge decomposition (7.9).   □

By Hodge duality, using Propositions 7.1.7(i) and 2.6.8(ii), we immediately obtain from Theorem 7.6.6 the following splitting using the five subspaces associated to normal boundary conditions.

Corollary 7.6.7 (Normal Hodge decomposition). *Let D be a bounded Lipschitz domain in Euclidean space. Then the spaces $\mathsf{R}(\underline{d})$ and $\mathsf{R}(\delta)$ of normal exact and coexact fields are closed subspaces of $L_2(D)$, and the normal cohomology space $\mathcal{C}_\perp(D)$ is finite-dimensional. Furthermore, we have an orthogonal splitting*

$$L_2(D) = \mathsf{R}(\underline{d}) \oplus \mathcal{C}_\perp(D) \oplus \mathsf{R}(\delta), \tag{7.10}$$

where $\mathsf{R}(\underline{d}) \oplus \mathcal{C}_\perp(D) = \mathsf{N}(\underline{d})$, $\mathcal{C}_\perp(D) \oplus \mathsf{R}(\delta) = \mathsf{N}(\delta)$, and $\mathcal{C}_\perp(D) = \mathsf{N}(\underline{d}) \cap \mathsf{N}(\delta)$. At the level of k-vector fields, each $L_2(D; \wedge^k V)$ splits in this way.

For applications, we note the following existence and uniqueness result for potentials, which the Hodge decompositions entail.

Corollary 7.6.8 (Hodge potentials). *Let D be a bounded Lipschitz domain in Euclidean space, and let $F \in L_2(D)$ be given. Then there exists $U \in L_2(D)$ such that*

$$F = \nabla \wedge U$$

if and only if $F \in \mathsf{R}(d)$, or equivalently if F is orthogonal to all tangential coclosed fields $G \in \mathsf{N}(\underline{\delta})$. The potential U is unique modulo $\mathsf{N}(d)$, and there exists a unique tangential coexact solution $U \in \mathsf{R}(\underline{\delta})$. This solution is characterized as the solution U with minimal L_2 norm. Similarly,

- *$F \in \mathsf{R}(\underline{d})$ has a unique coexact potential $U \in \mathsf{R}(\delta)$ such that $\nabla \wedge U = F$,*

- *$F \in \mathsf{R}(\delta)$ has a unique normal exact potential $U \in \mathsf{R}(\underline{d})$ such that $\nabla \lrcorner\, U = F$, and*

- *$F \in \mathsf{R}(\underline{\delta})$ has a unique exact potential $U \in \mathsf{R}(d)$ such that $\nabla \lrcorner\, U = F$,* *and we can characterize the existence of potentials as orthogonality of F to $\mathsf{N}(\delta)$, $\mathsf{N}(\underline{d})$, and $\mathsf{N}(d)$ respectively.*

Proof. The only claim that is not immediate from Theorem 7.6.6 is the uniqueness for U. Given $U \in L_2(D)$ such that $F = \nabla \wedge U$, we decompose U using (7.9) as

$$U = U_1 + U_2 + U_3.$$

We note that $\nabla \wedge U_1 = 0 = \nabla \wedge U_2$, and conclude that $\nabla \wedge U_3 = F$. This potential U_3 is the unique potential to F in $\mathsf{R}(\underline{\delta})$ since it is orthogonal to $\mathsf{N}(d)$. $\qquad\square$

Example 7.6.9 (Cohomology). By Theorem 7.6.6 we have

$$\mathsf{R}(d) \oplus \mathcal{C}_{\|}(D) = \mathsf{N}(d),$$

so the tangential cohomology subspace $\mathcal{C}_{\|}(D)$ is the gap between the closed and exact multivector fields in $L_2(D)$, consisting of the closed fields that are orthogonal to all the exact fields. As in Section 7.5, if D is star-shaped and therefore topologically trivial, then $\mathcal{C}_{\|}(D)$ consists only of the scalar and constant functions. For general domains, $\mathcal{C}_{\|}(D)$ contains information about the topology of D. A simple example of this is the annulus

$$D := \{x \in V \; ; \; 1 < |x| < 2\}$$

in the Euclidean plane V. Here we find that the vector field $F(x,y) := (-ye_1 + xe_2)/(x^2+y^2)$ is divergence- and curl-free, as well as tangential at ∂D. And indeed, we prove in Example 10.6.6 that for this annulus, $\mathcal{C}_{\|}(D)$ is two-dimensional and spanned by F along with the constant scalars. This happens because D is not topologically trivial but has one hole.

Similarly, for the corresponding three-dimensional annulus D, the bivector field $F(x) = *x/|x|^3$ belongs to $\mathcal{C}_{\|}(D; \wedge^2)$. In this case, D is simply connected and $\mathcal{C}_{\|}(D; \wedge^1) = \{0\}$.

Numerical examples of cohomology fields for more general planar domains are shown in Figures 7.3,7.4,10.2 and 10.3.

Example 7.6.10 (Helmholtz decompositions in $\mathbf{R}^3$). Let D be a bounded Lipschitz domain in three-dimensional oriented Euclidean space. The Hodge decomposition (7.9) amounts to the following splittings:

$$
\begin{aligned}
L_2(D; \wedge^3) &= \mathsf{R}(d; \wedge^3), \\
L_2(D; \wedge^2) &= \mathsf{R}(d; \wedge^2) \oplus \mathcal{C}_{\|}(D; \wedge^2) \oplus \mathsf{R}(\underline{\delta}; \wedge^2), \\
L_2(D; \wedge^1) &= \mathsf{R}(d; \wedge^1) \oplus \mathcal{C}_{\|}(D; \wedge^1) \oplus \mathsf{R}(\underline{\delta}; \wedge^1), \\
L_2(D; \wedge^0) &= \mathcal{C}_{\|}(D; \wedge^0) \oplus \mathsf{R}(\underline{\delta}; \wedge^0).
\end{aligned}
$$

Note that $\mathsf{R}(d; \wedge^0) = \{0\}$, since $\wedge^{-1}V = \{0\}$, and $\mathsf{N}(\underline{\delta}; \wedge^3) = \{0\}$, since the gradient with Dirichlet boundary conditions is injective.

We note that $b_0(D)$ is the number of connected components of D, since $\mathcal{C}_\parallel(D; \wedge^0)$ consists of the locally constant scalar functions on D. The Betti numbers $b_1(D)$ and $b_2(D)$ measures in a certain way the numbers of one- and two-dimensional topological obstructions in D. See Exercise 10.6.12.

In classical vector calculus notation, using the Hodge star to identify bivectors and vectors, as well as 3-vectors and scalars, we have $L_2(D; \wedge^0) = \mathsf{R}(\delta; \wedge^0)$, and the tangential Hodge splitting of bivector fields amounts to the normal Hodge splitting

$$L_2(D; \wedge^1) = \mathsf{R}(\underline{d}; \wedge^1) \oplus \mathcal{C}_\perp(D; \wedge^1) \oplus \mathsf{R}(\delta; \wedge^1)$$

of vector fields. From these splittings of vector fields, known as *Helmholtz decompositions* or the *fundamental theorem of vector calculus*, we obtain in particular existence and uniqueness results for the classical potentials.

(i) Given a vector field $F \in L_2(D; \wedge^1)$, we seek a scalar potential $U \in L_2(D; \wedge^0)$ such that

$$F = \nabla U.$$

This equation has a solution if and only if F is orthogonal to $\mathsf{N}(\underline{\delta}; \wedge^1)$, and in this case there exists a unique scalar potential U that is orthogonal to $\mathcal{C}_\parallel(D; \wedge^0)$. If we further assume that F is orthogonal to $\mathsf{N}(\delta; \wedge^1)$, then we have a unique potential $U \in L_2(D; \wedge^0)$ with Dirichlet boundary conditions $U = 0$ at ∂D.

(ii) Given a vector field $F \in L_2(D; \wedge^1)$, we seek a vector potential $U \in L_2(D; \wedge^1)$ such that

$$F = \nabla \times U = *(\nabla \wedge U).$$

This equation has a solution if and only if F is orthogonal to $\mathsf{N}(\underline{d}; \wedge^1)$, and in this case there exists a unique vector potential U that is orthogonal to $\mathsf{N}(d; \wedge^1)$. If we further assume that F is orthogonal to $\mathsf{N}(d; \wedge^1)$, meaning in particular that F is tangential at ∂D or equivalently that $*F$ is normal, then we have a unique potential that is normal at ∂D and orthogonal to $\mathsf{N}(\underline{d}; \wedge^1)$.

(iii) Given a scalar function $f \in L_2(D; \wedge^0)$, we seek a divergence potential $U \in L_2(D; \wedge^1)$ such that

$$f = \operatorname{div} U = \nabla \lrcorner U.$$

There always exists such a divergence potential that is orthogonal to $\mathsf{N}(\underline{d}; \wedge^1)$. If we further assume that f is orthogonal to $\mathcal{C}_\parallel(D; \wedge^0)$, that is, that f has mean value 0 on each connected component of D, then we have a unique potential U that is tangential at ∂D and orthogonal to $\mathsf{N}(d; \wedge^1)$.

Example 7.6.11 (FEM Helmholtz computation). The simplest nontrivial Hodge decomposition to compute numerically is the planar Helmholtz decomposition. Given a bounded domain $D \subset \mathbf{R}^2$ and a vector field $F \in L_2(D; \wedge^1 \mathbf{R}^2)$, we want to compute scalar functions U and V, and a cohomology field H, such that

$$F = \nabla U + H + \nabla \lrcorner (V j),$$

where $j = e_{12}$. We choose tangential boundary conditions, which amounts to imposing Dirichlet boundary conditions $V|_{\partial D} = 0$ on V, and on $H \in \mathcal{C}_\parallel(D)$. The natural algorithm for computing $U \in H^1(D)$, $V \in H_0^1(D)$, and $H \in \mathcal{C}_\parallel(D)$ is as follows. First we compute U, which is characterized by the property that $F - \nabla U \in \mathsf{N}(\underline{\delta})$, or equivalently $F - \nabla U$ is orthogonal to $\mathsf{R}(d)$. Therefore U can be computed using the finite element method (FEM) applied to

$$\int_D \langle \nabla U, \nabla \phi \rangle dx = \int_D \langle F, \nabla \phi \rangle dx, \qquad \phi \in H^1(D).$$

To compute V, we use the Hodge star maps to obtain

$$*F = \nabla V + *H - \nabla \lrcorner (Uj),$$

which leads to the variational equations

$$\int_D \langle \nabla V, \nabla \phi \rangle dx = \int_D \langle *F, \nabla \phi \rangle dx, \qquad \phi \in H_0^1(D),$$

for V. Finally the cohomology field H is obtained as

$$H = F - \nabla U - \nabla \lrcorner (Vj).$$

Figures 7.3, 7.4, 10.2, and 10.3 show numerical results for domains with different topology, and illustrate Example 7.6.9 and Exercise 7.6.2.

Example 7.6.12 (The Dirichlet problem). Consider the Dirichlet boundary value problem for the Laplace equation,

$$\begin{cases} \Delta u(x) = g(x), & x \in D, \\ u(x) = 0, & x \in \partial D, \end{cases}$$

interpreted in the weak sense that $u \in H_0^1(D)$, where $g \in L_2(D; \wedge^0)$ is given. The normal Hodge decomposition Corollary 7.6.7, and in particular the closedness of $\mathsf{R}(\underline{d})$ and $\mathsf{R}(\delta)$, implies the unique solvability of this boundary value problem. To see this, we rewrite $\Delta u = g$ as the first-order system

$$\operatorname{div} F = g, \tag{7.11}$$
$$\nabla u = F. \tag{7.12}$$

By Corollary 7.6.8, we have a unique solution $F \in \mathsf{R}(\underline{d}; \wedge^1)$ to (7.11). Again by Corollary 7.6.8, or more trivially by the definition of $\mathsf{R}(\underline{d}; \wedge^1)$, we have a unique solution u to (7.12) with Dirichlet boundary conditions.

Let us compare this solution of the Dirichlet boundary value problem to well-known variational formulations. The standard formulation is to seek $u \in H_0^1(D)$ such that

$$\int_D \langle \nabla u, \nabla \varphi \rangle dx = -\int_D g\varphi dx,$$

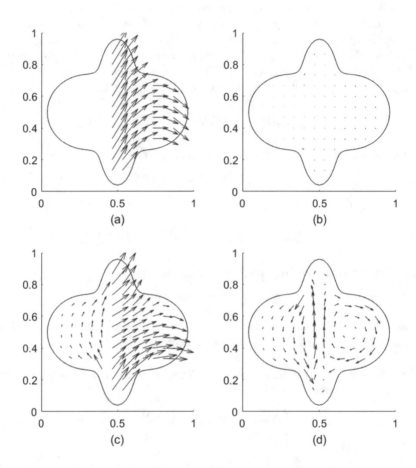

Figure 7.3: (a) Vector field F, with a jump at $x = 0.4$, to be decomposed. (b) The cohomology part H vanishes, modulo computational errors, since the domain is diffeomorphic to the disk. (c) The curl-free part ∇U of F. Note that only the e_1 component jumps at $x = 0.4$. (d) The divergence-free part $\nabla \lrcorner (Vj)$ of F, with tangential boundary conditions. Note that only the e_2 part jumps at $x = 0.4$.

for all $\varphi \in H_0^1(D)$. In other words, this means that we are looking for $F = \nabla U \in R(\underline{d})$ such that $\operatorname{div} F = g$.

Another variational formulation from mechanics is the principle of complementary energy, where we instead vary F over all vector fields such that $\operatorname{div} F = g$, and seek such a vector field satisfying

$$\int_D \langle F, V \rangle dx = 0$$

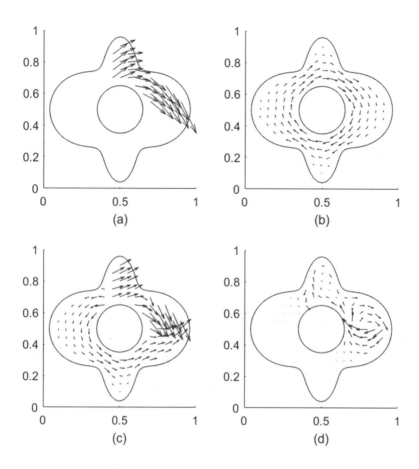

Figure 7.4: (a) Vector field F, with jumps at $x = 0.4$ and $y = 0.5$, to be decomposed. (b) The cohomology part H of F. Since D is diffeomorphic to an annulus, by Example 10.6.6, any other F will have a cohomology part parallel to this H. (c) The curl-free part ∇U of F. (d) The divergence-free part $\nabla \lrcorner (Vj)$ of F, with tangential boundary conditions.

for all divergence-free vector fields $V \in \mathsf{N}(\delta; \wedge^1)$. By Corollary 7.6.7 such F is a gradient vector field $F = \nabla u$ of a unique $u \in H_0^1(D)$.

Example 7.6.13 (The Neumann problem). Consider the Neumann boundary value problem for the Laplace equation,

$$\begin{cases} \Delta u(x) = g(x), & x \in D, \\ \langle \nu(x), \nabla u(x) \rangle = 0, & x \in \partial D, \end{cases}$$

interpreted in the weak sense that $u \in H^1(D)$ including the natural Neumann boundary condition, where $g \in L_2(D; \wedge^0)$ is given. Analogously to the Dirichlet problem, the tangential Hodge decomposition Theorem 7.6.6, and in particular the closedness of $\mathsf{R}(d)$ and $\mathsf{R}(\underline{\delta})$, implies the unique solvability of this boundary value problem. To see this, we again rewrite $\Delta u = g$ as the first-order system (7.11) and (7.12). By Corollary 7.6.8, we have a unique tangential vector field $F \in \mathsf{R}(d; \wedge^1)$ solving (7.11) if and only if g is orthogonal to $\mathcal{C}_{\|}(D)$, that is if it has mean value zero on each connected component of D. Note that F being tangential at ∂D is exactly the Neumann boundary condition we want. Moreover, since $F \in \mathsf{R}(d; \wedge^1)$, we can write $F = \nabla u$, where u is unique if we demand that it be orthogonal to the locally constant functions $\mathcal{C}_{\|}(D)$ just like the datum g.

Example 7.6.14 (Linear hydrodynamics). Consider steady-state flow of an incompressible fluid with viscosity $\mu > 0$. In the limit of small velocities v, the Navier–Stokes equations reduce to the linear Stokes equations

$$\begin{cases} \nabla p - \mu \Delta v = f, \\ \operatorname{div} v = 0. \end{cases}$$

We are given the external forces f, and we want to compute the velcity vector field v and the scalar pressure p. The equations express conservation of momentum and the incompressibility of the fluid respectively.

Assuming that the flow takes place inside a domain D, it is natural to demand in particular the boundary condition that the velocity vector field v is tangential at ∂D. We claim that the tangential Hodge decomposition Theorem 7.6.6 contains the solvability of this boundary value problem. To see this, we recall that with Clifford algebra we have $a^2 v = a \lrcorner (a \wedge v) + a\langle a, v \rangle$ for vectors a and v. Replacing a by ∇, we obtain by incompressibility that

$$\Delta v = \nabla \lrcorner (\nabla \wedge v).$$

Given a force vector field $f \in L_2(D; \wedge^1)$, we apply the tangential Hodge decomposition (7.9) and conclude when f is orthogonal to $\mathcal{C}_{\|}(D)$ that there exist a unique scalar function $p \in \mathsf{R}(\underline{\delta}; \wedge^0)$ and a tangential and exact bivector field $\omega \in \mathsf{R}(d; \wedge^2)$ such that

$$\nabla p - \mu \nabla \lrcorner \omega = f.$$

Again as in Corollary 7.6.8, there further exists a unique tangential vector field $v \in \mathsf{R}(\underline{\delta}; \wedge^1)$ such that

$$\omega = \nabla \wedge v.$$

Note in particular that v is divergence-free and tangential at ∂D, and that it is uniquely determined if we demand that it be orthogonal to $\mathcal{C}_{\|}(D; \wedge^1)$. The pressure p is unique modulo $\mathcal{C}_{\|}(D; \wedge^0)$. The curl $\omega = \nabla \wedge v$ is the vorticity of the flow, which in three-dimensional space can be represented by the Hodge dual vector field.

7.7 Comments and References

7.1 The standard terminology for k-covector fields is *differential forms*. The analysis of such differential forms was first developed by Élie Cartan, and they have become a standard tool in modern differential geometry.

The notation d and δ for the exterior and interior derivatives, when considered as partial differential operators, is standard in the literature. Beware, though, that it is standard to include an implicit minus sign in the notation δ, so that $\delta F = -\nabla \lrcorner F$ and $\delta = d^*$. In the literature it is also common to write df also for the total derivative $\underline{f} = \nabla \otimes f$. We reserve the notation df for the exterior derivative.

The nabla notations $\nabla \cdot F$ and $\nabla \times F$ are common in vector calculus. The generalizations $\nabla \wedge F$, $\nabla \lrcorner F$, $\nabla \otimes F$ of this notation, used in this book, are inspired by Hestenes and Sobczyk [57] and Hestenes [56].

Nilpotence of an object a usually means that $a^k = 0$ for some positive integer k. In this book, we consider only nilpotence of order 2.

7.2 The normalized pushforward of multivector fields is rarely found in the literature. In the special case of vector fields, this operation goes under the name *Piola transformation* in continuum mechanics.

Pullbacks and pushforwards by smooth maps act in a natural way on multivector-valued distributions. For maps that are not diffeomorphisms, the pullback acts on test functions and the normalized pushforward act on distributions. Using k-covector fields as test functions, following George de Rham one defines a k-current as a k-vector-valued distribution. A classical reference for the applications of currents to geometric measure theory is Federer [38].

7.3 The classical Stokes theorem from Example 7.3.10 dates back to Lord Kelvin and George Stokes in 1850. The general higher-dimensional result (7.5) was formulated by Élie Cartan in 1945. Standard notation with differential forms ω is

$$\int_M d\omega = \int_{\partial M} \omega,$$

where the oriented measure $\underline{dx}$ is implicit in the notation ω. The trivial extension to forms with values in a finite-dimensional linear space, as presented in Theorem 7.3.9, is rarely found in the literature. The numerous applications in this book, though, show the usefulness of having such a generalized Stokes formula ready to use.

7.4–7.5 The identity $\mathcal{L}_v^* \Theta = \nabla \wedge (v \lrcorner \Theta) + v \lrcorner (\nabla \wedge \Theta)$ is often referred to as *Cartan's magic formula* due to its usefulness.

A reference for the material in these two sections is Taylor [91], where results on ordinary differential equations, flows of vector fields, and the proof of Poincaré's theorem presented here can be found.

7.6 The theory of Hodge decompositions was developed by William Vallance Douglas Hodge in the 1930s, as a method for studying the cohomology of smooth manifold using PDEs. For further references for the presentation given here, which builds on the survey paper by Axelsson and McIntosh [14], we refer to Section 10.7. We find it convenient to work in the full exterior algebra $\wedge V$. However, since the exterior and interior products preserve homogeneity of multivector fields, we may rather state the results at each level $\wedge^k V$ to obtain Hodge splittings of k-vector fields, which is standard in the literature.

A reference for solving boundary value problems with Hodge decompositions, also on more general manifolds with boundary, but under the assumption that the boundary is smooth, can be found in Schwarz [85]. Two harmonic analysis works using multivector calculus, motivated by Example 7.6.14 and nonsmooth boundary value problems for the Stokes equations, are McIntosh and Monniaux [68] and Rosén [83].

Techniques from multivector calculus have also been used successfully in numerical analysis. A seminal paper on finite element exterior calculus is Arnold, Falk, and Winther [2].

Chapter 8

Hypercomplex Analysis

Prerequisites:

A solid background in analysis of one complex variable is required for this chapter, but no knowledge of analysis in several complex variables is needed. We make use of real-variable calculus, and build some on Chapter 7.

Road map:

We saw in Chapter 3 that even though it is natural to view Clifford algebras as a kind of hypercomplex numbers, the analogy fails in some important aspects, and it may be more appropriate to view Clifford algebras as matrix algebras, but from a geometric point of view. Nevertheless, a great deal of one-dimensional complex analysis does generalize to a noncommutative hypercomplex analysis in n-dimensional Euclidean space, replacing complex-valued functions by multivector fields. This yields a generalization of one-variable complex analysis that is fundamentally different from the commutative theory of several complex variables.

Recall from one-dimensional complex analysis the following equivalent characterizations of analytic/holomorphic functions $f : \mathbf{C} \to \mathbf{C}$ defined in a domain $D \subset \mathbf{C}$, where we assume that the total derivative f_z is injective. The analysis definition: f is analytic if the limit $f'(z) = \lim_{w \to 0}(f(z + w) - f(z))/w$ exists at each $z \in D$. The partial differential equation definition: f is analytic if it satisfies the *Cauchy–Riemann* system of partial differential equations

$$\partial_1 f_1(z) - \partial_2 f_2(z) = 0, \quad \partial_2 f_1(z) + \partial_1 f_2(z) = 0,$$

in D, where f_1, f_2 are the real component functions of $f = f_1 + if_2$. The algebra definition: f is analytic if around each point $z \in D$ it is locally the sum of a power series $f(w) = \sum_{k=0}^{\infty} a_k(w - z)^k$, convergent in $\{w \in D \; ; \; |w - z| < r(z)\}$ for some $r(z) > 0$. The geometry definition: f is analytic if it is an (orientation-preserving)

© Springer Nature Switzerland AG 2019

A. Rosén, *Geometric Multivector Analysis*, Birkhäuser Advanced Texts Basler Lehrbücher,
https://doi.org/10.1007/978-3-030-31411-8_8

conformal map, that is if at each $z \in D$ the derivative $\underline{f}_z$ is of the form

$$\underline{f}_z = \begin{bmatrix} a & -b \\ b & a \end{bmatrix},$$

where $a = \partial_1 f_1$ and $b = \partial_1 f_2$. This means that $\underline{f}_z$ is a nonzero multiple of a rotation matrix and can be expressed as complex multiplication by $f'(z)$.

In generalizing to a hypercomplex analysis in higher-dimensional Euclidean spaces, the partial differential equation definition turns out to be most successful, where the Cauchy–Riemann equations are replaced by a Dirac equation $\nabla_\triangle F(x) = 0$, using the nabla operator induced by the Clifford product. As in Example 7.3.13, we may express this Dirac equation in terms of an integral difference quotient. Behind the Dirac equation, a fundamental type of splitting of function spaces is lurking: splittings into *Hardy subspaces*. With a solid understanding of Clifford algebra, these are straightforward generalizations of the classical such splittings in the complex plane. Recall that in the complex plane, any function $f : \partial D \to \mathbf{C}$ on the boundary of a bounded domain D is in a unique way the sum

$$f = f^+ + f^-,$$

where f^+ is the restriction to ∂D of an analytic function in D, and f^- is the restriction to ∂D of an analytic function in $\mathbf{C} \setminus \overline{D}$ that vanishes at ∞. The two subspaces consisting of traces of analytic functions from the interior or the exterior domain are the Hardy subspaces, and the Cauchy integral formula

$$\frac{1}{2\pi i} \int_{\partial D} \frac{f(w)}{w - z} dw$$

provides the projection operators onto these subspaces. There is one important difference with the Hodge splitting from Section 7.6: the two Hardy spaces are in general not orthogonal subspaces of $L_2(\partial D)$, but the angle between them depends on the geometry of ∂D.

We show in Section 8.2 that the algebraic definition can be generalized to give power series expansions in higher dimensions. This is closely related to the classical theory of spherical harmonics. Later in Section 11.4, we shall see that the geometry definition does not generalize well. The higher dimensional conformal maps are very scarce indeed: the only ones are the Möbius maps!

Highlights:

- The higher dimensional Cauchy integral formula: 8.1.8

- Möbius pullbacks of monogenic fields: 8.1.14

- Splitting into spherical monogenics: 8.2.6

- Spherical Dirac operator: 8.2.15

- Splittings into Hardy subspaces: 8.3.6

8.1 Monogenic Multivector Fields

In this chapter we work in Euclidean space (X, V), and we study the following generalization of the Cauchy–Riemann equations.

Definition 8.1.1 (Monogenic fields). Let D be an open set in a Euclidean space X. If $F : D \to \triangle V$ is differentiable at $x \in D$, we define the Clifford derivative

$$\nabla \vartriangle F(x) = \sum_{i=1}^{n} e_i^* \vartriangle \partial_i F(x)$$

as in Definition 7.1.2, where $\{e_i^*\}$ is the basis dual to $\{e_i\}$, and ∂_i is the partial derivative with respect to the corresponding coordinate x_i.

The *Dirac operator* $\mathbf{D} : F \mapsto \nabla \vartriangle F$ is the nabla operator induced by Clifford multiplication. If F is a C^1 multivector field for which $\nabla \vartriangle F(x) = 0$ in all of D, then F is said to be a *monogenic* field in D.

Let $\{e_s\}$ be an induced ON-basis for $\triangle V$ and write

$$F(x) = \sum_s F_s(x) e_s.$$

If F is a monogenic field, then each scalar component function F_s is a harmonic function. To see this, we note that

$$0 = \mathbf{D}^2 F(x) = \sum_i \sum_j e_i e_j \partial_i \partial_j F(x) = \sum_i \partial_i^2 F(x) = \sum_s \left(\sum_i \partial_i^2 F_s(x) \right) e_s$$
$$= \sum_s (\Delta F_s(x)) e_s.$$

This is a consequence of the defining property $v^2 = |v|^2$ for the Clifford product, and it means that $\mathbf{D}$ is a first-order differential operator that is a square root of the componentwise Laplace operator. Similar to the situation for analytic functions, a monogenic multivector field consists of 2^n scalar harmonic functions, which are coupled in a certain sense described by the Dirac equation $\mathbf{D}F = 0$. In particular, monogenic fields are smooth, even real analytic.

The Dirac derivative further combines the exterior and interior derivative. Indeed, since $e_i^* \vartriangle w = e_i^* \lrcorner w + e_i^* \wedge w$, it is clear that

$$\mathbf{D}F(x) = \nabla \vartriangle F(x) = \nabla \lrcorner F(x) + \nabla \wedge F(x) = \delta F(x) + dF(x).$$

This means that

$$\mathbf{D}^2 = (d + \delta)^2 = d^2 + \delta^2 + d\delta + \delta d = d\delta + \delta d,$$

by nilpotence. Another way to see this is to put $v = \theta = \nabla$ in the anticommutation relation from Theorem 2.8.1.

As in Chapter 3, in using the Dirac operator, it is in general necessary to work within the full Clifford algebra, since typically $\mathbf{D}F$ will not be a homogeneous multivector field even if F is. However, in some applications the fields have values in the even subalgebra, or even are homogeneous k-vector fields.

Example 8.1.2 (Analytic functions). Let X be a two-dimensional Euclidean plane, and let $\mathbf{C} = \triangle^0 V \oplus \triangle^2 V$ be the standard geometric representation of the complex plane as in Section 3.2. Consider the Dirac equation $\nabla \vartriangle f = 0$ for a complex valued function $f = u + vj : \mathbf{C} \to \mathbf{C}$, where we have fixed an origin and ON-basis $\{e_1, e_2\}$ in $X = V$, giving the identification $V \leftrightarrow \mathbf{C}$, $e_1 \leftrightarrow 1$, $e_2 \leftrightarrow j = e_{12}$. Writing out the equation, we have

$$\nabla \vartriangle f = (e_1 \partial_1 + e_2 \partial_2) \vartriangle (u + e_{12} v) = (\partial_1 u - \partial_2 v)e_1 + (\partial_1 v + \partial_2 u)e_2.$$

Thus $\nabla \vartriangle f = 0$ coincides with the Cauchy–Riemann equations, and f is monogenic if and only if it is analytic. Note also that the only functions $f : \mathbf{C} \to \mathbf{C}$ that satisfy $\nabla \lrcorner f = 0 = \nabla \wedge f$ are the locally constant functions, since $\nabla \wedge f = \operatorname{grad} u$ and $\nabla \lrcorner f = -j \operatorname{grad} v$.

On the other hand, the complex function f corresponds to the vector field $F(x) = e_1 f(x)$ under the identification $V \leftrightarrow \mathbf{C}$. Reversing this relation gives $F(x) = \overline{F(x)} = \overline{f(x)}e_1$. Since the Clifford product is associative, it follows that $F(x)$ is a plane divergence and curl-free vector field if and only if

$$0 = \nabla \vartriangle F(x) = \nabla \vartriangle (\overline{f(x)} \vartriangle e_1) = (\nabla \vartriangle \overline{f(x)}) \vartriangle e_1,$$

that is, if f is antianalytic.

Example 8.1.3 (3D monogenic fields). Let F, G be vector fields and let f, g be scalar functions defined in an open set D in three-dimensional oriented Euclidean space. Then the multivector field $f(x) + F(x) + *G(x) + *g(x)$ is monogenic if and only if

$$\operatorname{div} F(x) = 0,$$
$$\nabla f(x) - \nabla \times G(x) = 0,$$
$$\nabla \times F(x) + \nabla g(x) = 0,$$
$$\operatorname{div} G(x) = 0.$$

We note that there is no restriction to assume that a monogenic field F takes values in the even subalgebra $\triangle^{\mathrm{ev}} V$. Indeed, if $F : D \to \triangle V$ is monogenic, we write $F = F^{\mathrm{ev}} + F^{\mathrm{od}}$ where $F^{\mathrm{ev}} : D \to \triangle^{\mathrm{ev}} V$ and $F^{\mathrm{od}} : D \to \triangle^{\mathrm{od}} V$. Then

$$0 = \nabla \vartriangle F(x) = \nabla \vartriangle F^{\mathrm{ev}} + \nabla \vartriangle F^{\mathrm{od}},$$

where $\nabla \vartriangle F^{\mathrm{ev}} : D \to \triangle^{\mathrm{od}} V$ and $\nabla \vartriangle F^{\mathrm{od}} : D \to \triangle^{\mathrm{ev}} V$, so we conclude that F^{ev} and F^{od} each are monogenic.

Example 8.1.4 (Stein–Weiss vector fields). If $F : D \to V = \triangle^1 V$ is a vector field in a general Euclidean space, then F is monogenic if and only if F is divergence- and curl-free. Thus, for vector fields the equation $\mathbf{D}F = 0$ is equivalent to the first-order system

$$\begin{cases} \operatorname{div} F(x) = 0, \\ \operatorname{curl} F(x) = 0. \end{cases}$$

This is a consequence of the fact that $\nabla \lrcorner F : D \to \triangle^0 V$ and $\nabla \wedge F : D \to \triangle^2 V$, where $\triangle^0 V \cap \triangle^2 V = \{0\}$.

We note that Example 8.1.4 generalizes as follows. When $F : D \to \triangle^k V$ is a homogeneous multivector field, then $\nabla \vartriangle F = 0$ if and only if

$$\begin{cases} \nabla \lrcorner F(x) = 0, \\ \nabla \wedge F(x) = 0. \end{cases}$$

The following proposition shows when all the homogeneous parts of a monogenic field are themselves monogenic.

Proposition 8.1.5 (Two-sided monogenicity). *Let $F : D \to \triangle V$ be a C^1 multivector field, and write $F = F_0 + \cdots + F_n$, where $F_j : D \to \triangle^j V$. Then the following are equivalent.*

(i) *All the homogeneous parts F_j are monogenic fields.*

(ii) *The field F satisfies both $\nabla \wedge F = 0$ and $\nabla \lrcorner F = 0$.*

(iii) *The field F is two-sided monogenic, that is, $\nabla \vartriangle F = 0$ and $F \vartriangle \nabla = \sum \partial_i F(x) \vartriangle e_i^* = 0$.*

Proof. (i) implies (iii): If F_j is monogenic, then $\nabla \wedge F_j = 0 = \nabla \lrcorner F_j$ as above, and therefore $\nabla \vartriangle F_j$ as well as $F_j \vartriangle \nabla = (-1)^j (\nabla \wedge F_j - \nabla \lrcorner F_j)$ is zero. Adding up all F_j proves (iii).

(iii) implies (ii): this is a consequence of the Riesz formulas, which show that

$$\nabla \wedge F = \tfrac{1}{2}(\nabla \vartriangle F - \widehat{F \vartriangle \nabla}) \quad \text{and} \quad \nabla \lrcorner F = \tfrac{1}{2}(\nabla \vartriangle F + \widehat{F \vartriangle \nabla}).$$

(ii) implies (i): If $\nabla \wedge F = 0$, then $0 = (\nabla \wedge F)_{j+1} = \nabla \wedge F_j$ for all j, since d maps j-vector fields to $(j+1)$-vector fields. Similarly $\nabla \lrcorner F_j = 0$ for all j. Thus $\nabla \vartriangle F_j = \nabla \lrcorner F_j + \nabla \wedge F_j = 0$. $\qquad \square$

We next consider the fundamental solution for the Dirac operator. In order to apply the Fourier transform componentwise as in Section 6.2, we complexify the Clifford algebra $\triangle V \subset \triangle V_c$. We note that the exterior, interior, and Clifford derivatives are the Fourier multipliers

$$\widehat{dF(x)} = i\xi \wedge \hat{F}(\xi),$$

$$\widehat{\delta F(x)} = i\xi \lrcorner \hat{F}(\xi), \quad \text{and}$$

$$\widehat{\mathbf{D}F(x)} = i\xi \vartriangle \hat{F}(\xi), \quad \xi \in V.$$

From this it follows that unlike d and δ, the Dirac operator $\mathbf{D}$ is elliptic and has a fundamental solution $\Psi(x)$ with Fourier transform

$$\hat{\Psi}(\xi) = (i\xi)^{-1} = -i\frac{\xi}{|\xi|^2}.$$

Using the formula for the fundamental solution Φ to the Laplace operator Δ from Example 6.3.1, where $\hat{\Phi}(\xi) = -1/|\xi|^2$, we obtain the following formula for $\Psi(x) = \nabla\Phi$. Note that unlike the situation for Φ, the two-dimensional case does not use any logarithm.

Definition 8.1.6 (Fundamental solution). The *fundamental solution* to the Dirac operator $\mathbf{D}$ in an n-dimensional Euclidean space with origin fixed, $n \geq 1$, is the vector field

$$\Psi(x) := \frac{1}{\sigma_{n-1}} \frac{x}{|x|^n},$$

where $\sigma_{n-1} := \int_{|x|=1} dx = 2\pi^{n/2}/\Gamma(n/2)$ is the measure of the unit sphere in V, and $\Gamma(z) := \int_0^\infty e^{-t}t^{z-1}dt$ is the gamma function, with $\Gamma(k) = (k-1)!$.

Exercise 8.1.7. Show by direct calculation that $\nabla \wedge \Psi(x) = 0$ and $\nabla \lrcorner \Psi(x) = \delta_0(x)$ in the distributional sense in V, where $\delta_0(x)$ is the Dirac delta distribution.

The following application of the general Stokes theorem is central to hypercomplex analysis.

Theorem 8.1.8 (Cauchy–Pompeiu formula for $\mathbf{D}$). *Let D be a bounded C^1-domain in Euclidean space X. If $F \in C^1(\overline{D}; \triangle V)$, then*

$$F(x) + \int_D \Psi(y-x)(\mathbf{D}F)(y)dy = \int_{\partial D} \Psi(y-x)\nu(y)F(y)\, dy,$$

for all $x \in D$, where $\nu(y)$ denotes the outward-pointing unit normal vector field on ∂D. In particular, for monogenic multivector fields F, we have the Cauchy reproducing formula

$$F(x) = \int_{\partial D} \Psi(y-x)\nu(y)F(y)\, dy, \quad x \in D. \tag{8.1}$$

Proof. For fixed $x \in D$, consider the linear 1-form

$$\theta(y, v) := \Psi(y-x) \wedge v \wedge F(y).$$

For $y \in D \setminus \{x\}$, its nabla derivative is

$$\theta(\dot{y}, \nabla) = \sum_{i=1}^n \partial_{y_i}\Big(\Psi(y-x) \wedge e_i \wedge F(y)\Big)$$

$$= (\Psi(\dot{y}-x) \wedge \nabla) \wedge F(y) + \Psi(y-x) \wedge (\nabla \wedge F(\dot{y})) = \Psi(y-x) \wedge (\mathbf{D}F)(y),$$

by associativity of the Clifford product and since $\Psi \vartriangle V = V \lrcorner \Psi - V \wedge \Psi = 0$ by Exercise 8.1.7.

To avoid using distribution theory, we consider the domain $D_\epsilon := D \backslash \overline{B(x, \epsilon)}$, obtained by removing a small ball around x. On $\partial B(x, \epsilon)$ the outward-pointing unit normal relative D_ϵ is $(x - y)/|x - y|$. The Stokes formula (7.4) gives

$$\int_{D_\epsilon} \Psi(y - x)(\mathbf{D}F)(y) dy$$

$$= \int_{\partial B(x,\epsilon)} \Psi(y - x) \frac{x - y}{|x - y|} F(y) \, dy + \int_{\partial D} \Psi(y - x)\nu(y)F(y) \, dy$$

$$= -\int_{\partial B(x,\epsilon)} \frac{1}{\epsilon^{n-1}\sigma_{n-1}} F(y) \, dy + \frac{1}{\sigma_{n-1}} \int_{\partial D} \frac{y - x}{|y - x|^n} \nu(y)F(y) \, dy.$$

Upon taking limits $\epsilon \to 0$, the first term on the right-hand side will converge to $-F(x)$, and the Cauchy–Pompeiu formula follows. □

Exercise 8.1.9 (Cauchy integral theorem). Apply Stokes's theorem and prove the general Cauchy theorem

$$\int_{\partial D} G(y) \vartriangle \nu(y) \vartriangle F(y) \, dy = 0$$

for a left monogenic field F and a right monogenic G, that is, $\nabla \vartriangle F = 0 = G(\dot{x})\vartriangle\nabla = 0$, in D. Deduce from this the classical Cauchy theorem $\int_{\partial D} f(w)dw = 0$ for an analytic function f from complex analysis. See Example 7.3.12.

Example 8.1.10 (Cauchy formula in $\mathbf{C}$). The Cauchy formula for analytic functions in the complex plane is a special case of Theorem 8.1.8. To see this, consider an analytic function $f(z)$ in a plane domain D. As in Example 8.1.2, we identify the vector $x \in V$ and the complex number $z = e_1 x \in \mathbf{C} = \triangle^{\mathrm{ev}}V$, $y \in V$ with $w = e_1 y \in \mathbf{C}$, and the normal vector ν with the complex number $n = e_1 \nu$. If $f(z) : D \to \mathbf{C} = \triangle^{\mathrm{ev}}V$ is analytic, thus monogenic, and if $x \in D$, then Theorem 8.1.8 shows that

$$f(z) = \frac{1}{\sigma_1} \int_{\partial D} \frac{e_1(w - z)}{|w - z|^2} (e_1 n(w)) f(w)|dw|$$

$$= \frac{1}{2\pi j} \int_{\partial D} \frac{e_1(w - z)e_1}{|w - z|^2} f(w)(jn(w)|dw|) = \frac{1}{2\pi j} \int_{\partial D} \frac{f(w)dw}{w - z}.$$

Here we have used that $e_1(w - z)e_1 = \overline{w - z}$, that complex numbers commute, and that jn is tangent to a positively oriented curve. We have written $|dw|$ for the scalar length measure on ∂D.

Note that unlike the situation for analytic functions, in higher dimensions the normal vector must be placed in the middle, between the fundamental solution and the monogenic field. This is because the Clifford product is noncommutative. For analytic functions, the normal infinitesimal element $\nu(y)dy$ corresponds to dw/j, which can be placed, for example, at the end of the expression, since complex numbers commute.

As in the complex plane, the Cauchy formula for monogenic fields has a number of important corollaries, of which we next consider a few.

Corollary 8.1.11 (Smoothness). *Let $F : D \to \triangle V$ be a monogenic field in a domain D. Then F is real analytic, and in particular a C^∞-regular field.*

Proof. Fix a ball $B(x_0, \epsilon)$ such that $B(x_0, 2\epsilon) \subset D$. Then

$$F(x) = \frac{1}{\sigma_{n-1}} \int_{|y-x_0|=\epsilon} \frac{y - x}{|y - x|^n} \nu(y)F(y)\, dy, \quad \text{for all } x \in B(x_0, \epsilon).$$

The stated regularity now follows from that of the fundamental solution $x \mapsto \frac{y-x}{|y-x|^n}$. $\qquad\square$

We also obtain a Liouville theorem for entire monogenic fields.

Corollary 8.1.12 (Liouville). *Let $F : X \to \triangle V$ be an entire monogenic field, that is monogenic on the whole Euclidean space X. If F is bounded, then F is a constant field.*

Proof. Let $x_0 \in X$. For all $R > 0$ and $1 \le k \le n$ we have

$$\partial_k F(0) = \frac{1}{\sigma_{n-1}} \int_{|y-x_0|=R} \left(\partial_{x_k} \frac{y - x}{|y - x|^n} \right)\Big|_{x=0} \nu(y)F(y)\, dy.$$

If F is bounded, the triangle inequality for integrals shows that

$$|\partial_k F(0)| \lesssim \int_{|y|=R} R^{-n} dy \lesssim 1/R.$$

Taking limits as $R \to 0$ shows that $\partial_k F(x_0) = 0$ for all k. Since x_0 was arbitrary, F must be constant. $\qquad\square$

Next we consider what further properties monogenic fields do and do not share with analytic functions. In contrast to analytic functions, monogenic fields do not form a multiplication algebra, that is, $F(x)$ and $G(x)$ being monogenic in D does not imply that $x \mapsto F(x) \vartriangle G(x)$ is monogenic. The obstacle here is the noncommutativity of the Clifford product, which causes $\mathbf{D}(FG) = (\mathbf{D}F)G + \mathbf{D}F\dot{G} \ne (\mathbf{D}F)G + F(\mathbf{D}G)$.

Although monogenic fields in general cannot be multiplied to form another monogenic field, we can do somewhat better than the real linear structure of monogenic fields. Recall that analytic functions form a complex linear space. This generalizes to monogenic fields as follows.

Proposition 8.1.13 (Right Clifford module). *Let D be an open set in a Euclidean space X. Then the monogenic fields in D form a right Clifford module, that is, if $F(x)$ is monogenic, then so is $x \mapsto F(x) \vartriangle w$ for every constant $w \in \triangle V$.*

Proof. This is a consequence of the associativity of the Clifford product, since

$$\nabla \vartriangle (F(x) \vartriangle w) = (\nabla \vartriangle F(x)) \vartriangle w = 0 \vartriangle w = 0. \qquad \square$$

In contrast to analytic functions, monogenic functions do not form a group under composition, that is, $F(y)$ and $G(x)$ being monogenic in appropriate domains does not imply that $x \mapsto F(G(x))$ is monogenic. Indeed, in general the composition is not even well defined, since the range space $\triangle V$ is not contained in V.

Although it does not make sense to compose monogenic fields, the situation is not that different in higher dimensions. Recall that in the complex plane, analytic functions are the same as conformal maps, at least for functions with invertible derivative. In higher dimensions one should generalize so that the inner function G is conformal and the outer function F is monogenic. In this way, we can do the following type of conformal change of variables that preserves monogenicity. Sections 4.5 and 11.4 are relevant here.

Proposition 8.1.14 (Conformal Kelvin transform). *Let $Tx = (ax + b)(cx + d)^{-1}$ be a fractional linear map of a Euclidean space $X = V$, and let $D \subset X$ be an open set such that $\infty \notin T(D)$. For a field $F : T(D) \to \triangle V$, define a pulled back field*

$$K_T^m F : D \to \triangle V : x \mapsto \frac{\overline{cx + d}}{|cx + d|^n} \vartriangle F(T(x)).$$

Then

$$\mathbf{D}(K_T^m F)(x) = \frac{\det_\triangle(T)}{|cx + d|^2} K_T^m(\mathbf{D}F)(x), \quad x \in D,$$

where $\det_\triangle(T) := a\overline{d} - b\overline{c}$. In particular, if F is monogenic, then so is $K_T^m F$.

Proof. Applying the product rule as in Example 7.1.9 shows that

$$\nabla \vartriangle (K_T F)(x) = \left(\nabla \frac{\overline{x + c^{-1}d}}{|x + c^{-1}d|^n} \right) \frac{\overline{c}}{|c|^n} F(T(x)) + \sum_{i=1}^{n} e_i \frac{\overline{cx + d}}{|cx + d|^n} \partial_i (F(T(x))).$$

The first term is zero, since the fundamental solution is monogenic outside the origin. For the second term, we note that since T is conformal, the derivative $\underline{T}_x$ will map the ON-basis $\{e_i\}$ onto a basis $\{e_i' = \underline{T}_x e_i\}$ of orthogonal vectors of equal length. By Exercise 4.5.18, we have

$$e_i' = (\det_\triangle(T)/\overline{cx + d})e_i(cx + d)^{-1}.$$

The dual basis is seen to be

$$e_i'^* = ((cx + d)/\det_\triangle(T))e_i \overline{cx + d},$$

so that

$$e_i \overline{cx + d} = (\det_\triangle(T)/(cx + d))e_i'^*.$$

According to the chain rule, the directional derivatives are

$$\partial_{e_i}(F(T(x))) = (\partial_{e_i'}F)(T(x)),$$

so

$$\nabla_\triangle(K_T^m F)(x) = \frac{\det_\triangle(T)}{|cx + d|^2} \frac{\overline{cx + d}}{|cx + d|^n} \sum_{i=1}^n e_i'^*(\partial_{e_i'}F)(T(x)) = \frac{\det_\triangle(T)}{|cx + d|^2} K_T^m(\mathbf{D}F)(x).$$

$\square$

Specializing to the inversion change of variables $Tx = 1/x$, we make the following definition.

Definition 8.1.15 (Kelvin transform). The *monogenic Kelvin transform* of a field $F : D \to \triangle V$ is the field

$$K^m F(x) := \frac{x}{|x|^n} \triangle F(1/x).$$

Similarly, using the fundamental solution for the Laplace operator, we define the *harmonic Kelvin transform* of a function $u : D \to \mathbf{R}$ to be

$$K^h u(x) := |x|^{2-n} u(1/x).$$

For the monogenic Kelvin transform, we have shown that

$$\mathbf{D}K^m = -|x|^{-2} K^m \mathbf{D}.$$

We now use this to obtain a similar result for the harmonic Kelvin transform.

Proposition 8.1.16. *The harmonic Kelvin transform satisfies the commutation relation*

$$\triangle(K^h u)(x) = |x|^{-4} K^h(\triangle u)(x).$$

In particular, the Kelvin transform of an harmonic function is harmonic.

Proof. We note that $\triangle = \mathbf{D}^2$ and $K^h u = x K^m u$. Thus

$$\begin{aligned}
\triangle K^h u &= \mathbf{D}\mathbf{D}(xK^m u) = \mathbf{D}(nK^m u + (2\partial_x - x\mathbf{D})K^m u) \\
&= n\mathbf{D}K^m u + 2(\mathbf{D}K^m u + \partial_x \mathbf{D}K^m u) - n\mathbf{D}K^m u - (2\partial_x - x\mathbf{D})\mathbf{D}K^m u \\
&= 2\mathbf{D}K^m u + x\mathbf{D}^2 K^m u = -2|x|^{-2}K^m \mathbf{D}u - x\mathbf{D}|x|^{-2}K^m \mathbf{D}u \\
&= -x^{-1}\mathbf{D}K^m \mathbf{D}u = x^{-1}|x|^{-2}K^m \mathbf{D}^2 u = |x|^{-4}K^h \triangle u.
\end{aligned}$$

Here $\partial_x f = \sum_{j=1}^n x_j \partial_j f = \langle x, \nabla\rangle f$ denotes the radial directional derivative, and we have used that $\mathbf{D}\partial_x f = \sum_{j=1}^n (\nabla x_j)\partial_j f + \partial_x \mathbf{D}f = (1 + \partial_x)\mathbf{D}f$ by the product rule.

$\square$

Exercise 8.1.17. Consider the special case of Proposition 8.1.14 in which

$$\rho(x) = Tx = \hat{q}xq^{-1}$$

is an isometry, where $q \in \text{Pin}(V)$. Investigate how the conformal Kelvin transform $K_T^m F$, the pullback $\rho^* F$, the pushforward $\rho_*^{-1} F$ and the normalized pushforward $\tilde{\rho}_*^{-1} F$ are related. Show that all these four fields are monogenic whenever F is monogenic, and relate this result to Proposition 8.1.13.

8.2 Spherical monogenics

In our n-dimensional Euclidean space $X = V$ with a fixed origin, we denote by

$$S := \{x \in X \; ; \; |x| = 1\}$$

the unit sphere. We generalize the well-known theory of Taylor series expansions of analytic functions in the plane. When $n = 2$, we know that a function analytic at 0 can be written as a convergent power series

$$f(x + iy) = \sum_{k=0}^{\infty} P_k(x, y), \quad x^2 + y^2 < \epsilon^2,$$

for some $\epsilon > 0$, where $P_k \in \mathcal{P}_k^m := \{a_k(x + iy)^k \; ; \; a_k \in \mathbf{C}\}$. A harmonic function can be written in the same way if we allow terms $P_k \in \mathcal{P}_k^h := \{a_k(x+iy)^k + b_k(x - iy)^k \; ; \; a_k, b_k \in \mathbf{C}\}$. Note that $\mathcal{P}_0^h$ and all $\mathcal{P}_k^m$ are one-dimensional complex linear spaces, whereas $\mathcal{P}_k^h$ are two-dimensional when $k \geq 1$. The spaces $\mathcal{P}_k^m$ and $\mathcal{P}_k^h$ are subspaces of the space $\mathcal{P}_k$ of all polynomials of order k, which has dimension $k+1$. A polynomial $P \in \mathcal{P}_k$ is in particular homogeneous of degree k in the sense that

$$P(rx) = r^k P(x), \quad \text{for all } r > 0, x \in \mathbf{R}^2.$$

This shows that P is uniquely determined by its restriction to the unit circle $|x| = 1$ if the degree of homogeneity is known. In the power series for f, the term P_k describes the kth-order approximation of f around the origin.

Next consider an n-dimensional space X, and the following generalization of the spaces above.

Definition 8.2.1 (Spherical harmonics and monogenics). Let $X = V$ be a Euclidean space, and let $k \in \mathbf{N}$ and $s \in \mathbf{R}$. Define function spaces

$$\mathcal{P}_k := \{P : X \to \triangle V \; ; \; \text{all component functions are homogeneous polynomials}$$
$$\text{of degree } k\},$$
$$\mathcal{P}_s^m := \{F : X \setminus \{0\} \to \triangle V \; ; \; \mathbf{D}F = 0, F(rx) = r^s F(x), x \neq 0, r > 0\},$$
$$\mathcal{P}_s^h := \{F : X \setminus \{0\} \to \triangle V \; ; \; \triangle F = 0, F(rx) = r^s F(x), x \neq 0, r > 0\}.$$

Let $\mathcal{P}_k^s \subset \mathcal{P}_k$ and $\mathcal{P}_s^{sh} \subset \mathcal{P}_s^h$ be the subspaces of scalar functions $F : X \setminus \{0\} \to \triangle^0 V = \mathbf{R}$, and let $\mathcal{P}_s^{em} \subset \mathcal{P}_s^m$ be the subspace of functions $F : X \setminus \{0\} \to \triangle^{\mathrm{ev}} V$ that take values in the even subalgebra.

Denote by $\mathcal{P}_s^h(S)$ the space of restrictions of functions $P \in \mathcal{P}_s^h$ to the unit sphere S. Denote by $\mathcal{P}_s^m(S)$ the space of restrictions of functions $P \in \mathcal{P}_s^m$ to the unit sphere S. We refer to these functions as (multivector-valued) *spherical harmonics* and *spherical monogenics* respectively.

Note that the spaces $\mathcal{P}_k$ and $\mathcal{P}_s^h$ essentially are spaces of scalar functions: Each function in these spaces has component functions that belong to the same space, since the conditions on the function do not involve any coupling between the component functions. Even if the definitions of $\mathcal{P}_s^m$ and $\mathcal{P}_s^h$ are quite liberal, these are essentially spaces of polynomials, as the following shows.

Proposition 8.2.2. *Let $n := \dim X$. The monogenic space $\mathcal{P}_s^m$ contains nonzero functions only if*

$$s \in \{\ldots, -(n+1), -n, -(n-1), 0, 1, 2, \ldots\}.$$

The harmonic space $\mathcal{P}_s^h$ contains nonzero functions only if

$$s \in \{\ldots, -(n+1), -n, -(n-1), -(n-2), 0, 1, 2, \ldots\}.$$

If $k \in \mathbf{N}$, then $\mathcal{P}_k^m \subset \mathcal{P}_k^h \subset \mathcal{P}_k$. The Kelvin transforms give self-inverse one-to-one correspondences

$$K^m : \mathcal{P}_s^m \to \mathcal{P}_{-(s+n-1)}^m, \quad K^h : \mathcal{P}_s^h \to \mathcal{P}_{-(s+n-2)}^h.$$

Proof. (i) First consider the monogenic spaces $\mathcal{P}_s^m$. Apply the Cauchy formula (8.1) to $P \in \mathcal{P}_s^m$ in the domain $D := B(0;1) \setminus B(0;\epsilon)$ for fixed $0 < \epsilon < 1$. For $x \in D$, we have

$$P(x) = \int_S \Psi(y-x)yP(y)dy + \int_{|y|=\epsilon} \Psi(y-x)\nu(y)P(y)dy.$$

For fixed $x \neq 0$, the second integral is dominated by $\epsilon^{n-1} \sup_{|y|=\epsilon} |P|$. Letting $\epsilon \to 0$, this tends to zero if $s > -(n-1)$, and it follows that 0 is a removable singularity of $P(x)$.

If $-(n-1) < s < 0$, Liouville's Theorem 8.1.12 shows that $P = 0$. Furthermore, generalizing the proof of Liouville's theorem by applying higher-order derivatives shows that if $s \geq 0$, then $P(x)$ must be a polynomial. Thus $\mathcal{P}_s^m \neq \{0\}$ only if $s \in \mathbf{N}$. That $K^m : \mathcal{P}_s^m \to \mathcal{P}_{-(s+n-1)}^m$ is bijective and self-inverse is straightforward to verify.

(ii) Next consider the harmonic spaces $\mathcal{P}_s^h$. If $P \in \mathcal{P}_s^h$, then $\mathbf{D}P \in \mathcal{P}_{s-1}^m$. If $s \notin \mathbf{Z}$ or $-(n-2) < s < 0$, then (i) shows that $\mathbf{D}P = 0$, so that $P \in \mathcal{P}_s^m$. Again by (i), we conclude that $P = 0$. If $s \in \mathbf{N}$, then the same argument shows that $\mathbf{D}P$ is a polynomial. Here we may assume that P is scalar-valued, so that $\mathbf{D}P = \nabla P$. Integrating we find that P is a polynomial as well. That $K^h : \mathcal{P}_s^h \to \mathcal{P}_{-(s+n-2)}^h$ is bijective and self-inverse is straightforward to verify. $\square$

We next examine the finite-dimensional linear spaces $\mathcal{P}_k^m$ and $\mathcal{P}_k^h$ for $k \in \mathbf{N}$. As we have seen, this also gives information about the spaces $\mathcal{P}_{-(k+n-1)}^m$ and $\mathcal{P}_{-(k+n-2)}^h$ via the Kelvin transforms. Note that unlike the situation in the plane, there is a gap $-(n-1) < s < 0$ and $-(n-2) < s < 0$ respectively between the nonzero spaces, and that this gap grows with dimension.

A polynomial $P(x) \in \mathcal{P}_k$, can be written

$$P(x) = \sum_{s \subset \overline{n}} \sum_{\alpha \in \mathbf{N}^k} P_{\alpha s} x^\alpha e_s.$$

Here we use multi-index notation $x^\alpha = x^{(\alpha_1, \ldots, \alpha_k)} := x_1^{\alpha_1} \cdots x_k^{\alpha_k}$, and we shall write $\delta_i := (0, \ldots, 0, 1, 0, \ldots, 0)$, where 1 is the ith coordinate. We introduce an auxiliary inner product

$$\langle P, Q \rangle_p := \sum_s \sum_\alpha \alpha! P_{\alpha s} Q_{\alpha s},$$

where $\alpha! = (\alpha_1, \ldots, \alpha_k)! := \alpha_1! \cdots \alpha_k!$.

Proposition 8.2.3. *With respect to the inner product $\langle \cdot, \cdot \rangle_p$ on $\mathcal{P}_k$, we have orthogonal splittings*

$$\mathcal{P}_k = \mathcal{P}_k^m \oplus x \mathcal{P}_{k-1}, \quad \mathcal{P}_k = \mathcal{P}_k^h \oplus x^2 \mathcal{P}_{k-2},$$

where $x \mathcal{P}_{k-1} := \{x \vartriangle P(x) ; P \in \mathcal{P}_{k-1}\}$, as well as

$$\mathcal{P}_k^h = \mathcal{P}_k^m \oplus x \mathcal{P}_{k-1}^m, \quad k \geq 1, \quad \mathcal{P}_0^h = \mathcal{P}_0^m.$$

Proof. (i) The key observation is that

$$\mathcal{P}_k \to \mathcal{P}_{k-1} : P(x) \mapsto \nabla \vartriangle P(x) \quad \text{and} \quad \mathcal{P}_{k-1} \to \mathcal{P}_k : P(x) \mapsto x \vartriangle P(x)$$

are adjoint maps with respect to $\langle \cdot, \cdot \rangle_p$. In fact, the inner product is designed for this purpose. To see this, write $P(x) = \sum_{s,\alpha} P_{\alpha,s} x^\alpha e_s$ and $Q(x) = \sum_{t,\beta} Q_{\beta,t} x^\beta e_t$. Then $\nabla \vartriangle P(x) = \sum_{i=1}^n \sum_{s,\alpha} P_{\alpha,s} \alpha_i x^{\alpha - \delta_i} \epsilon(i, s) e_{i \vartriangle s}$, so that

$$\langle \nabla \vartriangle P, Q \rangle_p = \sum_{i,s,\alpha,t,\beta} P_{\alpha,s} \alpha_i \epsilon(i, s) Q_{\beta,t} \langle x^{\alpha - \delta_i} e_{i \vartriangle s}, x^\beta e_t \rangle$$

$$= \sum_{i,s,\alpha} P_{\alpha,s} \alpha_i \epsilon(i, s) Q_{\alpha - \delta_i, i \vartriangle s} (\alpha - \delta_i)!$$

On the other hand, $x \vartriangle Q(x) = \sum_{i=1}^n \sum_{t,\beta} Q_{\beta,t} x^{\beta + \delta_i} \epsilon(i, t) e_{i \vartriangle t}$, so that

$$\langle P, x \vartriangle Q \rangle_p = \sum_{i,s,\alpha,t,\beta} P_{\alpha,s} Q_{\beta,t} \epsilon(i, t) \langle x^\alpha e_s, x^{\beta + \delta_i} e_{i \vartriangle t} \rangle$$

$$= \sum_{i,s,\alpha} P_{\alpha,s} Q_{\alpha - \delta_i, i \vartriangle s} \epsilon(i, i \vartriangle s) \alpha!.$$

Since $\alpha_i(\alpha - \delta_i)! = \alpha!$ and $\epsilon(i, s) = \epsilon(i, i \vartriangle s)$, the duality follows.

(ii) We note that $\mathcal{P}_k^m = \mathsf{N}(\nabla)$ and that $x\mathcal{P}_{k-1} = \mathsf{R}(x)$. Since the maps are adjoint, these subspaces are orthogonal complements in $\mathcal{P}_k$. Similarly, $\mathcal{P}_k = \mathcal{P}_k^h \oplus x^2 \mathcal{P}_{k-2}$, since $(\nabla^2)^* = x^2$. Finally, we consider the map $\mathcal{P}_k^h \to \mathcal{P}_{k-1}^m : P(x) \mapsto \nabla \vartriangle P(x)$. This is well defined, since $\nabla \vartriangle P$ is monogenic if P is harmonic. The adjoint operator will be $\mathcal{P}_{k-1}^m \to \mathcal{P}_k^h : Q(x) \mapsto x \vartriangle Q(x)$, provided $x \vartriangle Q$ is harmonic whenever Q is monogenic. To verify that this is indeed the case, we calculate as in the proof of Proposition 8.1.16 that

$$\mathbf{D}^2(xQ) = \mathbf{D}(nQ + (2\partial_x - x\mathbf{D})Q)$$
$$= n\mathbf{D}Q + 2(\mathbf{D} + \partial_x\mathbf{D})Q - (n\mathbf{D}Q + (2\partial_x - \mathbf{D})\mathbf{D}Q) = (2 + \mathbf{D})\mathbf{D}Q.$$

This proves that $\mathcal{P}_k^m = \mathsf{N}(\nabla)$ is the orthogonal complement to $x\mathcal{P}_{k-1}^m = \mathsf{R}(x)$ in $\mathcal{P}_k^h$. $\qquad\square$

Corollary 8.2.4 (Dimensions). *Let X be an n-dimensional Euclidean space. Then*

$$\dim \mathcal{P}_k^s = \binom{k + n - 1}{n - 1}, \quad \dim \mathcal{P}_k = 2^n \dim \mathcal{P}_k^s,$$

$$\dim \mathcal{P}_k^{em} = 2^{n-1}(\dim \mathcal{P}_k^s - \dim \mathcal{P}_{k-1}^s), \quad \dim \mathcal{P}_k^m = 2 \dim \mathcal{P}_k^{em},$$

$$\dim \mathcal{P}_k^{sh} = \dim \mathcal{P}_k^s - \dim \mathcal{P}_{k-2}^s, \quad \dim \mathcal{P}_k^h = 2^n \dim \mathcal{P}_k^{sh}.$$

Proof. To find $\dim \mathcal{P}_k^s$, note that this is the number of monomials of degree k in n variables. The standard combinatorial argument is as follows. Choose $n - 1$ of the numbers

$$1, 2, 3, \dots, k + n - 1,$$

say $1 \le m_1 < m_2 < \cdots < m_{n-1} \le k + n - 1$. This can be done in $\binom{k+n-1}{n-1}$ ways. Such choices $\{m_i\}$ are in one-to-one correspondence with monomials

$$x_1^{m_1-1} x_2^{m_2-m_1-1} x_3^{m_3-m_2-1} \cdots x_n^{k+n-1-m_{m-1}}.$$

From Proposition 8.2.3 the remaining formulas follow. $\qquad\square$

Exercise 8.2.5 (Two and three dimensions). Let V be a two-dimensional Euclidean space. In this case $\dim \mathcal{P}_k^{sh} = 2 = \dim \mathcal{P}_k^{em}$. Show that $\dim \mathcal{P}_k^{em}$ is a one-dimensional complex linear space with the geometric complex structure $j = e_{12} \in \Delta^2 V$. Find bases for these spaces using the complex powers $z^k = (x + jy)^k$. Identifying vectors and complex numbers as in Section 3.2, write the splitting $\mathcal{P}_k^h = \mathcal{P}_k^m \oplus x\mathcal{P}_{k-1}^m$ in complex notation.

Let V be a three-dimensional Euclidean space. In this case, $\dim \mathcal{P}_k^{sh} = 2k+1$ and $\dim \mathcal{P}_k^{em} = 4(k + 1)$. Find bases for the spherical harmonics $\mathcal{P}_k^{sh}$ and for the spherical monogenics $\mathcal{P}_k^{em}$. Note that $\mathcal{P}_k^{em}$ is a right vector space over $\mathbf{H}$ of dimension $k + 1$.

Recall from Fourier analysis that the trigonometric functions $\{e^{ik\theta}\}_{k\in\mathbf{Z}}$, suitably normalized, form an ON-basis for $L_2(S)$ on the unit circle $S \subset \mathbf{C} = \mathbf{R}^2$ in the complex plane. Thus every $f \in L_2(S)$ can be uniquely written

$$f(e^{i\theta}) = \sum_{k=-\infty}^{\infty} a_k e^{ik\theta}.$$

For $k \geq 0$, the function $e^{ik\theta}$ extends to the analytic function z^k on the disk $|z| < 1$. For $k < 0$, the function $e^{ik\theta}$ extends to the analytic function z^k on $|z| > 1$, which vanishes at ∞, or alternatively to the antianalytic and harmonic function $\bar{z}^{-k}$ on $|z| < 1$. In higher dimensions, we have the following analogue.

Theorem 8.2.6. *Let S be the unit sphere in an n-dimensional Euclidean space. The subspaces $\mathcal{P}_k^h(S)$, $k = 0, 1, 2, \ldots$, of spherical harmonics are pairwise orthogonal with respect to the $L_2(S)$ inner product*

$$\langle F, G \rangle := \int_S \langle F(x), G(x) \rangle dx.$$

Moreover, within each $\mathcal{P}_k^h(S)$, the two subspaces $\mathcal{P}_k^m(S)$ and $x\mathcal{P}_{k-1}^m(S)$ are orthogonal, and $x\mathcal{P}_{k-1}^m(S) = \mathcal{P}_{2-n-k}^m(S)$. The Hilbert space $L_2(S)$ splits into finite-dimensional subspaces as

$$L_2(S) = \bigoplus_{k=0}^{\infty} \mathcal{P}_k^h(S) = \bigoplus_{k=0}^{\infty} \mathcal{P}_k^m(S) \oplus \bigoplus_{k=-\infty}^{-(n-1)} \mathcal{P}_k^m(S).$$

Proof. Let $P \in \mathcal{P}_k^h(S)$ and $Q \in \mathcal{P}_l^h(S)$ with $k \neq l$. Green's second theorem, as in Example 7.3.11, shows that

$$\int_S \left(\langle \partial_x P(x), Q(x) \rangle - \langle P(x), \partial_x Q(x) \rangle \right) dx = \int_{|x|<1} \left(\langle \Delta P, Q \rangle - \langle P, \Delta Q \rangle \right) dx = 0.$$

But for every field $F(x)$ homogeneous of degree s, the radial derivative is $\partial_x F(x) = r\partial_r r^s F(x/r) = sr^s F(x/r) = sF(x)$. Thus we obtain $(k - l) \int_S \langle P, Q \rangle dx = 0$.

To prove the orthogonal splitting of $\mathcal{P}_k^h(S)$, assume that $P \in \mathcal{P}_k^m(S)$ and $Q \in x\mathcal{P}_k^m(S)$. The Stokes formula (7.4) with $\theta(x, v) = \langle P(x), v \vartriangle Q(x) \rangle$ shows that

$$\int_S \langle P(x), xQ(x) \rangle dx = \int_{|x|<1} \left(\langle P(x), \mathbf{D}Q(x) \rangle + \langle \mathbf{D}P(x), Q(x) \rangle \right) dx = 0,$$

since the normal vector is $\nu(x) = x$ on S.

We also note that the monogenic Kelvin transform of $P(x) \in \mathcal{P}_{k-1}$ is

$$K^m P(x) = \frac{x}{|x|^n} P(x/|x|^2).$$

The restriction to S coincides with $xP(x)$, which shows that $\mathcal{P}^m_{2-n-k}(S) = x\mathcal{P}^m_{k-1}(S)$.

What is left to prove is that the spherical harmonics $\mathcal{P}^h_k(S)$ span $L_2(S)$. Let $P \in \mathcal{P}_k$. According to Proposition 8.2.3, we can write

$$P(x) = P_k(x) + |x|^2 P_{k-2}(x) + |x|^4 P_{k-4}(x) + \cdots, \quad \text{where } P_{k-2j}(x) \in \mathcal{P}^h_{k-2j}.$$

Restricting to S, this shows that P, and therefore every inhomogeneous polynomial, can be written as a finite sum of spherical harmonics. By the Stone–Weierstrass theorem, the polynomials are dense in $L_2(S)$, which proves the theorem. $\qquad\square$

The spaces $\mathcal{P}^h_k(S)$ and $\mathcal{P}^m_k(S)$ can be quite large, although they are always finite-dimensional. We next construct higher-dimensional analogues of the cosine functions and the complex exponential functions respectively among all these spherical harmonics and monogenics.

Proposition 8.2.7 (Zonal functions). *For each $k \in \mathbf{N}$, there exists a unique function $Z_k : X \times X \to \Delta^0 V : (x,y) \mapsto Z_k(x,y)$ such that $Z_k(\cdot,x) \in \mathcal{P}^h_k$ for each $x \in X$ and*

$$P(x) = \int_S Z_k(y,x)P(y)dy, \quad \text{for all } P \in \mathcal{P}^h_k,\ x \in X.$$

For each $k \in \mathbf{N} \cup (-n-1-\mathbf{N})$, there exists a unique function $X_k : X \times X \to \Delta^{ev}V : (x,y) \mapsto X_k(x,y)$ such that $X_k(\cdot,x) \in \mathcal{P}^m_k$ for each $x \in X$ and

$$P(x) = \int_S \overline{X_k(y,x)}P(y)dy, \quad \text{for all } P \in \mathcal{P}^m_k,\ x \in X.$$

Extending these functions to k-homogeneous functions of the first argument, we have $Z_k(x,y) = Z_k(y,x)$ and $\overline{X_k(x,y)} = X_k(y,x)$ for all $x,y \in X$. The zonal functions $Z_k(y,x)$ and $X_k(y,x)$ are rotation invariant in the sense that

$$Z_k(qy\bar{q}, qx\bar{q}) = Z_k(y,x), \quad X_k(qy\bar{q}, qx\bar{q}) = qX_k(y,x)\bar{q},$$

for all $q \in \mathrm{Spin}(V), x,y \in X$.

Proof. To construct $Z_k(y,x)$, we consider the Euclidean space $\mathcal{P}^{sh}_k$ with inner product $\langle P,Q \rangle := \int_S P(x)Q(x)dx$. For fixed $x \in X$, consider the linear functional $\theta_x \in (\mathcal{P}^{sh}_k)^*$ defined by

$$\theta_x(P) := P(x).$$

By Proposition 1.2.3 there exists a unique $Z_k(\cdot,x) \in \mathcal{P}^{sh}_k$ that represents $\theta_x \in (\mathcal{P}^{sh}_k)^*$, as claimed. That the representation holds for all $P \in \mathcal{P}^h_k$ is seen by applying it to each component function of P.

Similarly, to construct $X_k(y,x)$ we consider the Euclidean space $\mathcal{P}^{em}_k$ with inner product $\langle P,Q \rangle := \int_S \langle P(x),Q(x) \rangle dx$. For fixed $x \in X$, consider the linear functional $\theta_x \in (\mathcal{P}^{em}_k)^*$ defined by

$$\theta_x(P) := \langle 1, P(x) \rangle.$$

By Proposition 1.2.3 there exists a unique $X_k(\cdot, x) \in \mathcal{P}_k^{em}$ such that $\theta_x(P) = \int_S \langle X_k(y,x), P(y) \rangle dy$ for all $P \in \mathcal{P}_k^{em}$. This gives

$$P(x) = \sum_s e_s \langle e_s, P(x) \rangle = \sum_s e_s \langle 1, P(x)\bar{e}_s \rangle = \sum_s e_s \int_S \langle X_k(y,x), P(y)\bar{e}_s \rangle dy$$

$$= \int_S \sum_s e_s \langle e_s, \overline{X_k(y,x)} P(y) \rangle dy = \int_S \overline{X_k(y,x)} P(y) dy,$$

for all $P \in \mathcal{P}_k^{em}$. That the representation holds for all $P \in \mathcal{P}_k^m$ is seen by writing $P = P_1 + P_2 v$, with $P_1, P_2 \in \mathcal{P}_k^{em}$ and $v \in V$ a unit vector.

To verify the symmetry property for $X_k(x,y)$, choose $P(x) = X_k(x,z)$. Then

$$\overline{X_k(x,z)} = \int_S \overline{X_k(y,x)} X_k(y,z) dy = \int_S \overline{X_k(y,z)} X_k(y,x) dy = \overline{X_k(z,x)}.$$

The proof for Z_k is similar but easier.

To show the rotation invariance, let $Tx = \bar{q}xq$ be a rotation. According to Proposition 8.1.14, the conformal Kelvin transform $K_T^m P(x) = qP(\bar{q}xq)$ preserves monogenic fields, and thus $\mathcal{P}_k^m$. Therefore $qP(\bar{q}xq) = \int_S X_k(x,y)qP(\bar{q}yq)dy$ for all $P \in \mathcal{P}_k^m$, that is,

$$P(x) = \int_S \bar{q} X_k(qx\bar{q}, qy\bar{q}) q P(y) dy.$$

Uniqueness of X_k shows that $\bar{q} X_k(qx\bar{q}, qy\bar{q}) q = X_k(x,y)$. The proof for Z_k is similar but simpler: $K_T^h P(x) = P(qx\bar{q})$ preserves $\mathcal{P}_k^h$. □

Definition 8.2.8. The function $Z_k(\cdot, x)$ is called the *zonal harmonic* of degree k with pole at x. The function $X_k(\cdot, x)$ is called the *zonal monogenic* of degree k with pole at x.

Exercise 8.2.9. In the complex plane, $n = 2$, use the bases $\{\cos(kt), \sin(kt)\}$ for $\mathcal{P}_k^{sh}(S)$ and $\{e^{jkt}, je^{jkt}\}$ for $\mathcal{P}_k^{em}(S)$, $j = e_{12}$, to deduce that

$$Z_k(e^{jt}, e^{js}) = \pi^{-1} \cos k(t-s), \ k \geq 1, \quad X_k(e^{jt}, e^{js}) = (2\pi)^{-1} e^{jk(t-s)}, \ k \geq 0.$$

Compare Corollary 8.2.11.

We next consider orthogonal projections onto the subspaces appearing in the splitting of $L_2(S)$ into spherical harmonics and monogenics from Theorem 8.2.6. Here Z_k provides the kernel for the projection onto $\mathcal{P}_k^h(S)$ and X_k provides the kernel for the projection onto $\mathcal{P}_k^m(S)$.

Proposition 8.2.10. *For $k \geq 1$, we have*

$$Z_k(x,y) = X_k(x,y) + X_{2-n-k}(x,y), \quad X_{2-n-k}(x,y) = x X_{k-1}(x,y) y, \quad x,y \in S,$$

whereas $Z_0(x,y) = X_0(x,y)$. *The Poisson and Cauchy kernels have absolutely convergent power series expansions*

$$\frac{1}{\sigma_{n-1}} \frac{1 - |x|^2}{|y - x|^n} = \sum_{k=0}^{\infty} Z_k(x,y), \quad \frac{1}{\sigma_{n-1}} \frac{1 - xy}{|y - x|^n} = \sum_{k=0}^{\infty} X_k(x,y),$$

for $|x| < 1, |y| = 1$. *The series converge uniformly on* $\{|x| < r\} \times S$ *for all* $r < 1$.

Proof. Consider the series of zonal monogenics. Note that $\frac{1}{\sigma_{n-1}} \frac{1-xy}{|y-x|^n} = \Psi(y-x)y$, where the normal to the sphere is $\nu(y) = y$, and consider

$$\int_S \Psi(y - x)yf(y)dy - \sum_{k=0}^{\infty} \int_S X_k(x,y)f(y)dy$$

for $|x| < 1$. If $f \in \mathcal{P}_j^m(S)$, $j \geq 0$, then this equals $f(x) - f(x) = 0$ by Theorem 8.1.8 and the orthogonality of the spaces of spherical monogenics. Note that only the term $k = j$ in the sum is nonzero. If $j \leq 1 - n$, then the difference equals $0 - 0 = 0$. It therefore remains to show that the series converges. We claim that

$$|X_k(x,y)| \lesssim (1 + k)^{n-2}|x|^k|y|^k,$$

from which the convergence follows. To see this, let $\{P_j\}$ be an ON-basis for $\mathcal{P}_k^m(S)$ and let $x \in S$. By the Pythagorean theorem,

$$\|X_k(\cdot, x)\|^2 = \sum_j |\langle X_k(\cdot, x), P_j \rangle|^2 \lesssim \sum_j |P_j(x)|^2,$$

since $\langle X_k(\cdot, x), P_j \rangle = \int_S \langle 1, \overline{X_k(y,x)}P_j(y) \rangle dy = \langle 1, P_j(x) \rangle$. But we have proved that X_k is rotation invariant, thus $\|X_k(\cdot, x)\|$ is independent of x, so integrating both sides shows that $\|X_k(\cdot, x)\|^2$ is dominated by the dimension of $\mathcal{P}_k^m$, which grows like $(1 + k)^{n-2}$ according to Corollary 8.2.4. This gives

$$|X_k(x,y)| = \left| \int_S X_k(x,z)X_k(z,y)dz \right| \lesssim \|X_k(\cdot, x)\|\|X_k(\cdot, y)\| \lesssim (1 + k)^{n-2}|x|^k|y|^k.$$

The proof for the Poisson kernel and the series of zonal harmonics is similar.

Since the Kelvin transform maps $\mathcal{P}_{k-1}^m(S) \to \mathcal{P}_{2-n-k}^m(S) : P(x) \mapsto xP(x)$, it follows from the uniqueness of the zonal monogenics that $X_{2-n-k}(x,y) = xX_{k-1}(x,y)y$. Similarly, writing $\mathcal{P}_k^h \ni P = P_1 + P_2 \in \mathcal{P}_k^m \oplus \mathcal{P}_{2-n-k}^m$, we see that

$$P(x) = P_1(x) + P_2(x) = \int_S X_k(x,y)P_1(y)dy + \int_S X_{2-n-k}(x,y)P_2(y)dy$$

$$= \int_S (X_k(x,y) + X_{2-n-k}(x,y))P(y)dy,$$

since $\int_S X_k(x,y)P_2(y)dy = 0 = \int_S X_{2-n-k}(x,y)P_1(y)dy$ by orthogonality. Uniqueness of $Z_k(x,y)$ shows that $Z_k(x,y) = X_k(x,y) + X_{2-n-k}(x,y)$. $\square$

whereas $Z_0(x, y) = X_0(x, y)$. The Poisson and Cauchy kernels have absolutely convergent power series expansions

$$\frac{1}{\sigma_{n-1}} \frac{1 - |x|^2}{|y - x|^n} = \sum_{k=0}^{\infty} Z_k(x, y), \qquad \frac{1}{\sigma_{n-1}} \frac{1 - xy}{|y - x|^n} = \sum_{k=0}^{\infty} X_k(x, y),$$

for $|x| < 1$, $|y| = 1$. The series converge uniformly on $\{|x| < r\} \times S$ for all $r < 1$.

Proof. Consider the series of zonal monogenics. Note that $\frac{1}{\sigma_{n-1}} \frac{1-xy}{|y-x|^n} = \Psi(y-x)y$, where the normal to the sphere is $\nu(y) = y$, and consider

$$\int_S \Psi(y - x)yf(y)dy - \sum_{k=0}^{\infty} \int_S X_k(x, y)f(y)dy$$

for $|x| < 1$. If $f \in \mathcal{P}_j^m(S)$, $j \geq 0$, then this equals $f(x) - f(x) = 0$ by Theorem 8.1.8 and the orthogonality of the spaces of spherical monogenics. Note that only the term $k = j$ in the sum is nonzero. If $j \leq 1 - n$, then the difference equals $0 - 0 = 0$. It therefore remains to show that the series converges. We claim that

$$|X_k(x, y)| \lesssim (1 + k)^{n-2}|x|^k|y|^k,$$

from which the convergence follows. To see this, let $\{P_j\}$ be an ON-basis for $\mathcal{P}_k^m(S)$ and let $x \in S$. By the Pythagorean theorem,

$$\|X_k(\cdot, x)\|^2 = \sum_j |\langle X_k(\cdot, x), P_j \rangle|^2 \lesssim \sum_j |P_j(x)|^2,$$

since $\langle X_k(\cdot, x), P_j \rangle = \int_S \langle 1, \overline{X_k(y, x)}P_j(y)\rangle dy = \langle 1, P_j(x)\rangle$. But we have proved that X_k is rotation invariant, thus $\|X_k(\cdot, x)\|$ is independent of x, so integrating both sides shows that $\|X_k(\cdot, x)\|^2$ is dominated by the dimension of $\mathcal{P}_k^m$, which grows like $(1 + k)^{n-2}$ according to Corollary 8.2.4. This gives

$$|X_k(x, y)| = \left| \int_S X_k(x, z)X_k(z, y)dz \right| \lesssim \|X_k(\cdot, x)\|\|X_k(\cdot, y)\| \lesssim (1 + k)^{n-2}|x|^k|y|^k.$$

The proof for the Poisson kernel and the series of zonal harmonics is similar.

Since the Kelvin transform maps $\mathcal{P}_{k-1}^m(S) \to \mathcal{P}_{2-n-k}^m(S) : P(x) \mapsto xP(x)$, it follows from the uniqueness of the zonal monogenics that $X_{2-n-k}(x, y) = xX_{k-1}(x, y)y$. Similarly, writing $\mathcal{P}_k^h \ni P = P_1 + P_2 \in \mathcal{P}_k^m \oplus \mathcal{P}_{2-n-k}^m$, we see that

$$P(x) = P_1(x) + P_2(x) = \int_S X_k(x, y)P_1(y)dy + \int_S X_{2-n-k}(x, y)P_2(y)dy$$

$$= \int_S (X_k(x, y) + X_{2-n-k}(x, y))P(y)dy,$$

since $\int_S X_k(x, y)P_2(y)dy = 0 = \int_S X_{2-n-k}(x, y)P_1(y)dy$ by orthogonality. Uniqueness of $Z_k(x, y)$ shows that $Z_k(x, y) = X_k(x, y) + X_{2-n-k}(x, y)$. $\square$

By Proposition 1.2.3 there exists a unique $X_k(\cdot, x) \in \mathcal{P}_k^{em}$ such that $\theta_x(P) = \int_S \langle X_k(y,x), P(y) \rangle dy$ for all $P \in \mathcal{P}_k^{em}$. This gives

$$P(x) = \sum_s e_s \langle e_s, P(x) \rangle = \sum_s e_s \langle 1, P(x)\overline{e}_s \rangle = \sum_s e_s \int_S \langle X_k(y,x), P(y)\overline{e}_s \rangle dy$$

$$= \int_S \sum_s e_s \langle e_s, \overline{X_k(y,x)}P(y) \rangle dy = \int_S \overline{X_k(y,x)}P(y)dy,$$

for all $P \in \mathcal{P}_k^{em}$. That the representation holds for all $P \in \mathcal{P}_k^m$ is seen by writing $P = P_1 + P_2 v$, with $P_1, P_2 \in \mathcal{P}_k^{em}$ and $v \in V$ a unit vector.

To verify the symmetry property for $X_k(x,y)$, choose $P(x) = X_k(x,z)$. Then

$$\overline{X_k(x,z)} = \int_S \overline{X_k(y,x)}X_k(y,z)dy = \int_S \overline{X_k(y,z)}X_k(y,x)dy = \overline{X_k(z,x)}.$$

The proof for Z_k is similar but easier.

To show the rotation invariance, let $Tx = \overline{q}xq$ be a rotation. According to Proposition 8.1.14, the conformal Kelvin transform $K_T^m P(x) = qP(\overline{q}xq)$ preserves monogenic fields, and thus $\mathcal{P}_k^m$. Therefore $qP(\overline{q}xq) = \int_S X_k(x,y)qP(\overline{q}yq)dy$ for all $P \in \mathcal{P}_k^m$, that is,

$$P(x) = \int_S \overline{q}X_k(qx\overline{q}, qy\overline{q})qP(y)dy.$$

Uniqueness of X_k shows that $\overline{q}X_k(qx\overline{q}, qy\overline{q})q = X_k(x,y)$. The proof for Z_k is similar but simpler: $K_T^h P(x) = P(qx\overline{q})$ preserves $\mathcal{P}_k^h$. $\qquad\square$

Definition 8.2.8. The function $Z_k(\cdot, x)$ is called the *zonal harmonic* of degree k with pole at x. The function $X_k(\cdot, x)$ is called the *zonal monogenic* of degree k with pole at x.

Exercise 8.2.9. In the complex plane, $n = 2$, use the bases $\{\cos(kt), \sin(kt)\}$ for $\mathcal{P}_k^{sh}(S)$ and $\{e^{jkt}, je^{jkt}\}$ for $\mathcal{P}_k^{em}(S)$, $j = e_{12}$, to deduce that

$$Z_k(e^{jt}, e^{js}) = \pi^{-1}\cos k(t-s), \quad k \geq 1, \qquad X_k(e^{jt}, e^{js}) = (2\pi)^{-1}e^{jk(t-s)}, \quad k \geq 0.$$

Compare Corollary 8.2.11.

We next consider orthogonal projections onto the subspaces appearing in the splitting of $L_2(S)$ into spherical harmonics and monogenics from Theorem 8.2.6. Here Z_k provides the kernel for the projection onto $\mathcal{P}_k^h(S)$ and X_k provides the kernel for the projection onto $\mathcal{P}_k^m(S)$.

Proposition 8.2.10. *For $k \geq 1$, we have*

$$Z_k(x,y) = X_k(x,y) + X_{2-n-k}(x,y), \quad X_{2-n-k}(x,y) = xX_{k-1}(x,y)y, \quad x,y \in S,$$

We can use these expansions of the Poisson and Cauchy kernels to calculate explicit expressions for the zonal functions. This uses the *Gegenbauer polynomials*

$$C_k^\lambda(t) := \sum_{j=0}^{[k/2]} \binom{-\lambda}{k-j}\binom{k-j}{j}(-2t)^{k-2j}, \quad k \in \mathbf{N}, \ \lambda \in \mathbf{R},$$

where $\binom{s}{m} := \frac{s(s-1)\cdots(s-(m-1))}{m!}$. In terms of generating functions, these appear in the expansion

$$\frac{1}{(1-2tr+r^2)^\lambda} = \sum_{k=0}^{\infty} C_k^\lambda(t)r^k$$

of the denominator in the kernels.

Corollary 8.2.11. *For $k \geq 0$, the zonal harmonics and monogenics equal*

$$Z_k(x,y) = \frac{1}{\sigma_{n-1}}\left(C_k^{n/2}(\langle x,y\rangle) - C_{k-2}^{n/2}(\langle x,y\rangle)\right), \quad x,y \in S,$$

$$X_k(x,y) = \frac{1}{\sigma_{n-1}}\left(C_k^{n/2}(\langle x,y\rangle) - xyC_{k-1}^{n/2}(\langle x,y\rangle)\right), \quad x,y \in S,$$

where we set $C_j^{n/2} = 0$ if $j < 0$. In particular $X_k(x,y) \in \mathrm{span}_\mathbf{R}\{1, x \wedge y\} \subset \triangle^0 V \oplus \triangle^2 V$ for all $x,y \in S$, $k \geq 0$.

Proof. We expand the Poisson and Cauchy kernels in power series and identify the homogeneous polynomials in these expansion, using

$$\frac{1}{|y-x|^n} = \frac{1}{(1-2\langle x/|x|,y\rangle|x|+|x|^2)^{n/2}} = \sum_{k=0}^{\infty} C_k^\lambda(\langle x/|x|,y\rangle)|x|^k,$$

$|x| < 1$, $|y| = 1$. Inserted in the series expansions in Proposition 8.2.10, this yields the stated formulas. $\square$

Exercise 8.2.12 (Legendre polynomials). If V is a three-dimensional Euclidean space, show that

$$Z_k(x,y) = \frac{2k+1}{4\pi^2}P_k(\langle x,y\rangle), \quad x,y \in S, k \in \mathbf{N},$$

where P_k denote the Legendre polynomials, that is, the Gegenbauer polynomials $P_k(t) = C_k^{1/2}(t)$.

We next consider the Dirac and Laplace equations in polar coordinates $x = ry$, $y \in S$, $r > 0$. Multiplying the Dirac equation $\nabla \vartriangle F(x) = 0$ by the radial vector field $x = \sum_i x_i e_i$ from the left, we obtain

$$\langle x, \nabla\rangle F(x) + (x \wedge \nabla) \vartriangle F(x) = 0.$$

The first term is the radial derivative $\partial_x F = r\partial_y F = r\partial_r F$ multiplied by r. Expressing ∇ in an ON-basis with one basis vector equal to y at the point x reveals that the term $(x \wedge \nabla) \vartriangle F(x)$ acts only by derivatives tangential to the sphere rS.

Definition 8.2.13 (Spherical Dirac). The *spherical Dirac operator* is the first order partial differential operator

$$\mathbf{D}_S F(x) := -(x \wedge \nabla) \vartriangle F(x), \quad x \in V.$$

Restricted to the sphere S, $\mathbf{D}_S$ defines an operator acting on fields $F : S \to \vartriangle V$.

Exercise 8.2.14. In two-dimensional Euclidean space $\mathbf{R}^2 = \mathbf{C}$, introduce polar coordinates (r, θ). Show that $\mathbf{D}_S = \frac{1}{j}\partial_\theta$, where $j = e_{12}$ is the geometric imaginary unit.

In three-dimensional oriented Euclidean space, relate $\mathbf{D}_S$ to the operator $-i\hbar x \times \nabla$ used in quantum mechanics to describe orbital angular momentum.

In particular, if $F \in \mathcal{P}_k^m$, then $\partial_x F(x) = kF(x)$ by homogeneity, so that

$$\mathbf{D}_S F = kF.$$

Thus the spaces $\mathcal{P}_k^m(S)$ of spherical monogenics are eigenspaces to the operator $\mathbf{D}_S$.

A better-known differential operator on the sphere is the *spherical Laplace operator* Δ_S. In polar coordinates for the Laplace operator, we have

$$\Delta u = \partial_r^2 u + (n-1)r^{-1}\partial_r u + r^{-2}\Delta_S u, \tag{8.2}$$

at $x = ry$, $y \in S$, $r > 0$. For the two spherical operators we have the following.

Proposition 8.2.15. *Let V be an n-dimensional Euclidean space, and consider the Hilbert space $L_2(S)$ on the unit sphere S. Then $\mathbf{D}_S$ defines a self-adjoint operator in $L_2(S)$ with spectrum*

$$\sigma(\mathbf{D}_S) = \mathbf{Z} \setminus \{-(n-2), \dots, -1\}.$$

The spherical Laplace operator equals

$$\Delta_S = \mathbf{D}_S(2 - n - \mathbf{D}_S).$$

In the splitting into spherical harmonics, $L_2(S) = \oplus_{k=0}^{\infty} \mathcal{P}_k^h(S)$, the spherical Laplace operator acts according to

$$\Delta_S\left(\sum_{k=0}^{\infty} f_k\right) = \sum_{k=0}^{\infty} k(2 - n - k)f_k,$$

whereas in the splitting into spherical monogenics,

$$L_2(S) = \bigoplus_{k=0}^{\infty} \mathcal{P}_k^m(S) \oplus \bigoplus_{k=-\infty}^{-(n-1)} \mathcal{P}_k^m(S),$$

the spherical Dirac operator acts according to

$$\mathbf{D}_S\left(\sum_{k=0}^{\infty} f_k + \sum_{k=-\infty}^{-(n-1)} f_k\right) = \sum_{k=0}^{\infty} k f_k + \sum_{k=-\infty}^{-(n-1)} k f_k.$$

Proof. It remains to prove that $\Delta_S = \mathbf{D}_S(2 - n - \mathbf{D}_S)$. Using polar coordinates $x = ry$, $y \in S$, we note that $\mathbf{D} = r^{-1}yx\mathbf{D} = r^{-1}y(\partial_x - \mathbf{D}_S) = y\partial_r - r^{-1}y\mathbf{D}_S$. Squaring this Euclidean Dirac operator, we get

$$\begin{aligned} \Delta = \mathbf{D}^2 &= (y\partial_r - r^{-1}y\mathbf{D}_S)^2 \\ &= y\partial_r y\partial_r - y\partial_r r^{-1}y\mathbf{D}_S - r^{-1}y\mathbf{D}_S y\partial_r + r^{-1}y\mathbf{D}_S r^{-1}y\mathbf{D}_S \\ &= \partial_r^2 - \partial_r r^{-1}\mathbf{D}_S - r^{-1}y\mathbf{D}_S y\partial_r + r^{-2}y\mathbf{D}_S y\mathbf{D}_S. \end{aligned}$$

Writing $[A, B] = AB - BA$ for the commutator of operators, we have used that $[\partial_r, y] = 0$ and $[\mathbf{D}_S, r] = 0$. To simplify further, we compute that $[\partial_r, r] = 1$ and

$$[\partial_x, \mathbf{D}_S] = [\partial_x, \partial_x - x\mathbf{D}] = -[\partial_x, x\mathbf{D}] = 0.$$

Thus $\partial_r r^{-1}\mathbf{D}_S = -r^{-2}\mathbf{D}_S + r^{-1}\mathbf{D}_S\partial_r$, so that

$$\Delta = \partial_r^2 - r^{-1}(\mathbf{D}_S + y\mathbf{D}_S y)\partial_r + r^{-2}(\mathbf{D}_S + y\mathbf{D}_S y\mathbf{D}_S).$$

Comparing this equation and (8.2), we see that $n - 1 = -(\mathbf{D}_S + y\mathbf{D}_S y)$ and

$$\Delta_S = \mathbf{D}_S + y\mathbf{D}_S y\mathbf{D}_S = \mathbf{D}_S + (1 - n - \mathbf{D}_S)\mathbf{D}_S = \mathbf{D}_S(2 - n - \mathbf{D}_S),$$

as claimed. $\qquad\square$

In three dimensions, it is standard to introduce spherical coordinates (r, θ, ϕ), and $\mathbf{D}_S$ and Δ_S can be expressed in terms of ∂_θ and ∂_ϕ. The classical expression for the spherical harmonics, obtained by separation of variables, is

$$r^k P_k^m(\cos\theta)e^{im\phi},$$

where the $P_k^m(t)$ denote the associated Legendre polynomials, $m = -k, \ldots, -1, 0, 1, \ldots, k$.

The optimal parametrization of the sphere S, though, uses stereographic projection, which is conformal and has only one singular point for the coordinate system.

Proposition 8.2.16 (Stereographic projection of $\mathbf{D}_S$). *Fix an $(n-1)$-dimensional subspace $V_S \subset V$ and a point $p \in S$ orthogonal to V_S, and consider the stereographic parametrization*

$$T : V_S \to S : y \mapsto T(y) = (py + 1)(y - p)^{-1},$$

as in (4.4). The monogenic Kelvin transform associated to the stereographic projection T defines an isometry of Hilbert spaces

$$2^{(n-1)/2} K_T^m : L_2(S) \to L_2(V_S),$$

and the spherical Dirac operator corresponds to

$$(K_T^m \mathbf{D}_S (K_T^m)^{-1}) G(y) = -\tfrac{1}{2} p((|y|^2 + 1)\mathbf{D}_y + (y - p)) G(y), \quad y \in V_S,$$

where $\mathbf{D}_y$ denotes the Dirac operator in the Euclidean space V_S.

Proof. According to Proposition 8.1.14, the Kelvin transform

$$K_T^m F(y) = \frac{y - p}{|y - p|^n} F((py + 1)(y - p)^{-1})$$

satisfies $\mathbf{D}(K_T^m F)(y) = -2|y - p|^{-2} K_T^m (\mathbf{D}F)(y)$. From the definition of $\mathbf{D}_S$ we get $K_T^m(\mathbf{D}_S F) = K_T^m(\partial_x F) - K_T^m(x\mathbf{D}F)$, where

$$\begin{aligned}
K_T^m(x\mathbf{D}F)(y) &= \frac{y - p}{|y - p|^n}(py + 1)(y - p)^{-1}(\mathbf{D}F)(T(y)) \\
&= (y - p)^{-1}(py + 1) K_T^m(\mathbf{D}F)(y) \\
&= -\tfrac{1}{2}(y - p)(py + 1)\mathbf{D}(K_T^m F)(y) = \tfrac{1}{2}(1 + |y|^2) p \mathbf{D}(K_T^m F)(y).
\end{aligned}$$

To rewrite $K_T^m(\partial_x F)$, we observe that the vertical derivative of the stereographic parametrization at $y \in V_S$ is

$$\underline{T}_y(p) = \frac{-2}{1 + |y|^2}(y - p)p(y - p)^{-1} = \frac{2}{1 + |y|^2} x \tag{8.3}$$

according to Exercise 4.5.18. Thus the chain and product rules give

$$\begin{aligned}
K_T^m(\partial_x F)(y) &= \frac{y - p}{|y - p|^n}(\partial_x F)(T(y)) = \frac{y - p}{|y - p|^n} \frac{1 + |y|^2}{2} \partial_{y_n}(F(T(y))) \\
&= \frac{1 + |y|^2}{2}\left(\partial_{y_n} K_T^m F(y) - \frac{p}{|y - p|^n} F(T(y))\right) \\
&= \frac{1 + |y|^2}{2}\partial_{y_n} K_T^m F(y) - \frac{p}{2}(y - p) K_T^m F(y).
\end{aligned}$$

Here ∂_{y_n} is the partial derivative in the direction p. Since $p\mathbf{D} = \partial_{y_n} + p\mathbf{D}_y$, we obtain the stated formula.

To show that the stated map is a Hilbert space isometry, note that by (8.3) the Jacobian is $J_T(y) = (2/(1 + |y|^2))^{n-1}$, since T is conformal. Thus

$$\int_S |F(x)|^2 dx = \int_{V_S} |F(T(y))|^2 \frac{2^{n-1} dy}{(1 + |y|^2)^{n-1}} = 2^{n-1} \int_{V_S} |K_T^m F(y)|^2 dy. \qquad \square$$

8.3 Hardy Space Splittings

Let $D = D^+$ be a bounded Lipschitz domain in Euclidean space X, with boundary ∂D separating it from the exterior unbounded domain $D^- = X \setminus \overline{D}$. Let ν denote the unit normal vector field on ∂D pointing into D^-. The main operator in this section is the *principal value Cauchy integral*

$$Eh(x) := 2\text{p.v.} \int_{\partial D} \Psi(y - x)\nu(y)h(y)\, dy = 2 \lim_{\epsilon \to 0} \int_{\partial D \setminus B(x;\epsilon)} \Psi(y - x)\nu(y)h(y)\, dy,$$

$x \in \partial D$, which appears when we let $x \in \partial D$, rather than $x \in D$, in the Cauchy reproducing formula from Theorem 8.1.8. Here we assume only suitable bounds on h, and in particular we do not assume that h is a restriction of a monogenic field. The factor 2 is a technicality that will ensure that $E^2 = I$. The singularity at $y = x$ in the integral is of order $|y - x|^{1-n}$ on the $(n-1)$-dimensional surface ∂D, which makes the definition and boundedness of E a nontrivial matter, and cancellations need to be taken into account. Due to the strong singularity at $y = x$, we also refer to E as the *Cauchy singular integral*. Ignoring these analytic problems for the moment, we first investigate by formal calculations how E is related to the two limits

$$E^+ h(x) := \lim_{z \in D^+, z \to x} \int_{\partial D} \Psi(y - z)\nu(y)h(y)\, dy, \quad x \in \partial D,$$

and

$$E^- h(x) := \lim_{z \in D^-, z \to x} \int_{\partial D} \Psi(y - z)(-\nu(y))h(y)\, dy, \quad x \in \partial D,$$

in the Cauchy reproducing formula (8.1) for D^+ and D^- respectively. Placing $z = x$ infinitesimally close, but interior, to ∂D, we have for E^+

$$E^+ h(x) = \lim_{\epsilon \to 0} \left(\int_{\Sigma_x^0} \Psi(y - x)\nu(y)h(x)dy + \int_{\Sigma_x^1} \Psi(y - x)\nu(y)h(y)dy \right)$$

$$= \tfrac{1}{2}h(x) + \tfrac{1}{2}Eh(x).$$

where $\Sigma_x^0 := \{y \in D^- \; ; \; |y - x| = \epsilon\}$ and $\Sigma_x^1 := \{y \in \partial D \; ; \; |y - x| > \epsilon\}$. We have here approximated $h(y) \approx h(x)$, changed the integration surface from $\partial D \setminus \Sigma_x^1$ to Σ_x^0 using Stokes's theorem, and used that $\Psi(y - x)\nu(y) = 1/(\sigma_{n-1}\epsilon^{n-1})$ on Σ_x^0 in the first integral. Thus the first term $h/2$ appears when we integrate around the singularity $y = x$ on an infinitesimal half-sphere.

Since $-\nu$ is outward pointing from D^-, a similar formal calculation indicates that $E^- h(x) = \tfrac{1}{2}h(x) - \tfrac{1}{2}Eh(x)$, and we deduce operator relations

$$\tfrac{1}{2}(I + E) = E^+,$$
$$\tfrac{1}{2}(I - E) = E^-,$$
$$E^+ + E^- = I,$$
$$E^+ - E^- = E.$$

Moreover, from Theorem 8.1.8 we conclude that

$$E^+E^+ = E^+,$$
$$E^-E^- = E^-,$$

since $E^{\pm}h$ by definition is the restriction of a monogenic field to ∂D, no matter what h is. This shows that E^+ and E^- are complementary projection operators. For a suitable space of multivector fields $\mathcal{H}$ on ∂D, these projections define a splitting

$$\mathcal{H} = E^+\mathcal{H} \oplus E^-\mathcal{H}.$$

This means that any given field h on ∂D can be uniquely written as a sum $h = h^+ + h^-$, where h^+ is the restriction to ∂D of a monogenic field in D^+ and h^- is the restriction to ∂D of a monogenic field in D^- that decays at ∞. We refer to $E^{\pm}\mathcal{H}$ as *Hardy subspaces*, and to $E^{\pm}$ as *Hardy projections*. Note also the structure of the Cauchy singular integral operator $E = E^+ - E^-$: it reflects the exterior Hardy subspace $E^-\mathcal{H}$ across the interior Hardy subspace $E^+\mathcal{H}$. In particular, $E^2 = I$, as claimed.

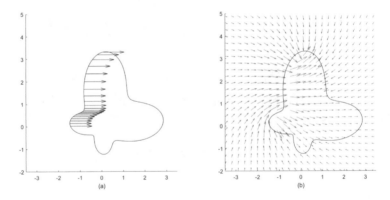

Figure 8.1: (a) The piecewise constant vector field $h : \partial D \to \triangle^1\mathbf{R}^2$ which equals e_1 in the second quadrant and vanishes on the rest of the curve ∂D. (b) The Hardy splitting of h as the sum of two traces of divergence- and curl-free vector fields.

Example 8.3.1 (Constant-curvature boundaries). The most natural space for the singular integral operator E is $\mathcal{H} = L_2(\partial D)$. In the simplest case, in which D is the upper complex half-plane, with ∂D the real axis, then E is a convolution singular integral, which under the Fourier transform corresponds to multiplication by

$$\text{sgn}(\xi) = \begin{cases} 1, & \xi > 0, \\ -1, & \xi < 0, \end{cases}$$

at least if h takes values in the even subalgebra and we use the geometric imaginary unit j as in Example 8.1.10.

The second simplest example is that in which D is the unit ball $|x| < 1$ as in Theorem 8.2.6. In this case, E^+ projects onto $\bigoplus_{k=0}^{\infty} \mathcal{P}_k^m(S)$, whereas E^- projects onto $\bigoplus_{k=-\infty}^{-(n-1)} \mathcal{P}_k^m(S)$.

In these examples the Hardy subspaces are orthogonal and $\|E^{\pm}\| = 1$. However, unlike Hodge splittings, the splitting into Hardy subspaces is not orthogonal for more general domains D. When ∂D has some smoothness beyond Lipschitz, Fourier methods apply to prove that E is a bounded operator on $L_2(\partial D)$, which geometrically means that the angle between the Hardy subspaces, although not straight, is always positive. A breakthrough in modern harmonic analysis was the discovery that this continues to hold for general Lipschitz domains.

Theorem 8.3.2 (Coifman–McIntosh–Meyer). *Let D be a bounded strongly Lipschitz domain. Then the principal value Cauchy integral $Eh(x)$ of any $h \in L_2(\partial D)$ is well defined for almost every $x \in \partial D$, and we have bounds*

$$\int_{\partial D} |Eh(x)|^2 dx \lesssim \int_{\partial D} |h(x)|^2 dx.$$

This is a deep result that is beyond the scope of this book. There exist many different proofs. A singular integral proof is to estimate the matrix for E in a wavelet basis for $L_2(\partial D)$ adapted to ∂D. A spectral proof is to identify $E^{\pm}$ as spectral projections of a Dirac-type operator on ∂D, generalizing the spherical Dirac operator $\mathbf{D}_S$ from Definition 8.2.13. The problem is that for general domains this operator is no longer self-adjoint, but rather has spectrum in a double sector around $\mathbf{R}$, and it becomes a nontrivial matter involving Carleson measures to estimate the spectral projections corresponding to the two sectors. See Section 8.4 for references and further comments. We remark only that from Theorem 8.3.2 one can prove that for $h \in L_2(\partial D)$, the *Cauchy extensions*

$$F^+(x) := \int_{\partial D} \Psi(y - x)\nu(y)h(y)\, dy, \quad x \in D^+, \tag{8.4}$$

and

$$F^-(x) := \int_{\partial D} \Psi(y - x)(-\nu(y))h(y)\, dy, \quad x \in D^-, \tag{8.5}$$

have limits as $x \to \partial D$ both in an $L_2(\partial D)$ sense and pointwise almost everywhere, provided that we approach ∂D in a nontangential way.

In the remainder of this section, we perform a rigorous analysis of the splitting of the space of Hölder continuous multivector fields

$$C^{\alpha}(\partial D) = C^{\alpha}(\partial D; \triangle V), \quad 0 < \alpha < 1,$$

from Example 6.4.1, into Hardy subspaces on a bounded C^1 surface ∂D. This setup is a good starting point for studying Hardy splittings that only requires straightforward estimates. We exclude the endpoint cases $\alpha = 0$, continuous functions, and $\alpha = 1$, Lipschitz continuous functions, for the reason that typically singular integral operators like E are not bounded on these spaces. It is also not bounded on $L_\infty(\partial D)$, something that could be seen in Figure 8.1 if we zoom in at the discontinuities of h.

Proposition 8.3.3 (Hardy projection bounds). *Let D be a bounded C^1 domain and $0 < \alpha < 1$, and assume that $h \in C^\alpha(\partial D)$. Define the Cauchy extensions $F^\pm$ in $D^\pm$ as in (8.4) and (8.5). Then F^+ is a monogenic field in D^+, and F^- is a monogenic field in D^- with decay $F^- = O(|x|^{-(n-1)})$ at ∞. At the boundary ∂D, the traces*

$$f^+(y) := \lim_{x \in D^+, x \to y} F^+(x), \quad f^-(y) := \lim_{x \in D^-, x \to y} F^-(x), \quad y \in \partial D,$$

exist, with estimates $\|f^+\|_\alpha \lesssim \|h\|_\alpha$ and $\|f^-\|_\alpha \lesssim \|h\|_\alpha$. In terms of operators, this means that the Hardy projections $E^\pm : h \mapsto f^\pm$ are bounded on $C^\alpha(\partial D)$.

Proof. (i) That F^+ and F^- are monogenic is a consequence of the associativity of the Clifford product. Indeed, applying the partial derivatives under the integral sign shows that

$$\nabla \vartriangle F^+(x) = \int_{\partial D} \nabla_x \vartriangle \big(\Psi(y-x) \vartriangle \nu(y) \vartriangle h(y)\big) dy$$

$$= \int_{\partial D} \big(\nabla_x \vartriangle (\Psi(y-x))\big) \vartriangle \nu(y) \vartriangle h(y) dy = 0,$$

when $x \notin \partial D$. The decay at infinity follows from the fact that ∂D and h are bounded and the decay of the fundamental solution Ψ.

(ii) We next consider the boundary trace of F^+. A similar argument applies to the trace of F^-. Note that in order to estimate $\|f^+\|_\alpha$, it suffices to estimate $|f^+(x) - f^+(y)| \lesssim |x - y|^\alpha \|h\|_\alpha$ for $|x - y| \le \delta$, provided that $|f^+(x)| \lesssim \|h\|_\alpha$ for all $x \in \partial D$, since ∂D is bounded. Thus we may localize to a neighborhood of a point $p \in \partial D$, in which we can assume that ∂D coincides with the graph of a C^1-function ϕ. We choose a coordinate system $\{x_i\}$ so that p is the origin and ∂D is given by $x_n = \phi(x')$, where $x' = (x_1, \ldots, x_{n-1})$, in the cylinder $|x'| < r$, $|x_n| < s$. Let $\delta < \min(r, s)$ and consider a point $x = (x', x_n) \in D \cap B(0, \delta)$. We claim that

$$|\partial_j F^+(x)| \lesssim \|h\|_\alpha (x_n - \phi(x'))^{\alpha-1}, \quad j = 1, \ldots, n. \tag{8.6}$$

To show this, consider the vertical projection $z = (x', \phi(x'))$ of x onto ∂D, and note that $F^+(x) - h(z) = \int_{\partial D} \Psi(y-x)\nu(y)(h(y) - h(z))dy$, since $\int \Psi(y-x)\nu(y)dy = 1$, according to the Cauchy formula. Thus differentiation with respect to x, with z fixed, gives

$$|\partial_j F^+(x)| \lesssim \|h\|_\alpha \int_{\partial D} |\partial_j \Psi(y-x)| \, |y - z|^\alpha dy = \|h\|_\alpha (I + II).$$

Here I denotes the part of the integral inside the cylinder, and II is the part outside. Since $|\partial_j \Psi(y - x)| \lesssim 1/|y - x|^n$, the term II is bounded. For the integral I, we change variable from $y = (y', \phi(y')) =: \rho(y') \in \partial D$ to $y' \in \mathbf{R}^{n-1}$. To find the change of $(n-1)$-volume, we calculate

$$\rho_{y'}(e_1 \wedge \cdots \wedge e_{n-1}) = (e_1 + (\partial_1 \phi)e_n) \wedge \cdots \wedge (e_{n-1} + (\partial_{n-1}\phi)e_n)$$

$$= e_{1\cdots(n-1)} + (\partial_1 \phi)e_{n2\cdots(n-1)} + (\partial_2 \phi)e_{1n3\cdots(n-1)} + \cdots + (\partial_{n-1}\phi)e_{123\cdots(n-2)},$$

the norm of which is $\sqrt{1 + |\nabla \phi|^2}$. Since the function ϕ is C^1, we conclude that $|\partial_j \Psi(y - x)| \approx 1/(|y' - x'| + t)^n$, $|y - z|^\alpha \approx |y' - x'|^\alpha$, and $dy \approx dy'$, where $t = x_n - \phi(x')$. Therefore

$$I \lesssim \int_{|y'|<r} (|y' - x'| + t)^{-n}|y' - x'|^\alpha dy' \lesssim \int_0^\infty (r + t)^{-n} r^\alpha r^{n-2} dr \lesssim t^{\alpha-1}.$$

(iii) We can now complete the proof, using (8.6). Let $y = (y', \phi(y')) \in \partial D \cap B(0, \delta)$, fix $r > \phi(x')$ and consider first the vertical limit $f^+(y)$. Since

$$|F^+(y', r) - F^+(y', \phi(y') + t)| \leq \int_{\phi(y')+t}^r |\partial_n F^+(y', s)| ds \lesssim \|h\|_\alpha \int_0^{r-\phi(y')} s^{\alpha-1} ds,$$

it is clear that this limit exists, since the integral is convergent. Moreover, we get the estimate $|f^+(y)| \lesssim \|h\|_\alpha$, since $|F^+(y', r)|$ is bounded by $\|h\|_\alpha$.

Next we aim to show that $\{F^+(x)\}$ converges when $x \to y$ from D^+ in general, and not only along the vertical direction. Let $x_1 = (x_1', t_1), x_2 = (x_2', t_2) \in D \cap B(y; \epsilon)$, and define $t := \max(t_1, t_2) + 2\epsilon(1 + \|\nabla\phi\|_\infty)$. Then

$$F^+(x_2) - F^+(x_1) = \int_\gamma \langle dx, \nabla \rangle F^+(x),$$

where γ is the piecewise straight line from x_1 to x_2 via (x_1', t) and (x_2', t). The first and last vertical line integrals are dominated by $\|h\|_\alpha \int_0^\epsilon t^{\alpha-1} dt$ as above, whereas in the middle horizontal line integral, the integrand is dominated by $\|h\|_\alpha \epsilon^{\alpha-1}$. In total we obtain the estimate

$$|F^+(x_2) - F^+(x_1)| \lesssim \|h\|_\alpha \epsilon^\alpha,$$

when $x_1, x_2 \in D \cap B(y, \epsilon)$. This shows the existence of the limit $f^+(y)$ as $x \to y$ from D^+. By taking $x_1, x_2 \in \partial D$, it also shows that $\|f^+\|_\alpha \lesssim \|h\|_\alpha$, which completes the proof. $\square$

Proposition 8.3.4 (Sokhotski–Plemelj jumps). *Let D be a bounded C^1 domain and $0 < \alpha < 1$. Then the Cauchy principal value integral $E : C^\alpha(\partial D) \to C^\alpha(\partial D)$ is a well-defined and bounded linear operator. The Hardy projections $E^\pm$ equal*

$$E^\pm h(x) = \tfrac{1}{2}h(x) \pm \text{p.v.} \int_{\partial D} \Psi(y - x)\nu(y)h(y)\, dy, \quad x \in \partial D.$$

In terms of operators, this means that $E^\pm = \tfrac{1}{2}(I \pm E)$.

Proof. We start by verifying the identity $E^+h(x) = \frac{1}{2}h(x) + \frac{1}{2}Eh(x)$ for $x \in \partial D$. As in the proof of Proposition 8.3.3, write $x = (x', \phi(x'))$ in a coordinate system in a cylinder around x. If $h \in C^\alpha(\partial D)$, the integrand of

$$\int_{\partial D} \Psi(y - (x + te_n))\nu(y)\big(h(y) - h(x)\big)dy$$

is seen to be bounded by $|y - x|^{\alpha - (n-1)}$, uniformly for $0 < t \leq t_0$. Here we view $h(x)$ as a constant function. Letting $t \to 0^+$ and applying the Lebesgue dominated convergence theorem, it follows that

$$E^+h(x) - h(x) = E^+(h - h(x))(x) = \int_{\partial D} \Psi(y - x)\nu(y)\big(h(y) - h(x)\big)dy$$

$$= \lim_{\epsilon \to 0} \int_{\partial D \setminus B(x;\epsilon)} \Psi(y - x)\nu(y)h(y)dy - \left(\lim_{\epsilon \to 0} \int_{\partial D \setminus B(x;\epsilon)} \Psi(y - x)\nu(y)dy \right) h(x).$$

The first equality follows from the fact that the Cauchy integral of the constant field $h(x)$ is the constant field $h(x)$ in D^+. It suffices to show that

$$\lim_{\epsilon \to 0} \int_{\partial D \setminus B(x;\epsilon)} \Psi(y - x)\nu(y)dy = \tfrac{1}{2}.$$

Applying the Cauchy formula for the domain $D^+ \setminus B(x; \epsilon)$ shows that it suffices to prove $\lim_{\epsilon \to 0} \int_{\partial B(x;\epsilon) \cap D^+} \Psi(y - x)\nu(y)dy = \frac{1}{2}$. But

$$\lim_{\epsilon \to 0} \int_{\partial B(x;\epsilon) \cap D^+} \Psi(y - x)\frac{y - x}{|y - x|}dy = \lim_{\epsilon \to 0} \frac{|\partial B(x;\epsilon) \cap D^+|}{|\partial B(x;\epsilon)|}, \qquad (8.7)$$

and on approximating ∂D by its tangent hyperplane at x, this limit is seen to be $1/2$, since ∂D is assumed to be C^1 regular at x.

To summarize, we have shown that $E^+ = \frac{1}{2}(1 + E)$, and Proposition 8.3.3 shows that $E = 2E^+ - I$ is a bounded and well-defined operator. Letting $t \to 0^-$ instead, we get $-E^-h(x) - 0 = Eh(x) - \frac{1}{2}h(x)$, which shows that $E^- = \frac{1}{2}(I - E)$. $\qquad\square$

Exercise 8.3.5. Generalize Proposition 8.3.3 to bounded Lipschitz domains. Show that Proposition 8.3.4 fails for bounded Lipschitz domains.

Summarizing the Hölder estimates in this section, we have the following main result.

Theorem 8.3.6 (Hardy subspace splitting). *Let D be a bounded C^1 domain and let $0 < \alpha < 1$. Then we have a splitting of the Hölder space $C^\alpha(\partial D)$ into Hardy subspaces*

$$C^\alpha(\partial D) = E^+C^\alpha \oplus E^-C^\alpha.$$

The Hardy subspaces are the ranges of the Hardy projections $E^\pm : C^\alpha(\partial D) \to C^\alpha(\partial D)$, which are the spectral projections $E^\pm = \frac{1}{2}(I \pm E)$ of the Cauchy singular integral operator E.

The interior Hardy subspace $E^+ C^\alpha_+$ consists of all traces $F^+|_{\partial D}$ of monogenic fields F^+ in D^+ that are Hölder continuous up to ∂D. The exterior Hardy subspace $E^- C^\alpha$ consists of all traces $F^-|_{\partial D}$ of monogenic fields F^- in D^- that are Hölder continuous up to ∂D and have limit $\lim_{x\to\infty} F^-(x) = 0$. In fact, all such F^- have decay $O(1/|x|^{n-1})$ as $x \to \infty$.

Proof. Proposition 8.3.3 shows that $E^\pm : C^\alpha(\partial D) \to C^\alpha(\partial D)$ are bounded projection operators. Proposition 8.3.4 shows in particular that they are complementary: $E^+ + E^- = I$. This shows that $C^\alpha(\partial D)$ splits into the two Hardy subspaces.

It is clear from the definition and Proposition 8.3.3 that the Hardy subspaces consist of traces of Hölder continuous monogenic fields $F^\pm$ in $D^\pm$ respectively. The decay of F^- at ∞ follows from that of Ψ. Conversely, the fact that the trace of every Hölder continuous monogenic field F^+ in D^+ belongs to $E^+ C^\alpha$ follows from Theorem 8.1.8. For the corresponding result for D^-, we apply the Cauchy reproducing formula to the bounded domain $D^-_R := D^- \cap B(0; R)$ for large R. We have

$$F^-(x) = - \int_{\partial D} \Psi(y - x)\nu(y)F^-(y)dy + \int_{|y|=R} \Psi(y - x)\nu(y)F^-(y)dy, \quad x \in D^-_R.$$

Since $|\partial B(0; R)|$ grows like R^{n-1} and $\Psi(x - y)$ decays like $1/R^{n-1}$ as $R \to \infty$, the last integral will vanish if $\lim_{x\to\infty} F^-(x) = 0$, showing that F^- is the Cauchy integral of $F^-|_{\partial D}$, so that $F^-|_{\partial D} \in E^- C^\alpha$. $\square$

8.4 Comments and References

8.1 The higher-dimensional complex analysis obtained from the Dirac equation and Clifford algebra has been developed since the 1980s. This research field is referred to as *Clifford analysis*. The pioneering work is Brackx, Delanghe, and Sommen [23]. Further references include Gilbert and Murray [42] and Delanghe, Sommen, and Soucek [33].

Div/curl systems like those in Example 8.1.4 have been used to define higher-dimensional harmonic conjugate functions in harmonic analysis. The seminal work is Stein and Weiss [89]

8.2 This material builds on the treatment by Axler, Bourdon, and Ramey [16] of spherical harmonics. We have generalized mutatis mutandis the theory for spherical harmonics to spherical monogenics.

8.3 The classical L_p-based Hardy spaces, named after G.H. Hardy, on the real axis or the unit circle in the complex plane where introduced by F. Riesz in

1923. The function space topologies for $p \leq 1$ that they provide are fundamental in modern harmonic analysis.

Theorem 8.3.2 was proved by R. Coifman, A. McIntosh, and Y. Meyer in [28] for general Lipschitz graphs in the complex plane. Earlier, A. Calderón had obtained a proof in the case of small Lipschitz constants. The higher-dimensional result in Theorem 8.3.2 is equivalent to the L_2 boundedness of the Riesz transforms on Lipschitz surfaces, and this was known already in [28] to follow from the one-dimensional case by a technique called Calderón's method of rotations. A direct proof using Clifford algebra is in [66]. From Calderón–Zygmund theory, also L_p boundedness for $1 < p < \infty$ follows.

A reference for wavelet theory, which is intimitely related to Theorem 8.3.2, is Meyer [69]. It is interesting to note that just like induced bases $\{e_s\}$ for multivectors, wavelet bases for function spaces also do not come with a linear order of the basis functions, but these are rather ordered as a tree. For Clifford algebras and wavelets, see Mitrea [71]. Unpublished lecture notes by the author containing the wavelet proof of Theorem 8.3.2 are [81]. The basic idea behind estimating singular integrals like the Cauchy integral using wavelets is simple: the matrices of such operators in a wavelet basis are almost diagonal in a certain sense. However, the nonlinear ordering of the basis elements and the details of the estimates make the proof rather technical.

There is also a much deeper extension to higher dimensions of the result in [28] that was known as the Kato square root problem. It was finally solved affirmatively by Auscher, Hofmann, Lacey, McIntosh, and Tchamitchian [6] 40 years after it was formulated by Kato, and 20 years after the one-dimensional case [28] was solved. As McIntosh used to tell the story, the works on linear operators by T. Kato and J.-L. Lions closed that field of research in the 1960s; only one problem remained open, and that was the Kato square root problem.

A reference for a spectral/functional calculus approach to Theorem 8.3.2 is Axelsson, Keith, and McIntosh [12]. See in particular [12, Consequence 3.6] for a proof of Theorem 8.3.2, and [12, Consequence 3.7] for a proof of the Kato square root problem. This paper illustrates well how Dirac operators and Hodge- and Hardy-type splittings can be used in modern research in harmonic analysis.

Chapter 9

Dirac Wave Equations

Prerequisites:

Some familiarity with electromagnetism and quantum mechanics is useful for Section 9.2. A background in partial differential equations, see Section 6.3, and boundary value problems is useful but not necessary for the later sections. For the operator theory that we use, the reader is referred to Section 6.4. Ideally, we would have liked to place Section 9.6 after Chapter 10. But since it belongs to the present chapter, we ask the reader to consult Chapter 10 for more on Hodge decompositions when needed.

Road map:

Acting with the nabla symbol through the Clifford product $\nabla \vartriangle F(x)$ on multivector fields, or through a representation $\nabla.\psi(x)$ on spinor fields, in Euclidean space we obtain first-order partial differential operators which are square roots of the Laplace operator Δ. However, Paul Dirac first discovered his original equation in 1928 for spin-1/2 massive particles, in the spacetime setting as a square root of the Klein–Gordon equation, that is, the wave equation with a zero-order term

$$\partial_x^2 \psi + \partial_y^2 \psi + \partial_z^2 \psi - c^{-2}\partial_t^2 \psi = \frac{m^2 c^2}{\hbar^2} \psi.$$

The resulting Dirac wave equation $\hbar \nabla.\psi = mc\psi$ describing the free evolution of the wave function for the particle, a spinor field $\psi : W \to \not\triangle W$ in physical spacetime, has been described as one of the most successful and beautiful equations ever. For example, it predicted the existence of antiparticles some years before these were experimentally found in 1932. In Section 9.2 we survey Dirac's equation, as well as Maxwell's equations from the early 1860s, which describes the evolution of the electrical and magnetic fields. We show how, in a very geometric way, the electromagnetic field is a multivector field and that the Maxwell equations, when written in terms of Clifford algebra, form a Dirac wave equation. The four classical

© Springer Nature Switzerland AG 2019
A. Rosén, *Geometric Multivector Analysis*, Birkhäuser Advanced Texts Basler Lehrbücher,
https://doi.org/10.1007/978-3-030-31411-8_9

equations correspond to the four spaces $\wedge^j V$ of homogeneous multivectors in three-dimensional Euclidean space V.

Motivated by applications to Maxwell's equations, Sections 9.3 to 9.7 develop a theory for boundary value problems (BVPs) for Dirac equations, and they show how it applies to electromagnetic scattering. We consider only time-harmonic waves at a fixed frequency. Our abstract setup for a BVP is to consider two splittings of a space $\mathcal{H}$ of functions on the boundary ∂D of the domain D. The first splitting,

$$\mathcal{H} = A^+\mathcal{H} \oplus A^-\mathcal{H},$$

encodes the differential equation and generalizes the splitting into Hardy subspaces $A^+\mathcal{H} = E^+\mathcal{H}$ and $A^-\mathcal{H} = E^-\mathcal{H}$ from Theorem 8.3.6. The second splitting,

$$\mathcal{H} = B^+\mathcal{H} \oplus B^-\mathcal{H},$$

encodes the boundary conditions. Typically the projections $B^\pm$ are pointwise and determined by the normal vector ν. Relating them to the classical boundary value problems for the Laplace operator, in that case B^+ would encode Dirichlet boundary conditions and B^- would encode Neumann boundary conditions. From this point of view of functional analysis, studying BVPs amounts to studying the geometry between these two different splittings. Well-posedness of BVPs will mean that the subspaces $A^\pm\mathcal{H}$ do not intersect the subspaces $B^\pm\mathcal{H}$, and in the optimal case, the two reflection operators A and B, where A generalizes the Cauchy principal value integral, anticommute.

In Section 9.5, we formulate integral equations for solving scattering problems for Dirac equations. The aim is to find singular but not hypersingular integral operators that are both bounded and invertible also on Lipschitz boundaries, whenever the scattering problem considered is well posed. A problem that we need to overcome to find an integral equation which is numerically useful is that we cannot easily discretize spaces like the Hardy spaces $E^\pm\mathcal{H}$, which are defined by a nonlocal integral constraint.

We obtain good integral equations on good function spaces for solving BVPs for the Dirac equation. To apply these to scattering problems for electromagnetic waves, we show in Sections 9.6 and 9.7 how we require a third splitting of the boundary function space $\mathcal{H}$: a boundary Hodge decomposition

$$\mathcal{H} = \mathsf{R}(\Gamma_k) \oplus \mathsf{R}(\Gamma_k^*),$$

where the Maxwell fields live in the Hodge component $\mathsf{R}(\Gamma_k)$. Embedding Maxwell's equations into the Dirac equation, and solving the BVP with a Dirac integral equation, we give examples in Section 9.7 of how this algorithm performs numerically.

Highlights:

- Boosting E and B using Clifford algebra: 9.2.4

- Discovery of antiparticles: 9.2.8

- Stratton–Chu as a Clifford–Cauchy integral: 9.3.8

- Well-posedness via operator Clifford algebra $\triangle \mathbf{R}^2$: 9.4.5

- Rellich spectral sector vs. Lipschitz geometry of boundary: 9.5.1

- Spin integral equation: 9.5.5

- Maxwell fields and boundary Hodge decompositions: 9.7.1

9.1 Wave and Spin Equations

The Dirac operator on a Euclidean space from Definition 8.1.1 generalizes in the obvious way to an inner product space of arbitrary signature.

Definition 9.1.1 ($\triangle$-Dirac operator). Let (X, V) be an inner product space. The $\triangle$-*Dirac operator* $\mathbf{D} = \mathbf{D}_V$ acting on multivector fields $F : D \to \triangle V$ defined on some domain $D \subset V$ is the nabla operator

$$(\mathbf{D}F)(x) := \nabla \vartriangle F(x) = \sum_{j=1}^{n} e_j^* \vartriangle \partial_j F(x)$$

induced by the Clifford product $V \times \triangle V \to \triangle V : (v, w) \mapsto v \vartriangle w$ as in Definition 7.1.2. Here ∂_j are partial derivatives with respect to coordinates in a basis $\{e_j\}$ for V, with dual basis $\{e_j^*\}$.

Since $\triangle$-Dirac operators are our main type of Dirac operator, we sometimes omit $\triangle$ in the notation. When V is a Euclidean space, we speak of a *harmonic Dirac operator*, while if $V = W$ is a spacetime, we speak of a *wave Dirac operator*. For a Euclidean space, we know from Chapter 8 that

$$\mathbf{D}^2 = \Delta.$$

We have seen that this means that multivector fields F solving $\mathbf{D}F = 0$ in particular have scalar component functions F_s that are harmonic. Turning to the wave Dirac operator, we now have

$$\mathbf{D}^2 = \square,$$

where $\square$ is the d'Alembertian from Section 6.3. Indeed, in an ON-basis $\{e_i\}$ we have

$$\mathbf{D}^2 F = (-e_0 \partial_0 + e_1 \partial_1 + \cdots + e_n \partial_n)^2 F = (-\partial_0^2 + \partial_1^2 + \cdots + \partial_n^2) F.$$

Similar to the Euclidean case, it follows that multivector fields solving the wave Dirac equation $\mathbf{D}F = 0$ have scalar component functions F_s that solve the wave equation $\square F_s = 0$.

The wave Dirac operator describes a wave propagation of multivector fields. For the harmonic Dirac operator we saw in Chapter 8 how the fundamental solution Φ to the Laplace operator yielded a fundamental solution $\Psi = \nabla\Phi$ to $\mathbf{D}$. Similarly, the fundamental solution to the wave equation encoded by the Riemann operators R_t from Proposition 6.2.2 and Example 6.3.2 now yield solution formulas for the wave Dirac operator.

Proposition 9.1.2 (Propagation of Dirac waves). *Fix a time-like unit vector e_0 in a spacetime W and let $V = [e_0]^{\perp}$ be the space-like complement. Consider the initial value problem for the wave Dirac equation*

$$\mathbf{D}_W F = G$$

for given initial data $F|_V = f$ and source G. We assume that f and $G_{x_0}(\cdot) = G(x_0, \cdot)$, for each fixed time x_0, belong to $L_2(V; \triangle W)$. Then the solution $F_{x_0}(x) = F(x_0, x)$ is given by

$$F(x_0, x) = M_{x_0} F_0(x) + \int_0^{x_0} M_{x_0 - s}(e_0 G_s)(x) ds, \quad x_0 > 0,$$

where the Fourier multiplier M_{x_0} on $L_2(V; \triangle W)$ is $M_{x_0} g := (\partial_0 - e_0 \mathbf{D}) R_{x_0} g$.

Proof. We apply the partial Fourier transform to $\mathbf{D}_W F = G$ in the x-variables, for each fixed x_0. We obtain the ODE $-e_0 \partial_0 \hat{F}_{x_0}(\xi) + i\xi \hat{F}_{x_0}(\xi) = \hat{G}_{x_0}(\xi)$, $\xi \in V$, with solution

$$\hat{F}_{x_0}(\xi) = \exp(-ie_0\xi x_0)\hat{f}(\xi) + \int_0^{x_0} \exp(-ie_0\xi(x_0 - s))e_0 \hat{G}_s(\xi) ds.$$

We have $\exp(-ie_0\xi x_0) = \cos(|\xi|x_0) - ie_0\xi \sin(|\xi|x_0)/|\xi|$ according to Exercise 1.1.5, which is seen to be the symbol of M_{x_0}. The inverse Fourier transformation yields the stated formula for F. $\square$

It follows that the time evolution of the wave Dirac equation is quite similar to that for the scalar second-order wave equation: with our scaling, the propagation speed is 1, and in odd dimensions there is a Huygens principle. However, although evolution backward in time is well posed, the wave Dirac equation is not symmetric in the time variable, unlike the scalar wave equation. Another difference is that the only inital datum that we need is $F(0, \cdot)$, and no normal derivative.

The second type of Dirac operator that we consider are the following spin-Dirac operators.

Definition 9.1.3 ($\not\triangle$- Dirac operator). Let (X, V) be an inner product space, with complex spinor space $\not\triangle V$. The $\not\triangle$-*Dirac operator* $\not{D} = \not{D}_V$ acting on spinor fields $\Psi : X \to \not\triangle V$ is the nabla operator

$$(\not{D}\Psi)(x) := \nabla.\Psi(x) = \sum_{j=1}^{n} e_j^* . \partial_j \Psi(x),$$

which is induced by the bilinear map $V \times \not{\!\!\Delta} V \to \not{\!\!\Delta} V : (\theta, \psi) \mapsto \theta.\psi$ as in Definition 7.1.2. Here ∂_j are partial derivatives with respect to coordinates in a basis $\{e_j\}$ for V, with dual basis $\{e_j^*\}$.

When V is a Euclidean space, we speak of a *harmonic $\not{\!\!\Delta}$-Dirac operator*, while if $V = W$ is a spacetime, we speak of a *wave $\not{\!\!\Delta}$-Dirac operator*. The $\not{\!\!\Delta}$-Dirac operators are best known for their representations as matrix first-order partial differential operators. Such expressions are straightforward to derive using representations of Clifford algebras. See Section 5.1.

Analogous to the $\triangle$-Dirac operator, for a Euclidean space, we have $\not{D}^2 = \Delta$, while in spacetime $\not{D}^2 = \Box$, and spinor fields solving $\not{D}\Psi = 0$ have harmonic functions and solutions to the wave equation as component functions, respectively.

Exercise 9.1.4 (Hypercomplex spin analysis). Show how the theory from Chapter 8 for the Cauchy integral, the monogenic Kelvin transform, and spherical monogenics generalizes in a natural way for solutions to the harmonic $\not{\!\!\Delta}$-Dirac operator. Explain why such solutions do not form a right Clifford module as in Proposition 8.1.13 and why the notion of two-sided monogenicity from Proposition 8.1.5 does not generalize.

Show how Proposition 9.1.2 generalize to describe the free wave evolution for wave $\not{\!\!\Delta}$-Dirac equations.

We next consider how the wave Dirac equations are related to the harmonic Dirac equations.

Proposition 9.1.5 ($\triangle V$ representation of $\mathbf{D}_W$). *Let W be a spacetime and fix a time-like unit vector e_0 and its Euclidean orthogonal complement $V = [e_0]^\perp$. Identify the even part $\triangle^{\mathrm{ev}} W$ of the spacetime Clifford algebra, and the Euclidean Clifford algebra $\triangle V$ via the isomorphism in Proposition 3.3.5. We write a general spacetime multivector $w \in \triangle W$ as $w = w^+ + e_0 w^-$, where $w^\pm \in \triangle^{\mathrm{ev}} W \leftrightarrow \triangle V$. Identifying a multivector field F in W in this way with a pair (F^+, F^-) of multivector fields in V, and similarly for G, the wave Dirac equation $\mathbf{D}_W F = G$ corresponds to*

$$\begin{cases} (\partial_0 + \mathbf{D}_V)F_+ = -G_-, \\ (\partial_0 - \mathbf{D}_V)F_- = G_+. \end{cases}$$

Proof. Since $\mathbf{D}_W$ swaps $\triangle^{\mathrm{ev}} W$ and $\triangle^{\mathrm{od}} W$ fields, we obtain in a spacetime ON-basis $\{e_i\}$ that $\mathbf{D}_W F = G$ is equivalent to

$$\begin{cases} (-e_0\partial_0 + \mathbf{D}')F_+ = e_0 G_-, \\ (-e_0\partial_0 + \mathbf{D}')(e_0 F_-) = G_+, \end{cases}$$

where $\mathbf{D}' := \sum_{j=1}^n e_j\partial_j$. Multiplying the first equation by e_0 and commuting e_0 to the left in the second equation establishes the claim. $\qquad\Box$

Changing notation, an argument as in the proof of Proposition 9.1.5 in the case that $\dim W$ is even also shows that the wave $\triangle$-Dirac equation $\mathbf{D}_W F = G$ corresponds to

$$\begin{cases} (\partial_0 + \mathbf{D}_V)F_+ = -G_-, \\ (\partial_0 - \mathbf{D}_V)F_- = G_+. \end{cases}$$

Here we use that $\triangle V = \triangle^+ W$ via the representation $\rho : V \to \triangle^+ W : v \mapsto e_0 v.(\cdot)$, and write the spacetime spinor field as $F = F_+ + e_0 F_-$ in the splitting $\triangle^+ W \oplus \triangle^- W$. Note that $e_0 : \triangle^+ W \to \triangle^- W$ is invertible.

We end this section by considering how these Dirac operators are related to exterior and interior derivative operators d and δ. For any inner product space, it is clear from the definitions that the $\triangle$-Dirac operators are

$$\mathbf{D} = d + \delta. \tag{9.1}$$

This holds true also for the wave $\triangle$-Dirac operator. We note the following refinement of Proposition 9.1.5.

Proposition 9.1.6 ($\wedge V$ representation of d_W, δ_W). *Let W be a spacetime and use notation as for $\mathbf{D}$ in Proposition 9.1.5. Then the differential equations $d_W F = G$ and $\delta_W F = G$ correspond to*

$$\begin{cases} (T^+\partial_0 + d_V)F_+ = -G_-, \\ (T^-\partial_0 - d_V)F_- = G_+, \end{cases} \qquad \begin{cases} (T^-\partial_0 + \delta_V)F_+ = -G_-, \\ (T^+\partial_0 - \delta_V)F_- = G_+, \end{cases}$$

respectively, where $T^+ f = \frac{1}{2}(f + \widehat{f})$ and $T^- f = \frac{1}{2}(f - \widehat{f})$ denote the projections onto $\wedge^{\mathrm{ev}} V$ and $\wedge^{\mathrm{od}} V$ respectively.

Proof. For example, we see that $d_W F = G$ is equivalent to

$$\begin{cases} -e_0 \wedge \partial_0 F_+ + d'F_+ = e_0 G_-, \\ -e_0 \wedge \partial_0(e_0 F_-) + d'(e_0 F_-) = G_+, \end{cases}$$

where $d'F := \sum_{j=1}^n e_j \wedge \partial_j F$. To relate the exterior product to the Clifford algebra isomorphism $\triangle^{\mathrm{ev}} W \approx \triangle V$, we use the Riesz formula (3.4). We also note that $w \mapsto -e_0 w e_0$ yields an automorphism of $\triangle^{\mathrm{ev}} W$ that negates $e_0 V$. Therefore $-e_0 w e_0$ corresponds to $\widehat{w}$ under the isomorphism $\triangle^{\mathrm{ev}} W \leftrightarrow \triangle V$. This yields for $F \in \triangle^{\mathrm{ev}} W \leftrightarrow \triangle V$,

$$e_0(e_0 \wedge F) = \tfrac{1}{2}(-F + e_0 F e_0) \leftrightarrow \tfrac{1}{2}(-F - \widehat{F}) = -T^+ F,$$

$$e_0 \wedge (e_0 F) = \tfrac{1}{2}(-F - e_0 F e_0) \leftrightarrow \tfrac{1}{2}(-F + \widehat{F}) = -T^- F,$$

and with nabla calculus using $\nabla' = \sum_{j=1}^n e_j \partial_j$ that

$$e_0(\nabla' \wedge F) = \tfrac{1}{2}((e_0 \nabla')F - e_0 F e_0(e_0 \nabla')) \leftrightarrow \tfrac{1}{2}(\nabla F + \widehat{F}\nabla) = d_V F,$$

$$\nabla' \wedge (e_0 F) = \tfrac{1}{2}(-(e_0 \nabla')F + e_0 F_0(e_0 \nabla')) \leftrightarrow \tfrac{1}{2}(-\nabla F - \widehat{F}\nabla) = -d_V F.$$

Similar calculations using the Riesz formula (3.3) prove the $\triangle V$ representation for $\delta_W F = G$. $\qquad\square$

The $\triangle$-Dirac operator in a general Euclidean or real inner product space, cannot be written as in (9.1). However, in the case of an even-dimensional Euclidean space with a complex structure given as in Example 5.1.5(i), we do have an invariant meaning of such exterior and interior derivative operators.

Given a Euclidean space V of dimension $n = 2m$ with an isometric complex structure J, consider the complex exterior algebra $\wedge \mathcal{V}$ for the complex vector space $\mathcal{V} = (V, J)$, which comes with the corresponding Hermitian inner product $(\cdot, \cdot)$. As in Example 5.1.5(i), the real linear map

$$V \to \mathcal{L}(\wedge \mathcal{V}) : v \mapsto v \lrcorner_* (\cdot) + v \wedge (\cdot)$$

gives a representation of the complex spinor space $\triangle \mathcal{V} = \wedge \mathcal{V}$. But the two terms induce separately the nabla operators

$$\Gamma_1 \psi := \nabla \wedge \psi \quad \text{and} \quad \Gamma_2 \psi := \nabla \lrcorner_* \psi$$

acting on spinor fields $\psi : V \to \triangle \mathcal{V}$ and $\mathbf{D} = \Gamma_1 + \Gamma_2$. Fixing a complex ON-basis $\{e_j\}_{j=1}^m$ for $\mathcal{V}$ and writing x_j for the real coordinates along e_j and y_j for the real coordinates along Je_j, we have, since $\{e_j\} \cup \{Je_j\}$ form a real ON-basis for V from Proposition 7.1.3, that

$$\Gamma_1 \psi = \sum_{j=1}^m e_j \wedge \partial_{x_j} \psi + \sum_{j=1}^m ie_j \wedge \partial_{y_j} \psi = \sum_{j=1}^m e_j \wedge \partial_{z_j^c} \psi$$

and

$$\Gamma_2 \psi = \sum_{j=1}^m e_j \lrcorner_* \partial_{x_j} \psi + \sum_{j=1}^m (ie_j) \lrcorner_* \partial_{y_j} \psi = \sum_{j=1}^m e_j \lrcorner_* \partial_{z_j} \psi,$$

since $Je_j = ie_j$ in $\mathcal{V}$ and $(ie_j)\lrcorner_* w = -i(e_j \lrcorner_* w)$ by sesquilinearity. Here we used the classical complex analysis operators $\partial_{z_j} := \partial_{x_j} - i\partial_{y_j}$ and $\partial_{z_j^c} := \partial_{x_j} + i\partial_{y_j}$. Since $\Gamma_1^* = -\Gamma_2$, one can develop a complex version of the theory of Hodge decomposition similar to the real theory in Chapter 10.

9.2 Dirac Equations in Physics

The aim of this section is to briefly review how Dirac equations appear in electromagnetic theory and quantum mechanics in physics. We model our universe by spacetime W with three space dimensions, as in special relativity. See Section 1.3. The unit of length is the meter [m]. Fixing a future-pointing time-like vector e_0 with $e_0^2 = -1$, we write V for the three-dimensional Euclidean space $[e_0]^\perp$. We write the coordinate along e_0 as

$$x_0 = ct,$$

where t is time measured in seconds [s] and $c \approx 2.998 \cdot 10^8$ [m/s] is the speed of light. Our discussion uses SI units. Out of the seven SI base units, we need the meter [m] for length, kilogram [kg] for mass, second [s] for time and ampere [A] for electric current. From this we have the SI derived units newton [N= kg·m/s²] for force, coulomb [C=A·s] for electric charge, volt [V= N·m/C] for electric potential, joule [J= Nm] for energy.

We consider first Maxwell's equations, which describe the time evolution of the electric and magnetic fields, which mediate the forces that electric charges in motion exert on each other. The charges that generate the electric and magnetic fields are described by a *charge density* and *electric current density*

$$\rho(t,x) \in \wedge^0 V \quad \text{and} \quad J(t,x) \in \wedge^1 V,$$

measured in units [C/m³] and [A/m²] respectively. This means that a given domain $D \subset V$ contains the charge $\int_D \rho dx$, and the electric current through a 2-surface S in V is $\int_S \langle J, *\underline{dy} \rangle$, at time t. Here $[\underline{dy}]$ is the tangent plane to S and $*\underline{dy}$ is an infinitesimal vector normal to S, in the direction in which we measure the current.

Maxwell's four equations, which we discuss below, describe how ρ and J generate a vector field

$$E(t,x) \in \wedge^1 V, \quad \text{measured in units } [\text{N/C} = \text{V/m}],$$

which is called the *electric field*, and a bivector field

$$B(t,x) \in \wedge^2 V, \quad \text{measured in units of tesla } [\text{T} = \text{N}/(\text{A} \cdot \text{m})],$$

which we refer to as the *magnetic field*. The way we measure these fields is by placing a test charge with charge q_0 at the point moving with velocity $v_0 \in \wedge^1 V$. The electric and magnetic fields will then exert a force on this test charge given by the Lorentz force

$$F = q_0 E + q_0 B \llcorner v_0. \tag{9.2}$$

Experiments show that the magnetic force is orthogonal to the velocity, and thus is described by a skew symmetric map. Recalling Proposition 4.2.3, this demonstrates that the magnetic field is a bivector field rather than a vector field. In classical vector notation, the magnetic field is described by the Hodge dual vector field $*B$, in which case the magnetic force is given by the vector product

$$q_0 v_0 \times (*B).$$

The three-dimensional exterior algebra

$$\wedge V = \wedge^0 V \oplus \wedge^1 V \oplus \wedge^2 V \oplus \wedge^3 V$$

provides a natural framework for expressing the four Maxwell equations, or more precisely eight scalar equations, for determining $E \in \wedge^1 V$ and $B \in \wedge^2 V$ from $\rho \in \wedge^0 V$ and $J \in \wedge^1 V$. The constants of proportionality appearing are the *permittivity of free space* $\epsilon_0 \approx 8.854 \cdot 10^{-12}$ [C/(V· m)] and *permeability of free space* $\mu_0 = 4\pi \cdot 10^{-7}$ [V· s/(A· m)].

$\wedge^0$ Gauss's law for the electric field states that the flow of the electric field out through the boundary of a domain D is proportional to the charge $Q = \int_D \rho dx$ contained in the domain:

$$\int_{\partial D} \langle E, *\underline{dy} \rangle = \epsilon_0^{-1} Q.$$

By Stokes's theorem, Gauss's law is equivalent to the $\wedge^0 V$-valued differential equation

$$\epsilon_0 \nabla \lrcorner E = \rho.$$

In classical vector notation this reads $\epsilon_0 \langle \nabla, E \rangle = \rho$.

$\wedge^1$ The Ampère–Maxwell law states that

$$\mu_0^{-1} \int_{\partial S} \langle B, *\underline{dy} \rangle = \int_S \langle J, *\underline{dx} \rangle + \epsilon_0 \partial_t \int_S \langle E, *\underline{dx} \rangle$$

for every 2-surface S. In the stationary case that ρ, J, E, and B are time-independent, it reduces to Ampère's law, which shows that an electric current $I := \int_S \langle J, *\underline{dx} \rangle$ through S produces a magnetic field with circulation $\int_{\partial S} \langle B, *\underline{dy} \rangle = \mu_0 I$. In the nonstationary case, Maxwell added the necessary additional term $\epsilon_0 \mu_0 \partial_t \int_S \langle E, *\underline{dx} \rangle$ to the equation.

By Stokes's theorem, Ampère–Maxwell's law is equivalent to the $\wedge^1 V$-valued differential equation

$$\epsilon_0 \partial_t E + \mu_0^{-1} \nabla \lrcorner B = -J.$$

In classical vector notation this reads $\epsilon_0 \partial_t E - \mu_0^{-1} \nabla \times (*B) = -J$.

$\wedge^2$ Faraday's law of induction states that a change of the integral of the magnetic field B over a 2-surface S induces an electric field around the boundary curve:

$$\int_{\partial S} \langle E, \underline{dy} \rangle = -\partial_t \int_S \langle B, \underline{dx} \rangle.$$

By Stokes's theorem, Faraday's law is equivalent to the $\wedge^2 V$-valued differential equation

$$\partial_t B + \nabla \wedge E = 0.$$

In classical vector notation this reads $\partial_t (*B) + \nabla \times E = 0$.

$\wedge^3$ Gauss's law for magnetic fields states that the integral of a magnetic field over the boundary of a domain D vanishes:

$$\int_{\partial D} \langle B, \underline{dy} \rangle = 0.$$

By Stokes's theorem, the magnetic Gauss's law is equivalent to the $\wedge^3 V$-valued differential equation

$$\nabla \wedge B = 0.$$

In classical vector notation this reads $\langle \nabla, *B \rangle = 0$.

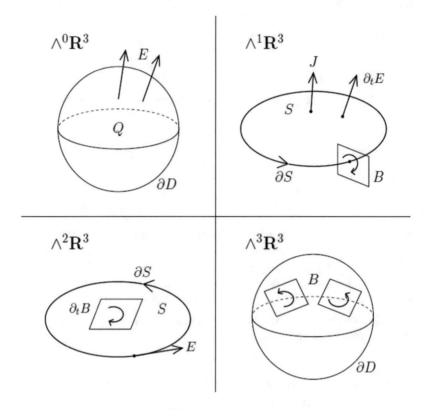

Figure 9.1: Maxwell's equations for the electric vector field E and the magnetic bivector field B in $\wedge \mathbf{R}^3$.

Since the electric and magnetic fields take values in the two different sub-spaces $\wedge^1 V$ and $\wedge^2 V$ of the exterior algebra, we can add them to obtain a six-dimensional total electromagnetic multivector field F. The most natural scaling is such that $|F|^2$ is an energy density, with dimension $[\mathrm{J/m}^3]$. We set

$$F := \epsilon_0^{1/2} E + \mu_0^{-1/2} B \in \wedge^1 V \oplus \wedge^2 V. \tag{9.3}$$

Collecting and rescaling Maxwell's equations, we have

$$\nabla \wedge (\mu_0^{-1/2} B) = 0,$$

$$c^{-1}\partial_t(\mu_0^{-1/2} B) + \nabla \wedge (\epsilon_0^{1/2} E) = 0,$$

$$c^{-1}\partial_t(\epsilon_0^{1/2} E) + \nabla \lrcorner (\mu_0^{-1/2} B) = -\mu_0^{1/2} J,$$

$$\nabla \lrcorner (\epsilon_0^{1/2} E) = \epsilon_0^{-1/2} \rho,$$

where $c = (\epsilon_0 \mu_0)^{-1/2}$. Adding these four equations, we see that Maxwell's equations are equivalent to the Dirac equation

$$c^{-1}\partial_t F + \nabla \vartriangle F = G, \qquad (9.4)$$

since Maxwell's equations take values in the different homogeneous subspaces of $\wedge V$. Here $G := \epsilon_0^{-1/2} \rho - \mu_0^{1/2} J$ is a $\wedge^0 V \oplus \wedge^1 V$-valued multivector field, which we refer to as the *electric four-current*. From (9.4) it is clear that Maxwell's equation is a wave Dirac equation for the $\wedge^1 V \oplus \wedge^2 V$-valued electromagnetic field F.

Example 9.2.1 (Static electromagnetic field). Assume that the sources ρ and J and the electromagnetic field are constant with respect to time, and that J is divergence-free. Then Maxwell's equations reduce to the inhomogeneous Dirac equation $\nabla \vartriangle F = G$, which by the Cauchy–Pompeiu formula from Theorem 8.1.8 has solution $F(x) = \Psi(x) * G(x)$ if G decays as $x \to \infty$. This amounts to

$$E(x) = \frac{1}{4\pi\epsilon_0} \int_V \rho(x - y) \frac{y}{|y|^3} dy,$$

$$B(x) = \frac{\mu_0}{4\pi} \int_V J(x - y) \wedge \frac{y}{|y|^3} dy.$$

Thus E is the Coulomb field from charge density ρ, and B is determined from J by the law of Biot–Savart.

Exercise 9.2.2 (Pauli representation). Using an ON-basis $\{e_1, e_2, e_3\}$ for V, write $\epsilon_0^{1/2} E = \tilde{E}_1 e_1 + \tilde{E}_2 e_2 + \tilde{E}_3 e_3$, $\mu_0^{-1/2} B = \tilde{B}_1 e_{23} + \tilde{B}_2 e_{31} + \tilde{B}_3 e_{12}$, $\mu_0^{1/2} J = \tilde{J}_1 e_1 + \tilde{J}_2 e_2 + \tilde{J}_3 e_3$, and $\epsilon_0^{-1/2} \rho = \tilde{\rho}$. Represent the basis vectors $\{e_1, e_2, e_3\}$ by the Pauli matrices from Example 3.4.19 and show that Maxwell's equations become

$$\begin{bmatrix} c^{-1}\partial_t + \partial_3 & \partial_1 - i\partial_2 \\ \partial_1 + i\partial_2 & c^{-1}\partial_t - \partial_3 \end{bmatrix} \begin{bmatrix} \tilde{E}_3 + i\tilde{B}_3 & \tilde{E}_1 - i\tilde{E}_2 + i\tilde{B}_1 + \tilde{B}_2 \\ \tilde{E}_1 + i\tilde{E}_2 + i\tilde{B}_1 - \tilde{B}_2 & -\tilde{E}_3 - i\tilde{B}_3 \end{bmatrix}$$
$$= \begin{bmatrix} \tilde{\rho} - \tilde{J}_3 & -\tilde{J}_1 + i\tilde{J}_2 \\ -\tilde{J}_1 - i\tilde{J}_2 & \tilde{\rho} + \tilde{J}_3 \end{bmatrix}.$$

Note that this representation requires that the components of the fields be real-valued, since we use a real algebra isomorphism $\triangle_{\mathbf{R}} V \leftrightarrow \mathbf{C}(2)$.

For time-dependent electromagnetic fields we can obtain a spacetime Dirac formulation of Maxwell's equation from Proposition 9.1.5. Namely, the electromagnetic field is really the spacetime bivector field

$$F_W := \epsilon_0^{1/2} e_0 \wedge E + \mu_0^{-1/2} B \in \wedge^2 W,$$

solving the spacetime Dirac equation

$$\mathbf{D}_W F_W = -G_W, \qquad (9.5)$$

where $G_W := \epsilon_0^{-1/2}\rho e_0 + \mu_0^{1/2}J \in \wedge^1 W$ is the spacetime representation of the electric four-current. Since G_W is a spacetime vector field and F_W is a spacetime bivector field, Maxwell's equations can equivalently be written as the system

$$\begin{cases} d_W F_W = 0, \\ \delta_W F_W = -G_W, \end{cases}$$

by the mapping properties of $\mathbf{D}_W = d_W + \delta_W$.

The difference between Maxwell's equations and the Dirac equation is a constraint similar to the one described in Proposition 8.1.5.

Proposition 9.2.3 (Maxwell = Dirac + constraint). *Let* $G_W = \epsilon_0^{-1/2}\rho e_0 + \mu_0^{1/2}J \in \wedge^1 W$. *If* F_W *is a* $\wedge^2 W$-*valued solution to* (9.5)*, then* ρ *and* J *satisfy the continuity equation*

$$\partial_t \rho + \operatorname{div} J = 0.$$

Conversely, if this continuity equation holds, then the multivector field F_W *solving the wave Dirac equation* (9.5) *described in Proposition 9.1.2 is* $\wedge^2 W$-*valued at all times, provided it is so at* $t = 0$ *with* $\operatorname{div} E = \rho/\epsilon_0$ *and* $\nabla \wedge B = 0$.

Recall that the continuity equation $\partial_t \rho + \operatorname{div} J = 0$ expresses the fact that total charge is conserved. By Gauss's theorem it shows that

$$\partial_t \int_D \rho\, dx = - \int_{\partial D} \langle J, *\underline{dy} \rangle$$

for every domain $D \subset V$.

Proof. The necessity of the continuity equation follows from

$$\delta_W G_W = -\delta_W^2 F_W = 0,$$

by the nilpotence of the spacetime interior derivative.

For the converse, we investigate the proof of Proposition 9.1.2 and compute the $\wedge^0 W$ and $\wedge^4 W$ parts of $\hat{F}_W = \hat{F}$. The $\wedge^4 W$ part is

$$\mu_0^{-1/2}\frac{\sin(|\xi|ct)}{|\xi|}i\xi \wedge \hat{B}_0 = 0,$$

since $\nabla \wedge B_0 = 0$. The $\wedge^0 W$ part is

$$i\frac{\sin(|\xi|ct)}{|\xi|}e_0 \lrcorner (\xi \lrcorner \hat{F}_0)$$
$$+ c\int_0^t \left(\epsilon_0^{-1/2}\cos(|\xi|c(t-s))\hat{\rho}_s - \mu_0^{1/2}\frac{\sin(|\xi|c(t-s))}{|\xi|}i\xi \lrcorner \hat{J}_s \right)ds$$
$$= \frac{\sin(|\xi|ct)}{|\xi|}(\epsilon_0^{1/2}i\xi \lrcorner \hat{E}_0 - \epsilon_0^{-1/2}\hat{\rho}_0)$$
$$- c\int_0^t \frac{\sin(|\xi|c(t-s))}{|\xi|}(c^{-1}\epsilon_0^{-1/2}\partial_s\hat{\rho}_s + \mu_0^{1/2}i\xi \lrcorner \hat{J}_s)ds,$$

which vanishes, since div $E_0 = \rho_0/\epsilon_0$ and $\partial_t \rho + \mathrm{div}\, J = 0$. This shows that F_W is a homogeneous spacetime bivector field for all times. □

Example 9.2.4 (Lorentz transformation of E and B). From the spacetime representation F_W of the electromagnetic field, we can find how the electric and magnetic fields transform under a change of inertial system. Consider two inertial observers O and O', with ON-bases $\{e_0, e_1, e_2, e_3\}$ and $\{e'_0, e'_1, e'_2, e'_3\}$ respectively. Assume that O sees O' traveling in direction e_3 at speed v. As in Example 4.4.1, the Lorentz boost that maps $\{e_i\}$ to $\{e'_i\}$ is

$$Tx = \exp(\phi e_{03}/2)x \exp(-\phi e_{03}/2), \quad \tanh \phi = v/c.$$

In $\wedge^2 W$ we have the electromagnetic field

$$\mu_0^{1/2} F = c^{-1}e_0 \wedge E + B = c^{-1}e'_0 \wedge E' + B',$$

where $E = E_1 e_1 + E_2 e_2 + E_3 e_3$ and $B = B_1 e_{23} + B_2 e_{31} + B_3 e_{12}$ are the fields measured by O and $E' = E'_1 e'_1 + E'_2 e'_2 + E'_3 e'_3$ and $B' = B'_1 e'_{23} + B'_2 e'_{31} + B'_3 e'_{12}$ are the fields measured by O'. We now compare the two measurements by identifying the two bases as in the discussion above Example 4.6.5, letting $\tilde{E} = E'_1 e_1 + E'_2 e_2 + E'_3 e_3$ and $\tilde{B} = B'_1 e_{23} + B'_2 e_{31} + B'_3 e_{12}$. Then

$$c^{-1}e_0 \wedge E + B = \exp(\phi e_{03}/2)(c^{-1}e_0 \wedge \tilde{E} + \tilde{B}) \exp(-\phi e_{03}/2).$$

Applying the isomorphism $\triangle^{\mathrm{ev}}W \approx \triangle V$ from Proposition 3.3.5, we have equivalently

$$c^{-1}\tilde{E} + \tilde{B} = \exp(-\phi e_3/2)(c^{-1}E + B) \exp(\phi e_3/2).$$

Computing the action of $x \mapsto \exp(-\phi e_3/2)x \exp(\phi e_3/2)$ on $e_1, e_2, e_3, e_{23}, e_{31}, e_{12}$, we get

$$\begin{cases} E'_1 = (E_1 - vB_2)/\sqrt{1 - v^2/c^2}, \\ E'_2 = (E_2 + vB_1)/\sqrt{1 - v^2/c^2}, \\ E'_3 = E_3, \\ B'_1 = (B_1 + (v/c^2)B_2)/\sqrt{1 - v^2/c^2}, \\ B'_2 = (B_2 - (v/c^2)E_1)/\sqrt{1 - v^2/c^2}, \\ B'_3 = B_3. \end{cases}$$

From this we see that for speeds v comparable to the speed of light c, there is a significant mixing of E and B, which shows that indeed it is correct to speak of the electromagnetic field rather than electric and magnetic fields only, since the latter two depend on the inertial frame.

We have seen that Maxwell's equations can be written as a $\triangle$-Dirac wave equation $\mathbf{D}_W F_W = 0$. However, the electromagnetic field F_W is not a general spacetime multivector field, but a bivector field. This means that Maxwell's equations are not identical to the $\triangle$-Dirac equation, but rather that we can embed

Maxwell's equations in a Dirac equation. We show in the remaining sections of this chapter that this is a very useful technique, since in some respects, Dirac equations are better behaved than the Maxwell equations.

An equation from physics that truly is a Dirac equation is Dirac's original equation for the relativistic motion of spin-1/2 particles in quantum mechanics, such as electrons and quarks. With our notation this is a wave $\!\!\!\!/\!\!\!\!A$-Dirac equation in physical spacetime with a lower-order mass term. Without any external potential, the free Dirac equation reads

$$\hbar \not{D} \psi = mc\psi. \qquad (9.6)$$

Here c is the speed of light and $\hbar = 6.626 \cdot 10^{-34}$ [Js] is Planck's constant. The parameter m is the mass of the particle, which in the case of the electron is $m \approx 9.109 \cdot 10^{-31}$ [kg]. Dirac's original approach was to look for a first-order differential equation that is a square root of the Klein–Gordon equation, that is, the wave equation with a mass term

$$\hbar^2 \square \psi = m^2 c^2 \psi, \qquad (9.7)$$

which is obtained from the relativistic energy–momentum relation $E^2 c^{-2} - p^2 = m^2 c^2$ by substituting $E \to i\hbar\partial_t$ and $p \to -i\hbar\nabla$. Such a scalar first-order differential equation does not exist, but Dirac succeeded by allowing matrix coefficients. Having multivectors and spinors at our disposal, we already know that the $\!\!\!\!/\!\!\!\!A$-Dirac equation (9.6) for spacetime spinor fields $\psi : W \to \!\!\!\!/\!\!\!\!A W$ has an invariant geometric meaning.

Exercise 9.2.5 (Matrix representation). Fix an ON-basis $\{e_0, e_1, e_2, e_3\}$ for spacetime, and represent the dual basis $\{-e_0, e_1, e_2, e_3\}$ by the imaginary Dirac matrices $\{i\gamma^0, i\gamma^1, i\gamma^2, i\gamma^3\}$, where γ^k are Dirac's gamma matrices as in Example 5.1.9. Show that Dirac's equation reads

$$i\hbar \begin{bmatrix} \partial_0 & 0 & \partial_3 & \partial_1 - i\partial_2 \\ 0 & \partial_0 & \partial_1 + i\partial_2 & -\partial_3 \\ -\partial_3 & -\partial_1 + i\partial_2 & -\partial_0 & 0 \\ -\partial_1 - i\partial_2 & \partial_3 & 0 & -\partial_0 \end{bmatrix} \begin{bmatrix} \psi_1 \\ \psi_2 \\ \psi_3 \\ \psi_4 \end{bmatrix} = mc \begin{bmatrix} \psi_1 \\ \psi_2 \\ \psi_3 \\ \psi_4 \end{bmatrix}.$$

The physical interpretation of complex-valued wave functions ψ in quantum mechanics is that $|\psi|^2$ represents a probability density for the position of the particle. For the spinor-valued wave function $\psi : W \to \!\!\!\!/\!\!\!\!A W$, we require an inner product on the spinor space $\!\!\!\!/\!\!\!\!A V$. The following is a version of Proposition 5.3.1 for physical spacetime.

Proposition 9.2.6 (Inner product). *Let W be four-dimensional spacetime, with chosen future time direction fixed and complex spinor space $\!\!\!\!/\!\!\!\!A W$. Then there exists a complex inner product $(\cdot, \cdot)$ on $\!\!\!\!/\!\!\!\!A W$ such that*

$$(\psi_1, v.\psi_2) = -(v.\psi_1, \psi_2)$$

for all $\psi_1, \psi_2 \in \cancel{A}W$ and $v \in W$, and

$$-i\langle \psi, v.\psi \rangle > 0$$

for all $\psi \in \cancel{A}W \setminus \{0\}$ and $v \in W_{t+}$. If $(\cdot, \cdot)'$ is any other such inner product, then there is a constant $\lambda > 0$ such that $(\psi_1, \psi_2)' = \lambda(\psi_1, \psi_2)$ for all $\psi_1, \psi_2 \in \cancel{A}W$.

Proof. The proof is analogous to that of Proposition 5.3.1. We look for a matrix M such that

$$M\rho(v) = -\rho(v)^* M,$$

which exists, unique up to complex nonzero multiples, by Theorem 5.2.3, since $(-\rho(v)^*)^2 = \rho(v^2)^* = \langle v \rangle^2 I$. Using the representation ρ from Example 5.1.9, we see that we have $M = \lambda\rho(e_0)$, $\lambda \in \mathbf{C} \setminus \{0\}$, where e_0 is a fixed future-pointing time-like unit vector. For the duality to be an inner product, that is, symmetric, we must choose $\mathrm{Re}\,\lambda = 0$, and to have $-i\langle \psi, e_0.\psi \rangle > 0$, we must have $\mathrm{Im}\,\lambda < 0$. This shows uniqueness. Choosing $\lambda = -i$ and $v = e_0 + v'$, $\langle e_0, v' \rangle = 0$, we have

$$-i\langle \psi, v.\psi \rangle = \psi^*(1 - \rho(e_0 v'))\psi > 0$$

if $|v'| < 1$, since $\rho(e_0 v')$ is $|v'|$ times a $\mathbf{C}^4$ isometry. This completes the existence proof. □

This spacetime spinor inner product is used as follows. Given a wave function solving Dirac's equation, a spinor field

$$\psi : W \to \cancel{A}W$$

in spacetime, we define uniquely a vector field $j_p : W \to \wedge^1 W$ by demanding

$$\langle j_p, v \rangle = i\langle \psi, v.\psi \rangle$$

for all $v \in W$. This exists by Proposition 1.2.3 and is referred to as the *probability four-current*. Fixing a future time direction e_0 and writing

$$j_p = \rho_p e_0 + c^{-1} J_p,$$

$J_p \in [e_0]^\perp$, it follows from the properties of $(\cdot, \cdot)$ that j_p is a real vector field with time component $\rho_p \geq 0$. This represents the probability density for the position of the particle. That J_p defines a *probability current* is clear from the continuity equation

$$\partial_t \rho_p + \mathrm{div}_V J_p = 0.$$

This holds whenever ψ solves Dirac's equation (9.6), since

$$c^{-1}(\partial_t \rho_p + \mathrm{div}_V J_p) = \delta_W j_p = -\partial_0 \langle j_p, e_0 \rangle + \sum_1^3 \partial_k \langle j_p, e_k \rangle$$

$$= i \sum_0^3 \partial_k \langle \psi, e_k^*.\psi \rangle = i\langle \psi, \cancel{D}\psi \rangle - i\langle \cancel{D}\psi, \psi \rangle = 0.$$

Recall the main reflector from Definition 5.2.1, which for physical spacetime we choose as
$$w_4 = ie_{0123} \in \triangle^4 W.$$

In physics $\rho(w_4)$ is referred to as the *chiral operator*, and spinors in its eigenspaces $\triangle^{\pm}W$ are called *right-* and *left-handed spinors* respectively. To obtain a Euclidean formulation of Dirac's equation, we fix a future time direction e_0 and rewrite (9.6) as a coupled system of Euclidean $\triangle$-Dirac equations for the right- and left-handed components of the wave function. As in the discussion after Proposition 9.1.5, we obtain
$$\begin{cases} (c^{-1}\partial_t + \slashed{D})\psi^+ = -\tilde{m}\psi^-, \\ (c^{-1}\partial_t - \slashed{D})\psi^- = \tilde{m}\psi^+, \end{cases}$$

where $\psi^{\pm}(t,x) \in \triangle V \leftrightarrow \triangle^+ W$, $k = 1, 2$, $\tilde{m} := mc/\hbar$, and $\slashed{D} = \slashed{D}_V$ is the $\triangle$-Dirac operator for the Euclidean three-dimensional space $V = [e_0]^{\perp}$.

Exercise 9.2.7. Under the algebra isomorphism
$$\triangle V^2 \ni (\psi^+, \psi^-) \leftrightarrow \psi^+ + e_0\psi^- \in \triangle^+ W \oplus \triangle^- W = \triangle W,$$

show by uniqueness, with suitable normalization $\lambda > 0$, that the spacetime spinor inner product of $\psi_1 = \psi_1^+ + e_0\psi_1^-$ and $\psi_2 = \psi_2^+ + e_0\psi_2^-$ from Proposition 9.2.6 corresponds to
$$i((\psi_1^+, \psi_2^-) - (\psi_1^-, \psi_2^+)),$$

where $(\cdot, \cdot)$ denotes the Hermitian spinor inner product on $\triangle V$.

Using the Hermitian inner product $(\psi_1^+, \psi_2^+) + (\psi_1^-, \psi_2^-)$ on $\triangle V^2$, we have the following Hilbert space result.

Proposition 9.2.8 (Antiparticles and time evolution). *Write Dirac's equation as* $i\hbar\partial_t\psi = H_0\psi$, *where*
$$H_0 := -i\hbar c \begin{bmatrix} \slashed{D} & \tilde{m} \\ -\tilde{m} & -\slashed{D} \end{bmatrix} : L_2(V; \triangle V^2) \to L_2(V; \triangle V^2).$$

Then the free Dirac Hamiltonian H_0 has spectrum $\sigma(H_0) = (-\infty, -mc^2] \cup [mc^2, \infty)$. We have an orthogonal splitting of L_2 into spectral subspaces
$$L_2(V; \triangle V^2) = L_2^+(V; \triangle V^2) \oplus L_2^-(V; \triangle V^2),$$

where
$$L_2^{\pm}(V; \triangle V^2) = \left\{ \begin{bmatrix} -i\tilde{m}\psi & i\slashed{D}\psi \pm \sqrt{\tilde{m}^2 - \Delta}\psi \end{bmatrix}^t ; \psi \in \triangle V \right\}$$

are the spectral subspaces for the energy intervals $[mc^2, \infty)$ and $(-\infty, -mc^2]$ respectively.

The solution to the inital value problem for the wave Dirac equation is
$$\psi(t,x) = (c^{-1}\partial_t + H_0/(\hbar c))R_{ct}^{\tilde{m}}\psi(0,x), \quad \psi(t,\cdot) \in L_2(V; \triangle V^2),$$

where $R_{ct}^{\tilde{m}}$ denotes the Klein–Gordon Riemann function from Corollary 6.2.3, act-ing component-wise.

Splitting the wave function $\psi = \psi^+ + \psi^-$, where $\psi^\pm \in L_2^\pm(V; \not{A}V^2)$, the parts ψ^+ and ψ^- of positive and negative energy describe a particle and an antiparticle respectively. Note that time evolution by H_0 preserves the subspaces $L_2^\pm(V; \not{A}V^2)$. It follows from Corollary 6.2.3 that Dirac's equation has finite propagation speed $\leq c$. However, unlike the massless case in Proposition 9.1.2, the Huygens principle is not valid for Dirac's equation in three spatial dimensions. Compare also this time evolution for Dirac's equation to that for Schrödinger's equation in Example 6.3.6, where instead of finite propagation speed, we have the evolution given by an oscillatory quadratic exponential.

Proof. Applying the Fourier transform in V, Dirac's equation is turned into the ordinary differential equation

$$\partial_t \psi = -c \begin{bmatrix} i\rho(\xi) & \tilde{m} \\ -\tilde{m} & -i\rho(\xi) \end{bmatrix} \psi,$$

where $\rho(\xi) \in \mathcal{L}(\not{A}V)$. For the matrix we obtain eigenvalues $\pm i\sqrt{|\xi|^2 + \tilde{m}^2}$ and eigenvectors $\begin{bmatrix} -i\tilde{m}\psi & -\rho(\xi)\psi \pm \sqrt{|\xi|^2 + \tilde{m}^2}\psi \end{bmatrix}^t$. Applying the inverse Fourier transform, this translates to the stated splitting.

To calculate the time evolution, we write $j := \frac{1}{\sqrt{|\xi|^2 + \tilde{m}^2}} \begin{bmatrix} i\rho(\xi) & \tilde{m} \\ -\tilde{m} & -i\rho(\xi) \end{bmatrix}$ and note that $j^2 = -I$. It follows from Exercise 1.1.5 that

$$\exp(-c\sqrt{|\xi|^2 + \tilde{m}^2}j) = \cos(ct\sqrt{|\xi|^2 + \tilde{m}^2}) - j\sin(ct\sqrt{|\xi|^2 + \tilde{m}^2}),$$

which under the Fourier transform is equivalent to the stated evolution formula. □

Example 9.2.9 (Foldy–Wouthuysen transformation). The particle and antiparticle splitting of a solution $\psi : W \to \not{A}W$ to Dirac's equation (9.6) is independent of the inertial frame for W. Indeed, since $H_0\psi = i\hbar\partial_t\psi$, we have

$$ic^{-1}\partial_t\psi = \pm\sqrt{\tilde{m}^2 - \Delta}\psi, \tag{9.8}$$

with sign $+1$ for particles, that is, $\psi^- = 0$, and sign -1 for antiparticles, that is, $\psi^+ = 0$. Note that (9.6) is a differential equation that is a square root of the Klein–Gordon equation (9.7), and that (9.8) are also square roots of (9.7) although not differential equations. Using the spacetime Fourier transform, we see that the Fourier transforms of wave functions, in the distributional sense, for particles and antiparticles are supported on the two branches of the hyperboloid $\langle\xi\rangle^2 + \tilde{m}^2 = 0$. In particular, this shows the claimed relativistic invariance.

Exercise 9.2.10 (Charge conjugation). Consider the spinor space $\triangle W$ of physical spacetime, with spinor inner product as in Proposition 9.2.6. Show by generalizing the Euclidean theory from Section 5.3 that there exists an antilinear spinor conjugation $\triangle W \to \triangle W : \psi \mapsto \psi^\dagger$ such that

$$(v.\psi)^\dagger = v.\psi^\dagger, \quad v \in W, \; \psi \in \triangle W,$$

and $(\psi^\dagger)^\dagger = \psi$, $\psi \in \triangle W$, and that this is unique modulo a complex factor $|\lambda| = 1$. Show further that in the representation from Example 5.1.9, we can choose $\psi^\dagger = (\rho(e_2)\psi)^c$ and that this spinor conjugation is compatible with the spinor inner product as in Lemma 5.3.4.

For Dirac's equation, the operation $\psi \mapsto \psi^\dagger$ represents *charge conjugation* in physics, an operation that switches particles and antiparticles, which is readily seen from (9.8). Mathematically, note that since $(\psi^\dagger)^\dagger = \psi$, the spinor conjugation yields a real structure on the spinor space of physical spacetime. This agrees with the fact that with our sign convention, the Clifford algebra $\triangle W$ is isomorphic to $\mathbf{R}(4)$ by Theorem 3.4.13.

Recall that a classical particle in an electromagnetic field is acted upon by the Lorentz force (9.2). For a quantum spin-1/2 particle in an electromagnetic field, the Dirac equation is modified by adding a source term and reads

$$\hbar \slashed{D} \psi = mc\psi + iqA_W.\psi. \tag{9.9}$$

The vector field $A_W : W \to \wedge^1 W$ is a four-potential of the electromagnetic field $F_W = d_W A_W$ and q is the charge of the particle, which in case of the electron is $q \approx -1.602 \cdot 10^{-19}$ [C]. A geometric interpretation of (9.9) is that A_W provides Christoffel symbols for a covariant derivative as in Definition 11.1.5.

The Faraday and magnetic Gauss laws show that the electromagnetic field F_W is a closed spacetime bivector field, that is, $d_W F_W = 0$. Poincaré's theorem (Theorem 7.5.2) shows that locally this is equivalent to the existence of a spacetime vector field A_W such that $F_W = d_W A_W$. As we have seen, at least in the Euclidean setting, in Section 7.6, globally there can be topological obstructions preventing every closed field from being exact. And indeed, the famous Aharonov–Bohm experiment shows that in fact, F_W being an exact bivector field is the correct physical law, and not $d_W F_W = 0$. Writing $A_W = \epsilon_0^{1/2}\Phi e_0 + \mu_0^{-1/2}A$ to obtain a Euclidean expression for potential, where $\Phi : W \to \mathbf{R}$ and $A : W \to V$ are scalar and vector potentials of the electromagnetic field, we have

$$\begin{cases} E = -\nabla\Phi - \partial_t A, \\ B = \nabla \wedge A. \end{cases}$$

Returning to (9.9), we note that a solution ψ still yields a probability four-current j_p satisfying the continuity equation, as a consequence of A_W being a real spacetime vector field. As in the free case, (9.9) describes the time evolution of the wave functions for a particle and antiparticle pair. What is, however, not immediately clear is how the nonuniqueness of A_W influences the solution ψ. To explain this, consider an exact spacetime bivector field $F_W : W \to \triangle^2 W$ representing the electromagnetic field, and let $A_W, \tilde{A}_W : W \to \triangle^1 W$ be two different vector potentials, so that

$$F_W = d_W A_W = d_W \tilde{A}_W.$$

Another application of Poincaré's theorem (Theorem 7.5.2) shows that locally, the closed vector field $\tilde{A}_W - A_W$ is exact, so that

$$\tilde{A}_W = A_W + \nabla U,$$

for some scalar potential $U : W \to \triangle^0 W = \mathbf{R}$. From the product rule, we deduce that $\hbar \mathbf{D}\psi = mc\psi + iqA_W.\psi$ if and only if

$$\hbar \mathbf{D}(e^{iqU/\hbar}\psi) = (mc + iq\tilde{A}_W.)(e^{iqU/\hbar}\psi).$$

Therefore $\tilde{\psi} := e^{iqU/\hbar}\psi$ is the wave function of the particle in the electromagnetic field with potential $\tilde{A}_W$. However, since $(\tilde{\psi}, v.\tilde{\psi}) = (\psi, v.\psi)$ by sesquilinearity, the wave functions for the two choices of electromagnetic four-potential yield the same probability four-current j_p. Therefore the physical effects are independent of the choice of electromagnetic four-potential A_W.

9.3 Time-Harmonic Waves

Let W be a spacetime and fix a future pointing time-like unit vector e_0, and let $V = [e_0]^\perp$. For the remainder of this chapter, we study time-harmonic solutions to the wave $\triangle$-Dirac equation $\mathbf{D}_W F = \mathbf{D}_V F - e_0 c^{-1}\partial_t F = 0$. We use the complexified spacetime Clifford algebra $\triangle W_c$, where the component functions $F_s(x)$ belong to $\mathbf{C}$. With a representation of the time-harmonic field as in Example 1.5.2, the Dirac equation reads

$$(\mathbf{D} + ike_0)F(x) = 0,$$

with a wave number $k := \omega/c \in \mathbf{C}$. This is now an elliptic equation with a zero-order term ike_0 added to $\mathbf{D} = \mathbf{D}_V$, rather than a hyperbolic equation. Since even the inner product on the real algebra $\triangle W$ is indefinite, we require the following modified Hermitian inner product for the analysis and estimates to come.

Definition 9.3.1 (Hermitian inner product). With $V = [e_0]^\perp \subset W$ as above, define the auxiliary inner product

$$\langle w_1, w_2 \rangle_V := \langle e_0 \widehat{w}_1 e_0^{-1}, w_2 \rangle, \quad w_1, w_2 \in \triangle W.$$

We complexify both the standard indefinite inner product $\langle \cdot, \cdot \rangle$ on $\triangle W$ and $\langle \cdot, \cdot \rangle_V$ to sesquilinear inner products $(\cdot, \cdot)$ and $(\cdot, \cdot)_V$ on $\triangle W_c$ respectively.

We note that if $w = u + e_0 v$, with $u, v \in \triangle V$, then $e_0 \widehat{w} e_0^{-1} = u - e_0 v$. It follows that $(\cdot, \cdot)_V$ is a Hermitian inner product in which the induced basis $\{e_s\}$ is an ON-basis for $\triangle W_c$ whenever $\{e_j\}_{j=1}^n$ is an ON-basis for V. We use the L_2 norm $\|f\|_{L_2}^2 = \int (f(x), f(x))_V dx$ of complex spacetime multivector fields f.

The aim of this section is to generalize Section 8.3 from the static case $k = 0$ to $k \in \mathbf{C}$. Note that

$$(\mathbf{D} \pm ike_0)^2 = \Delta + k^2. \tag{9.10}$$

Definition 9.3.2 (Fundamental solution). Let Φ_k be the fundamental solution to the Helmholtz equation from Corollary 6.2.4 for $\operatorname{Im} k \geq 0$. Define fundamental solutions

$$\Psi_k^\pm = (\mathbf{D} \pm ike_0)\Phi_k$$

to the Dirac operators $\mathbf{D} \pm ike_0$.

Note the relation

$$\Psi_k^-(x) = -\Psi_k^+(-x)$$

between these two families of fundamental solutions, and that $\Psi_0^+ = \Psi_0^-$ equals Ψ from Definition 8.1.6. It is clear from Corollary 6.2.4 that $\Psi_k^\pm$ in general can be expressed in terms of Hankel functions $H_\nu^{(1)}$, which in odd dimensions are elementary functions involving the exponential function $e^{ik|x|}$.

Exercise 9.3.3 (Asymptotics). Show that in three dimensions,

$$\Psi_k^\pm(x) = \left(\frac{x}{|x|^3} - \frac{ik}{|x|} \left(\tfrac{x}{|x|} \pm e_0 \right) \right) \frac{e^{ik|x|}}{4\pi}.$$

Note that $\Psi_k^\pm \approx \Psi$ near $x = 0$, while $\Psi_k^\pm \in \triangle W_c$ is almost in the direction of the light-like vector $\frac{x}{|x|} \pm e_0$ near $x = \infty$.

Show that in dimension $\dim V = n \geq 2$ we have

$$\Psi_k^\pm(x) - \Psi(x) = O(|x|^{-(n-2)}), \quad \text{as } x \to 0,$$

as well as $\nabla \otimes (\Psi_k^\pm(x) - \Psi(x)) = O(|x|^{-(n-1)})$ as $x \to 0$, and that

$$\Psi_k^\pm(x)e^{-ik|x|} - \tfrac{1}{2}e^{-i\frac{\pi}{2}\frac{n-1}{2}} \left(\frac{k}{2\pi} \right)^{(n-1)/2} \frac{\frac{x}{|x|} \pm e_0}{|x|^{(n-1)/2}} = O(|x|^{-(n+1)/2}), \quad \text{as } x \to \infty.$$

Theorem 9.3.4. *Let $D \subset V$ be a bounded C^1-domain. If $F : D \to \triangle W_c$ solves $(\mathbf{D} + ike_0)F = 0$ in D and is continuous up to ∂D, then*

$$F(x) = \int_{\partial D} \Psi_k^-(y - x)\nu(y)F(y)\, dy, \quad \text{for all } x \in D. \tag{9.11}$$

Note that since $\Psi^-(x) = -\Psi_k^+(-x)$, we can write the reproducing formula equivalently as

$$F(x) = - \int_{\partial D} \Psi_k^+(x - y)\nu(y)F(y)\, dy.$$

Proof. The proof is analogous to that of Theorem 8.1.8. We define the linear 1-form

$$D \setminus \{x\} \times V \to \triangle W_c : (y, v) \mapsto \theta(y, v) := \Psi_k^-(y - x) v F(y).$$

For $y \neq x$, its exterior derivative is

$$\theta(\dot{y}, \nabla) = \sum_{i=1}^{n} \partial_{y_i} \left(\Psi_k^-(y - x) \vartriangle e_i \vartriangle F(y) \right)$$

$$= (\Psi_k^-(\dot{y} - x) \vartriangle \nabla) \vartriangle F(y) + \Psi_k^-(y - x) \vartriangle (\nabla \vartriangle F(\dot{y})).$$

Since $\mathbf{D}F = -ike_0 F$ and

$$\Psi_k^-(\dot{x}) \vartriangle \nabla = (\nabla - ike_0) \vartriangle \Phi_k^-(\dot{x}) \vartriangle \nabla$$

$$= \Phi_k^-(\dot{x})((\nabla - ike_0)^2 + (\nabla - ike_0) \vartriangle ike_0) = \Psi_k^-(x) \vartriangle ike_0,$$

we obtain $\theta(\dot{y}, \nabla) = 0$.

Applying the Stokes formula on the domain $D_\epsilon := D \setminus \overline{B(x, \epsilon)}$ and using the asymptotics of Ψ^- near the origin from Exercise 9.3.3, the rest of the proof follows as for Theorem 8.1.8. $\qquad \square$

It is essential in Theorem 9.3.4 that the domain $D = D^+$ is bounded. In the exterior domain $D^- = V \setminus \overline{D}$, we need appropriate decay of F at ∞. When $k \neq 0$, this takes the form of a radiation condition as follows.

Definition 9.3.5 (Radiating fields). Let F be a multivector field that solves $(\mathbf{D} + ike_0)F = 0$ in D^-. We say that F *radiates* at ∞ if

$$\lim_{R \to \infty} \int_{|y| = R} \Psi_k^-(y - x) \nu(y) F(y) \, dy = 0,$$

for every $x \in D^-$.

Note that by applying Theorem 9.3.4 to the annulus $R_1 < |x| < R_2$, the limit is trivial since the integrals are constant for $R > |x|$. We need an explicit description of this radiation condition. Note that

$$\left(\tfrac{x}{|x|} + e_0 \right)^2 = 0.$$

Proposition 9.3.6 (Radiation conditions). *Let F be a multivector field that solves $(\mathbf{D} + ike_0)F = 0$ in D^- and is continuous up to ∂D, and assume that $\operatorname{Im} k \geq 0$ and $k \neq 0$. If*

$$\left(\tfrac{x}{|x|} + e_0 \right) F = o(|x|^{-(n-1)/2} e^{(\operatorname{Im} k)|x|})$$

as $x \to \infty$, then F radiates. Conversely, if F radiates, then

$$F(x) = \int_{\partial D} \Psi_k^+(x - y) \nu(y) F(y) \, dy, \tag{9.12}$$

for all $x \in D^-$. In particular,

$$F = O(|x|^{-(n-1)/2}e^{-(\operatorname{Im} k)|x|}) \quad \text{and} \quad (\tfrac{x}{|x|} + e_0)F = O(|x|^{-(n+1)/2}e^{-(\operatorname{Im} k)|x|})$$

as $x \to \infty$.

Not only does this give an explicit description of the radiation condition, but it also bootstraps it in that the necessary condition is stronger than the sufficient condition.

Proof. Assuming the decay condition on $(\tfrac{x}{|x|} + e_0)F$, it suffices to prove that

$$\int_{|x|=R} |F|_V^2 \, dx = O(e^{2\operatorname{Im} kR}) \tag{9.13}$$

as $R \to \infty$. Indeed, the Cauchy–Schwarz inequality and the asymptotics for Ψ_k^- from Exercise 9.3.3 then show that F radiates. To estimate $|F|_V$, we note that

$$|(\tfrac{x}{|x|} + e_0)F|_V^2 = (F, (\tfrac{x}{|x|} - e_0)(\tfrac{x}{|x|} + e_0)F)_V = 2|F|_V^2 + 2(F, \tfrac{x}{|x|}e_0 F)_V.$$

Applying Stokes's theorem for bodies on the domain $D_R^- := \{x \in D^- \; ; \; |x| < R\}$, we obtain

$$\int_{|x|=R} (F, \tfrac{x}{|x|}e_0 F)_V \, dx = \int_{\partial D} (F, \nu e_0 F)_V \, dx$$

$$+ \int_{D_R^-} \left((-ike_0 F, e_0 F)_V + (F, -e_0(-ike_0 F))_V \right) dx.$$

In total, this shows that

$$\int_{|x|=R} |F|_V^2 \, dx = \frac{1}{2}\int_{|x|=R} |(\tfrac{x}{|x|} + e_0)F|_V^2 \, dx - \int_{\partial D} (F, \nu e_0 F)_V \, dx - 2\operatorname{Im} k \int_{D_R^-} |F|_V^2 \, dx,$$

from which (9.13) follows from the hypothesis, since ∂D is independent of R and since $\operatorname{Im} k \geq 0$.

The converse is an immediate consequence of Theorem 9.3.4 applied in D_R^-, and the asymptotics for Ψ_k^- from Exercise 9.3.3. □

There are two important applications of the Dirac equation $(\mathbf{D} + ike_0)F = 0$ to classical differential equations, namely time-harmonic acoustic and electromagnetic waves.

Example 9.3.7 (Helmholtz's equation). For a scalar function u, define

$$F = \nabla u + ike_0 u.$$

Then u solves the Helmholtz equation $\Delta u + k^2 u = 0$ from Example 6.3.4 if and only if F solves the Dirac equation $\mathbf{D}F + ike_0 F = 0$. However, note that F is not a general solution to this equation: it is a vector field $F(x) \in \triangle^1 W_c$.

To investigate the reproducing formula (9.11) for this vector field F, we evaluate the time-like and space-like parts of the equation, and get

$$u(x) = \int_{\partial D} \Big(\partial_\nu \Phi_k(y-x)u(y) - \Phi_k(y-x)\partial_\nu u(y) \Big) dy, \tag{9.14}$$

$$\nabla u(x) = \int_{\partial D} \Big(\nabla \Phi_k(y-x) \vartriangle \nu(y) \vartriangle \nabla u(y) + k^2 \Phi_k(y-x)u(y)\nu(y) \Big) dy, \tag{9.15}$$

for $x \in D$, where ∂_ν denotes the derivative in the normal direction. Equation (9.14) we recognise as the Green second identity for solutions to the Helmholtz equation, whereas (9.15) is an analogue of this for the gradient. This latter equation can be further refined by expanding the triple Clifford vector product as

$$\nabla \Phi_k \vartriangle \nu \vartriangle \nabla u = (\partial_\nu \Phi_k)\nabla u + \nabla \Phi_k(\partial_\nu u) - \langle \nabla \Phi_k, \nabla u\rangle \nu + \nabla \Phi_k \wedge \nu \wedge \nabla u.$$

Evaluating the vector part of (9.14), we obtain

$$\nabla u(x) = \int_{\partial D} \Big(\partial_\nu \Phi_k(y-x)\nabla u(y) + \nabla \Phi_k(y-x)\partial_\nu u(y)$$
$$- \langle \nabla \Phi_k(y-x), \nabla u(y)\rangle \nu(y) + k^2 \Phi_k(y-x)u(y)\nu(y) \Big) dy, \quad x \in D.$$

For solutions to the Helmholtz equation $\Delta u + k^2 u = 0$, the classical decay condition at ∞ is the *Sommerfield radiation condition*

$$\partial_r u - iku = o(|x|^{-(n-1)/2} e^{(\operatorname{Im} k)|x|}),$$

with ∂_r denoting the radial derivative. To see its relation to the radiation condition for $\mathbf{D} + ike_0$, we compute

$$(\tfrac{x}{|x|} + e_0)F = (\tfrac{x}{|x|} + e_0)(\nabla u + ike_0 u) = (1 + e_0 \wedge \tfrac{x}{|x|})(\partial_r u - iku) + (\tfrac{x}{|x|} + e_0) \wedge \nabla_S u,$$

where $\nabla_S u := \tfrac{x}{|x|} \lrcorner (\tfrac{x}{|x|} \wedge \nabla u)$ is the angular derivative. By considering the scalar part of this identity, we see that the Dirac radiation condition entails the Sommerfield radiation condition. In fact the two conditions are equivalent. To see this, we can argue similarly to the proof of Proposition 9.3.6 to show that Green's second identity (9.14) holds for the exterior domain D^-. This will yield an estimate on $\nabla_S u$, given the Sommerfield radiation condition.

Example 9.3.8 (Time-harmonic Maxwell's equations). Consider a time-harmonic electromagnetic wave F in a spacetime with three space dimensions. As in Section 9.2 we have $F = \epsilon_0^{1/2} e_0 \wedge E + \mu_0^{-1/2} B \in \triangle^2 W_c$ solving

$$(\mathbf{D} + ike_0)F = 0,$$

with wave number $k := \omega/c = \omega\sqrt{\epsilon_0\mu_0}$. To investigate the reproducing formula (9.11) for this bivector field F, we evaluate the time-like and space-like bivector parts of the equation, and obtain two classical equations known as the *Stratton–Chu formulas*:

$$E(x) = \nabla \times \int_{\partial U} \Phi_k(x - y)\,\nu(y) \times E(y)\,dy$$

$$- \nabla \int_{\partial U} \Phi_k(x - y)\,\nu(y) \cdot E(y)\,dy$$

$$+ ikc \int_{\partial U} \Phi_k(x - y)\,\nu(y) \times (*B)(y)\,dy,$$

$$*B(x) = \nabla \times \int_{\partial U} \Phi_k(x - y)\,\nu(y) \times (*B)(y)\,dy$$

$$- \nabla \int_{\partial U} \Phi_k(x - y)\,\nu(y) \cdot (*B)(y)\,dy$$

$$- ik/c \int_{\partial U} \Phi_k(x - y)\,\nu(y) \times E(y)\,dy.$$

For solutions to the time-harmonic Maxwell's equations, the classical decay condition at ∞ is the *Silver–Müller radiation condition*

$$\begin{cases} \frac{x}{|x|} \times E(x) - c(*B)(x) = o(|x|^{-(n-1)/2} e^{(\operatorname{Im} k)|x|}), \\ c\frac{x}{|x|} \times (*B)(x) + E(x) = o(|x|^{-(n-1)/2} e^{(\operatorname{Im} k)|x|}). \end{cases}$$

Since

$$\left(\tfrac{x}{|x|} + e_0\right)\left(e_0 E + cB\right) = e_0\left(cB - \tfrac{x}{|x|}E\right) + \left(-E + c\tfrac{x}{|x|}B\right),$$

we see that the Dirac radiation condition for the electromagnetic field is equivalent to the Silver–Müller radiation condition. Note that both radiation conditions also give decay of the radial parts of the vector fields E and $*B$.

Given the Cauchy reproducing formulas for $\mathbf{D} + ike_0$, we can extend the theory of Hardy subspaces from Section 8.3 to the case $k \neq 0$. Acting on functions $h : \partial D \to \triangle W_c$ we define traces of Cauchy integrals

$$E_k^+ h(x) := \lim_{z \to x, z \in D^+} \int_{\partial D} \Psi_k^-(y - z)\nu(y)h(y)dy,$$

$$E_k^- h(x) := - \lim_{z \to x, z \in D^-} \int_{\partial D} \Psi_k^-(y - z)\nu(y)h(y)dy, \quad x \in \partial D,$$

and the principal value Cauchy integral

$$E_k h(x) := \lim_{\epsilon \to 0^+} \int_{\partial D \setminus B(x;\epsilon)} \Psi_k^-(y - x)\nu(y)h(y)dy, \quad x \in \partial D.$$

As in the static case $k = 0$, we limit ourselves to proving splittings of Hölder spaces of multivector fields into Hardy subspaces.

Theorem 9.3.9 (Hardy wave subspace splitting). *Let $D = D^+ \subset V$ be a bounded C^1 domain, with exterior domain $D^- = V \backslash \overline{D}$, and let $\text{Im } k \geq 0$. Consider the function space $C^\alpha = C^\alpha(\partial D; \triangle W_c)$, for fixed regularity $0 < \alpha < 1$. Then the operators E_k^+, E_k^-, and E_k are well defined and bounded on $C^\alpha(\partial D)$. The operators $E_k^\pm$ are complementary projections, with $E_k^\pm = \frac{1}{2}(I \pm E_k)$, and they split $C^\alpha(\partial D)$ into Hardy subspaces*

$$C^\alpha(\partial D) = E_k^+ C^\alpha \oplus E_k^- C^\alpha.$$

There is a one-to-one correspondence, furnished by the Cauchy integral (9.11) and the trace map, between fields in the interior Hardy subspace $E_k^+ C^\alpha$ and fields in D^+ solving $\mathbf{D}F + ike_0F = 0$ and Hölder continuous up to ∂D. Likewise, there is a one-to-one correspondence, furnished by the Cauchy integral (9.12) and the trace map, between fields in the exterior Hardy subspace $E_k^- C^\alpha$ and fields in D^- solving $\mathbf{D}F + ike_0F = 0$, Hölder continuous up to ∂D and radiating at ∞.

Proof. Define the operator

$$R_k h(x) = \int_{\partial D} (\Psi_k^-(y - x) - \Psi(y - x))\nu(y)h(y)dy, \quad x \in V.$$

By the asymptotics at $x = 0$ from Exercise 9.3.3, $R_k h(x)$ is a well-defined convergent integral for all $x \in V$. Furthermore, by differentiating under the integral sign, we have

$$|\partial_j R_k h(x)| \lesssim \|h\|_\infty \int_{\partial D} \frac{dy}{|y - x|^{n-1}} \lesssim \|h\|_\infty \ln(1/\text{dist }(x, \partial D)), \qquad (9.16)$$

for $0 < \text{dist }(x, \partial D) < 1/2$ and $j = 1, \ldots, n$, where $\|h\|_\infty := \sup_{\partial D} |h|$. Integrating (9.16) and (8.6) as in Proposition 8.3.3, it follows that $E_k^\pm$ are bounded on C^α, since

$$E_k^\pm = E^\pm \pm R_k.$$

Note that integrating (9.16) in fact shows that $R_k h$ is Hölder continuous across ∂D. Therefore it follows from Proposition 8.3.4 that

$$E_k^\pm h = E^\pm h \pm R_k h = \tfrac{1}{2}h \pm \tfrac{1}{2}(Eh + 2R_k h) = \tfrac{1}{2}(h \pm E_k h)$$

and that $E_k = E + 2R_k$ is a bounded operator on C^α.

As in Theorem 8.3.6, we conclude that $E_k^+ + E_k^- = I$ and that $E^\pm$ are projections by Theorem 9.3.4 and Proposition 9.3.6 respectively. This proves the splitting into Hardy subspaces for $\mathbf{D} + ike_0$. $\square$

9.4 Boundary Value Problems

For the remainder of this chapter, we study boundary value problems (BVPs) for Dirac operators, where our problem is to find a solution F to

$$\mathbf{D}F(x) + ike_0F(x) = 0$$

in a domain D that satisfies a suitable condition on the trace $F|_{\partial D}$. To make the problem precise, one needs to state assumptions on ∂D: How smooth is it? Is it bounded or unbounded? We also need to specify the space of functions on ∂D in which we consider $F|_{\partial D}$, and in what sense the boundary trace $F|_{\partial D}$ is meant. To start with, we postpone these details, and assume only given a Banach space $\mathcal{H}$ of functions on ∂D. A concrete example is $\mathcal{H} = C^\alpha(\partial D)$ from Theorem 9.3.9. We assume that the Cauchy integral operator E_k acts as a bounded operator in $\mathcal{H}$, and we recall that E_k is a reflection operator, $E_k^2 = I$, and it induces a splitting of $\mathcal{H}$ into Hardy wave subspaces. Solutions F to $\mathbf{D}F + ike_0F = 0$ in $D = D^+$ are in one-to-one correspondence with $f = F|_{\partial D}$ in $E_k^+\mathcal{H}$. A formulation of a Dirac BVP is

$$\begin{cases} \mathbf{D}F + ike_0F = 0, & \text{in } D, \\ Tf = g, & \text{on } \partial D. \end{cases}$$

Here $T : \mathcal{H} \to \mathcal{Y}$ is a given bounded and linear operator onto an auxiliary Banach function space $\mathcal{Y}$, which contains the boundary datum g to the BVP. In such an operator formulation, well-posedness of the BVP means that the restricted map

$$T : E_k^+\mathcal{H} \to \mathcal{Y} \tag{9.17}$$

is an isomorphism. Indeed, if so, then for every data $g \in \mathcal{Y}$ we have a unique solution $f \in E_k^+\mathcal{H}$, or equivalently a solution F to $\mathbf{D}F + ike_0F = 0$ in D, which depends continuously on g. The main goal in studying BVPs is to prove such well-posedness. Almost as good is to prove well-posedness in the Fredholm sense, meaning that T is a Fredholm map. In this case, g needs to satisfy a finite number of linear constraints for f to exist, and this is unique only modulo a finite-dimensional subspace.

Proposition 9.4.1. *Let* $T : \mathcal{H} \to \mathcal{Y}$ *be a surjective bounded linear operator. Then the restriction* $T : E_k^+\mathcal{H} \to \mathcal{Y}$ *is an isomorphism if and only if we have a splitting*

$$E_k^+\mathcal{H} \oplus \mathsf{N}(T) = \mathcal{H}.$$

Proof. If $T : E_k^+\mathcal{H} \to \mathcal{Y}$ is an isomorphism, denote its inverse by $T_0 : \mathcal{Y} \to E_k^+\mathcal{H}$. Then $P := T_0T : \mathcal{H} \to \mathcal{H}$ is a projection with null space $\mathsf{N}(T)$ and range $E_k^+\mathcal{H}$, which proves the splitting. Conversely, if we have a splitting $E_k^+\mathcal{H} \oplus \mathsf{N}(T) = \mathcal{H}$, then clearly $T : E_k^+\mathcal{H} \to \mathcal{Y}$ is injective and surjective. $\qquad\square$

Without much loss of generality, we assume from now on that T is a bounded projection on $\mathcal{H}$ with range $\mathcal{Y} \subset \mathcal{H}$. We consider the following abstract formulation of the BVP, in terms of two bounded reflection operators A and B on $\mathcal{H}$:

$$A^2 = I \quad \text{and} \quad B^2 = I.$$

The operator A plays the role of the Cauchy integral E_k, so that $A^+ = \frac{1}{2}(I + A)$ projects onto traces of solutions to the differential equation in D^+ and $A^- =$

$\frac{1}{2}(I - A)$ projects onto traces of solutions to the differential equation in D^-, with appropriate decay at infinity. The operator B encodes two complementary boundary conditions: either $T = B^+ = \frac{1}{2}(I + B)$ or $T = B^- = \frac{1}{2}(I - B)$ can be used to define boundary conditions. Note that we have null spaces $\mathsf{N}(B^+) = B^-\mathcal{H}$ and $\mathsf{N}(B^-) = B^+\mathcal{H}$. We note that the algebra for each of the operators A and B is similar to that of E_k in Theorem 9.3.9. We have two different splittings of $\mathcal{H}$:

$$\mathcal{H} = A^+\mathcal{H} \oplus A^-\mathcal{H} \quad \text{and} \quad \mathcal{H} = B^+\mathcal{H} \oplus B^-\mathcal{H},$$

and $A = A^+ - A^-$ and $B = B^+ - B^-$. The core problem in the study of BVPs is to understand the geometry between on the one hand the subspaces $A^\pm\mathcal{H}$ related to the differential equation, and on the other hand the subspaces $B^\pm\mathcal{H}$ related to the boundary conditions.

Example 9.4.2 (BVP = operator $\triangle\mathbf{R}^2$). The algebra of two reflection operators A and B can be viewed as an operator version of the Clifford algebra $\triangle\mathbf{R}^2$ for the Euclidean plane $\mathbf{R}^2$. Indeed, consider two unit vectors $a, b \in V$. Since $a^2 = b^2 = 1$ in $\triangle\mathbf{R}^2$, we have here a very simple example of an abstract BVP. The geometry of a and b is described by the angle ϕ between the vectors. We recall that this angle can be calculated from the anticommutator

$$\tfrac{1}{2}(ab + ba) = \cos\phi,$$

or from the exponential

$$ab = e^{\phi j},$$

where j is the unit bivector with orientation of $a \wedge b$.

Definition 9.4.3 (Well-posedness). Let $A, B : \mathcal{H} \to \mathcal{H}$ be two reflection operators on a Banach space $\mathcal{H}$. Define the *cosine operator*

$$\tfrac{1}{2}(AB + BA)$$

and the *rotation operators*

$$AB \quad \text{and} \quad BA = (AB)^{-1}.$$

We say that the AB boundary value problems are *well posed* (*in the Fredholm sense*) if the four restricted projections $B^\pm : A^\pm\mathcal{H} \to B^\pm\mathcal{H}$ are all isomorphisms (Fredholm operators).

Exercise 9.4.4 (Simplest abstract BVP). Let $\mathcal{H} = \mathbf{C}^2$ and consider the two orthogonal reflection operators

$$A = \begin{bmatrix} 1 & 0 \\ 0 & -1 \end{bmatrix} \quad \text{and} \quad B = \begin{bmatrix} \cos(2\alpha) & \sin(2\alpha) \\ \sin(2\alpha) & -\cos(2\alpha) \end{bmatrix},$$

for some $0 \leq \alpha \leq \pi/2$. Compute the cosine and rotation operators and show that the AB BVPs are well posed if and only if $0 < \alpha < \pi/2$. Show that we have spectra $\sigma(\frac{1}{2}(AB + BA)) = \{\cos(2\alpha)\}$ and $\sigma(AB) = \{e^{i2\alpha}, e^{-i2\alpha}\}$, and that the AB BVPs fail to be well posed exactly when these spectra hit $\{+1, -1\}$.

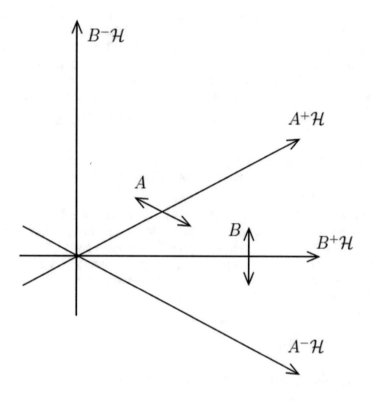

Figure 9.2: The two splittings encoding an abstract BVP, with associated reflection operators.

For two general reflection operators A and B, the associated cosine and rotation operators each contain the necessary information to conclude well-posedness of the AB BVPs. Useful identities include the following, which are straightforward to verify:

$$\tfrac{1}{2}(I + BA) = B^+A^+ + B^-A^-, \tag{9.18}$$

$$\tfrac{1}{2}(I - BA) = B^+A^- + B^-A^+, \tag{9.19}$$

$$2(I + C) = (I + BA)B(I + BA)B, \tag{9.20}$$

$$2(I - C) = (I - BA)B(I - BA)B. \tag{9.21}$$

Proposition 9.4.5 (Well-posedness and spectra). *Let $A, B : \mathcal{H} \to \mathcal{H}$ be two reflection operators on a Banach space $\mathcal{H}$. Then the following are equivalent:*

(i) *The AB BVPs are well posed.*

(ii) *The spectrum of the rotation operator BA does not contain $+1$ or -1.*

(iii) *The spectrum of the cosine operator $C = \frac{1}{2}(AB + BA)$ does not contain $+1$ or -1.*

Similarly, the AB BVPs are well posed in the Fredholm sense if and only if $I \pm BA$ are Fredholm operators, if and only if $I \pm C$ are Fredholm operators.

Proof. We note that $B^+ A^+ + B^- A^-$ is invertible if and only if the BVPs $B^+ : A^+ \mathcal{H} \to B^+ \mathcal{H}$ and $B^- : A^- \mathcal{H} \to B^- \mathcal{H}$ are well posed, and similarly for $B^+ A^- + B^- A^+$. Also $((I + BA)B)^2$ is invertible if and only if $I + BA$ is invertible, and similarly for $I - BA$. The equivalences follow. $\square$

With this general setup, and Proposition 9.4.5 as our main tool for proving well-posedness of Dirac BVPs, we now consider the two main examples that we have in mind. The boundary condition B, unlike A, is typically a pointwise defined multiplier, derived from the orientation of the tangent space to ∂D, described by the normal vector ν. For the remainder of this section we assume that D is a bounded C^2 domain. In this case, we note that ν is a C^1 smooth vector field on ∂D. We will see below that the cosine operators for such smooth BVPs tend to be compact, leading directly to BVPs that are Fredholm well posed by Proposition 9.4.5. Indeed, by the general Fredholm theory outlined in Section 6.4, the operators $I \pm C$ will then be Fredholm operators with index zero. The cosine operators typically are generalizations of the following classical integral operator from potential theory.

Exercise 9.4.6 (Double layer potential). Consider the integral operator

$$Kf(x) := \int_{\partial D} \langle \Psi(y - x), \nu(y) \rangle f(y) dy, \quad x \in \partial D,$$

with kernel $k(x, y) = \langle \Psi(y - x), \nu(y) \rangle$. In three dimensions, a physical interpretation of $k(x, y)$ is that of the electric potential from a dipole at y, in the direction $\nu(y)$, and for this reason K is called the *double layer potential operator*. The operator K is weakly singular on smooth domains. More precisely, show that on a C^2 boundary ∂D of dimension $n - 1$, we have kernel estimates $|k(x, y)| \lesssim |x - y|^{2-n}$ and $|\nabla'_x k(x, y)| \lesssim |x - y|^{1-n}$, $x \neq y$, $x, y \in \partial D$, where ∇'_x denotes the tangential gradient in the x-variable.

Lemma 9.4.7 (Weakly singular = compact). *Let*

$$Tf(x) = \int_{\partial D} k(x, y) f(y) dy, \quad x \in \partial D,$$

be a weakly singular integral operator with kernel estimates $|k(x, y)| \lesssim |x - y|^{2-n}$ and $|\nabla'_x k(x, y)| \lesssim |x - y|^{1-n}$, $x, y \in \partial D$. Here ∇'_x denotes the tangential gradient along ∂D in the variable x. Then T is a compact operator on $C^\alpha(\partial D)$ for all $0 < \alpha < 1$.

Proof. Assume that $x, x' \in \partial D$ with $|x - x'| = \epsilon$. Write

$$
\begin{aligned}
&Tf(x') - Tf(x) \\
&= \int_{|y-x|\leq 2\epsilon} (k(x',y) - k(x,y))f(y)dy + \int_{|y-x|>2\epsilon} (k(x',y) - k(x,y))f(y)dy \\
&=: I_0 + I_1.
\end{aligned}
$$

For I_1, we obtain from the mean value theorem the estimate $|k(x',y) - k(x,y)| \lesssim \epsilon/|y - x|^{1-n}$ when $|y - x| > 2\epsilon$. This yields

$$
|I_1| \lesssim \epsilon \ln \epsilon^{-1} \|f\|_{L_\infty}.
$$

For I_0, we estimate $|k(x',y) - k(x,y)| \lesssim |y - x|^{2-n}$ and obtain $|I_0| \lesssim \epsilon \|f\|_{L_\infty}$. It follows that $T : C^\alpha(\partial D) \to C^\beta(\partial D)$ is bounded for all $\beta < 1$, and that $T : C^\alpha(\partial D) \to C^\alpha(\partial D)$ is a compact operator. $\qquad\square$

Example 9.4.8 (Normal/tangential BVP). Our main example of a Dirac BVP occurs when the differential equation is $\mathbf{D}F + ike_0F = 0$, that is, $A = E_k$, for a fixed wave number $k \in \mathbf{C}$, and the boundary conditions are encoded by the reflection operator $B = N$ given by

$$
Nf(x) := \nu(x) \vartriangle \widehat{f(x)} \vartriangle \nu(x), \quad x \in \partial D.
$$

We know from Section 4.1 that N reflects the multivector $f(x)$ across the tangent plane to ∂D at x, and assuming that $\nu \in C^1$, we have that N is bounded on $C^\alpha(\partial D)$. The projection N^+ in this case will yield a boundary condition that specifies the part of $f(x)$ tangential to ∂D in the sense of Definition 2.8.6. This can be verified using the Riesz formula (3.4), as

$$
N^+f = \tfrac{1}{2}(f + \nu \widehat{f} \nu) = \nu \tfrac{1}{2}(\nu f + \widehat{f}\nu) = \nu(\nu \wedge f) = \nu \lrcorner (\nu \wedge f).
$$

The corresponding calculation using (3.3) shows that $N^-f = \nu \wedge (\nu \lrcorner f)$ yields the boundary condition that specifies the normal part of $f(x)$.

   The four E_kN BVPs consist of two BVPs for solutions to $\mathbf{D}F + ike_0F = 0$ in the interior domain D^+, where the tangential or normal part of $F|_{\partial D}$ is specified, and two BVPs for solutions to $\mathbf{D}F + ike_0F = 0$ in the exterior domain D^-, where the tangential or normal part of $F|_{\partial D}$ is specified. By Proposition 9.4.5, the well-posedness of these four BVPs may be studied via the associated cosine operator $E_kN + NE_k = (E_k + NE_kN)N$. When $k = 0$, we calculate using $\Psi\nu = 2\langle \Psi, \nu \rangle - \nu\Psi$

that

$$\tfrac{1}{2}(E + NEN)f(x) = \text{p.v.} \int_{\partial D} \Big(\Psi(y - x)\nu(y)f(y) + \nu(x)\Psi(y - x)f(y)\nu(y)\nu(x) \Big) dy$$

$$= 2\text{p.v.} \int_{\partial D} \langle \Psi(y - x), \nu(y) \rangle f(y) dy$$

$$+ \text{p.v.} \int_{\partial D} (\nu(x) - \nu(y)) \Psi(y - x) f(y) dy$$

$$+ \nu(x) \left(\text{p.v.} \int_{\partial D} \Psi(y - x) f(y)(\nu(y) - \nu(x)) dy \right) \nu(x).$$

Assume now that D is a bounded C^2 domain. We can then apply Lemma 9.4.7 to each of these three terms, showing that $EN + NE$ is a compact operator on C^α. Moreover, the compactness of $E_k - E$ on C^α follows by yet another application of Lemma 9.4.7. We conclude that the $E_k N$ BVPs are well posed in the sense of Fredholm in $C^\alpha(\partial D)$ for C^2 domains D.

Example 9.4.9 (Spin BVP). The second example of a Dirac BVP that we shall consider is that in which the boundary conditions are induced by left Clifford multiplication by the normal vector ν. For technical reasons we study boundary conditions encoded by the reflection operator $B = S$ given by

$$Sf(x) := e_0 \vartriangle \nu(x) \vartriangle f(x), \quad x \in \partial D.$$

Note that $(e_0 n)^2 = -e_0^2 n^2 = 1$, so indeed S is a reflection operator, and it is bounded on C^α, since ν is C^1 regular. The factor e_0 is motivated by Proposition 3.3.5 as in Proposition 9.1.5, and makes $\triangle W^{\text{ev}}$ invariant under S.

As before, we study the differential equation $\mathbf{D}F + ike_0 F = 0$ encoded by the reflection operator $A = E_k$. It would be more natural to consider the operators E_k and S acting on spinor fields $\partial D \to \cancel{\triangle} W$, though, since both operators use only left multiplication by multivectors. So the true nature of the $E_k S$ BVPs are BVPs for the $\cancel{\triangle}$-Dirac operator. However, we here consider the $\triangle$-Dirac operator, since we aim to combine the $E_k S$ and the $E_k N$ BVPs in Section 9.5. The ranges of the projections

$$S^+ f = \tfrac{1}{2}(1 + e_0 \nu)f \quad \text{and} \quad S^- f = \tfrac{1}{2}(1 - e_0 \nu)f$$

are seen to be the subspaces of multivector fields containing left Clifford factors that are respectively the light-like vectors $\nu \pm e_0$. The advantage of the S boundary conditions is that in some sense, the $E_k S$ BVPs are the best local BVPs possible for the differential equation $\mathbf{D}F + ike_0 F = 0$. We will see several indications of this below.

For the cosine operator $\frac{1}{2}(ES + SE)$, we calculate

$$\tfrac{1}{2}(E + SES)f(x) = \text{p.v.} \int_{\partial D} \Big(\Psi(y - x)\nu(y)f(y) + \nu(x)\Psi(y - x)f(y) \Big) dy$$

$$= 2\text{p.v.} \int_{\partial D} \langle \Psi(y - x), \nu(y) \rangle f(y) dy$$

$$+ \text{p.v.} \int_{\partial D} (\nu(x) - \nu(y)) \Psi(y - x) f(y) dy,$$

since e_0 anticommutes with the space-like vectors ν and Ψ. As in Example 9.4.8, we conclude from this, using Lemma 9.4.7, that the $E_k S$ BVPs are well posed in the sense of Fredholm in C^α on C^2 domains.

Having established well-posedness in the Fredholm sense for the $E_k N$ and $E_k S$ BVPs, we know that the BVP maps (9.17) are Fredholm operators, so that the null spaces are finite-dimensional and the ranges are closed subspaces of finite codimension. It remains to prove injectivity and surjectivity, whenever possible.

Proposition 9.4.10 (Injectivity). *Let $0 < \alpha < 1$ and $\operatorname{Im} k \geq 0$.*

- *For the $E_k N$ BVPs we have*

$$E_k^+ C^\alpha \cap N^+ C^\alpha = E_k^+ C^\alpha \cap N^- C^\alpha = E_k^- C^\alpha \cap N^+ C^\alpha = E_k^- C^\alpha \cap N^- C^\alpha = \{0\}$$

  if $\operatorname{Im} k > 0$. Moreover, if D^- is a connected domain and $k \in \mathbf{R} \setminus \{0\}$, then $E_k^- C^\alpha \cap N^+ C^\alpha = E_k^- C^\alpha \cap N^- C^\alpha = 0$.

- *For the $E_k S$ BVPs we have*

$$E_k^+ C^\alpha \cap S^+ C^\alpha = E_k^- C^\alpha \cap S^- C^\alpha = \{0\}$$

  whenever $\operatorname{Im} k \geq 0$.

Proof. For the estimates we require the Hermitian inner product $(w_1, w_2)_V := (e_0 \widehat{w}_1 e_0^{-1}, w_2)$ on $\triangle W_c$ from Definition 9.3.1. Consider first the interior BVPs. Given $f = F|_{\partial D} \in E_k^+ C^\alpha$, we define the linear 1-form

$$\overline{D} \times V \to \mathbf{C} : (y, v) \to (e_0 v F(y), F(y))_V,$$

which has nabla derivative

$$(e_0 \nabla F(\dot{y}), F(\dot{y}))_V = (e_0 (\nabla F), F)_V - (e_0 F, \nabla F)_V$$
$$= (e_0(-ik e_0 F), F)_V - (e_0 F, -ik e_0 F)_V = -2\operatorname{Im} k |F|_V^2.$$

From the Stokes formula (7.4), it follows that

$$\int_{\partial D} (Sf, f)_V dy = -2\operatorname{Im} k \int_{D^+} |F|_V^2 dx.$$

If $f \in N^{\pm}C^{\alpha}$, then $(Sf, f)_V = 0$, and we conclude that $F = 0$ if $\operatorname{Im} k > 0$. So in this case, $E_k^+ C^{\alpha} \cap N^{\pm} C^{\alpha} = \{0\}$. If $f \in S^+ C^{\alpha}$, then $(Sf, f)_V = |f|_V^2$, and we conclude that $f = 0$ whenever $\operatorname{Im} k \geq 0$, so $E_k^+ C^{\alpha} \cap S^+ C^{\alpha} = \{0\}$.

Consider next the exterior BVPs. Let $f = F|_{\partial D} \in E_k^- C^{\alpha}$, and fix a large radius R. From Stokes's theorem applied to the domain $D_R^- := D^- \cap \{|x| < R\}$, we have

$$\int_{|x|=R} (e_0 \tfrac{x}{|x|} F, F)_V \, dx - \int_{\partial D} (Sf, f)_V \, dy = -2\operatorname{Im} k \int_{D_R^-} |F|_V^2 \, dx.$$

Furthermore, on the sphere $|x| = R$, we note that

$$|(\tfrac{x}{|x|} + e_0)F|_V^2 = 2|F|_V^2 - 2(e_0 \tfrac{x}{|x|} F, F)_V,$$

and obtain the identity

$$\int_{|x|=R} (|F|_V^2 - \tfrac{1}{2}|(\tfrac{x}{|x|} + e_0)F|_V^2) \, dx - \int_{\partial D} (Sf, f)_V \, dy = -2\operatorname{Im} k \int_{D_R^-} |F|_V^2 \, dx.$$

Using Proposition 9.3.6, we have $\lim_{R \to \infty} \int_{|x|=R} |(\tfrac{x}{|x|} + e_0)F|_V^2 \, dx = 0$ for all $\operatorname{Im} k \geq 0$. If $f \in S^- C^{\alpha}$, then $(Sf, f)_V = -|f|_V^2$, and we again conclude that $f = 0$, so $E_k^- C^{\alpha} \cap S^- C^{\alpha} = \{0\}$. If $f \in N^{\pm} C^{\alpha}$, then $(Sf, f)_V = 0$, and we have

$$\int_{|x|=R} |F|_V^2 \, dx + 2\operatorname{Im} k \int_{D_R^-} |F|_V^2 \, dx = \tfrac{1}{2} \int_{|x|=R} |(\tfrac{x}{|x|} + e_0)F|_V^2 \, dx \to 0, \quad R \to \infty.$$

When $\operatorname{Im} k > 0$, this shows that $F = 0$. When $k \in \mathbf{R} \setminus \{0\}$, we have

$$\lim_{R \to \infty} \int_{|x|=R} |F|_V^2 \, dy = 0.$$

Applying Rellich's lemma (Lemma 6.3.5) to the component functions F_s of F, which satisfy Helmholtz's equation $\Delta F_s + k^2 F_s = 0$, we also in this case conclude that $F = 0$, so in either case, $E_k^- C^{\alpha} \cap N^{\pm} C^{\alpha} = \{0\}$. $\square$

Summarizing our findings, we have obtained the following well-posedness results.

Theorem 9.4.11 (C^{α} well-posedness). *For the Dirac BVPs with boundary function space $C^{\alpha}(\partial D)$, $0 < \alpha < 1$, on domains with C^2 regular boundary ∂D, we have the following well-posedness results.*

The four BVPs $N^{\pm} : E_k^{\pm} C^{\alpha} \to N^{\pm} C^{\alpha}$ are well posed when $\operatorname{Im} k > 0$. If the exterior domain D^- is connected, then the exterior BVPs $N^{\pm} : E_k^- C^{\alpha} \to N^{\pm} C^{\alpha}$ are well posed for all nonzero $\operatorname{Im} k \geq 0$.

The two spin-Dirac BVPs $S^- : E_k^+ C^{\alpha} \to S^- C^{\alpha}$ and $S^+ : E_k^- C^{\alpha} \to S^+ C^{\alpha}$ are well posed for all $\operatorname{Im} k \geq 0$.

We remark that by applying analytic Fredholm theory, one can prove that in fact, also the interior $E_k N$ BVPs are well posed for $k \in \mathbf{R}$, except for a discrete set of resonances.

Proof. We make use of the Fredholm theory outlined in Section 6.4. By Example 9.4.8 and Proposition 9.4.5, the $E_k N$ BVPs are well posed in the Fredholm sense for all k. By Proposition 9.4.10 the four maps $N^\pm : E_k^\pm C^\alpha \to N^\pm C^\alpha$ are injective when $\operatorname{Im} k > 0$. We conclude that $I \pm \frac{1}{2}(E_k N + N E_k)$ are injective Fredholm operators with index zero, and therefore invertible. So the $E_k N$ BVPs are well posed when $\operatorname{Im} k > 0$.

For $k \in \mathbf{R} \backslash \{0\}$, we have injective semi-Fredholm maps $N^\pm : E_k^- C^\alpha \to N^\pm C^\alpha$ by Proposition 9.4.10. By perturbing E_k^- to $\operatorname{Im} k > 0$, Lemma 9.4.12 below proves that they are invertible.

The well-posedness of $S^- : E_k^+ C^\alpha \to S^- C^\alpha$ and $S^+ : E_k^- C^\alpha \to S^+ C^\alpha$ follows from Example 9.4.8 and Proposition 9.4.10, using Proposition 9.4.5. Note that $I - \frac{1}{2}(E_k S + S E_k) = ((S^- E_k^+ + S^+ E_k^-)S)^2$ is an injective Fredholm operator with index zero, and hence invertible. $\qquad\square$

The following two techniques for proving existence of solutions to BVPs turn out to be useful.

Lemma 9.4.12 (Perturbation of domains). *Let A_t, $t \in [0,1]$, and B be reflection operators on a Banach space $\mathcal{H}$, and consider the family of BVPs described by $B^+ : A_t^+ \mathcal{H} \to B^+ \mathcal{H}$. If these are all semi-Fredholm maps and if $t \mapsto A_t$ is continuous, then the indices of $B^+ : A_0^+ \mathcal{H} \to B^+ \mathcal{H}$ and $B^+ : A_1^+ \mathcal{H} \to B^+ \mathcal{H}$ are equal.*

Proof. We parametrize the domains $A_t^+ \mathcal{H}$ by the fixed space $A_0^+ \mathcal{H}$. Considering $\tilde{A}_t^+ := A_t^+ : A_0^+ \mathcal{H} \to A_t^+ \mathcal{H}$ as one of the four abstract $A_0 A_t$ BVPs, we note that

$$I + A_t A_0 = 2I + (A_t - A_0) A_0.$$

If $\|A_t - A_0\| \le 1/\|A_0\|$, it follows that $I + A_t A_0$ is invertible, and from (9.18) we see in particular for $0 \le t \le \epsilon$, that $\tilde{A}_t^+$ is invertible. Let $\tilde{B}_t^+ := B^+ : A_t^+ \mathcal{H} \to B^+ \mathcal{H}$. Applying the method of continuity to $\tilde{B}_t^+ \tilde{A}_t^+$, we conclude that $\operatorname{Ind}(\tilde{B}_\epsilon^+ \tilde{A}_\epsilon^+) = \operatorname{Ind}(\tilde{B}_0^+ \tilde{A}_0^+)$. Since $\tilde{A}_t^+$ are invertible, we obtain $\operatorname{Ind}(\tilde{B}_\epsilon^+) = \operatorname{Ind}(\tilde{B}_0^+)$. Repeating this argument a finite number of times, we conclude that $\operatorname{Ind}(\tilde{B}_1^+) = \operatorname{Ind}(\tilde{B}_0^+)$. $\qquad\square$

Lemma 9.4.13 (Subspace duality). *Let A and B be two reflection operators on a Banach space $\mathcal{H}$, and consider the BVP described by $B^+ : A^+ \mathcal{H} \to B^+ \mathcal{H}$. This map is surjective if and only if the dual BVP described by $(B^*)^- : (A^*)^- \mathcal{H}^* \to (B^*)^- \mathcal{H}^*$ is an injective map.*

Proof. Note that A^* and B^* are reflection operators in $\mathcal{H}^*$. By duality as in Section 6.4, we have $(A^+ \mathcal{H})^\perp = \mathsf{R}(A^+)^\perp = \mathsf{N}((A^*)^+) = \mathsf{R}((A^*)^-) = (A^*)^- \mathcal{H}^*$ and similarly $(B^- \mathcal{H})^\perp = (B^*)^+ \mathcal{H}^*$. Similarly to Proposition 9.4.1, since

$$(A^+ \mathcal{H} + B^- \mathcal{H})^\perp = (A^+ \mathcal{H})^\perp \cap (B^- \mathcal{H})^\perp,$$

this translates to the claim. $\qquad\square$

We end this section with two applications of the techniques in this section to Dirac's equation.

Example 9.4.14 (The MIT bag model). Consider Dirac's equation

$$i\hbar\partial_t\psi = H_0\psi$$

from Proposition 9.2.8 on a bounded domain $D \subset V$. The MIT bag model is used in physics to describe the quarks in a nucleon, that is, a proton or neutron. The bag D represents the nucleon, and the boundary condition is

$$\nu.\psi = \psi,$$

or in the $\triangle V^2$ representation $e_0\nu.\psi = \psi$. This boundary condition implies in particular that the probability current

$$\langle j_p, \nu\rangle = i(\psi, \nu.\psi) = i(\psi, \psi)$$

across ∂D vanishes, since j_p is a real spacetime vector field. We see that with suitable modifications, such BVPs for time-harmonic solutions to Dirac's equation can be studied with the methods described in this section.

Example 9.4.15 (Chirality of (anti-)particles). What we refer to here as abstract BVPs, namely the algebra of two reflection operators describing the geometry between two splittings of a function space, appear in many places independent of any BVPs. One of many such examples we saw in connection to Proposition 9.2.8. Consider the Hilbert space $\mathcal{H} := L_2(V; \triangle V_c^2)$, where we saw two different splittings. The reflection operator $B = \begin{bmatrix} I & 0 \\ 0 & -I \end{bmatrix}$ encodes the Chiral subspaces of right- and left-handed spinors, whereas

$$A = \operatorname{sgn}(H_0) = \frac{-i}{\sqrt{\tilde{m}^2 - \Delta}} \begin{bmatrix} \slashed{D} & \tilde{m} \\ -\tilde{m} & -\slashed{D} \end{bmatrix}.$$

Using, for example, the representation of $\triangle V$ by Pauli matrices, the Fourier multiplier of the rotation operator AB at frequency $\xi \in V$ is seen to have the four eigenvalues

$$\lambda = (\pm|\xi| \pm i\tilde{m})/\sqrt{|\xi|^2 + \tilde{m}^2}.$$

Therefore the spectrum of AB is precisely the unit circle $|\lambda| = 1$. We conclude that although the spectral subspaces $L_2^\pm(V; \triangle V^2)$ do not intersect the chiral subspaces, the angle between them is zero. The problem occurs at high frequencies: particles or antiparticles of high energy may be almost right- or left-handed.

9.5 Integral Equations

The aim of this section is to use the somewhat abstract theory from Section 9.4 to derive integral equations for solving Dirac BVPs, with good numerical properties, that have recently been discovered.

- It is desirable to extend the theory to nonsmooth domains, which have boundaries that may have corners and edges, as is often the case in applications. Ideally, one would like to be able to handle general Lipschitz domains.

- To solve a given BVP, we want to have an equivalent integral formulation

$$\int_{\partial D} k(x,y)f(y)dy = g(x), \quad x \in \partial D,$$

where the boundary datum gives g and the integral equation is uniquely solvable for f if and only if the BVP to solve is well posed. Ideally we want to have a function space without any constraints, meaning a space of functions $\partial D \to L$ with values in a fixed linear space L and coordinate functions in some classical function space.

In this section we let D be a bounded strongly Lipschitz domain. At this generality, the normal vector field ν is only a measurable function without any further smoothness. To extend the theory from Section 9.4 and keep the basic operators E_k, N and S bounded, we shall use

$$L_2 = L_2(\partial D; \triangle W_c),$$

which is the most fundamental space to use for singular integral operators like E_k. Indeed, the singular integral operator E_k is bounded on $L_2(\partial D)$ for every Lipschitz domain D, by Theorem 8.3.2 and Exercises 9.3.3 and 6.4.3.

We first consider Fredholm well-posedness of the $E_k N$ BVPs in L_2 on bounded strongly Lipschitz domains. On such nonsmooth domains, it is not true in general that $E_k N + N E_k$, or even the classical double layer potential from Exercise 9.4.6, is compact. However, we recall from Proposition 9.4.5 that it suffices to show that the spectrum of $\frac{1}{2}(E_k N + N E_k)$ does not contain ± 1.

Theorem 9.5.1 (Rellich estimates). *Let D be a bounded strongly Lipschitz domain, and let θ be a smooth compactly supported field that is transversal to ∂D as in Exercise 6.1.8. Define the local Lipschitz constant $L := \sup_{\partial D}(|\theta \wedge \nu|/\langle \theta, \nu \rangle)$ for ∂D. Then*

$$\lambda I + E_k N$$

is a Fredholm operator on $L_2(\partial D)$ of index zero whenever $\lambda = \lambda_1 + i\lambda_2$, $|\lambda_2| < |\lambda_1|/L$, $\lambda_1, \lambda_2 \in \mathbf{R}$.

Note that since $E_k N$ and $(E_k N)^{-1} = N E_k$ are bounded, we also know that the spectrum of $E_k N$ is contained in an annulus around 0. Furthermore, since

$$((\lambda I + E_k N)E_k)^2 = \lambda(\lambda + \lambda^{-1} + E_k N + N E_k),$$

the resolvent set of the cosine operator contains the hyperbolic regions onto which $\lambda \mapsto \frac{1}{2}(\lambda + \lambda^{-1})$ maps the double cone $|\lambda_2| < |\lambda_1|/L$. And for $\lambda = \pm 1$ it follows in particular that the $E_k N$ BVPs are well posed in the Fredholm sense in $L_2(\partial D)$.

Proof. To motivate the calculations to come, we consider first the BVP described by $N^+ : E_k^+ L_2 \to N^+ L_2$. To estimate $\|f\|_{L_2}$ in terms of $\|N^+ f\|_{L_2}$, we insert the factor $\langle \theta, \nu \rangle$ and express it with the Clifford product as

$$\|f\|_{L_2}^2 \approx \int_{\partial D} |f|_V^2 \langle \theta, \nu \rangle dy = \frac{1}{2} \int_{\partial D} \langle f, f(\theta\nu + \nu\theta) \rangle_V dy = \mathrm{Re} \int_{\partial D} \langle f\nu, f\theta \rangle_V dy.$$

We next use the reversed twin of the Riesz formula (3.4) to write $f\nu = 2f \wedge \nu - \nu\widehat{f}$. We estimate the last term so obtained by applying Stokes's theorem with the linear 1-form $(y, v) \mapsto \langle v\widehat{f(y)}, f(y)\theta(y) \rangle_V$, giving

$$\int_{\partial D} \langle \nu\widehat{f}, f\theta \rangle_V dy = \int_D \left(\langle -ike_0\widehat{f}, f\theta \rangle_V + \langle \widehat{f}, (-ike_0 f)\theta \rangle_V + \sum_{j=1}^n \langle \widehat{f}, e_j f(\partial_j \theta) \rangle_V \right) dy \tag{9.22}$$

Combining and estimating, we get

$$\|f\|_{L_2(\partial D)}^2 \lesssim \|f \wedge \nu\|_{L_2(\partial D)} \|f\|_{L_2(\partial D)} + \|F\|_{L_2(D_\theta)}^2,$$

where $D_\theta := D \cap \mathrm{supp}\,\theta$. The Cauchy integral $L_2(\partial D) \to L_2(D_\theta) : f \mapsto F$ can be shown to be a bounded operator by generalizing the Schur estimates from Exercise 6.4.3 to integral operators from ∂D to D_θ. Moreover, such estimates show by truncation of the kernel that this Cauchy integral is the norm limit of Hilbert–Schmidt operators, and hence compact. On the first term we can use the absorption inequality $\|N^+ f\| \|f\| \leq \frac{1}{2\epsilon} \|N^+ f\|^2 + \frac{\epsilon}{2} \|f\|^2$. Choosing ϵ small leads to a lower bound, showing that $N^+ : E_k^+ L_2 \to N^+ L_2$ is a semi-Fredholm operator.

Next consider the integral equation $\lambda h + E_k Nh = g$, where we need to estimate $\|h\|_{L_2}$ in terms of $\|g\|_{L_2}$. To this end, we note that $E_k Nh = g - \lambda h$, so that

$$E_k^\pm L_2 \ni f^\pm := 2E_k^\pm Nh = Nh \pm (g - \lambda h).$$

Applying (9.22) to f^+ and the corresponding application of Stokes's theorem to f^-, we obtain estimates

$$\left| \int_{\partial D} \langle \nu\widehat{f^\pm}, f^\pm\theta \rangle_V dy \right| \lesssim \|F\|_{L_2(\mathrm{supp}\,\theta)}^2.$$

We now expand the bilinear expressions on the left, writing $f^\pm = Nh \mp \lambda h \pm g$, and observe that the integrals

$$\int_{\partial D} \langle \nu\widehat{h}, h\theta \rangle_V dy \quad \text{and} \quad \int_{\partial D} \langle \nu\widehat{Nh}, Nh\theta \rangle_V dy$$

are bad in the sense that we have only an upper estimate by $\|h\|_{L_2}^2$, whereas the terms

$$\int_{\partial D} \langle \nu\widehat{Nh}, \lambda h\theta \rangle_V dy = \lambda \int_{\partial D} \langle h, h\theta\nu \rangle_V dy$$

and $\int_{\partial D}(\nu\lambda\widehat{h}, Nh\theta)_V \, dy = \left(\lambda \int_{\partial D}(h, h\theta\nu)_V \, dy\right)^c$ are good in the sense that they are comparable to $\|h\|_{L_2}^2$. To avoid the bad terms, we subtract identities and obtain

$$\|F\|_{L_2(\operatorname{supp}\theta)}^2 \gtrsim \left|\int_{\partial D}\left((\nu\widehat{f}^+, f^+\theta)_V - (\nu\widehat{f}^-, f^-\theta)_V\right) dy\right|$$

$$\gtrsim \left|2\operatorname{Re}\left(\lambda \int_{\partial D}(h, h\theta\nu)_V \, dy\right)\right| - \|h\|_{L_2}\|g\|_{L_2} - \|g\|_{L_2}^2.$$

Writing $\theta\nu = \langle\theta,\nu\rangle + \theta\wedge\nu$, we know that $|\theta\wedge\nu| \le L\langle\theta,\nu\rangle$ for some $L < \infty$. It follows that $\left|2\operatorname{Re}\left(\lambda \int_{\partial D}(h, h\theta\nu)_V \, dy\right)\right| \gtrsim \|h\|_{L_2}^2$ if $|\lambda_2| < |\lambda_1|/L$, and we conclude that in this case, $\lambda I + E_k N$ is a semi-Fredholm operator. That it is a Fredholm operator with index zero follows from the method of continuity, by perturbing λ, for example, to 0, where $E_k N$ is an invertible operator. $\square$

Theorem 9.5.2 (L_2 well-posedness for $E_k N$). *For the $E_k N$ Dirac BVPs with boundary regularity $L_2(\partial D)$ on bounded strongly Lipschitz domains D, we have the following well-posedness results.*

The four BVPs $N^\pm : E_k^\pm L_2 \to N^\pm L_2$ are well posed for all $\operatorname{Im} k \ge 0$ except for a discrete set of real $k \ge 0$. If the exterior domain D^- is connected, then the exterior BVPs $N^\pm : E_k^- L_2 \to N^\pm L_2$ are well posed for all nonzero $\operatorname{Im} k \ge 0$.

Proof. By Theorem 9.5.1 and Proposition 9.4.5, the $E_k N$ BVPs are well posed in the Fredholm sense for all k. Proposition 9.4.10 can be verified when the C^α topology is replaced by L_2. The proof can now be completed as in Theorem 9.4.11. $\square$

For the remainder of this section we consider the second problem posed above, namely how to formulate a given Dirac BVP as an integral equation that is good for numerical applications. As a concrete example, we take the exterior BVP with prescribed tangential part, that is,

$$N^+ : E_k^- L_2 \to N^+ L_2. \tag{9.23}$$

This BVP has important applications, since a solution of it yields an algorithm for computing, for example, how acoustic and electromagnetic waves are scattered by an object D. Assuming that the exterior domain D^- is connected and $\operatorname{Im} k \ge 0$, $k \ne 0$, we know that this BVP is well posed. Although it is an invertible linear equation by which we can solve the BVP, it is not useful for numerical applications. The reason is that the solution space $E_k^- L_2$ is defined by a nonlocal constraint on $f \in L_2$. What we need is an ansatz, meaning some operator

$$U : \mathcal{Y} \to E_k^- L_2,$$

where $\mathcal{Y}$ is a function space that is good for numerical purposes and U has good invertibility properties. Using such a U, we can solve the BVP (9.23) by solving

$$N^+ U h = g$$

for $h \in \mathcal{Y}$. This gives the solution $f = Uh \in E_k^- L_2$.

As a first try, we swap the roles of E_k and N and consider

$$U = E_k^- : N^+ L_2 \to E_k^- L_2.$$

This leads to the operator $N^+ E_k^-|_{N^+ L_2}$, which can be shown to be closely related to the double layer potential operator from Exercise 9.4.6. The function space $\mathcal{Y} = N^+ L_2$ is good, but although this U is a Fredholm operator, it fails to be invertible for a discrete set of real k. Indeed, $\mathsf{N}(U) = N^+ L_2 \cap E_k^+ L_2$ will contain the eigenvalues of the self-adjoint operator $-ie_0 \mathbf{D}$ with tangential boundary conditions on the bounded domain D^+. This explains a well-known problem in the numerical solution of BVPs by integral equations: the existence of *spurious interior resonances* k, where the integral equation fails to be invertible, even though the BVP it is used to solve is itself well posed.

A better try, which should be more or less optimal, comes from the $E_k S$ BVPs. Swapping the roles of E_k and S, we consider

$$U = E_k^- : S^+ L_2 \to E_k^- L_2.$$

Similarly, a good ansatz for an interior Dirac BVP is $U = E_k^+ : S^- L_2 \to E_k^+ L_2$. It is important not to swap S^+ and S^- in these ansatzes. Maybe the best way to see that the $E_k S$ BVPs have well-posedness properties superior to those for the $E_k N$ BVPs on L_2, even in the Fredholm sense and in particular on Lipschitz domains, is to consider the rotation operator

$$E_k S f(x) = 2\text{p.v.} \int_{\partial D} \Psi_k^-(y - x) f(y) dy, \quad x \in \partial D.$$

Note that we used that $\nu^2 = 1$. Since $E_k - E$ is a weakly singular integral operator, it is compact on $L_2(\partial D)$, and when $k = 0$, we note that ES is a skew-symmetric operator, since Ψ is a space-like vector depending skew-symmetrically on x and y. In particular, this means that the spectrum of ES is on the imaginary axis and the operators $I \pm ES$ are invertible with $\|(I \pm ES)^{-1}\| \le 1$. By the identities from Proposition 9.4.5, this means, for example, that

$$\|E^- h\| = \tfrac{1}{2}\|(I - E_k S)h\| \ge \tfrac{1}{2}\|h\|, \quad h \in S^+ L_2.$$

For general k, we note that there is still a major difference in well-posedness properties of the $E_k S$ BVPs as compared to those for the $E_k N$ BVPs. The operator $\lambda I + E_k S$ can fail to be Fredholm only when $\operatorname{Re} \lambda = 0$, whereas $\lambda I + E_k N$ can fail to be Fredholm whenever $|\operatorname{Re} \lambda| \le L|\operatorname{Im} \lambda|$, not far away from $\lambda = \pm 1$ for large L. So, as compared to the $E_k N$ BVPs, the well-posedness properties for the $E_k S$ BVPs do not essentially depend on the Lipschitz geometry of ∂D.

Theorem 9.5.3 (L_2 well-posedness for $E_k S$). *For the $E_k S$ Dirac BVPs with boundary regularity $L_2(\partial D)$ on bounded strongly Lipschitz domains D, we have the following well-posedness results. The two spin-Dirac BVPs $S^- : E_k^+ L_2 \to S^- L_2$ and*

$S^+ : E_k^- L_2 \to S^+ L_2$ *are well posed for all* $\operatorname{Im} k \geq 0$. *Equivalently, the ansatzes* $E_k^- : S^+ L_2 \to E_k^- L_2$ *and* $E_k^+ : S^- L_2 \to E_k^+ L_2$ *are invertible for all* $\operatorname{Im} k \geq 0$.

Proof. As before, we note the identity $\frac{1}{2}(I - E_k S) = E_k^+ S^- + E_k^- S^+$ and its twin $S^+ E_k^- + S^- E_k^+ = \frac{1}{2}(I - S E_k) = \frac{1}{2}(E_k S - I) S E_k$. From the discussion above it follows that $I - E_k S$ is a Fredholm operator of index 0, which directly shows that the two BVPs and the two ansatzes are Fredholm maps. By Proposition 9.4.10 adapted to L_2, the four maps are injective for all $\operatorname{Im} k \geq 0$. Therefore $I - E_k N$ is injective, hence surjective. We conclude that the two BVPs and the two ansatzes are invertible. □

Example 9.5.4 (Asymptotic APS BVPs). Consider the Cauchy reflection operator $A = E_k$ encoding the Dirac equation $\mathbf{D}F + ike_0 F = 0$, together with the abstract boundary conditions $B = E_l$, where $k, l \in \mathbf{C}$. Clearly not all four $E_k E_l$ BVPs are well posed, since $E_l - E_k$ is a compact operator. However, since

$$\frac{1}{2}(I + E_l E_k) = E_l^+ E_k^+ + E_l^- E_k^-$$

clearly is a Fredholm operator with index zero, the two BVPs $E_l^+ : E_k^+ L_2 \to E_l^+ L_2$ and $E_l^- : E_k^- L_2 \to E_l^- L_2$ are Fredholm operators. Such BVPs with nonlocal boundary conditions defined by the differential equation itself are essentially the boundary conditions employed by Atiyah, Patodi, and Singer (APS) in their work on index theory for manifolds with boundary.

We next let $l \to \infty$ along the upper imaginary axis. The operators E_l are not norm convergent, but for a fixed function h, one can show that

$$E_l h \to -Sh.$$

Note from the formula for Ψ_l^- how the singular integral operators E_l localize to the pointwise multiplier $-S$. This shows that indeed, the operator S is related to the differential equation as a local asymptotic Cauchy singular integral, and to some extent explains why the $E_k S$ BVPs are so remarkably well posed.

Example 9.5.5 (Spin integral equation). We now return to the exterior Dirac BVP (9.23) with prescribed tangential parts, which we know is well posed whenever $\operatorname{Im} k \geq 0$, $k \neq 0$, and D^- is connected. Using the invertible ansatz $E_k^- : S^+ L_2 \to E_k^- L_2$ from Theorem 9.5.3, we can solve the BVP (9.23), given datum $g \in N^+ L_2$, by solving

$$N^+ E_k^- h = g \qquad (9.24)$$

for $h \in S^+ L_2$, giving the solution $f = E_k^- h \in E_k^- L_2$ and

$$F(x) = \int_{\partial D} \Psi_k^-(y - x)(-\nu(y)) h(y) dy, \quad x \in D^-,$$

solving $\mathbf{D}F + ike_0 F = 0$ in D^- with $N^+ F|_{\partial D} = g$. This is certainly numerically doable, since both spaces $N^+ L_2$ and $S^+ L_2$ are defined by a simple pointwise constraint determined by the normal ν. However, we can enhance the integral equation somewhat as follows. Consider the reflection operator T given by

$$Tf = -e_0 \widehat{f} e_0.$$

We note that, similarly to N, replacing ν by the time-like vector e_0, indeed $T^2 = I$ and T reflects time-like multivectors in the subspace of space-like multivectors. Computing relevant cosine operators, we have

$$(TS + ST)f = -e_0(\widehat{e_0 \nu f})e_0 + e_0 \nu (-e_0 \widehat{f} e_0) = 0,$$
$$(TN + NT)f = -e_0 \nu f \nu e_0 - \nu e_0 f e_0 \nu \neq 0,$$
$$(NS + SN)f = \nu(\widehat{e_0 \nu f})\nu + e_0 \nu(\nu \widehat{f} \nu) = 0.$$

By Proposition 9.4.5, this means that we have optimally well posed abstract BVPs TS and NS. In particular, this allows us to parametrize the domain space $S^+ L_2$ of the integral equation (9.24) for example by

$$T^+ L_2 = L_2(\partial D; \triangle V_c),$$

the space of space-like multivector fields, which is an ideal space for applications. In fact, we verify that $S^+ : T^+ L_2 \to S^+ L_2$ is $1/\sqrt{2}$ times an isometry.

Since $TN + NT \neq 0$, we cannot directly parametrize the range space $N^+ L_2$ of (9.24) by $T^+ L_2$. However, we can go via the splitting $L_2 = S^+ L_2 \oplus S^- L_2$, since for example,

$$T^+ S^+ : N^+ L_2 \to T^+ L_2$$

is invertible. In fact, both $S^+ : N^+ L_2 \to S^+ L_2$ and $T^+ : S^+ L_2 \to T^+ L_2$ are $1/\sqrt{2}$ times isometries.

To summarize, we propose that the exterior BVP (9.23) with prescribed tangential part is best solved using the integral equation

$$T^+ S^+ N^+ E_k^- S^+ h = T^+ S^+ g,$$

for $h \in T^+ L_2$. Indeed, the derivation above shows that this integral equation is uniquely solvable, and the function space for the variable h and the datum $T^+ S^+ g$ is simply $T^+ L_2 = L_2(\partial D; \triangle V_c)$. To write out this equation more explicitly, we compute that

$$T^+ S^+ g = \tfrac{1}{2}(g_0 + \nu g_1), \quad \text{when } g = g_0 + e_0 g_1$$

and $g_1, g_2 \in N^+ L_2 \cap T^+ L_2$, so the time-like part is mapped onto a normal part when the original multivector is tangential. We also compute that $T^+ S^+ N^+ S^+ T^+ = \tfrac{1}{4} T^+$. Writing $E_k^- = \tfrac{1}{2}(I - E_k)$, the integral equation for $h \in L_2(\partial D; \triangle V_c)$ becomes

$$\tfrac{1}{2} h(x) + M(x) \text{p.v.} \int_{\partial D} \Psi_k^-(y - x)(\nu(y) - e_0) h(y) dy = 2M(x) g(x), \quad x \in \partial D.$$

Here M denotes the multiplier that projects onto tangential multivectors and maps tangential time-like multivectors onto normal space-like multivectors by replacing a left factor e_0 into ν. We refer to this integral equation as a *spin integral equation* for solving the BVP (9.23), since the key feature is that it uses an ansatz derived from the $E_k S$ BVPs, which, as we have discussed in Example 9.4.9, really are BVPs for the $\triangle$-Dirac equation $\not{D}\psi + ike_0\psi = 0$.

Example 9.5.6 (Transmission problems). Transmission problems generalize boundary value problems in that we look for a pair of fields $F^+ : D^+ \to \triangle W_c$ and $F^- : D^- \to \triangle W_c$ such that

$$\begin{cases} \mathbf{D}F^+ + ik_2e_0F^+ = 0, & \text{in } D^+, \\ \mathbf{D}F^- + ik_1e_0F^- = 0, & \text{in } D^-, \\ Mf^+ = f^- + g, & \text{on } \partial D. \end{cases} \tag{9.25}$$

Here the wave numbers $k_1, k_2 \in \mathbf{C}$ are different in the two domains, with $\operatorname{Im} k_1 \geq 0$ and $\operatorname{Im} k_2 \geq 0$. The relation between the traces $f^+ = F^+|_{\partial D}$ and $f^- = F^-|_{\partial D}$ on ∂D is described by a multiplier $M \in \mathcal{L}(L_2)$ and a given source $g \in L_2$.

For solving the transmission problem (9.25), unlike in the case of BVPs, we have a good ansatz directly available, namely

$$U : L_2 \to E_{k_2}^+ L_2 \oplus E_{k_1}^- L_2 : h \mapsto (E_{k_2}^+ h, E_{k_1}^- h).$$

In the case $k_1 = k_2$, it is clear from the L_2 analogue of Theorem 9.3.9 that U is invertible. What is somewhat surprising is that U is invertible for all $\operatorname{Im} k_1 \geq 0$ and $\operatorname{Im} k_2 \geq 0$. To prove this, it suffices by the method of continuity to show that U is injective. To this end, note that $Uh = 0$ means that

$$h = F^+|_{\partial D} = F^-|_{\partial D},$$

where $\mathbf{D}F^+ + ik_1e_0F^+ = 0$ in D^+ and $\mathbf{D}F^- + ik_2e_0F^- = 0$ in D^-. Applying Stokes's theorem twice to $\int_{\partial D}(e_0\nu h, h)_V dy$, computations as in the proof of Proposition 9.4.10 give

$$2\operatorname{Im} k_1 \int_{D^+} |F^+|_V^2 dx + 2\operatorname{Im} k_2 \int_{D_R^-} |F^-|_V^2 dx + \int_{|x|=R} |F^-|_V^2 dx$$

$$= \tfrac{1}{2} \int_{|x|=R} |(\tfrac{x}{|x|} + e_0)F^-|_V^2 dx.$$

Using radiation conditions and jumps, this shows that $F^+ = F^- = 0$ and therefore $h = 0$.

Using this invertible ansatz U, we can now solve the transmission problem (9.25) by solving the integral equation

$$(ME_{k_2}^+ - E_{k_1}^-)h = g$$

for $h \in L_2$. Note that this is an integral equation in $L_2(\partial D; \triangle W_c)$ without any constraints. From the solution h, we finally compute the field

$$F^+(x) = \int_{\partial D} \Psi^-_{k_2}(y - x)\nu(y)h(y)dy,$$

$$F^-(x) = -\int_{\partial D} \Psi^-_{k_1}(y - x)\nu(y)h(y)dy,$$

solving the transmission problem. In Section 9.7, we apply this integral equation for Dirac transmission problems to solve scattering problems for electromagnetic waves.

9.6 Boundary Hodge Decompositions

We have considered Dirac BVPs in the previous sections and how to solve them by integral equations. Returning to Examples 9.3.7 and 9.3.8, one important issue remains. We saw there that both the Helmholtz equation and Maxwell's equations can be viewed as special cases of the Dirac equation $\mathbf{D}F + ike_0F = 0$. However, in these examples F is a vector field and a bivector field respectively, and not a general multivector field. If we intend, for example, to solve BVPs for Helmholtz's or Maxwell's equations by a spin integral equation as in Example 9.5.5 or a transmission problem with a Dirac integral equation as in Example 9.5.6, then we need a tool to ensure that the solution multivector field F is in fact a vector or bivector field. It turns out that there exists an exterior/interior derivative operator acting on multivector fields $\partial D \to \triangle W_c$, which we shall denote by Γ_k, which is the tool needed. Applications to Maxwell scattering are found in Section 9.7. The point of departure for our explanations is Proposition 8.1.5, where we noted that for a monogenic field $\nabla \vartriangle F = 0$, each of its homogeneous component functions F_j is monogenic if and only if $\nabla \wedge F = 0 = \nabla \lrcorner F$. Generalizing this to time-harmonic waves with wave number $k \in \mathbf{C}$, we have the following.

Lemma 9.6.1 (Two-sided k-monogenic fields). *Assume that $F : D \to \triangle W_c$ solves $\mathbf{D}F + ike_0F = 0$ in some open set $D \subset V$. Write $F = F_0 + F_1 + \cdots + F_{n+1}$, where $F_j : D \to \triangle^j W_c$. Then $\mathbf{D}F_j + ike_0F_j = 0$ in D for all $0 \le j \le n+1$ if and only if*

$$\begin{cases} dF + ike_0 \wedge F = 0, \\ \delta F + ike_0 \lrcorner F = 0. \end{cases}$$

The way we use this result is that if we construct F solving $\mathbf{D}F + ike_0$ and some BVP, and if $dF + ike_0 \wedge F = 0$, then we can conclude, for example, that F_2 is a bivector field solving the Dirac equation, since the homogeneous parts of F decouple, and thus F_2 is an electromagnetic field satisfying Maxwell's equations.

Proof. If $(\nabla + ike_0) \wedge F = 0 = (\nabla + ike_0) \lrcorner F$, then

$$(\nabla + ike_0) \wedge F_j = ((\nabla + ike_0) \wedge F)_{j+1} = 0$$

and $(\nabla + ike_0) \lrcorner F_j = ((\nabla + ike_0) \lrcorner F)_{j-1} = 0$, and so $(\nabla + ike_0) \vartriangle F_j = (\nabla + ike_0) \lrcorner F_j + (\nabla + ike_0) \wedge F_j = 0$ for all j. Conversely, if $(\nabla + ike_0) \vartriangle F_j = 0$ for all j, then

$$(\nabla + ike_0) \wedge F_j = ((\nabla + ike_0) \vartriangle F_j)_{j+1} = 0$$

and $(\nabla + ike_0) \lrcorner F_j = ((\nabla + ike_0) \vartriangle F_j)_{j-1} = 0$. Summing over j, we obtain $(\nabla + ike_0) \wedge F = 0 = (\nabla + ike_0) \lrcorner F$. $\qquad\square$

To proceed with the analysis, we need to choose a function space. Since our theory for Hodge decompositions as well as for spin integral equations is set in Hilbert spaces, we choose $L_2(\partial D)$.

Definition 9.6.2 (Boundary Γ_k operator). Consider the Hardy space splitting $L_2(\partial D) = E_k^+ L_2 \oplus E_k^- L_2$ on a strongly Lipschitz domain. Define the operator Γ_k by

$$\Gamma_k f := g^+ + g^-,$$

where $f = E_k^+ f + E_k^- f$, $F^\pm$ denote the Cauchy integrals of f in $D^\pm$ so that $E_k^\pm f = F^\pm|_{\partial D}$, and $g^\pm = G^\pm|_{\partial D} \in E_k^\pm L_2$ are such that their Cauchy integrals equal

$$G^\pm = (\nabla + ike_0) \wedge F^\pm \quad \text{in } D^\pm.$$

The domain of Γ_k is the set of f for which such $g^\pm$ exist.

In a series of lemmas, we derive below a more concrete expression for this unbounded operator Γ_k as a tangential differential operator on $L_2(\partial D)$. It turns out that Γ_k acts by exterior differentiation along ∂D on tangential fields and by interior differentiation along ∂D on normal fields, modulo zero order terms determined by k.

Definition 9.6.3 (Tangential derivatives). Consider the Lipschitz boundary $M = \partial D$, which is a Lipschitz manifold in the sense that the transition maps, as in Section 6.1, are Lipschitz regular. As in Definitions 11.2, 11.2.6, 12.1.1 and extending to Lipschitz regularity as in Section 10.2, we define *tangential exterior and interior derivative operators* d' and δ' in $L_2(M; \wedge M)$, such that $(d')^* = -\delta'$.

In the notation of this chapter, complexifying the bundle $\wedge M$ to $\wedge M_c$, we have

$$N^+ L_2 = \{f_1 + e_0 \wedge f_2 \; ; \; f_1, f_2 \in L_2(M; \wedge M_c)\},$$

and extending d' and δ' to operators in $N^+ L_2$ acting as

$$d' f := d' f_1 - e_0 \wedge (d' f_2),$$
$$\delta' f := \delta' f_1 - e_0 \wedge (\delta' f_2),$$

on $f = f_1 + e_0 \wedge f_2$, with $f_1, f_2 \in L_2(M; \wedge M)$.

The reader is kindly advised to consult the relevant sections of the following chapters, as indicated in Definition 9.6.3, for further details. Note that the minus sign in the actions on $N^+ L_2$ occurs because the time-like e_0 and the formally space-like tangential ∇' anticommute.

Lemma 9.6.4 ($E_k^{\pm}L_2$ to N^+L_2). *If $f \in E_k^+ L_2 \cap D(\Gamma_k)$, then $N^+ f \in D(d')$ and*

$$d'(N^+ f) + ike_0 \wedge (N^+ f) = N^+(\Gamma_k f).$$

The same holds for $f \in E_k^- L_2 \cap D(\Gamma_k)$.

Proof. Let $f = F|_{\partial D}$, where F is the Cauchy extension of f. Write $f = f_1 + e_0 \wedge f_2$ and $F = F_1 + e_0 \wedge F_2$, where f_j and F_j are space-like fields, $j = 1, 2$. Generalizing Exercise 11.2.3, with methods as in Lemma 10.2.4, to Lipschitz regular hypersurfaces, we have $N^+ f_j = \rho^* F_j$, where $\rho : \partial D \to V$ denotes the embedding of ∂D into V. The commutation theorem shows that

$$d' \rho^* F_j = \rho^*(dF_j),$$

giving $d'(N^+ f) = d' \rho^* F_1 - e_0 \wedge d' \rho^* F_2 = N^+(dF)$. This proves the first statement, since $e_0 \wedge N^+ f = N^+(e_0 \wedge f)$, and the proof for $E_k^- L_2$ is similar. $\qquad \square$

Using Hodge star dualities, we next derive the corresponding result for the normal part. This uses left Clifford multiplication by ν, which is an isometry between $N^- L_2$ and $N^+ L_2$.

Lemma 9.6.5 ($E_k^{\pm}L_2$ to $N^- L_2$). *If $f \in E_k^+ L_2 \cap D(\Gamma_k)$, then $\nu \lrcorner f = \nu N^- f \in D(\delta')$ and*

$$\delta'(\nu \lrcorner f) + ike_0 \lrcorner (\nu \lrcorner f) = \nu \lrcorner (\Gamma_k f).$$

The same holds for $f \in E_k^- L_2 \cap D(\Gamma_k)$.

Proof. Using nabla calculus with $\nabla_k := \nabla + ike_0$, given $f = F|_{\partial D} \in E_k^+ L_2 \cap D(\Gamma_k)$, we write for example $\mathbf{D}F + ike_0 F = 0$ as $\nabla_k F = \nabla_k \vartriangle F = 0$. Extending Proposition 8.1.13, such solutions form a right Clifford module, so

$$G = Fw = \overline{F}* = \overline{F} \lrcorner w,$$

writing $w = e_{012\cdots n}$ for the spacetime volume element, with dual volume element $w^* = -w \in \triangle^n W^*$, is also a solution to $\nabla_k G = 0$ in D^+. Moreover,

$$*(\nabla_k \wedge G) = -w \llcorner (\nabla_k \wedge G) = (-w \llcorner G) \llcorner \nabla_k = \overline{F} \llcorner \nabla_k = \overline{\nabla_k \lrcorner F},$$

making use of the algebra from Section 2.6. By Lemma 9.6.4 applied to G, we have $N^+(\nabla_k \wedge G)|_{\partial D} = \nabla_k' \wedge (N^+ g)$ with $g = G|_{\partial D}$, writing d' formally with nabla calculus using $\nabla_k' = \nabla' + ike_0$ along ∂D. The spacetime Hodge dual of the left-hand side is

$$*(N^+(\nabla_k \wedge G)|_{\partial D}) = N^-(\overline{\nabla_k \lrcorner F})|_{\partial D}.$$

For the right hand side, we note for $h := \nabla_k' \wedge (N^+ g) \in N^+ L_2$ that $*h = -(\nu w')\overline{h} = -\nu(w'\overline{h}) = -\nu(w' \llcorner h)$, where $w' := \nu \lrcorner w$. We used here Corollary 3.1.10 and $\langle w \rangle^2 = -1$. We get

$$*(\nabla_k' \wedge (N^+ g)) = -\nu((w' \llcorner N^+ g) \llcorner \nabla_k') = \nu((\nu \lrcorner \overline{f}) \llcorner \nabla_k') = \nu \overline{\nabla_k' \lrcorner (f \llcorner \nu)}.$$

Note that the first step uses a nonsmooth extension of Exercise 11.2.7. Reversing these two equations, multiplying them from the left by ν, and equating them yields

$$\nu \lrcorner (\nabla_k \lrcorner F)|_{\partial D} = \nu(\nabla'_k \lrcorner (f \llcorner \nu))\nu = \nabla'_k \lrcorner (\widehat{f} \llcorner \nu) = -\nabla'_k \lrcorner (\nu \lrcorner f).$$

In the second step we used that $\nu h = \widehat{h}\nu$ whenever $h \in N^+L_2$, and in the last step we applied the commutation relation from Proposition 2.6.3. This proves the lemma for $E_k^+L_2$, since $\Gamma_k f = -(\nabla'_k \lrcorner F)|_{\partial D}$. The proof for $E_k^-L_2$ is similar. $\square$

We next show the converses of Lemmas 9.6.4 and 9.6.5.

Lemma 9.6.6 ($N^\pm L_2$ to $E_k^\pm L_2$). *If* $f \in N^+L_2$ *and* $f \in D(d')$, *then* $f \in D(\Gamma_k)$ *with*

$$\Gamma_k f = d'f + ike_0 \wedge f.$$

Similarly, if $f \in N^-L_2$ *and* $\nu \lrcorner f \in D(\delta')$, *then* $f \in D(\Gamma_k)$ *with* $\Gamma_k f = \nu \wedge (\delta' + ike_0\lrcorner)(\nu \lrcorner f)$.

Proof. Let $f \in N^+L_2$ and define Cauchy extensions

$$F^\pm(x) = \pm \int' \Psi_k^-(y - x)\nu(y)f(y)dy, \quad x \in D^\pm.$$

Differentiating under the integral sign, with notation as in the proof of Lemma 9.6.5, we have

$$\nabla_k \wedge F(x) = \mp \int_{\partial D} \Psi_k^-(y - \dot{x})(\nabla_k \wedge \nu(y) \wedge f(y))dy$$

$$= \pm \int_{\partial D} \Psi_k^-(\dot{y} - x)(\nabla_{-k} \wedge \nu(y) \wedge f(y))dy,$$

where we have used the algebraic anticommutation relation

$$\nabla_k \wedge (\Psi_k^- h) = (\nabla_k, \Psi_k^-)h - \Psi_k^-(\nabla_k \wedge h)$$

and the first term vanishes, since $\Psi_k^+(\cdot - y) = -\Psi_k^-(y - \cdot)$ is a fundamental solution to $\mathbf{D} + ike_0$. Aiming to apply a nonsmooth extension of Exercise 11.2.7, we form the inner product with a fixed multivector $w \in \triangle W_c$, and obtain

$$\langle w, \nabla_k \wedge F(x)\rangle = \pm \int_{\partial D} \langle \nu(y) \lrcorner (\nabla_k \lrcorner (\Psi_k^+(\dot{y} - x)w)), f(y)\rangle dy.$$

We choose to use the complex bilinear pairing on $\triangle W_c$, but this is not important. By Lemma 9.6.5, we have $\nu(y)\lrcorner(\nabla_k\lrcorner(\Psi_k^+(\dot{y}-x)w)) = -(\delta'+ike_0\lrcorner)(\nu(y)\lrcorner(\Psi_k^+(y-x)w))$. Note that F in the proof of Lemma 9.6.5 need not solve a Dirac equation for such a trace result to be true. Duality yields

$$\langle w, \nabla_k \wedge F(x)\rangle = \pm \int_{\partial D} \langle w, \Psi_k^-(y - x)\nu(y)(d'f + ike_0 \wedge f)\rangle dy.$$

Since w is arbitrary, this proves the lemma for N^+L_2.

The proof for $f \in N^-L_2$ is similar. We calculate

$$\nabla_k \wedge F(x) = -\nabla_k \lrcorner F(x) = \mp \int_{\partial D} \Psi_k^-(\dot{y} - x)(\nabla_{-k} \lrcorner (\nu(y) \lrcorner f(y)))dy.$$

Pairing with w gives

$$\langle w, \nabla_k \wedge F(x)\rangle = \mp \int_{\partial D} \langle \nabla_k \wedge (\Psi_k^+(\dot{y} - x)w), \nu(y) \lrcorner f(y)\rangle dy$$

$$= \mp \int_{\partial D} \langle (d' + ike_0 \wedge)N^+(\Psi_k^+(\dot{y} - x)w), \nu(y) \lrcorner f(y)\rangle dy$$

$$= \pm \int_{\partial D} \langle w, \Psi_k^-(y - x)\nu(y)(\nu(y) \wedge ((\delta' + ike_0 \lrcorner)(\nu(y) \lrcorner f(y))))\rangle dy.$$

The second equality uses that $\nu \lrcorner f \in N^+L_2$ and Lemma 9.6.4. Since w is arbitrary, this proves the lemma for N^-L_2. $\qquad\square$

Summarizing the above results, we obtain the following concrete expression for Γ_k. Given Lemmas 9.6.4, 9.6.5, and 9.6.6, the proof is straightforward.

Proposition 9.6.7. *The operator Γ_k is a nilpotent operator in $L_2(\partial D)$ in the sense of Definition 10.1.1. Its domain equals*

$$D(\Gamma_k) = \{f \in L_2(\partial D) \; ; \; N^+f \in D(d') \text{ and } \nu \lrcorner f \in D(\delta')\}$$

and

$$\Gamma_k f = (d' + ike_0 \wedge)N^+f + \nu \wedge (\delta' + ike_0 \lrcorner)(\nu \lrcorner f), \quad f \in D(\Gamma_k).$$

The operator Γ_k commutes with E_k and with N.

Having uncovered this nilpotent operator Γ_k, we now investigate the Hodge splitting of $L_2(\partial D)$ that it induces. We need a Hermitian inner product on $L_2(\partial D)$, and we choose

$$(f, g) = \int_{\partial D} (f(x), g(x))_V dx.$$

Proposition 9.6.8 (Boundary Hodge decomposition)**.** *When $k \neq 0$, the nilpotent operator Γ_k induces an exact Hodge splitting*

$$L_2(\partial D) = R(\Gamma_k) \oplus R(\Gamma_k^*),$$

where the ranges $R(\Gamma_k) = N(\Gamma_k)$ and $R(\Gamma_k^*) = N(\Gamma_k^*)$ are closed. When $k = 0$, the ranges are still closed, but the finite-dimensional cohomology space $\mathcal{H}(\Gamma_k) = N(\Gamma_k) \cap N(\Gamma_k^*)$ will be nontrivial.

Proof. Proposition 10.1.2 shows that $\Gamma = \Gamma_k$ induces an orthogonal splitting

$$L_2(\partial D) = \overline{R(\Gamma_k)} \oplus \mathcal{H}(\Gamma_k) \oplus \overline{R(\Gamma_k^*)}.$$

When D is smooth and $k = 0$, it follows from Propositions 12.1.3 and 10.1.6 that the ranges are closed and that the cohomology space is finite-dimensional for $\Gamma = \Gamma_k$. Adapting the methods from Theorem 10.3.1 to the manifold setting, this result can be extended to the case that D is merely a Lipschitz domain. However, on nonsmooth boundaries ∂D we do not have $\mathsf{D}(d') \cap \mathsf{D}(\delta') = H^1(\partial D)$, but still $\mathsf{D}(d') \cap \mathsf{D}(\delta')$ is compactly embedded in $L_2(\partial D)$.

Assume next that $k \neq 0$ and define the nilpotent operator

$$\mu f = ike_0 \wedge N^+ f + \nu \wedge (ike_0 \lrcorner (\nu \lrcorner f))$$

so that $\Gamma_k = \Gamma + \mu$. We compute

$$\mu^* f = ik^c e_0 \lrcorner N^+ f + \nu \wedge (ik^c e_0 \wedge (\nu \lrcorner f)).$$

As in Example 10.1.7, we note that $\mathsf{N}(\mu) \cap \mathsf{N}(\mu^*) = \{0\}$. Consider the abstract Dirac operators

$$\Gamma_k + \Gamma_k^* = (\Gamma + \Gamma^*) + (\mu + \mu^*) : \mathsf{D}(\Gamma) \cap \mathsf{D}(\Gamma^*) \to L_2.$$

Since $\Gamma + \Gamma^* : \mathsf{D}(\Gamma) \cap \mathsf{D}(\Gamma^*) \to L_2$ is a Fredholm operator and $\mu + \mu^* : \mathsf{D}(\Gamma) \cap \mathsf{D}(\Gamma^*) \to L_2$ is a compact operator, it follows from Proposition 10.1.6 that $\Gamma_k + \Gamma_k^* : \mathsf{D}(\Gamma) \cap \mathsf{D}(\Gamma^*) \to L_2$ is a Fredholm operator. Thus the ranges are closed. To prove that in fact the cohomology space $\mathsf{N}(\Gamma_k) \cap \mathsf{N}(\Gamma_k^*)$ in fact is trivial, we note that

$$\Gamma \mu^* + \mu^* \Gamma = 0.$$

Thus, if $\Gamma f + \mu f = 0 = \Gamma^* f + \mu^* f$, then

$$0 = (f, (\Gamma \mu^* + \mu^* \Gamma) f)_V = (\Gamma^* f, \mu^* f)_V + (\mu f, \Gamma f)_V = -\|\mu^* f\|^2 - \|\mu f\|^2.$$

This shows that $f = 0$ and completes the proof.   □

Exercise 9.6.9. Roughly speaking, Γ_k^* acts as interior derivative on $N^+ L_2$ and as exterior derivative on $N^- L_2$. Write down the details of this, and show that Γ_k^* commutes with N, but not with E_k in general.

9.7 Maxwell Scattering

In this section, we demonstrate how classical Helmholtz and Maxwell boundary value and transmission problems can be solved using the operators E_k, N, and Γ_k. Recall that E_k is the reflection operator for the Hardy space splitting from Theorem 9.3.9, that N is the reflection operator for the splitting into normal and tangential fields from Example 9.4.8, and that Γ_k is the nilpotent operator for the boundary Hodge decomposition from Proposition 9.6.8. The basic operator algebra is that

$$E_k^2 = N^2 = I, \quad E_k N \neq N E_k,$$
$$\Gamma_k^2 = 0, \quad \Gamma_k E_k = E_k \Gamma_k, \quad \Gamma_k N = N \Gamma_k.$$

The $E_k N$ BVPs are essentially well posed, so by Proposition 9.4.5, roughly speaking E_k and N are closer to anticommuting than to commuting.

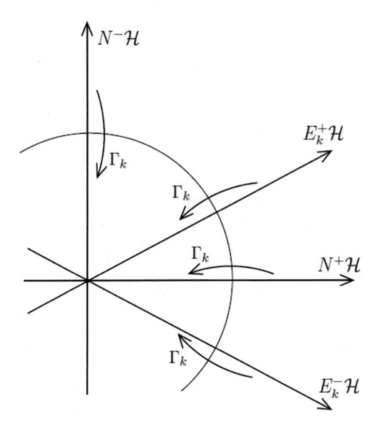

Figure 9.3: A rough sketch of the splittings involved in a Dirac BVP. The splitting into $E_k^{\pm}\mathcal{H}$ encodes the Dirac equation. The splitting into $N^{\pm}\mathcal{H}$ encodes the boundary conditions. The circle indicates the boundary Hodge splitting, with the interior of the circle illustrating $\mathsf{N}(\Gamma_k)$ where the Maxwell BVP takes place.

We note that the operator S from Example 9.4.9 does not commute with Γ_k, but we will not need this, since we use only S as a computational tool for solving an $E_k N$ BVP.

We consider as an example the exterior Dirac BVP (9.23) with prescribed tangential part. The other three $E_k N$ BVPs can be analyzed similarly. Using the operator Γ_k, we have three relevant $L_2(\partial D; \triangle W_c)$-based function spaces, namely

$$\mathcal{H} = L_2, \quad \mathcal{H} = \mathsf{D}(\Gamma), \quad \text{and } \mathcal{H} = \mathsf{N}(\Gamma_k).$$

Note that $\mathsf{D}(\Gamma) = \mathsf{D}(\Gamma_k)$ is a dense subspace of L_2, which does not depend on k, although the equivalent norms $\|f\|_{\mathsf{D}(\Gamma_k)} = (\|f\|_{L_2}^2 + \|\Gamma_k f\|_{L_2}^2)^{1/2}$ do depend on k. Further note that $\mathsf{N}(\Gamma_k)$ is a closed subspace of L_2, as well as of $\mathsf{D}(\Gamma)$, which is roughly speaking half of the latter spaces by Hodge decomposition. Since E_k and N commute with Γ_k, they act as bounded linear operators in each of the three function spaces $\mathcal{H}$, and in each case we see that $E_k^2 = N^2 = I$. Therefore we can consider the BVP (9.23), expressed as the restricted projection

$$N^+ : E_k^- \mathcal{H} \to N^+ \mathcal{H},$$

in each of the three function spaces $\mathcal{H}$. Our aim in this section is to solve BVPs in $\mathcal{H} = \mathsf{N}(\Gamma_k)$. This, however, is a function space defined by a differential constraint, which we may want to avoid numerically. For this reason, we prefer to enlarge the function space to either L_2 or to the function space $\mathsf{D}(\Gamma)$ in which roughly speaking half of the functions have Sobolev regularity H^1, since Γ_k is nilpotent, and to solve the integral equation in such a space.

Proposition 9.7.1 (Constrained Dirac BVPs). *Consider the exterior Dirac BVP $N^+ : E_k^- L_2 \to N^+ L_2$ with prescribed tangential part at ∂D, and assume that $\operatorname{Im} k \geq 0$ and $k \neq 0$, so we have L_2 well-posedness of this BVP by Theorem 9.5.2. Then the restricted map*

$$N^+ : E_k^- \mathcal{H} \to N^+ \mathcal{H}$$

is also invertible for each of the function spaces $\mathcal{H} = \mathsf{D}(\Gamma)$ and $\mathcal{H} = \mathsf{N}(\Gamma_k)$.

For the solution $f \in E_k^- L_2$ to the BVP with datum $g = N^+ f \in L_2$, the following holds. If $g \in \mathsf{D}(\Gamma)$, then $f \in \mathsf{D}(\Gamma)$. If $\Gamma_k g = 0$, then $\Gamma_k f = 0$. If $\Gamma_k g = 0$ and g is a j-vector field, then f is a j-vector field.

Note that if $g \in N^+ L_2$ is a j-vector field, then in general the solution $f \in E_k^- L_2$ to the BVP will not be a homogeneous j-vector field. The constraint $\Gamma_k g = 0$ is crucial.

Proof. (i) Lower bounds for

$$N^+ : E_k^- \mathsf{D}(\Gamma) \to N^+ \mathsf{D}(\Gamma) \tag{9.26}$$

hold, since

$$\|f\|_{\mathsf{D}(\Gamma)} \approx \|f\|_{L_2} + \|\Gamma_k f\|_{L_2} \lesssim \|N^+ f\|_{L_2} + \|N^+ \Gamma_k f\|_{L_2}$$
$$= \|N^+ f\|_{L_2} + \|\Gamma_k (N^+ f)\|_{L_2} \approx \|N^+ f\|_{\mathsf{D}(\Gamma)}.$$

To show surjectivity, we can proceed as follows. First apply Lemma 9.4.12 with $A = E_k$, $B = N$, $\mathcal{H} = \mathsf{D}(\Gamma)$ and perturb k into $\operatorname{Im} k > 0$. This shows that it suffices to show surjectivity for $\operatorname{Im} k > 0$. Then we use that N and E_k commute with Γ_k, and similarly to the above, we derive lower bounds for

$$\lambda I + E_k N : \mathsf{D}(\Gamma) \to \mathsf{D}(\Gamma),$$

when $|\lambda_1| > L|\lambda_2|$, from Theorem 9.5.1. Therefore the method of continuity shows that $I \pm E_k N$ are Fredholm operators of index zero on $\mathsf{D}(\Gamma)$. The argument in Proposition 9.4.10 shows that all four $E_k N$ BVPs are injective when $\operatorname{Im} k > 0$, and so it follows from (9.19) that (9.26) is surjective.

If $f \in E_k^- L_2$ solves the BVP with datum $g \in \mathsf{D}(\Gamma)$, then let $\tilde{f} \in E_k^- \mathsf{D}(\Gamma)$ be the solution to the well-posed BVP described by (9.26). By uniqueness of the solutions to the L_2 BVP, we conclude that $f = \tilde{f} \in \mathsf{D}(\Gamma)$.

(ii) Next consider $N^+ : E_k^- \mathsf{N}(\Gamma_k) \to N^+ \mathsf{N}(\Gamma_k)$. This map is clearly bounded and injective with a lower bound. To show surjectivity, let $g \in N^+ \mathsf{N}(\Gamma) \subset N^+ \mathsf{D}(\Gamma)$. By (i) there exists $f \in E_k^- \mathsf{D}(\Gamma)$ such that $N^+ f = g$. Since $N^+(\Gamma_k f) = \Gamma_k(N^+ f) = \Gamma_k g = 0$, it follows from L_2 well-posedness that $f \in E_k^- \mathsf{N}(\Gamma_k)$.

If furthermore $g \in \mathsf{N}(\Gamma_k)$ is a j-vector field, then the solution f satisfies $\Gamma_k f = 0$, and we conclude from Lemma 9.6.1 that each homogeneous component function f_m belongs to $E_k^+ L_2$. Since $N^+ f_m = g_m = 0$ if $m \neq j$, it follows in this case that $f_m = 0$ by uniqueness of solutions to the BVP. Therefore $f = f_j$ is a j-vector field. □

Example 9.7.2 (Helmholtz BVPs). In Example 9.3.7 we saw how the Helmholtz equation for a scalar acoustic wave u is equivalent to the vector field $F = \nabla u + ike_0 u$ solving the Dirac equation $\mathbf{D}F + ike_0 F = 0$.

(i) The Neumann BVP for u amounts to specifying the normal part

$$N^- f = (\partial_\nu u)\nu$$

of $f = F|_{\partial D}$. In this case, by Proposition 9.6.7 the condition $\Gamma_k(N^- f) = 0$ is automatic for a vector field f, since $\wedge^{-1} W = \{0\}$. Therefore, solving the Dirac BVP for F with this prescribed datum on ∂D will produce a vector field F according to Proposition 9.7.1. From Proposition 9.6.8 it follows that $F \in \mathsf{R}(\Gamma_k)$, which means that there exists a scalar function u such that $F = \nabla u + ike_0 u$. In particular, u solves $\Delta u + k^2 u = 0$ with prescribed Neumann datum.

(ii) The Dirichlet BVP for u amounts to specifying the tangential part

$$N^+ f = \nabla' u + ike_0 u$$

of $f = F|_{\partial D}$. For a given tangential vector field $g = g_1 + e_0 g_0$, where g_1 is a space-like vector field and g_0 is a scalar function, we note that $\Gamma_k g = \nabla' \wedge g_1 + e_0 \wedge (-\nabla' g_0 + ik g_1)$, so $g \in \mathsf{N}(\Gamma_k)$ amounts to

$$ik g_1 = \nabla' g_0.$$

Therefore, solving the Dirac BVP for F with such a tangential vector field $g \in \mathsf{N}(\Gamma_k)$ on ∂D as datum will produce a vector field of the form $F = \nabla u + ike_0 u$ by Proposition 9.7.1, where u solves the Helmholtz Dirichlet problem.

Example 9.7.3 (Maxwell BVPs). In Example 9.3.8 we saw how Maxwell's equations for an electromagnetic wave F are equivalent to the bivector field $F = \epsilon_0^{1/2} e_0 \wedge E + \mu_0^{-1/2} B$ solving the Dirac equation $\mathbf{D}F + ike_0 F = 0$. We now assume that the interior domain $D^+ \subset \mathbf{R}^3$ is a perfect electric conductor, so that $E = B = 0$ in D^+. If Maxwell's equations are to hold in the distributional sense in all $\mathbf{R}^3$, by the vanishing right-hand sides in the Faraday and magnetic Gauss laws, we need $N^+ f = 0$ for $f = F|_{\partial D}$. If the electromagnetic wave in D^- is the superposition of an incoming wave f_0 and a reflected wave $f_1 \in E_k^- L_2$, then f_1 needs to solve the BVP where $N^+ f_1$ is specified to cancel the datum $N^+ f_0$. Note that for the classical vector fields E and $*B$, the tangential part $N^+ f$ corresponds to the tangential part of E and the normal part of $*B$.

For a given tangential bivector field $g = e_0 \wedge g_1 + g_2$, where g_1 is a space-like tangential vector field and g_2 is a space-like tangential bivector field, we note that $\Gamma_k g = e_0 \wedge (-\nabla' \wedge g_1 + ikg_2) + \nabla' \wedge g_2$, so $g \in \mathsf{N}(\Gamma_k)$ amounts to

$$ikg_2 = \nabla' \wedge g_1.$$

In terms of the electric and magnetic fields, the tangential part of B is given by the tangential curl of the tangential part of E.

From Proposition 9.7.1 it follows that if we solve the Dirac BVP (9.23) with such a tangential bivector field $g \in \mathsf{N}(\Gamma_k)$ on ∂D as datum, then the solution f will indeed be a bivector field representing an electromagnetic field.

Example 9.7.4 (Maxwell transmission problems). When an electromagnetic wave propagates in a material and not in vacuum, we account for the material's response to the field by replacing ϵ_0 and μ_0 by permittivity and permeability constants ϵ and μ depending on the material properties. These may in general be variable as well as matrices, but we limit ourselves to homogeneous and isotropic materials for which ϵ and μ are constant complex numbers. Similar to (9.3), we define the electromagnetic field

$$F := \epsilon^{1/2} e_0 \wedge E + \mu^{-1/2} B.$$

Maxwell's equations in such a material read $\mathbf{D}F + ike_0 F = 0$, with $k = \omega\sqrt{\epsilon\mu}$.

Consider the following transmission problem. We assume that the exterior domain D^- consists of a material with electromagnetic properties described by ϵ_1, μ_1, giving a wave number $k_1 := \omega\sqrt{\epsilon_1\mu_1}$, and that the interior domain D^+ consists of a material with electromagnetic properties described by ϵ_2, μ_2, giving a wave number $k_2 := \omega\sqrt{\epsilon_2\mu_2}$. We obtain a transmission problem of the form (9.25) for a pair of electromagnetic fields

$$F^\pm : D^\pm \to \triangle^2 W_c.$$

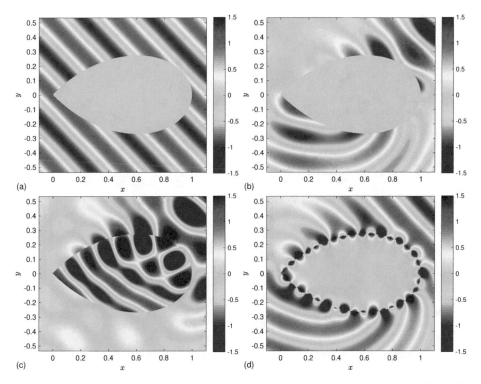

Figure 9.4: TM magnetic waves $U = B_{12}$. $\partial\Omega$ parametrized by $\sin(\pi s)\exp(i(s-1/2)$ $\pi/2)$, $0 \le s \le 1$. (a) Incoming wave $U_0 = \exp(18i(x+y)/\sqrt{2})$ from south-west. (b) Wave reflected by a perfect electric conductor, computed with the spin integral equation in Example 9.5.5. (c) Waves reflected into Ω^- and transmitted into a dielectric object Ω^+, computed with a tweaked version of the Dirac integral equation in Example 9.5.6. Wave numbers $k_1 = 18$ and $k_2 = 27$ as in Example 9.7.4. (d) As in (c), but Ω^+ is now a conducting object described by the Drude model and an imaginary wave number $k_2 = i18\sqrt{1.1838}$. Here the wave decays exponentially into Ω^+ and surface plasmon waves, excited by the corner singularity, appear near $\partial\Omega$.

The jump condition $Mf^+ = f^- + g$ is found by returning to the original formulation of Maxwell's equations for E and B. For these to hold in the distributional sense across ∂D, Faraday's law and the magnetic Gauss laws dictate that $\nu \wedge E$ and $\nu \wedge B$ do not jump across ∂D. Furthermore, assuming that we do not have any electric charges and current except for those induced in the material described by ϵ and μ, the Ampère and Gauss laws require that $\nu \lrcorner (\mu^{-1}B)$ and $\nu \lrcorner (\epsilon E)$ not jump across ∂D. If we translate this to spacetime multivector algebra, this specifies the multiplier

$$M = \sqrt{\tfrac{\mu_2}{\mu_1}}N^+T^+ + \sqrt{\tfrac{\epsilon_1}{\epsilon_2}}N^+T^- + \sqrt{\tfrac{\mu_1}{\mu_2}}N^-T^+ + \sqrt{\tfrac{\epsilon_2}{\epsilon_1}}N^-T^-,$$

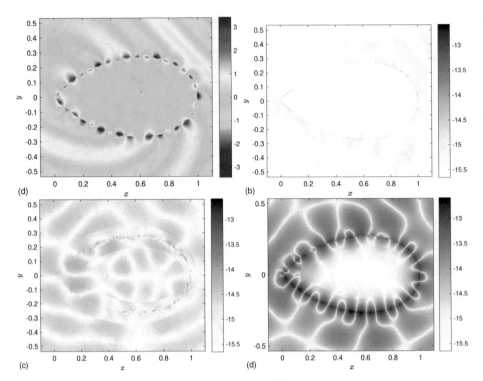

Figure 9.5: Upper left (d) is same as Figure 9.4(d), but scaled so the peaks of the plasmon wave are visible. (b), (c) and (d) show $\log_{10}$ of the estimated absolute error for the three scattering computations. (d) indicates the numerical challenge in computing surface plasmon waves. Here the parameters hit the essential spectrum, where the integral equation fails to be Fredholm.

using the normal reflection operator N and the time reflection operator T. Note how the two commuting reflection operators N and T split the electromagnetic field into these four parts.

With this formulation, and with the datum g being the boundary trace $g = F_0|_{\partial D}$ of an incoming electromagnetic wave F_0 in D^-, we can use the Dirac integral equation proposed in Example 9.5.6 to compute the transmitted wave F^+ in D^+ and the reflected wave F^- in D^-.

We end this chapter with some examples of how the integral equations from Examples 9.5.5 and 9.5.6 perform numerically when applied to scattering problems for electromagnetic fields as in Examples 9.7.3 and 9.7.4. Results are shown in Figures 9.4 and 9.5. For simplicity, we consider a two-dimensional scattering problem in which the object represented by the domain $D^+ \subset \mathbf{R}^3$ is a cylinder

$$D^+ = \Omega^+ \times \mathbf{R}$$

along the z-axis over the base $\Omega^+ \subset \mathbf{R}^2 = [e_{12}]$ in the xy-plane, and the field is transversal magnetic. This means that we assume that

$$F = F(x,y) = \sqrt{\epsilon}e_0 \wedge (E_1(x,y)e_1 + E_2(x,y)e_2) + \tfrac{1}{\sqrt{\mu}}B_{12}(x,y)e_{12}.$$

In classical vector calculus notation this means that E is parallel to $\mathbf{R}^2$ and the vector field $*B$ is orthogonal to $\mathbf{R}^2$, explaining the terminology. Maxwell's equations, after dividing F by $\sqrt{\epsilon}$, read

$$(\nabla + ike_0)(e_0 E + cB) = (c\nabla B - ikE) + e_0(-\nabla E + i\omega B) = 0,$$

where $\nabla = e_1\partial_1 + e_2\partial_2$ is the nabla symbol for $\mathbf{R}^2$. From the space- and time-like parts of this equation, we get

$$\Delta B = \nabla(\nabla B) = (ik/c)\nabla E = (ik/c)i\omega B = -k^2 B,$$

that is, $U := B_{12}$ solves the Helmholtz equation, and

$$E = (c/ik)\nabla B = (c/ik)(\nabla U)e_{12}.$$

This means that Maxwell's equations for transversal magnetic fields F are equivalent to the Helmholtz equation for $U = B_{12}$ and that E is obtained from the gradient ∇U by rotation and scaling. In particular, it follows that for transversal magnetic fields F, the tangential boundary datum N^+f corresponds to the Neumann data $\partial_\nu U$ for U.

9.8 Comments and References

9.2 Building on the work of Michael Faraday and André-Marie Ampère, James Clerk Maxwell (1831–1879) collected and completed the system of equations governing electromagnetic theory in the early 1860s. His *Treatise on Electricity and Magnetism* was published in 1873. The equations that he obtained showed that electric and magnetic fields propagate at the speed of light, and they were relativistically correct decades before Einstein formulated relativity theory.

The fundamental equation of quantum mechanics, the Schrödinger equation from Example 6.3.6, was first discovered in 1925 and describes physics at small scales. The famous Stern–Gerlach experiment from 1922 showed that the intrinsic angular momentum of particles is quantized. The Pauli equation from 1927 is a modification of the Schrödinger equation that takes this spin phenomenon into account, but neither of these is the correct equation at high speeds, that is, they are not relativistically correct. The Klein–Gordon equation from 1926 is a relativistically correct version of the Schrödinger equation, but it does not incorporate spin. Paul Dirac finally succeeded in

1928 in finding the equation that is correct from the point of view of both
quantum mechanics and relativity theory, as well as correctly describing spin-
1/2 particles, which include all the elementary particles constituting ordinary
matter.

The classical derivation of the Dirac equation is to seek matrices γ_0, γ_1,
γ_2, γ_3 by which one can factorize the Klein–Gordon equation into a first-order
wave equation. This amounts to using a matrix representation of the spacetime Clifford algebra, something that the pioneers of quantum mechanics
were unaware of. Starting from the 1960s there has been a renewed interest in Clifford's geometric algebra, where in particular, David Hestenes [55],
Hestenes and Sobczyk [57], and Hestenes [56] have advocated geometric algebra as the preferred mathematical framework for physics. In particular,
[55] is a reference for using Clifford algebra to study Maxwell's and Dirac's
equations. The formulations (9.4) and (9.5) of Maxwell's equations as wave
$\triangle$-Dirac equations go back to M. Riesz. A further reference for the use of
multivectors in electromagnetic theory is Jancewicz [60].

A standard mathematics reference for the analysis of Dirac's equation
is Thaller [93]. Further references on Dirac operators and spinors in physics
include Benn and Tucker [19] and Hitchin [58].

9.3-9.6 The material covered in these sections, which aim to solve Maxwell BVPs
using multivector calculus, builds on the author's PhD thesis and publications
[8, 9, 7, 14, 10].

The first basic idea for solving boundary value problems for Maxwell's
equations is to embed it into a Dirac equation as in Example 9.3.8. This was
first used by McIntosh and M. Mitrea in [67] in connection with BVPs on
Lipschitz domains.

The second basic idea is to formulate Dirac boundary value problems
in terms of Hardy projections $E_k^{\pm}$ and projections $N^{\pm}$ encoding boundary
conditions, and to show that these subspaces are transversal. This was first
worked out by Axelsson, Grognard, Hogan, and McIntosh [11].

The third main idea is to extract a Maxwell solution from the Dirac solution as in Proposition 9.7.1, using the Hodge decomposition on the boundary
defined by the operator Γ_k from Section 9.6. This was worked out in detail
in [9].

We have chosen to use the spacetime formulation, but as in Propositions 9.1.5 and 9.1.6, we can equally well use a $\triangle V$ formulation in which
the Dirac equation reads $\mathbf{D}F = ikF$ for $F : D \to \triangle V_c$. The main reason
for our choice is that the operator Γ_k in Section 9.6 is difficult, although not
impossible, to handle using the latter formalism. To minimize the algebra,
the $\triangle V_c$ formulation was used in [84, 80], where the spin integral equation
from Example 9.5.5 was first introduced.

A main philosophy in [9] and associated publications is to handle the boundary value problems by first-order operators. It is clear what this means for the differential operators: in (9.10) the second-order Helmholtz operator is factored by the first-order Dirac operator. But we also have corresponding factorizations of the boundary integral operators. In the abstract formulation with Proposition 9.4.5, the second-order cosine operator is factored by the first-order rotation operator in (9.20)–(9.21). We think of the rotation operators being as of first order, since they essentially are direct sums of two restricted projections as in (9.18)–(9.19). Similarly, the cosine operator can be seen to be essentially the direct sum of compositions of two restricted projections, hence of second order.

A reference for Bessel functions and Exercise 9.3.3 is Watson [95].

Standard references for the classical double and single layer potential integral equations are Colton and Kress [29, 30] and Kress [62]. The method to prove semi-Fredholm estimates of singular integral operators on Lipschitz domains as in Theorem 9.5.1 using Stokes's theorem and a smooth transversal vector field as in Exercise 6.1.8 goes back to Verchota [94]. The spectral estimates in Theorem 9.5.1 are from [7].

9.7 Figures 9.4 and 9.5 have been produced by Johan Helsing using the spin and tweaked Dirac integral equations. The state-of-the-art numerical algorithm RCIP, recursively compressed inverse preconditioning that he uses is described in [51], with applications to Helmholtz scattering in [52] and [53]. Since the Dirac equation is more general than the Helmholtz and Maxwell equations that it embeds, the spin and Dirac integral equations cannot quite compete with the most efficient Kleinman–Martin type integral equation [53, eq. 45] in terms of computational economy. In terms of achievable numerical accuracy in the solution, however, the two systems of integral equations perform almost on par with each other. Moreover, the spin and Dirac integral equations apply equally well to Maxwell scattering in three dimensions, where the present understanding of integral formulations for Maxwell's equations is incomplete.

Chapter 10

Hodge Decompositions

Prerequisites:

The reader is assumed to have read Sections 7.5 and 7.6, which this chapter develops further. A good understanding of unbounded Hilbert space operators and the material in Section 6.4 is desirable. Some exposure to distribution theory and algebraic topology helps, but is not necessary.

Road map:

We saw in Section 7.6 that every multivector field F on a domain D can be decomposed into three canonical parts

$$F = \nabla \wedge U + H + \nabla \lrcorner V,$$

where $\nabla \wedge H = 0 = \nabla \lrcorner H$, and H and the potential V are tangential on ∂D. This is the Hodge decomposition of the multivector field F, which amounts to a splitting of the space of all multivector fields F into two subspaces $\mathsf{R}(d)$ and $\mathsf{R}(\delta)$ of exact and coexact fields respectively, and a small subspace $\mathcal{C}_{\parallel}(D)$ of closed and coclosed fields, all with appropriate boundary conditions. Alternatively, we can instead demand that H and the potential U be normal on ∂D. At least four types of questions arise.

(i) Are the subspaces $\mathsf{R}(d)$ and $\mathsf{R}(\delta)$ transversal, that is do they intersect only at 0 and at a positive angle? This would mean that these subspaces give a splitting of the function space $\mathcal{H}$ that we consider, modulo $\mathcal{C}_{\parallel}(D)$. In the case of $\mathcal{H} = L_2(D)$ which we only consider here, these subspaces are in fact orthogonal, but more generally this problem amounts to estimating singular integral operators realizing the Hodge projections onto these subspaces. We touch on this problem in Proposition 10.1.5 and Example 10.1.8.

(ii) Are the ranges $\mathsf{R}(d)$ and $\mathsf{R}(\delta)$ closed subspaces? This is a main problem that we address in this chapter, and we show that this is indeed the case for

© Springer Nature Switzerland AG 2019
A. Rosén, *Geometric Multivector Analysis*, Birkhäuser Advanced Texts Basler Lehrbücher,
https://doi.org/10.1007/978-3-030-31411-8_10

bounded domains D. See Section 10.3 and Example 10.1.8. We saw in Section 7.6 that such closedness yields well-posedness results for boundary value problems.

(iii) What properties, in particular regularity, of the potentials U and V do we have? Note that the parts $\nabla \wedge U$ and $\nabla \lrcorner V$ are uniquely determined by F, but not so for the potentials U and V. We show in Section 10.4 that the most obvious choice, the Hodge potentials of minimal L_2 norm, are not always the best choices. Even more surprising is the fact that there exist Bogovskiĭ potentials V, for which we have full Dirichlet boundary conditions $V|_{\partial D} = 0$.

(iv) Is the cohomology space $\mathcal{C}_\parallel(D)$ finite-dimensional? More exactly, how do we go about calculating the dimension of this subspace for a given domain D? As compared to the first three questions, which belong to analysis, this fourth question belongs to algebraic topology and is addressed in Section 10.6.

In the analysis of Hodge decompositions on domains, the regularity and curvature of the boundary play an important role through Weitzenböck formulas. Hodge decompositions can also be considered on manifolds, and in this case also the curvature of the manifold in the interior of the domain enters the picture. This will be a central idea in Chapters 11 and 12. In the present chapter we avoid the technicalities of vector bundles and limit the discussion to domains in affine spaces.

Highlights:

- Compactness and Hodge decomposition: 10.1.6

- Natural boundary conditions for d and δ: 10.2.3

- Weitzenböck boundary curvature: 10.3.6

- Bogovskiĭ and Poincaré potentials: 10.4.3

- Čech computation of Betti numbers: 10.6.5

10.1   Nilpotent operators

In terms of operators, a splitting of a function space corresponds to a projection P, along with its complementary projection $I - P$. Somewhat similarly, we show in this section how Hilbert space operators Γ with the property that $\Gamma^2 = 0$ induce splittings of the function space in a natural way, generalizing Hodge decompositions. Usually the condition $\Gamma^k = 0$ for some $k \in \mathbf{Z}_+$ defines nilpotence, but we shall always assume index $k = 2$.

Definition 10.1.1 (Nilpotent). A linear, possibly unbounded, operator $\Gamma : \mathcal{H} \to \mathcal{H}$ in a Hilbert space $\mathcal{H}$ is said to be *nilpotent* (with index 2) if it is densely defined, closed, and if $\mathsf{R}(\Gamma) \subset \mathsf{N}(\Gamma)$. In particular, $\Gamma^2 f = 0$ for all $f \in \mathsf{D}(\Gamma)$. We say that a nilpotent operator Γ is *exact* if $\overline{\mathsf{R}(\Gamma)} = \mathsf{N}(\Gamma)$.   .

Recall that the null space $N(\Gamma)$ is always closed if Γ is closed but that in general, the range $R(\Gamma)$ is not a closed subspace. If Γ is nilpotent, then we have inclusions

$$R(\Gamma) \subset \overline{R(\Gamma)} \subset N(\Gamma) \subset D(\Gamma) \subset \mathcal{H}.$$

Let $\mathcal{H}_0$ denote any closed subspace complementary to $N(\Gamma)$, for example $\mathcal{H}_0 = N(\Gamma)^\perp$, so that $\mathcal{H} = N(\Gamma) \oplus \mathcal{H}_0$. Then the restricted map

$$\Gamma : \mathcal{H}_0 \to R(\Gamma) \subset N(\Gamma)$$

is injective, which roughly speaking means that $N(\Gamma)$ is at least half of $\mathcal{H}$. For this reason it is natural to combine a nilpotent operator Γ_1 with a "complementary" nilpotent operator Γ_2. Ideally one would like to have a splitting of the Hilbert space

$$\mathcal{H} = R(\Gamma_1) \oplus R(\Gamma_2),$$

where $R(\Gamma_1) = N(\Gamma_1)$ and $R(\Gamma_2) = N(\Gamma_2)$. Since $N(\Gamma_1^*) = R(\Gamma_1)^\perp$, the natural choice in a Hilbert space is $\Gamma_2 = \Gamma_1^*$.

Proposition 10.1.2 (Abstract Hodge decomposition). *Let Γ be a nilpotent operator in a Hilbert space $\mathcal{H}$. Then so is Γ^*, and there is an orthogonal splitting into closed subspaces*

$$\mathcal{H} = \overline{R(\Gamma)} \oplus \mathcal{C}(\Gamma) \oplus \overline{R(\Gamma^*)}, \tag{10.1}$$

where

$$\mathcal{C}(\Gamma) := N(\Gamma) \cap N(\Gamma^*),$$
$$N(\Gamma) = \overline{R(\Gamma)} \oplus \mathcal{C}(\Gamma), \quad and$$
$$N(\Gamma^*) = \mathcal{C}(\Gamma) \oplus \overline{R(\Gamma^*)}.$$

Note that $\mathcal{C}(\Gamma) = \{0\}$ if and only if Γ is exact.

Proof. If T is a densely defined and closed operator in $\mathcal{H}$, then $R(T)^\perp = N(T^*)$ and therefore $\overline{R(T)} = N(T^*)^\perp$. This proves that $\overline{R(\Gamma^*)} = N(\Gamma)^\perp \subset R(\Gamma)^\perp = N(\Gamma^*)$, showing that Γ^* is nilpotent and that we have orthogonal splittings

$$\mathcal{H} = N(\Gamma) \oplus \overline{R(\Gamma^*)} = \overline{R(\Gamma)} \oplus N(\Gamma^*).$$

But $R(\Gamma) \subset N(\Gamma)$, since Γ is nilpotent, so using the second splitting in the first, we get

$$N(\Gamma) = \overline{R(\Gamma)} \oplus \left(N(\Gamma) \cap \overline{R(\Gamma)}^\perp \right) = \overline{R(\Gamma)} \oplus \left(N(\Gamma) \cap N(\Gamma^*) \right),$$

which proves the stated splitting. $\qquad\square$

The mapping properties of Γ and Γ^* are as follows. In the Hodge decomposition (10.1), the operator Γ is zero on $\overline{R(\Gamma)} \oplus \left(N(\Gamma) \cap N(\Gamma^*) \right) = N(\Gamma)$, and Γ^* is zero on $\left(N(\Gamma^*) \cap N(\Gamma) \right) \oplus \overline{R(\Gamma^*)} = N(\Gamma^*)$. On the other hand, we see that the restrictions $\Gamma : \overline{R(\Gamma^*)} \to \overline{R(\Gamma)}$ and $\Gamma^* : \overline{R(\Gamma)} \to \overline{R(\Gamma^*)}$ are injective and have dense ranges:

$$\mathcal{H} \;=\; \overline{R(\Gamma)} \quad \oplus \quad N(\Gamma) \cap N(\Gamma^*) \quad \oplus \quad \overline{R(\Gamma^*)} \tag{10.2}$$

$$\mathcal{H} \;=\; \overline{R(\Gamma)} \quad \oplus \quad N(\Gamma) \cap N(\Gamma^*) \quad \oplus \quad \overline{R(\Gamma^*)}$$

We have been using the formally skew-adjoint Dirac operator $\mathbf{D} = d + \delta$ in Chapters 8 and 9. Using instead the anti-Euclidean Clifford product leads to a formally self-adjoint Dirac operator $d - \delta$. For the following results we can use either the abstract Dirac operator $\Gamma - \Gamma^*$ or its self-adjoint analogue $\Gamma + \Gamma^*$. To be able to use resolvents without complexifying the space, we choose to work with $\Gamma - \Gamma^*$. Note from the mapping properties of Γ and Γ^* that such operators swap the subspaces $\overline{R(\Gamma)}$ and $\overline{R(\Gamma^*)}$.

Proposition 10.1.3 (Abstract Hodge–Dirac operators). *Let Γ be a nilpotent operator in a Hilbert space $\mathcal{H}$. Consider the operator $\Pi := \Gamma - \Gamma^*$ with domain $D(\Pi) := D(\Gamma) \cap D(\Gamma^*)$. Then Π is skew-adjoint, that is, $\Pi^* = -\Pi$ in the sense of unbounded operators, with $N(\Pi) = \mathcal{C}(\Gamma)$ and $R(\Pi) = R(\Gamma) + R(\Gamma^*)$.*

We refer to operators $\Pi = \Gamma - \Gamma^*$, derived from a nilpotent operator Γ, as an abstract *Hodge–Dirac operator*. Note that in Euclidean spaces, the $\triangle$-Dirac operator $\mathbf{D}$ from Definition 9.1.1 is an example of a Hodge–Dirac operator, whereas to have the $\slashed{\triangle}$-Dirac operator $\slashed{D}$ from Definition 9.1.3 as a Hodge–Dirac operator requires a complex structure on our Euclidean space, as discussed at the end of Section 9.1.

Proof. We use the Hodge decomposition from Proposition 10.1.2. If $\Gamma u + \Gamma^* u = 0$, then $\Gamma u = 0 = \Gamma^* u$ by orthogonality, from which $N(\Pi) = \mathcal{C}(\Gamma)$ follows. If $f = \Gamma u_1 + \Gamma^* u_2$, then $f = \Pi(P_{\Gamma^*} u_1 + P_{\Gamma} u_2)$, from which $R(\Pi) = R(\Gamma) + R(\Gamma^*)$ follows. Note that $u_1 - P_{\Gamma^*} u_1 \in N(\Gamma) \subset D(\Gamma)$ and similarly for u_2.

It is clear that $-\Pi$ is the formal adjoint of Π. It remains to prove that if

$$\langle f, \Pi g \rangle + \langle f', g \rangle = 0$$

for all $g \in D(\Pi)$, then $f \in D(\Pi)$ and $f' = \Pi f$. Writing $f = f_1 + f_2 + f_3$ in the Hodge splitting, and similarly for f', we have

$$\langle f_1, \Gamma g \rangle + \langle f_3', g \rangle = 0,$$
$$0 + \langle f_2', g \rangle = 0,$$
$$\langle f_3, -\Gamma^* g \rangle + \langle f_1', g \rangle = 0,$$

by choosing $g \in \overline{R(\Gamma^*)} \cap D(\Gamma)$, $g \in \mathcal{C}(\Gamma)$ and $g \in \overline{R(\Gamma)} \cap D(\Gamma^*)$ respectively. Since Γ and Γ^* are adjoint in the sense of unbounded operators, we conclude that $f_1 \in D(\Gamma^*)$, $f_3' = -\Gamma^* f_1$, $f_2' = 0$, $f_3 \in D(\Gamma)$ and $f_1' = \Gamma f_3$. This shows that $f \in D(\Pi)$ and $f' = \Pi f$. $\qquad\square$

Definition 10.1.4 (Hodge projections). Let Γ be a nilpotent operator in a Hilbert space $\mathcal{H}$. The associated *Hodge projections* are the orthogonal projections P_Γ and P_{Γ^*} onto the subspaces $\overline{R(\Gamma)}$ and $\overline{R(\Gamma^*)}$ respectively. The orthogonal projection $P_{\mathcal{C}(\Gamma)}$ onto the Γ-*cohomology space* $\mathcal{C}(\Gamma)$ is $P_{\mathcal{C}(\Gamma)} = I - P_\Gamma - P_{\Gamma^*}$.

Proposition 10.1.5 (Formulas for Hodge projections). *Let Γ be a nilpotent operator in a Hilbert space $\mathcal{H}$. If Γ is exact, then*

$$P_\Gamma f = \Gamma\Pi^{-1}f = -\Pi^{-1}\Gamma^*f = -\Gamma\Pi^{-2}\Gamma^*f,$$
$$P_{\Gamma^*}f = -\Gamma^*\Pi^{-1}f = \Pi^{-1}\Gamma f = -\Gamma^*\Pi^{-2}\Gamma f,$$

for $f \in D(\Pi) \cap R(\Pi)$.

If Γ is not exact, let $\epsilon \in \mathbf{R}\setminus\{0\}$. Then we have $P_{\mathcal{C}(\Gamma)}f = \lim_{\epsilon\to 0} \epsilon(\epsilon I + \Pi)^{-1}f$, and the Hodge projections are

$$P_\Gamma f = \lim_{\epsilon\to 0} \Gamma(\epsilon I + \Pi)^{-1}f,$$
$$P_{\Gamma^*}f = -\lim_{\epsilon\to 0} \Gamma^*(\epsilon I + \Pi)^{-1}f,$$

with convergence in $\mathcal{H}$, for $f \in \mathcal{H}$. We also have $P_\Gamma f = -\lim(\epsilon I + \Pi)^{-1}\Gamma^* f$ for $f \in D(\Gamma^*)$ and $P_{\Gamma^*}f = \lim(\epsilon I + \Pi)^{-1}\Gamma f$ for $f \in D(\Gamma)$.

Proof. The formulas for exact operators Γ involving Π^{-1} are immediate from (10.11), and the final second-order formulas follow since $P_\Gamma = P_\Gamma^2$ and $P_{\Gamma^*} = P_{\Gamma^*}^2$.

For nonexact Γ, consider first $P_{\mathcal{C}(\Gamma)}f$. If $f \in \mathcal{C}(\Gamma)$, then $\epsilon(\epsilon I + \Pi)^{-1}f = f$. If $f = \Pi u \in R(\Pi)$, then

$$\epsilon(\epsilon I + \Pi)^{-1}\Pi u = \epsilon u - \epsilon^2(\epsilon I + \Pi)^{-1}u \to 0$$

as $\epsilon \to 0$. We have used the skew-adjointness of Π, which implies that $\|\epsilon(\epsilon I + \Pi)^{-1}\| \le 1$. These uniform bounds also allow us to conclude that $\epsilon(\epsilon I + \Pi)^{-1}f \to 0$ also for all $f \in \overline{R(\Pi)}$. This proves the formula for $P_{\mathcal{C}(\Gamma)}f$, from which it immediately follows that

$$\Gamma(\epsilon I + \Pi)^{-1}f = P_\Gamma\Pi(\epsilon I + \Pi)^{-1}f \to P_\Gamma(f - P_{\mathcal{C}(\Gamma)}f) = P_\Gamma f,$$

and similarly for P_{Γ^*}. Alternatively, for $f \in D(\Gamma^*)$, we have

$$-(\epsilon I + \Pi)^{-1}\Gamma^* f = (\epsilon I + \Pi)^{-1}\Pi P_\Gamma f \to (I - P_{\mathcal{C}(\Gamma)})P_\Gamma f = P_\Gamma f,$$

and similarly for P_{Γ^*}. $\qquad\square$

The following result describes an important property that a nilpotent operator may have, which we will establish for d and δ on bounded Lipschitz domains.

Proposition 10.1.6 (Compact potential maps). *For a nilpotent operator Γ in a Hilbert space $\mathcal{H}$, the following are equivalent.*

(i) *The subspaces $R(\Gamma)$ and $R(\Gamma^*)$ are closed and $C(\Gamma)$ is finite-dimensional, and the inverses of $\Gamma : R(\Gamma^*) \to R(\Gamma)$ and $\Gamma^* : R(\Gamma) \to R(\Gamma^*)$ are compact.*

(ii) *There exist compact operators K_0, $K_1 : \mathcal{H} \to \mathcal{H}$, with $R(K_1) \subset D(\Gamma)$, such that the homotopy relation*

$$\Gamma K_1 f + K_1 \Gamma f + K_0 f = f$$

holds for all $f \in D(\Gamma)$.

(iii) *The Hilbert space $D(\Gamma) \cap D(\Gamma^*)$, equipped with the norm $(\|f\|^2 + \|\Gamma f\|^2 + \|\Gamma^* f\|^2)^{1/2}$, is compactly embedded in $\mathcal{H}$.*

Carefully note that unlike (i) and (iii), property (ii) does not involve the adjoint Γ^*. We exploit this in Theorem 10.3.1 below to reduce the problem of existence of potentials, from Lipschitz domains to smooth domains.

Also note for (i) that when the ranges are closed, $\Gamma : R(\Gamma^*) \to R(\Gamma)$ has a compact inverse if and only if there exists a compact operator $K_\Gamma : R(\Gamma) \to \mathcal{H}$ such that $\Gamma K_\Gamma = I_{R(\Gamma)}$. Indeed, if we have such K_Γ, then $P_\Gamma K_\Gamma$ is a compact operator giving a potential $u \in R(\Gamma^*)$.

Proof. Assume (i). Define compact operators $K_0 := P_{C(\Gamma)}$, and

$$K_1 f := \begin{cases} \Gamma^{-1} f \in R(\Gamma^*), & f \in R(\Gamma), \\ 0, & f \in N(\Gamma^*). \end{cases}$$

It is straightforward to verify that $\Gamma K_1 = P_\Gamma$ and $K_1 \Gamma = P_{\Gamma^*}$, from which (ii) follows.

Assume (ii). Let $(f_j)_{j=1}^\infty$ be a sequence such that f_j, Γf_j and $\Gamma^* f_j$ all are bounded sequences in $\mathcal{H}$. We have

$$(I - P_\Gamma) f_j = (I - P_\Gamma)(\Gamma K_1 f_j + K_1(\Gamma f_j) + K_0 f_j) = (I - P_\Gamma) K_1(\Gamma f_j) + (I - P_\Gamma) K_0 f_j.$$

By duality, we also obtain from the homotopy relation that

$$P_\Gamma f_j = P_\Gamma(\Gamma^* K_1^* f_j + K_1^*(\Gamma^* f_j) + K_0^* f_j) = P_\Gamma K_1^*(\Gamma^* f_j) + P_\Gamma K_0^* f_j.$$

This shows that $(P_{\Gamma^*} f_j)_{j=1}^\infty$, $(P_{C(\Gamma)} f_j)_{j=1}^\infty$ and $(P_{\Gamma^*} f_j)_{j=1}^\infty$ have subsequences that converge in $\mathcal{H}$, and (iii) follows.

Assume (iii). The operator $I + \Pi$ is an isometry between the Hilbert spaces $D(\Gamma) \cap D(\Gamma^*)$ and $\mathcal{H}$, since Π is a skew-adjoint operator. Since I is compact between these spaces, perturbation theory shows that

$$\Pi : D(\Gamma) \cap D(\Gamma^*) \to \mathcal{H}$$

is a Fredholm operator, and (i) follows. $\square$

Nilpotent operators appear naturally from the exterior and interior products, since $v \wedge v \wedge w = 0$ and $v \lrcorner (v \lrcorner w) = 0$.

Example 10.1.7 (Algebraic Hodge decomposition). Fix a unit vector $v \in V$ in an n-dimensional Euclidean space and define nilpotent linear maps

$$\mu(w) := v \wedge w, \qquad \mu^*(w) := v \lrcorner w, \qquad w \in \wedge V.$$

We apply the abstract theory above to $\Gamma = \mu$ and $\mathcal{H}$, the finite-dimensional Hilbert space $\wedge V$. Lemma 2.2.7 shows that $\mathsf{R}(\mu) = \mathsf{N}(\mu)$, so in this case μ is exact and the Hodge decomposition reads

$$\wedge V = \mathsf{R}(\mu) \oplus \mathsf{R}(\mu^*),$$

where $\mathsf{R}(\mu)$ are the multivectors normal to and $\mathsf{R}(\mu^*)$ are the multivectors tangential to the hyperplane $[v]^\perp$, in the sense of Definition 2.8.6. We have $(\mu - \mu^*)^2 = -1$, and the Hodge projections are $\mu\mu^*$ onto normal multivectors, and $\mu^*\mu$ onto tangential multivectors.

Note that $\mathsf{R}(\mu)$ and $\mathsf{R}(\mu^*)$, for the full algebra $\wedge V$, both have dimension 2^{n-1}. However, this is not true in general for the restrictions to $\wedge^k V$. For example, the space $\mathsf{R}(\mu) \cap \wedge^1 V$ of vectors normal to $[v]^\perp$ is one-dimensional, whereas the space $\mathsf{R}(\mu^*) \cap \wedge^1 V$ of vectors tangential to $[v]^\perp$ has dimension $n - 1$. The smaller k is, the more tangential k-vectors exist as compared to normal k-vectors. At the ends, all scalars are tangential and all n-vectors are normal.

Example 10.1.8 ($\mathbf{R}^n$ Hodge decomposition). Consider the exterior and interior derivative operators

$$dF(x) = \nabla \wedge F(x) \quad \text{and} \quad \delta F(x) = \nabla \lrcorner F(x)$$

in the Hilbert space $\mathcal{H} = L_2(V; \wedge V_c)$ on the whole Euclidean space $X = V$, where we complexify the exterior algebra in order to use the Fourier transform. These two nilpotent operators are the Fourier multipliers

$$\widehat{dF}(\xi) = \sum_{k=1}^n e_i \wedge \int_X \partial_i F(x) e^{-i\langle \xi, x \rangle} dx = i\xi \wedge \hat{F}(\xi),$$

$$\widehat{\delta F}(\xi) = \sum_{k=1}^n e_i \lrcorner \int_X \partial_i F(x) e^{-i\langle \xi, x \rangle} dx = i\xi \lrcorner \hat{F}(\xi) = (-i\xi) \lrcorner F(\xi),$$

defining the interior product as the sesquilinear adjoint of the exterior product. Define the pointwise multiplication operators

$$\mu_\xi(\hat{F}(\xi)) := \xi \wedge \hat{F}(\xi), \qquad \mu_\xi^*(\hat{F}(\xi)) := \xi \lrcorner \hat{F}(\xi).$$

We view $\mu_\xi, \mu_\xi^* : L_2(X; \wedge V_c) \to L_2(X; \wedge V_c)$ as multiplication operators by the radial vector field $X \ni \xi \mapsto \xi = \sum_{k=1}^n \xi_k e_k \in V$. Thus we have

$$\mathcal{F}(dF) = i\mu_\xi(\mathcal{F}(F)), \qquad \mathcal{F}(\delta F) = i\mu_\xi^*(\mathcal{F}(F)).$$

In particular, F is closed if and only if $\hat{F}$ is a *radial* multivector field, that is, $\xi \wedge \hat{F} = 0$, and F is coclosed if and only if $\hat{F}$ is an *angular* multivector field, that is, $\xi \lrcorner \hat{F} = 0$.

From Plancherel's theorem it is clear that $\Gamma = d$ and $\Gamma^* = -\delta$, with domains $D(\Gamma) := \{F \in L_2 \; ; \; \xi \wedge \hat{F} \in L_2\}$ and $D(\Gamma^*) := \{F \in L_2 \; ; \; \xi \lrcorner \hat{F} \in L_2\}$, are nilpotent operators in $\mathcal{H}$, and that $d = -\delta^*$ in the sense of unbounded operators. In this case, the Hodge decomposition reads

$$L_2(V; \wedge V_c) = \overline{R(d)} \oplus \overline{R(\delta)}.$$

That d is exact is a consequence of μ_ξ being exact for each $\xi \in V \setminus \{0\}$. By considering $\hat{F}$ near $\xi = 0$, we see that the ranges are not closed, which is a consequence of the domain X not being bounded. Using the formulas from Proposition 10.1.5, we see that the Hodge projections are the singular integrals

$$P_d F(x) = \nabla \wedge \int_X \Psi(x - y) \lrcorner F(y) dy$$
$$= \frac{k}{n} F(x) + \text{p.v.} \int_X \nabla \wedge (\Psi(\dot{x} - y) \lrcorner F(y)) dy,$$
$$P_\delta F(x) = \nabla \lrcorner \int_X \Psi(x - y) \wedge F(y) dy$$
$$= \frac{n - k}{n} F(x) + \text{p.v.} \int_X \nabla \lrcorner (\Psi(\dot{x} - y) \wedge F(y)) dy,$$

for k-vector fields $F \in L_2(X; \wedge^k V_c)$. We have used the distributional derivative $\partial_i \Psi(x) = e_i \delta(x)/n + \text{p.v.} \partial_i \Psi(x)$.

10.2   Half-Elliptic Boundary Conditions

For the remainder of this chapter, we study the nilpotent operators d and δ on bounded domains D, at least Lipschitz regular, in Euclidean space X. The main idea in this section is to use the commutation theorem (Theorem 7.2.9) and reduce the problems to smooth domains. Realizing the operators that are implicit in Definition 7.6.1 as unbounded nilpotent operators, we have the following.

Definition 10.2.1 (d and δ on domains). Let D be a bounded Lipschitz domain in a Euclidean space (X, V). Define unbounded linear operators $d, \underline{d}, \delta, \underline{\delta}$ in $L_2(D) = L_2(D; \wedge V)$ as follows.

Assume that $F, F' \in L_2(D)$ and consider the equation

$$\int_D \Big(\langle F'(x), \phi(x) \rangle + \langle F(x), \nabla \lrcorner \phi(x) \rangle \Big) dx = 0.$$

If this holds for all $\phi \in C_0^\infty(D)$, then we define $F \in D(d)$ and $dF := F'$. If this holds for all $\phi \in C^\infty(\overline{D})$, then we define $F \in D(\underline{d})$ and $\underline{d}F := F'$.

Assume that $F, F' \in L_2(D)$ and consider the equation

$$\int_D \Big(\langle F'(x), \phi(x) \rangle + \langle F(x), \nabla \wedge \phi(x) \rangle \Big) dx = 0.$$

If this holds for all $\phi \in C_0^\infty(D)$, then we define $F \in \mathsf{D}(\delta)$ and $\delta F := F'$. If this holds for all $\phi \in C^\infty(\overline{D})$, then we define $F \in \mathsf{D}(\underline{\delta})$ and $\underline{\delta} F := F'$.

We recall from Section 7.6 that by Stokes's theorem we interpret $F \in \mathsf{D}(\underline{d})$ as being normal at ∂D in a weak sense, and $F \in \mathsf{D}(\underline{\delta})$ as being tangential at ∂D in a weak sense. Basic properties of these operators are the following.

Proposition 10.2.2 (Nilpotence). *Let D be a bounded Lipschitz domain in a Euclidean space. Then the operators $d, \underline{d}, \delta, \underline{\delta}$ are well-defined nilpotent operators on $L_2(D)$. In particular, they are linear, closed, and densely defined. With the pointwise Hodge star and involution maps, we have*

$$\delta(F*) = (d\widehat{F})*, \quad F \in \mathsf{D}(d),$$
$$\underline{d}(F*) = (\underline{\delta}\widehat{F})*, \quad F \in \mathsf{D}(\underline{\delta}).$$

Proof. Consider $\underline{d}$. The other proofs are similar. That $\underline{d}$ is defined on $C_0^\infty(D)$, linear, and closed is clear. It is well defined, since $F = 0$ implies $F' = 0$, since $C^\infty(\overline{D})$ is dense in $L_2(D)$. To show nilpotence, assume $F \in \mathsf{D}(\underline{d})$. Then

$$\int_D \Big(0 + \langle \underline{d}F(x), \nabla \lrcorner \phi(x) \rangle \Big) dx = - \int_D \langle F(x), \nabla \lrcorner (\nabla \lrcorner \phi(x)) \rangle dx = 0$$

for all $\phi \in C^\infty(\overline{D})$, which shows that $\underline{d}(\underline{d}F) = 0$.

The relation between $d, \underline{\delta}$ and $\delta, \underline{d}$ follows from Proposition 7.1.7(i). $\qquad\square$

The goal of this section is to prove the following duality. Recall the definition (6.4) of adjointness in the sense of unbounded operators.

Proposition 10.2.3 (Duality). *Let D be a bounded Lipschitz domain in a Euclidean space. Then d and $-\underline{\delta}$ are adjoint in the sense of unbounded operators. Similarly, $\underline{d}$ and $-\delta$ are adjoint in the sense of unbounded operators.*

From Propositions 10.1.2 and 10.2.3 we obtain a Hodge decomposition with tangential boundary conditions

$$L_2(D) = \overline{\mathsf{R}(d)} \oplus \mathcal{C}_\|(D) \oplus \overline{\mathsf{R}(\underline{\delta})},$$

where $\mathcal{C}_\|(D) := \mathsf{N}(d) \cap \mathsf{N}(\underline{\delta})$, and a Hodge decomposition with normal boundary conditions

$$L_2(D) = \overline{\mathsf{R}(\underline{d})} \oplus \mathcal{C}_\perp(D) \oplus \overline{\mathsf{R}(\delta)},$$

where $\mathcal{C}_\perp(D) := \mathsf{N}(\underline{d}) \cap \mathsf{N}(\delta)$. We will prove in Section 10.3 that the ranges of the four operators are closed, so the closures here are redundant. For the proof of Proposition 10.2.3, we need the following results.

Lemma 10.2.4 (Local nonsmooth commutation theorem). *Let $\rho : D_1 \to D_2$ be a Lipschitz diffeomorphism between domains D_1 and D_2 in Euclidean space. If $F \in D(d)$ in D_2 with supp $F \subset D_2$, then $\rho^* F \in D(d)$ with*

$$d(\rho^* F) = \rho^*(dF)$$

in D_1. Similarly, if $F \in D(\delta)$ in D_1 with supp $F \subset D_1$, then $\tilde{\rho}_* F \in D(\delta)$ with $\delta(\tilde{\rho}_* F) = \tilde{\rho}_*(\delta F)$ in D_2.

We recall, for example, that supp $F \subset D_2$ means that $F = 0$ in a neighborhood of ∂D_2. Note that for general Lipschitz changes of variables, $\rho^* F$ and $\tilde{\rho}_* F$ are defined almost everywhere by Rademacher's theorem.

Proof. By Proposition 7.2.7 it suffices to prove the first statement. Consider first $F \in C_0^\infty(D_2)$. We mollify and approximate ρ by

$$\rho_t(x) := \eta_t * \rho(x),$$

where $\eta \in C_0^\infty(X; \mathbf{R})$ with $\int \eta = 1$ and $\eta_t(x) := t^{-n}\eta(x/t)$. Note that ρ_t is well defined on every compact subset of D_1 for small t. It follows that $\rho_t \in C^\infty$ and

$$d(\rho_t^* F) = \rho_t^*(dF)$$

holds by Theorem 7.2.9. From the dominated convergence theorem we conclude that $\rho_t^* F \to \rho^* F$ in $L_2(D_1)$. Since for the same reason $\rho_t^*(dF) \to \rho^*(dF)$, and d is a closed operator, it follows that $\rho^* F \in D(d)$ and $d(\rho^* F) = \rho^*(dF)$.

Next consider general $F \in D(d)$ with compact support in D_2. Similarly to above, we now mollify and approximate F by $F_n \in C_0^\infty(D_2)$, with $F_n \to F$ and $dF_n \to dF$ in $L_2(D_2)$. We have shown above that $d(\rho^* F_n) = \rho^*(dF_n)$. Using that $\rho^* : L_2(D_2) \to L_2(D_1)$ is bounded and that d is closed, it follows that $\rho^* F \in D(d)$ and $d(\rho^* F) = \rho^*(dF)$. $\square$

The following shows that the normal and tangential boundary conditions for d and δ are obtained by closure from C_0^∞.

Proposition 10.2.5 (Half Dirichlet conditions). *Let D be a bounded Lipschitz domain in a Euclidean space. If $F \in D(\underline{d})$, then there exists $F_t \in C_0^\infty(D)$ such that*

$$F_t \to F \quad and \quad \underline{d} F_t \to \underline{d} F$$

in $L_2(D)$ as $t \to 0$. Similarly, if $F \in D(\underline{\delta})$, then there exists $F_t \in C_0^\infty(D)$ such that $F_t \to F$ and $\underline{\delta} F_t \to \underline{\delta} F$ in $L_2(D)$ as $t \to 0$.

Proof. By Hodge star duality it suffices to consider $\underline{d}$. By the compactness of D, we can localize and assume that supp $F \subset D_p \cap \overline{D}$ near $p \in \partial D$ as in Definition 6.1.4. We note from Definition 10.2.1 that extending F by 0 outside D, we have $F \in D(d)$ on X as in Example 10.1.8. Pulling back by the local parametrization ρ, Lemma 10.2.4 shows that $\rho^* F \in D(d)$ on $\mathbf{R}^n$. We translate $\rho^* F$ up into Ω_p and pull back by ρ^{-1} to define $\tilde{F}_t := (\rho^*)^{-1}(\rho^* F(x', x_n - t))$. This yields $\tilde{F}_t \in \mathsf{D}(d)$ with supp $\tilde{F}_t \subset D$. Finally, we mollify and approximate $\tilde{F}_t$ by

$$F_t(x) := \eta_t(x) * \tilde{F}_t(x), \quad x \in D,$$

where $\eta \in C_0^\infty(X; \mathbf{R})$ with $\int \eta = 1$, supp $\eta \subset B(0, r)$, and $\eta_t(x) := t^{-n}\eta(x/t)$. If $0 < r < t$ is chosen small enough depending on the Lipschitz geometry, we obtain $F_t \in C_0^\infty(D)$ and can verify that F_t and dF_t converge to F and dF respectively. $\quad\square$

Proof of Proposition 10.2.3. Consider the equation

$$\int_D \left(\langle F'(x), \phi(x) \rangle + \langle F(x), \nabla \lrcorner \phi(x) \rangle \right) dx = 0.$$

This holds for all $F \in \mathsf{D}(d)$ with $F' = dF$, and all $\phi \in C_0^\infty(D)$ by Definition 10.2.1. By Proposition 10.2.5 and a limiting argument, this continues to hold for $\phi \in \mathsf{D}(\underline{\delta})$. This shows that d and $-\underline{\delta}$ are formally adjoint. Furthermore, assume that the equation holds for some F and $F' \in L_2(D)$ and all $\phi \in \mathsf{D}(\delta)$. In particular, it holds for all $\phi \in C_0^\infty(D)$, and it follows by definition that $F \in \mathsf{D}(d)$ and $F' = dF$. This shows that d and $-\underline{\delta}$ are adjoint in the sense of unbounded operators. The proof that $\underline{d}$ and $-\delta$ are adjoint in the sense of unbounded operators is similar. $\quad\square$

We next remove the assumption of compact support in Lemma 10.2.4.

Lemma 10.2.6 (Nonsmooth commutation theorem). *Let $\rho : D_1 \to D_2$ be a Lipschitz diffeomorphism between bounded Lipschitz domains D_1 and D_2 in Euclidean space. If $F \in \mathsf{D}(d)$ on D_2, then $\rho^* F \in \mathsf{D}(d)$ on $\mathsf{D}(D_1)$ with*

$$d(\rho^* F) = \rho^*(dF)$$

in D_1. Similarly, if $F \in \mathsf{D}(\delta)$ on D_1, then $\tilde{\rho}_* F \in \mathsf{D}(\delta)$ with $\delta(\tilde{\rho}_* F) = \tilde{\rho}_*(\delta F)$ on D_2.

Proof. By Proposition 10.2.2, it suffices to consider d. In this case, we must show that

$$\int_{D_1} \left(\langle \rho^*(dF), \phi \rangle + \langle \rho^* F, \nabla \lrcorner \phi \rangle \right) dx = 0$$

for $\phi \in C_0^\infty(D_1)$. By the Lipschitz change of variables formula (6.2), see Section 6.5, and Lemma 10.2.4, this is equivalent to

$$\int_{D_2} \left(\langle dF, \tilde{\rho}_* \phi \rangle + \langle F, \nabla \lrcorner (\tilde{\rho}_* \phi) \rangle \right) dx = 0,$$

which holds by Proposition 10.2.3. $\quad\square$

It is clear from the definition that $\mathsf{D}(\underline{d})$ on D can be viewed as a subspace of $\mathsf{D}(d)$ on X, by extending F on D by zero to all X. The following existence of extension maps shows that $\mathsf{D}(d)$ on D can be identified with the quotient space $\mathsf{D}(d_X)/\mathsf{D}(\underline{d}_{X \setminus \overline{D}})$.

Proposition 10.2.7 (Extensions for d and δ). *Let D be a bounded Lipschitz domain in a Euclidean space X.*

Assume that $F \in D(d)$ on D. Then there exists $\tilde{F} \in D(d)$ on X such that $\tilde{F}|_D = F$. Furthermore, there exists $F_t \in C^\infty(\overline{D})$ such that $F_t \to F$ and $dF_t \to dF$ in $L_2(D)$ as $t \to 0$.

Similarly, assume that $F \in D(\delta)$ on D. Then there exists $\tilde{F} \in D(\delta)$ on X such that $\tilde{F}|_D = F$. Furthermore, there exists $F_t \in C^\infty(\overline{D})$ such that $F_t \to F$ and $\delta F_t \to \delta F$ in $L_2(D)$ as $t \to 0$.

Proof. As in the proof of Proposition 10.2.5, it suffices to consider d, and we may assume that supp $F \subset D_p \cap \overline{D}$, a small neighborhood of $p \in \partial D$. By Lemma 10.2.6 we have $\rho^* F \in D(d)$ on $\Omega_p \cap \{x_n > 0\}$. Define

$$G(x) := \begin{cases} \rho^* F(x), & x_n > 0, \\ R^* \rho^* F(x), & x_n < 0, \end{cases}$$

where $R(x', x_n) := (x', -x_n)$ denotes reflection in $\mathbf{R}^{n-1}$. We claim that $G \in D(d)$ on all Ω_p across $\mathbf{R}^{n-1}$. To see this, for $\phi \in C_0^\infty(\Omega_p)$, we calculate

$$\int_{x_n > 0} \Big(\langle d\rho^* F, \phi \rangle + \langle \rho^* F, \nabla \lrcorner \phi \rangle \Big) dx + \int_{x_n < 0} \Big(\langle dR^* \rho^* F, \phi \rangle + \langle R^* \rho^* F, \nabla \lrcorner \phi \rangle \Big) dx$$

$$= \int_{x_n > 0} \Big(\langle d\rho^* F, \phi + R_* \phi \rangle + \langle \rho^* F, \nabla \lrcorner (\phi + R_* \phi) \rangle \Big) dx.$$

Since $\phi + R_* \phi$ is tangential on $\mathbf{R}^{n-1}$, we have $\phi + R_* \phi \in D(\underline{\delta})$ on $\Omega_p \cap \mathbf{R}_+^n$, so by Proposition 10.2.3, the integral vanishes.

By Lemma 10.2.4, the field $\tilde{F} := (\rho^*)^{-1} G \in D(d)$ on X is an extension of F, and if we mollify and approximate $\tilde{F}$ by

$$F_t(x) := \eta_t * \tilde{F}(x), \quad x \in D,$$

as above, we obtain $F_t \in C^\infty(\overline{D})$ and can verify that F_t and dF_t converge to F and dF respectively. $\qquad \square$

10.3   Hodge Potentials

Our main result on Hodge decompositions is the following.

Theorem 10.3.1 (Hodge decompositions on Lipschitz domains). *Let D be a bounded Lipschitz domain in a Euclidean space X. Then the operators d, δ, $\underline{d}$, $\underline{\delta}$ in $L_2(D; \wedge V)$ all have closed ranges, the cohomology spaces $C_{\|}(D) = N(d) \cap N(\underline{\delta})$ and $C_{\perp}(D) = N(\underline{d}) \cap N(\delta)$ are finite-dimensional, and we have Hodge decompositions*

$$L_2(D; \wedge V) = R(d) \oplus C_{\|}(D) \oplus R(\underline{\delta}) = R(\underline{d}) \oplus C_{\perp}(D) \oplus R(\delta).$$

Moreover, the inverses of $d : R(\underline{\delta}) \to R(d)$, $\underline{\delta} : R(d) \to R(\underline{\delta})$, $\underline{d} : R(\delta) \to R(\underline{d})$, and $\delta : R(\underline{d}) \to R(\delta)$ are all L_2 compact.

The proof follows from the following reduction and Theorem 10.3.3 below.

Reduction of Theorem 10.3.1 to a ball. We prove that there are compact operators K_0 and K_1 on $L_2(D)$ such that $dK_1F + K_1dF + K_0F = F$ for all $F \in \mathsf{D}(d)$. By Propositions 10.1.6 and 10.2.2, this will prove Theorem 10.3.1.

By Definition 6.1.4 we have a finite covering $D = \bigcup_\alpha D_\alpha$, with Lipschitz diffeomorphisms $\rho_\alpha : B \to D_\alpha$ from the unit ball B. Moreover, we have a partition of unity $\eta_\alpha \in C^\infty(\overline{D})$ subordinate to this covering.

By Theorem 10.3.3 for the ball B, we have compact maps K_1^B and K_0^B on $L_2(B)$ such that $dK_1^B F + K_1^B dF + K_0^B F = F$. Note that we need only part (i) in the proof of Theorem 10.3.3 for this. Define

$$K_1F := \sum_\alpha \eta_\alpha (\rho_\alpha^*)^{-1} K_1^B (\rho_\alpha^* F|_{D_\alpha}),$$

which is seen to be compact on $L_2(D)$. We calculate

$$dK_1F = \sum_\alpha \eta_\alpha (\rho_\alpha^*)^{-1}(I - K_1^B d - K_0^B)(\rho_\alpha^* F|_{D_\alpha}) + \sum_\alpha \nabla\eta_\alpha \wedge (\rho_\alpha^*)^{-1} K_1^B (\rho_\alpha^* F|_{D_\alpha})$$

$$= F - K_1 dF - K_0 F,$$

where

$$K_0F := \sum_\alpha \eta_\alpha (\rho_\alpha^*)^{-1} K_0^B (\rho_\alpha^* F|_{D_\alpha}) - \sum_\alpha \nabla\eta_\alpha \wedge (\rho_\alpha^*)^{-1} K_1^B (\rho_\alpha^* F|_{D_\alpha})$$

is seen to be compact on $L_2(D)$. Note the critical use of Theorem 7.2.9. This proves Theorem 10.3.1 for Lipschitz domains D. $\qquad\square$

In the proof of Theorem 10.3.1 we used Proposition 10.1.6(ii). As for the characterization (iii), it is natural to ask whether $\mathsf{D}(d) \cap \mathsf{D}(\delta) \subset H^1(D)$, that is, whether the total derivative $\nabla \otimes F$ belongs to $L_2(D)$, whenever $F, dF, \delta F \in L_2(D)$. This is not true for general Lipschitz domains, where the irregularities of ∂D may prevent $F \in \mathsf{D}(d) \cap \mathsf{D}(\delta)$ from having full Sobolev H^1 regularity, but it does hold for smooth domains.

Example 10.3.2 (Nonconvex corner). Let $D_\alpha \subset \mathbf{R}^2$ be a bounded domain that is smooth except at 0, in a neighborhood of which D_α coincides with the sector $\{re^{i\phi} ; r > 0, 0 < \phi < \alpha\}$. Define a scalar function $u : D_\alpha \to \mathbf{R}$ such that

$$u = r^{\pi/\alpha} \sin(\pi\phi/\alpha)\eta,$$

where $\eta \in C_0^\infty(\mathbf{R}^2)$, $\eta = 1$ in a neighborhood of 0, and $\eta = 0$ where D_α differs from the sector.

Consider the gradient vector field $F := \nabla u \in \mathsf{R}(d)$. Using the estimate $|F| \lesssim r^{\pi/\alpha-1}$, we verify that $F \in \mathsf{D}(\underline{d}) \cap \mathsf{D}(\delta)$. However,

$$\int_D |\nabla \otimes F|^2 dxdy \gtrsim \int_0^1 (r^{\pi/\alpha-2})^2 rdr.$$

Therefore, when D_α is not convex, that is, when $\alpha > \pi$, then $F \notin H^1(D)$. Figure 10.1 shows the case $\alpha = 3\pi/2$.

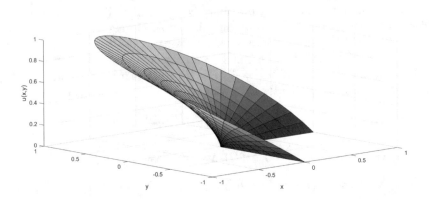

Figure 10.1: The harmonic function $r^{2/3} \sin(2\phi/3)$ in quadrants 1–3 in the unit circle, with Dirichlet boundary conditions but infinite gradient at the origin.

Theorem 10.3.3 (Full regularity of Hodge potentials). *Let D be a bounded C^2 domain. Then*

$$D(d) \cap D(\underline{\delta}) = H^1_\parallel(D) := \{F \in H^1(D) \; ; \; \nu \lrcorner F|_{\partial D} = 0\} \quad and$$
$$D(\underline{d}) \cap D(\delta) = H^1_\perp(D) := \{F \in H^1(D) \; ; \; \nu \wedge F|_{\partial D} = 0\}.$$

For the proof of Theorem 10.3.3 we shall prove a Weitzenböck identity for d and δ on D, involving a boundary curvature term. This requires the following definitions from differential geometry and uses that the boundary ∂D is C^2 regular. In this case, the unit normal vector field ν on ∂D is C^1, and the curvature of the boundary is a continuous function.

Proposition 10.3.4 (Derivative of normal). *Let D be a bounded C^2 domain, with outward-pointing unit normal vector field ν on ∂D. At $p \in \partial D$, let $T_p(\partial D)$ denote the tangent hyperplane. Then the map*

$$S^p_{\partial D} : T_p(\partial D) \to T_p(\partial D) : v \mapsto \partial_v \nu,$$

is linear and symmetric. Moreover, for any tangential C^1 vector fields u and v on ∂D, at each $p \in \partial D$ we have

$$\langle u, S^p_{\partial D} v \rangle = -\langle \partial_u v, \nu \rangle = \langle S^p_{\partial D} u, v \rangle.$$

Proof. We have $0 = \partial_v |\nu|^2 = 2\langle \partial_v \nu, \nu\rangle$, since $|\nu| = 1$ on ∂D, which shows that $S_{\partial D}^p(v)$ is a tangential vector. To show the symmetry of $S_{\partial D}^p$ at $p \in \partial D$, we note that

$$0 = \partial_u \langle v, \nu\rangle = \langle \partial_u v, \nu\rangle + \langle v, \partial_u \nu\rangle \quad \text{and}$$
$$0 = \partial_v \langle u, \nu\rangle = \langle \partial_v u, \nu\rangle + \langle u, \partial_v \nu\rangle,$$

since u and v are tangential on ∂D. The symmetry of $S_{\partial D}^p$ now follows, since the Lie bracket $\partial_u v - \partial_v u = [u, v] = \mathcal{L}_u v$ is tangential. $\qquad\square$

Definition 10.3.5 (Second fundamental form). Let D be a bounded C^2 domain. The symmetric bilinear form

$$B_{\partial D}^p : T_p(\partial D) \times T_p(\partial D) \to \mathbf{R} : (u, v) \mapsto -\langle \partial_u v, \nu\rangle$$

from Proposition 10.3.4 is called *the second fundamental form* for ∂D. The associated symmetric map

$$S_{\partial D}^p : T_p(\partial D) \to T_p(\partial D) : v \mapsto \partial_v \nu$$

from Proposition 10.3.4 is called the *Weingarten map*, or *shape operator*, for ∂D. The eigenvalues $\{\kappa_1, \ldots, \kappa_{n-1}\}$ of $S_{\partial D}^p$ are called the *principal curvatures* of ∂D at p, and a corresponding ON-basis $\{e_1', \ldots, e_{n-1}'\}$ for $T_p(\partial D)$ of eigenvectors to $S_{\partial D}^p$ is referred to as the *principal directions of curvatures* at p.

Note that if D is a convex domain, then $\kappa_j \geq 0$.

Theorem 10.3.6 (Weitzenböck identities). *Let D be a bounded C^2 domain, and let e_j' denote the principal directions of curvatures κ_j, $j = 1, \ldots, n-1$. Then*

$$\int_D |\nabla \otimes F|^2 dx = \int_D (|dF|^2 + |\delta F|^2) dx - \sum_{j=1}^{n-1} \int_{\partial D} \kappa_j |e_j' \wedge F|^2 dy, \quad F \in H_\perp^1(D),$$

$$\int_D |\nabla \otimes F|^2 dx = \int_D (|dF|^2 + |\delta F|^2) dx - \sum_{j=1}^{n-1} \int_{\partial D} \kappa_j |e_j' \lrcorner F|^2 dy, \quad F \in H_\|^1(D),$$

where $|\nabla \otimes F|^2 = \sum_{j=1}^n |\partial_j F|^2$.

Example 10.3.7 (Kadlec's formula). Consider a scalar function $U : D \to \mathbf{R}$ satisfying Poisson's equation $\Delta U = f$ in D, with Dirichlet boundary conditions $U|_{\partial D} = 0$. This means that its gradient vector field $F = \nabla U$ is normal at the boundary. Assuming that $F \in H_\perp^1(D)$, we have *Kadlec's formula*

$$\sum_{i,j=1}^n \int_D |\partial_i \partial_j U|^2 dx = \int_D |f|^2 dx - (n-1) \int_{\partial D} H(y) |\nabla U|^2 dy,$$

where $H(y) := \text{Tr}(S^y_{\partial D})/(n-1)$ is the *mean curvature* of the boundary. Note that Lagrange's identity Proposition 3.1.1 shows that $|e'_j \wedge F|^2 = |e'_j|^2 |F|^2 - |\langle e'_j, F \rangle|^2 = |F|^2$.

If instead U satisfies Neumann boundary conditions $\langle \nu, \nabla U \rangle = 0$, then we get a similar identity

$$\sum_{i,j=1}^n \int_D |\partial_i \partial_j U|^2 dx = \int_D |f|^2 dx - \sum_{j=1}^n \int_{\partial D} \kappa_j |\langle e'_j, \nabla U \rangle|^2 dy,$$

but where all the principal curvatures appear and not only the mean curvature.

For convex domains, the Weitzenböck identities imply that

$$\int_D |\nabla \otimes F|^2 dx \leq \int_D (|dF|^2 + |\delta F|^2) dx, \qquad \text{for all } F \in H^1_\perp(D) \cup H^1_\parallel(D),$$

since in this case all $\kappa_j \geq 0$. In general, we have the following estimates.

Corollary 10.3.8 (Gaffney's inequality). *Let D be a bounded C^2 domain. Then*

$$\int_D |\nabla \otimes F|^2 dx \lesssim \int_D (|dF|^2 + |\delta F|^2 + |F|^2) dx, \quad \text{for all } F \in H^1_\perp(D) \cup H^1_\parallel(D).$$

Proof. For a C^2 domain, we note that the principal curvatures κ_j are bounded functions, which shows that the boundary integral terms are $\lesssim \|F\|^2_{L_2(\partial D)}$. To replace this by a term $\|F\|^2_{L_2(D)}$, we apply Stokes's theorem to obtain a standard trace estimate as follows. Let $\theta \in C_0^\infty(X; V)$ be a vector field such that $\inf_{y \in \partial D} \langle \theta(y), \nu(y) \rangle > 0$, that is, θ is uniformly outward pointing on ∂D. Stokes's theorem gives

$$\int_{\partial D} |F|^2 \langle \theta, \nu \rangle dy = \int_D \left(2\langle \partial_\theta F, F \rangle + |F|^2 \, \text{div} \, \theta \right) dx.$$

Estimating, this shows that

$$\|F\|^2_{L_2(\partial D)} \lesssim \int_{\partial D} |F|^2 \langle \theta, \nu \rangle dy \lesssim \int_D \left(|\nabla \otimes F||F| + |F|^2 \right) dx.$$

It follows from the Weitzenböck identities that

$$\int_D |\nabla \otimes F|^2 dx \leq \int_D (|dF|^2 + |\delta F|^2) dx + C \int_D |\nabla \otimes F||F| dx + C \int_D |F|^2 dx, \quad (10.3)$$

for some constant $C < \infty$. We next use an estimate technique called the *absorption inequality*, which is

$$ab \leq \tfrac{\epsilon}{2} a^2 + \tfrac{1}{2\epsilon} b^2.$$

This is, of course, nothing deeper than $(\sqrt{\epsilon}a - b/\sqrt{\epsilon})^2 \geq 0$. To use this, we take $a = |\nabla \otimes F(x)|$, $b = |F(x)|$, and $\epsilon = C^{-1}$. This shows that the second term on the right-hand side in (10.3) is

$$C \int_D |\nabla \otimes F||F|dx \leq \tfrac{1}{2} \int_D |\nabla \otimes F|^2 dx + \tfrac{C^2}{2} \int_D |F|^2 dx,$$

where the first term can be moved to the left-hand side in (10.3) and be absorbed there. Gaffney's inequality follows. $\qquad\qquad\qquad\qquad\square$

Proof of Theorem 10.3.6. (i) Let first $F \in C^2(\overline{D})$ and consider the 1-form

$$\theta(x,v) := \sum_{j=1}^{n} \langle v, e_j \rangle \langle F(x), \partial_j F(x) \rangle - \langle v \wedge F(x), dF(x) \rangle - \langle v \lrcorner F(x), \delta F(x) \rangle,$$

for $x \in \overline{D}$, $v \in V$. We calculate its exterior derivative

$$\theta(\dot{x}, \nabla) = (|\nabla \otimes F|^2 + \langle F, \Delta F \rangle) - (|dF|^2 + \langle F, \delta dF \rangle) - (|\delta F|^2 + \langle F, d\delta F \rangle)$$
$$= |\nabla \otimes F|^2 - |dF|^2 - |\delta F|^2,$$

since $\Delta = \delta d + d\delta$. The Stokes' formula (7.4) gives

$$\int_D \Big(|\nabla \otimes F|^2 - |dF|^2 - |\delta F|^2 \Big) dx = \int_{\partial D} \Big(\langle F, \partial_\nu F \rangle - \langle \nu \wedge F, dF \rangle - \langle \nu \lrcorner F, \delta F \rangle \Big) dy.$$

We continue and rewrite the right-hand side with nabla calculus as

$$\langle F, \langle \nu, \nabla \rangle F \rangle - \langle F, \nu \wedge (\nabla \lrcorner F) \rangle - \langle \nu \wedge F, \nabla \wedge F \rangle$$
$$= \langle F, \nabla \lrcorner (n \wedge \dot{F}) \rangle - \langle \nu \wedge F, \nabla \wedge F \rangle \qquad (10.4)$$
$$= -\langle F, \nabla \lrcorner (\dot{n} \wedge F) \rangle + \langle F, \nabla \lrcorner (n \wedge F) \rangle - \langle \nu \wedge F, \nabla \wedge F \rangle, \qquad (10.5)$$

where $n \in C^1(X; V)$ denotes an extension of ν. The first step uses the algebraic anticommutation relation $\nu \wedge (\nabla \lrcorner F) = \langle \nu, \nabla \rangle F - \nabla \lrcorner (n \wedge \dot{F})$, and the second step uses the analytic product rule $\nabla \lrcorner (n \wedge F) = \nabla \lrcorner (\dot{n} \wedge F) + \nabla \lrcorner (n \wedge \dot{F})$. At $p \in \partial D$, we calculate the first term in the ON-basis $\{e'_1, \ldots, e'_{n-1}, \nu\}$ and get

$$\langle F, \nabla \lrcorner (\dot{n} \wedge F) \rangle = \sum_{j=1}^{n-1} \kappa_j |e'_j \wedge F|^2 + \langle \nu \wedge F, (\partial_\nu n) \wedge F \rangle.$$

On the other hand, the normal derivatives in the last two terms in (10.4) are

$$\langle F, \nu \lrcorner \partial_\nu (n \wedge F) \rangle - \langle \nu \wedge F, \nu \wedge \partial_\nu F \rangle = \langle \nu \wedge F, (\partial_\nu n) \wedge F \rangle.$$

Therefore these three terms involving the normal derivatives cancel, and we obtain the identity

$$\int_D (|\nabla \otimes F|^2 - |dF|^2 - |\delta F|^2)dx$$

$$= -\sum_{j=1}^{n-1} \int_{\partial D} \kappa_j |e'_j \wedge F|^2 dy + \int_{\partial D} (\langle F, \nabla' \lrcorner (\nu \wedge F)\rangle - \langle \nu \wedge F, \nabla' \wedge F\rangle)dy, \quad (10.6)$$

where $\nabla' := \nu \lrcorner (\nu \wedge \nabla) = \sum_{j=1}^{n-1} e'_j \partial_{e'_j}$.

(ii) Next consider $F \in H^1_\perp(D)$. To obtain the first Weitzenböck identity, we use the fact that $C^2(\overline{D})$ is dense in $H^1(D)$ and take $F_j \in C^2(\overline{D})$ such that $F_j \to F$ and $\nabla \otimes F_j \to \nabla \otimes F$ in $L_2(D)$. On the C^2 manifold ∂D, we use the Sobolev spaces $H^{1/2}(\partial D)$ and $H^{-1/2}(\partial D)$, as discussed in Example 6.4.1, where $H^{1/2} \subset L_2 \subset H^{-1/2}$. As usual, we allow the functions to be multivector fields, and require that each component function be in such Sobolev space. We need the following well-known facts.

- The trace map $H^1(D) \to H^{1/2}(\partial D) : F \mapsto F|_{\partial D}$ is a bounded linear operator.

- The tangential derivative ∇' defines a bounded linear operator $\nabla' : H^{1/2}(\partial D) \to H^{-1/2}(\partial D)$.

- Multiplication by a C^1 function like ν is a bounded operation on $H^{1/2}(\partial D)$.

- The spaces $H^{1/2}(\partial D)$ and $H^{-1/2}(\partial D)$ are dual; in particular, we have the estimate
$$\left| \int_{\partial D} \langle F, G\rangle dx \right| \lesssim \|F\|_{H^{1/2}(\partial D)} \|G\|_{H^{-1/2}(\partial D)}.$$

Given this, we apply (10.6) to F_j and take the limit as $j \to \infty$. Since $\nu \wedge F_j \to \nu \wedge F = 0$ in $H^{1/2}(\partial D)$, we obtain the Weitzenböck identity for $F \in H^1_\perp(D)$.

(iii) To obtain the Weitzenböck identity for $F \in H^1_\parallel(D)$, we instead rewrite θ as

$$\theta(x, \nu) = \langle F, \langle \nu, \nabla\rangle F\rangle - \langle F, \nu \lrcorner (\nabla \wedge F)\rangle - \langle \nu \lrcorner F, \nabla \lrcorner F\rangle$$
$$= \langle F, \nabla \wedge (n \lrcorner \dot{F})\rangle - \langle \nu \lrcorner F, \nabla \lrcorner F\rangle$$
$$= -\langle F, \nabla \wedge (\dot{n} \lrcorner F)\rangle + \langle F, \nabla \wedge (n \lrcorner F)\rangle - \langle \nu \lrcorner F, \nabla \lrcorner F\rangle,$$

and proceed as in (i) and (ii).   □

We finally prove Theorem 10.3.3. From the Weitzenböck identities, the Gaffney inequalities show that on C^2 domains we have $H^1_\parallel(D) \subset D(d) \cap D(\delta)$ and $H^1_\perp(D) \subset D(\underline{d}) \cap D(\delta)$ and that

$$\|\nabla \otimes F\|^2 + \|F\|^2 \approx \|\nabla \wedge F\|^2 + \|\nabla \lrcorner F\|^2 + \|F\|^2$$

in $L_2(D)$ norm, for all $F \in H^1_\parallel(D)$ and all $F \in H^1_\perp(D)$. It is important to note that this equivalence of norms, without further work, does not imply that $\mathsf{D}(d) \cap \mathsf{D}(\underline{\delta}) \subset H^1_\parallel(D)$ and $\mathsf{D}(\underline{d}) \cap \mathsf{D}(\delta) \subset H^1_\perp(D)$. It particular, the absorption technique in the proof of Corollary 10.3.8 fails to prove this.

Proof of Theorem 10.3.3. It remains to prove $\mathsf{D}(d) \cap \mathsf{D}(\underline{\delta}) \subset H^1_\perp(D)$. By Hodge duality as in Proposition 10.2.2, this will imply the corresponding result for normal boundary conditions.

(i) Consider first the case that D is the unit ball $B := \{x \in V \; ; \; |x| < 1\}$ and let $F \in \mathsf{D}(d) \cap \mathsf{D}(\underline{\delta})$. Using a partition of unity, we write $F = F_0 + F_1$, $F_0, F_1 \in \mathsf{D}(d) \cap \mathsf{D}(\underline{\delta})$, where $F_0(x) = 0$ when $|x| > 1/2$ and $F_1(x) = 0$ when $|x| < 1/3$. We use inversion $R(x) = 1/x$ in the unit sphere, with derivative $\underline{R}_x h = -x^{-1} h x^{-1}$ to extend F_1 to

$$\tilde{F}_1(x) := \begin{cases} F_1(x), & |x| < 1, \\ R^* F_1(x), & |x| > 1. \end{cases}$$

Arguing as in the proof of Proposition 10.2.7, replacing $\mathbf{R}^{n-1}$ by the sphere $|x| = 1$, we conclude that $\tilde{F} \in \mathsf{D}(d)$ on X. Moreover, R is a conformal map and $R^* = |x|^{2(n-1)} \tilde{R}_*^{-1}$. From this it follows that $R^* F_1 \in \mathsf{D}(\delta)$ on X, with

$$\nabla \lrcorner R^* F_1(x) = |x|^{2(n-2)} x \lrcorner \tilde{R}_*^{-1} F_1(x) + |x|^{2(n-1)} \tilde{R}_*^{-1}(\nabla \lrcorner F_1)(x)$$

for $|x| > 1$. Recall that F_1, extended by 0 for $|x| > 1$ belongs to $\mathsf{D}(\delta)$.

We obtain an extension $\tilde{F} := F_0 + \tilde{F}_1$ of F to X, with $\tilde{F} = 0$ for $|x| > 3$ and $\tilde{F}$, $d\tilde{F}$, and $\delta\tilde{F}$ all belonging to $L_2(X)$. By Plancherel's theorem and Langrange's identity, we get

$$(2\pi)^n \int_X |\nabla \otimes \tilde{F}|^2 dx = \int_X |\mathcal{F}(\tilde{F})|^2 |\xi|^2 d\xi = \int_X \left(|\xi \wedge \mathcal{F}(\tilde{F})|^2 + |\xi \lrcorner \mathcal{F}(\tilde{F})|^2 \right) d\xi < \infty.$$

Recall from Example 10.1.8 that d and δ on X are the Fourier multipliers $i\mu_\xi$ and $i\mu_\xi^*$. This shows that $F \in H^1(D)$.

(ii) Next consider a general bounded C^2 domain D. Localizing the problem with a partition of unity, we may assume that D is C^2 diffeomorphic to B. Moreover, we may assume that we have a C^2 map

$$\rho : [0, 1] \times B \to X$$

such that $\rho_t = \rho(t, \cdot)$ defines a C^2 diffeomorphism $B \to \rho_t(B) =: D_t$, with $D_0 = B$ and $D_1 = D$. For fixed $t \in [0, 1]$, we consider the inclusion $H^1_\parallel(D_t) \subset \mathsf{D}(d) \cap \mathsf{D}(\underline{\delta})$ on the C^2 domain D_t. We note from Proposition 10.1.3 that $I + d + \underline{\delta} : \mathsf{D}(d) \cap \mathsf{D}(\underline{\delta}) \to L_2(D_t)$ is an invertible isometry, so the inclusion amounts to

$$I + d + \delta : H^1_\parallel(D_t) \to L_2(D_t)$$

being an injective semi-Fredholm operator. See Definition 6.4.9. From (i), we know that it is surjective for the ball at $t = 0$. To apply the method of continuity and conclude that it is surjective for all t, and in particular for D at $t = 1$, we note that the normalized pushforward $\tilde{\rho}_{t*}$ defines invertible maps $H^1_{\|}(B) \to H^1_{\|}(D_t)$ and $L_2(B) \to L_2(D_t)$. The method of continuity therefore applies to the family of semi-Fredholm operators

$$(\tilde{\rho}_{t*})^{-1}(I + d + \delta)\tilde{\rho}_{t*} : H^1_{\|}(B) \to L_2(B).$$

We conclude that $I + d + \delta : H^1_{\|}(D_t) \to L_2(D_t)$ is invertible, which shows that $\mathsf{D}(d) \cap \mathsf{D}(\underline{\delta}) = H^1_{\|}(D)$ and completes the proof of Theorem 10.3.3. $\qquad\square$

10.4   Bogovskiĭ and Poincaré Potentials

Recall that exterior and interior potentials in general are highly nonunique. In this section we prove the following surprising results about potentials on strongly Lipschitz domains D.

- We have seen in Example 10.3.2 that in contrast to smooth domains, the potential U in the subspace $\mathsf{R}(\underline{\delta})$ to $F = dU \in \mathsf{R}(d)$ may not belong to $H^1(D)$. We refer to this potential U as the *Hodge potential* for F, which is characterized by its minimal L_2 norm.

  It follows from Theorem 10.4.3 below that every exact field $F \in \mathsf{R}(d)$ on any bounded and strongly Lipschitz domain D nevertheless has a potential $\tilde{U}$, in general different from the Hodge potential, such that $\tilde{U} \in H^1(D)$ and $dU = F$. We refer to such potentials as (regularized) *Poincaré potentials* for F.

- We have seen that the Hodge potential $U \in \mathsf{R}(d)$ to $F = \underline{\delta}U \in \mathsf{R}(\underline{\delta})$ is tangential on ∂D, meaning that half of the component functions of U vanish there. Theorem 10.4.3 below show that every field $F \in \mathsf{R}(\underline{\delta})$ on any bounded and strongly Lipschitz domain D in fact has a potential $\tilde{U}$, in general different from the Hodge potential, such that $\tilde{U} \in H^1_0(D)$ and $\underline{\delta}U = F$. This means that all component functions of $\tilde{U}$ vanish on ∂D, and we note that this is a nontrivial result also for smooth domains. We refer to such potentials as *Bogovskiĭ potentials* for F.

Similarly, and related by the Hodge star, there exist Poincaré potentials $\tilde{U} \in H^1(D)$ for $F \in \mathsf{R}(\delta)$, and Bogovskiĭ potentials $\tilde{U} \in H^1_0$ for $F \in \mathsf{R}(\underline{d})$. We will formulate the results only for d and $\underline{\delta}$, and leave it to the reader to translate the results in this section to $\underline{d}$ and δ.

First consider a star-shaped domain D. In what follows, we extend the operators initially defined on k-vector fields, by linearity to act on general multivector fields. The method we use to construct a Poincaré potential U to a given field $F \in \mathsf{R}(d)$ on D builds on Poincaré's Theorem 7.5.2. If D is shar-shaped with respect to $p_0 \in D$, then this gives the potential

$$T_{p_0}(F)(x) := (x - p_0) \lrcorner \int_0^1 F(p_0 + t(x - p_0))\, t^{k-1} dt, \quad x \in D, \qquad (10.7)$$

provided $k \geq 1$ and F is a smooth k-vector field. For a scalar function $F : D \to \wedge^0 \mathbf{R}$, we let $T_{p_0} F = 0$. We would like to extend (10.7) to fields that are square integrable, without any assumption on regularity. To obtain a bounded operator, we need to average the formula (10.7) over base points p around p_0. In what follows, we assume that D is star-shaped not only with respect to a point, but to a whole ball. We fix $\theta \in C_0^\infty(B(p_0; \epsilon))$ and assume that D is star-shaped with respect to each $p \in B(p_0; \epsilon)$, where $\epsilon > 0$ and $\int \theta dx = 1$. Then define the averaged operator

$$T_D F(x) := \int_{|p-p_0| \leq \epsilon} \theta(p) T_p F(x) dp, \quad x \in D. \tag{10.8}$$

We rewrite this formula by changing the variables p and t to $y := p + t(x-p)$ and $s = 1/(1-t) - 1$. This gives

$$T_D F(x) = \int_D (x-y) \lrcorner F(y) \, k_\theta(x,y) dy, \tag{10.9}$$

where

$$k_\theta(x,y) := \int_0^\infty \theta(y + s(y-x)) s^{k-1}(1+s)^{n-k} ds.$$

This operator T_D constructs the regularized Poincaré potential for an exact k-vector field on a bounded domain that is star-shaped with respect to $B(p_0; \epsilon)$.

Exercise 10.4.1 (Kernel support). Show that $k_\theta(x,y) \neq 0$ is possible only when y lies on the straight line between x and a point $p \in \operatorname{supp} \eta$, and that we have estimates

$$|k_\theta(x,y)| \lesssim \frac{1}{|x-y|^n}, \quad x,y \in D,$$

so that T_D is a weakly singular integral operator. Note how by averaging with θ we have replaced the line integral for T_{p_0} by a volume integral over a conical region for T_D.

The adjoint operator

$$T_D^* F(x) = \int_D (y-x) \wedge F(y) \, k_\theta(y,x) dy \tag{10.10}$$

constructs the Bogovskiĭ potential for a $(k-1)$-vector field $F \in \mathsf{R}(\delta)$ on the star-shaped domain D. We see from Exercise 10.4.1 that $T_D^* F|_{\partial D} = 0$, since for T_D^* we integrate over a cone starting at x, away from $B(p_0, \epsilon)$.

For domains D that are C^2 diffeomorphic to a domain that is star-shaped with respect to a ball, we can pull back and push forward these operators T_D and T_D^* to obtain Bogovskiĭ and regularized Poincaré potentials. Next we extend these constructions to general strongly Lipschitz domains, and provide the necessary analysis.

Definition 10.4.2 (Bogovskiĭ and Poincaré maps). Let D be a bounded and strongly Lipschitz domain. Fix a finite cover $D = \bigcup_\alpha D_\alpha$ by domains D_α that are star-shaped with respect to balls $B(p_\alpha; \epsilon)$. Further fix $\theta_\alpha \in C_0^\infty(B(p_\alpha; \epsilon))$ with $\int \theta_\alpha dx = 1$ and a partition of unity $\eta_\alpha \in C^\infty(\overline{D})$ subordinate to the covering D_α. We assume that $\eta_\alpha = 1$ on a neighborhood of $\mathrm{supp}\,\theta_\alpha$. The *regularized Poincaré map* with these choices $D_\alpha, \theta_\alpha, \eta_\alpha$, for d on D is

$$T_D F(x) = \sum_\alpha \eta_\alpha(x) T_{D_\alpha}(F|_{D_\alpha})(x), \quad x \in D.$$

The *Bogovskiĭ map*, with these choices $D_\alpha, \theta_\alpha, \eta_\alpha$, for $\underline{\delta}$ on D is

$$T_D^* F(x) = \sum_\alpha T_{D_\alpha}^*(\eta_\alpha F|_{D_\alpha})(x), \quad x \in D.$$

Here T_{D_α} and $T_{D_\alpha}^*$ are the Poincaré and Bogovskiĭ maps on the star-shaped domains D_α, constructed as above.

Unlike the star-shaped case, these Bogovskiĭ and regularized Poincaré maps on general strongly Lipschitz domains do not straight away give potentials for (co-)exact fields. We proceed as follows.

Theorem 10.4.3 (Bogovskiĭ and Poincaré homotopies). *Let D be a bounded and strongly Lipschitz domain. The regularized Poincaré potential map from Definition 10.4.2, maps $T_D : C^\infty(\overline{D}) \to C^\infty(\overline{D})$ and extends by continuity to a bounded operator*

$$T_D : L_2(D) \to H^1(D).$$

The Bogovskiĭ potential map from Definition 10.4.2 maps $T_D^* : C_0^\infty(D) \to C_0^\infty(D)$ and extends by continuity to a bounded operator

$$T_D^* : L_2(D) \to H_0^1(D).$$

We have homotopy relations

$$d(T_D F) + T_D(dF) + K_D F = F, \quad F \in D(d),$$
$$-\underline{\delta}(T_D^* F) - T^*(\underline{\delta} F) + K_D^* F = F, \quad F \in D(\underline{\delta}),$$

with perturbation terms

$$K_D F(x) = \sum_\alpha \eta_\alpha(x) \int \theta_\alpha F_0 dy + \sum_\alpha \nabla \eta_\alpha(x) \wedge T_{D_\alpha}(F|_{D_\alpha})(x),$$

$$K_D^* F(x) = \sum_\alpha \theta_\alpha(x) \int \eta_\alpha F_0 dy + \sum_\alpha T_{D_\alpha}^*(\nabla \eta_\alpha \lrcorner F|_{D_\alpha})(x),$$

which are bounded, $K_D : L_2(D) \to H^1(D)$ and $K_D^* : L_2(D) \to H_0^1(D)$. Here F_0 denotes the $\wedge^0 V$ part of F.

To see how Theorem 10.4.3 implies the existence of Bogovskiĭ and Poincaré potentials, we consider the following Hodge decomposition:

$$L_2(D) \quad = \quad \mathsf{R}(d) \quad \oplus \quad \mathcal{C}_\|(D) \quad \oplus \quad \mathsf{R}(\underline{\delta}) \tag{10.11}$$

$$L_2(D) \quad = \quad \mathsf{R}(d) \quad \oplus \quad \mathcal{C}_\|(D) \quad \oplus \quad \mathsf{R}(\underline{\delta})$$

Given $F \in \mathsf{R}(d)$, we apply the homotopy relation to the Hodge potential $U \in \mathsf{R}(\underline{\delta})$, with $dU = F$, to obtain $U = dT_D U + T_D dU + K_D U$, and in particular,

$$F = dU = d(T_D F + K_D U).$$

Therefore the field $\tilde{U} := T_D F + K_D U \in H^1(D)$ is a regularized Poincaré potential for F.

Similarly, for $F \in \mathsf{R}(\underline{\delta})$ we apply the homotopy relation to the Hodge potential $U \in \mathsf{R}(d)$, with $\underline{\delta} U = F$, to obtain

$$F = \underline{\delta} U = \underline{\delta}(-T_D^* F + K_D^* U),$$

where the field $\tilde{U} := -T_D^* F + K_D^* U \in H_0^1(D)$ is a Bogovskiĭ potential for F.

Proof of Theorem 10.4.3. (i) Let $F \in C^\infty(\overline{D})$. Then $F|_{D_\alpha} \in C^\infty(\overline{D}_\alpha)$, and we see from (10.8) for the star-shaped domain D_α that $T_{D_\alpha}(F|_{D_\alpha}) \in C^\infty(\overline{D}_\alpha)$. Note that T_{D_α} acts on $C^\infty(X)$, but the values $T_{D_\alpha} F(x)$, for $x \in D_\alpha$, depend only on $F|_D$. With the partition of unity η_α, we obtain $T_D F \in C^\infty(\overline{D})$.

Let $F \in C_0^\infty(D)$. Then $\eta_\alpha F|_{D_\alpha} \in C_0^\infty(D_\alpha)$, and we see from Exercise 10.4.1 that $\operatorname{supp} T_{D_\alpha}^*(\eta_\alpha F|_{D_\alpha})$ is compactly contained in D_α. To verify smoothness, we write

$$T_{D_\alpha}^* G(x) = -\int_{D_\alpha} z \wedge G(x - z) \left(\int_0^\infty \theta_\alpha(x + sz) s^{k-1}(1 + s)^{n-k} ds \right) dz.$$

Differentiation with respect to x shows that $T_{D_\alpha}^*(\eta_\alpha F|_{D_\alpha}) \in C_0^\infty(D_\alpha)$, and therefore that $T_D^* F \in C_0^\infty(D)$.

Averaging the homotopy relation in Exercise 7.5.6, we obtain

$$d(T_{D_\alpha} F) + T_{D_\alpha}(dF) + K_{D_\alpha} F = F$$

on D_α, with

$$K_{D_\alpha} F := \int \theta_\alpha F_0 dx.$$

As in the proof of Theorem 10.3.1, the product rule yields $d(T_D F) + T_D(dF) + K_D F = F$ on D, and duality yields the stated formulas for $\underline{\delta}$.

(ii) It remains to establish H^1 bounds for T_{D_α} and $T_{D_\alpha}^*$. To this end, assume that $D = D_\alpha$ is star-shaped with respect to a ball and consider the operators (10.9)

and (10.10). By Exercise 10.4.1, these are weakly singular integral operators, and
Schur estimates as in Exercise 6.4.3 show that T_D is bounded on $L_2(D)$. Expanding
$(1+s)^{n-k}$ with the binomial theorem, we may further replace $k_\theta(x,y)$ by

$$\int_0^\infty \theta(y+s(y-x))s^{n-1}ds.$$

Indeed, in estimating $\|\nabla \otimes T_D F\|_{L_2}$ the difference will be a weakly singular operator
that is bounded as above, and similarly for $\|\nabla \otimes T_D^* F\|_{L_2}$.

Make the change of variables $t = s|y-x|$, fix a coordinate $1 \le i \le n$, and
define

$$k(x,z) := \left(\int_0^\infty \eta\left(x+t\frac{z}{|z|}\right) t^{n-1} dt \right) \frac{z_i}{|z|^n}.$$

Estimating the multivector fields componentwise, we see that it is enough to consider the operators

$$Sf(x) := \int_D k(y,y-x)f(y)\,dy \quad \text{and} \quad S^*f(x) := \int_D k(x,x-y)f(y)\,dy,$$

and prove bounds on $\|\nabla Sf\|_{L_2}$ and $\|\nabla S^*f\|_{L_2}$. We note that $k(x,z)$ is homogeneous of degree $-n+1$ with respect to z. For fixed x, we expand $k(x,z/|z|)$ in a series of spherical harmonics on the unit sphere S. We get

$$k(x,z) = \frac{1}{|z|^{n-1}} \sum_{j=0}^\infty \sum_{m=1}^{h_j} k_{jm}(x)Y_{jm}(z/|z|) = \sum_{j=0}^\infty \sum_{m=1}^{h_j} k_{jm}(x)\frac{Y_{jm}(z)}{|z|^{n-1+j}},$$

where $\{Y_{jm}\}_{m=1}^{h_j}$ denotes an ON-basis for the space $\mathcal{P}_j^{sh}(S)$ of scalar-valued spherical harmonics, for $j \in \mathbf{N}$. See Section 8.2. In particular the coefficients are
$k_{jm}(x) := \int_S k(x,z)Y_{jm}(z)\,dz$. Define weakly singular convolution integral operators

$$S_{jm}(x) := \int_D \frac{Y_{jm}(x-y)}{|x-y|^{n-1+j}} f(y)\,dy.$$

With k_{jm} as multipliers we have

$$Sf(x) = \sum_{j=0}^\infty (-1)^j \sum_{m=1}^{h_j} S_{jm}(k_{jm}f)(x), \quad S^*f(x) = \sum_{j=0}^\infty \sum_{m=1}^{h_j} k_{jm}(x)S_{jm}f(x).$$

The main estimate we need is

$$\|S_{jm}\|_{L_2(D) \to H^1(D)} \lesssim (1+j)^{n-2}. \tag{10.12}$$

To see this, we use zonal harmonics as in Section 8.2 to estimate

$$|Y_{jm}(z)| = \left| \int_S Z_j(z,y)Y_{jm}(y)dy \right| \le \|Z_j(z,\cdot)\|_{L_2(S)}\|Y_{jm}\|_{L_2(S)} \lesssim (1+j)^{n-2}|z|^j,$$

which yields the L_2 estimate. To bound $\nabla S_{jm}f$ on $L_2(X)$, we use Proposition 6.2.1 to see that ∇S_{jm} is a Fourier multiplier with estimates

$$\left| \xi 2c \frac{\Gamma((1+j)/2)}{\Gamma((n-1+j)/2)} Y_{jm}(\xi)/|\xi|^{1+j} \right| \lesssim (1+j)^{n-2}, \quad \xi \in X,$$

of the symbol. This proves (10.12).

To estimate the multipliers

$$k_{jm}(x) = \int_S k(x,z) Y_{jm}(z)dz,$$

we use that $k(x,\cdot)$ is smooth on S, while Y_{jm} becomes more oscillatory as j grows, to show that k_{jm} decays with j as follows. By Proposition 8.2.15, the spherical Laplace operator Δ_S is a self-adjoint operator in $L_2(S)$ with

$$\Delta_S Y_{jm} = (2-n-j)j Y_{jm}.$$

Using self-adjointness N times shows that

$$k_{jm}(x) = \frac{1}{(2-n-j)^N j^N} \int_S (\Delta_S^N k(x,z)) Y_{jm}(z)dz.$$

Since $\Delta_S^N k(x,\cdot)$, for any fixed N, is bounded, we get the estimate

$$|k_{jm}(x)| \lesssim (1+j)^{-2N}.$$

Similarly, we bound

$$\nabla k_{jm}(x) = \int_S \nabla_x k(x,z) Y_{jm}(z)dz$$

uniformly by $(1+j)^{-2N}$. Collecting our estimates, we obtain

$$\|Sf\|_{H^1(D)} \lesssim \sum_{j=0}^{\infty} h_j (1+j)^{n-2} (1+j)^{-N} \|f\|_{L_2(D)} \lesssim \|f\|_{L_2(D)},$$

$$\|S^*f\|_{H^1(D)} \lesssim \sum_{j=0}^{\infty} h_j (1+j)^{-N} (1+j)^{n-2} \|f\|_{L_2(D)} \lesssim \|f\|_{L_2(D)},$$

provided we fix large enough N. This completes the proof. □

10.5 Čech Cohomology

In this section we collect some tools from algebraic topology that we use in Section 10.6 to calculate the dimensions of the finite-dimensional cohomology space

$N(d) \cap N(\delta)$, more precisely the Betti numbers $b_k(D)$, from Definition 7.6.3. We also use these tools in Chapters 11 and 12.

Our starting point is the notion of a sheaf, where we only use the following simplified version of this concept. We consider some set D and some fixed finite covering of it by subsets $D_1, \ldots, D_N$, so that

$$D = D_1 \cup \cdots \cup D_N.$$

By a *sheaf* $\mathcal{F}$ on D we mean a collection of linear spaces $\mathcal{F}(D')$, one for each intersection D' of the subsets D_j. In fact, it is only the additive structure of sheaves that is relevant, and in Chapter 11 we shall use Čech cohomology, where the spaces $\mathcal{F}(D')$ are the smallest additive group $\mathbf{Z}_2 = \{0, 1\}$. The linear spaces that we use in this chapter are supposed to behave like spaces of functions defined on D' in the sense that we require that there exist linear *restriction maps* $\mathcal{F}(D') \to \mathcal{F}(D'') : f \mapsto f|_{D''}$ whenever $D'' \subset D' \subset D$. If an intersection D' is empty, then we require that the linear space $\mathcal{F}(D')$ be trivial, that is, $\mathcal{F}(D') = \{0\}$. The intersections

$$D_s = D_{s_1} \cap \cdots \cap D_{s_k},$$

of distinct subsets D_{s_j} that we consider, are indexed by the 2^N subsets $s = \{s_1, \ldots, s_k\} \subset \{1, \ldots, N\}$. Since the Čech algebra that we are about to construct is alternating, we choose below to index the intersections not by s, but by auxiliary basis multivectors e_s in $\wedge \mathbf{R}^N$. This is only a formal notation, which turns out to be useful, since it allows us to recycle some, by now well known to us, exterior algebra.

Definition 10.5.1 (*k-cochains*). Let $\mathcal{F}$ be a sheaf on D as above, with covering $\underline{D} = \{D_1, \ldots, D_N\}$. A *Čech k-cochain* f associates to each $(k+1)$-fold intersection D_s, $|s| = k+1$, an element in the linear space $\mathcal{F}(D')$, which we denote by $\langle f, e_s \rangle \in \mathcal{F}(D_s)$. This is not an inner product, but only a convenient notation for the value of f on D_s. We also extend the definition of f homogeneously by letting $\langle f, \alpha e_s \rangle := \alpha \langle f, e_s \rangle$, for $\alpha \in \mathbf{R}$.

The space of all Čech k-cochains f on $\underline{D}$ with values in $\mathcal{F}$ is denoted by $C^k(\underline{D}; \mathcal{F})$. Viewing $C^k(\underline{D}; \mathcal{F})$ as $\oplus_{s:|s|=k+1} \mathcal{F}(D_s)$, it is clear that this is a linear space. For $k < 0$ and $k \geq N$ we let $C^k(\underline{D}; \mathcal{F}) := \{0\}$.

The Čech *coboundary operator* $\partial_k : C^k(\underline{D}; \mathcal{F}) \to C^{k+1}(\underline{D}; \mathcal{F})$ is the linear map defined by

$$\langle \partial_k f, e_s \rangle := \sum_{j=1}^{N} \langle f, e_j \lrcorner e_s \rangle|_{D_s}, \quad |s| = k+2, \ f \in C^k(\underline{D}; \mathcal{F}).$$

For $k < 0$ and $k \geq N - 1$, we let $\partial_k = 0$.

We will see that Čech k-cochains and ∂_k behave in many ways like k-covector fields and the exterior derivative d. We need some terminology.

Definition 10.5.2 (Complex of spaces). A *complex* of linear spaces is a sequence of linear maps between linear spaces

$$\overset{\partial_{j-3}}{\to} V_{j-2} \overset{\partial_{j-2}}{\to} V_{j-1} \overset{\partial_{j-1}}{\to} V_j \overset{\partial_j}{\to} V_{j+1} \overset{\partial_{j+1}}{\to} V_{j+2} \overset{\partial_{j+2}}{\to}$$

such that $\mathsf{R}(\partial_{j-1}) \subset \mathsf{N}(\partial_j)$ in V_j. The complex is said to be *exact at* V_j if $\mathsf{R}(\partial_{j-1}) = \mathsf{N}(\partial_j)$. If it is exact at all V_j, we say that the complex is *exact*. More generally, the *cohomology* of the complex at V_j is the quotient space $H^j(V) := \mathsf{N}(\partial_j)/\mathsf{R}(\partial_{j-1})$.

An important special case occurs when $V_j = \{0\}$ for some j, so that $\partial_j = \partial_{j-1} = 0$. In this case, exactness at V_{j+1} means that ∂_{j+1} is injective, and exactness at V_{j-1} means that ∂_{j-2} is surjective.

Lemma 10.5.3. *If (V_j, ∂_j) is a an exact complex of finite-dimensional linear spaces and $V_{j_1} = V_{j_2} = 0$, then*

$$\sum_{j_1 < j < j_2} (-1)^j \dim V_j = 0.$$

Proof. The dimension theorem for linear maps shows that $\dim V_j = \dim \mathsf{N}(\partial_j) + \dim \mathsf{R}(\partial_j)$. Since $\mathsf{R}(\partial_j) = \mathsf{N}(\partial_{j+1})$, we get a telescoping sum

$$\sum_{j_1 < j < j_2} (-1)^j \dim V_j = \sum_{j_1 < j < j_2} (-1)^j (\dim \mathsf{N}(\partial_j) + \dim \mathsf{N}(\partial_{j+1}))$$

$$= (-1)^{j_1+1} \dim \mathsf{R}(\partial_{j_1}) + (-1)^{j_2-1} \dim \mathsf{N}(\partial_{j_2}) = 0. \qquad \square$$

Lemma 10.5.4 (Čech complex). *The Čech sequence*

$$\overset{\partial_{k-2}}{\to} C^{k-1}(\underline{D}; \mathcal{F}) \overset{\partial_{k-1}}{\to} C^k(\underline{D}; \mathcal{F}) \overset{\partial_k}{\to} C^{k+1}(\underline{D}; \mathcal{F}) \overset{\partial_{k+1}}{\to} C^{k+2}(\underline{D}; \mathcal{F}) \overset{\partial_{k+2}}{\to}$$

is a complex of linear spaces.

Proof. Let $f \in C^k(\underline{D}; \mathcal{F})$ and $|s| = k + 3$. Then

$$\langle \partial_{k+1} \partial_k f, e_s \rangle = \sum_j \langle \partial_k f, e_j \lrcorner e_s \rangle|_{D_s} = \sum_j \left(\sum_i \langle f, e_i \lrcorner (e_j \lrcorner e_s) \rangle|_{D_{s \setminus \{j\}}} \right)\Big|_{D_s}$$

$$= \sum_{i,j} \langle f, (e_j \wedge e_i) \lrcorner e_s \rangle|_{D_s} = 0,$$

since $e_i \wedge e_j = -e_j \wedge e_i$ on performing the sum. $\qquad \square$

We denote the Čech cohomology spaces associated with this complex by $H^k(\underline{D}; \mathcal{F})$. A key result that we now prove is, roughly speaking, that sheaves of functions defined without any constraints will have trivial cohomology spaces. A constraint here could mean that we consider functions that are constant or that satisfy some differential equation. More precisely, a sheaf $\mathcal{F}$ is defined to be a *fine sheaf* if every sufficiently smooth cutoff function $\eta : D \to \mathbf{R}$ gives well-defined multiplication operators $f \mapsto \eta f$ on each of the linear spaces $\mathcal{F}(D')$. In particular, if $\mathrm{supp}\, \eta \subset D'$, then ηf is supposed to be extendable by zero to a function $\eta f \in \mathcal{F}(D)$ on all D. When restricted to some D'', this defines $\eta f \in \mathcal{F}(D'')$.

Proposition 10.5.5 (Cohomology of fine sheaves). *If $\mathcal{F}$ is a fine sheaf on D, then $H^k(\underline{D}; \mathcal{F}) = \{0\}$ when $k \geq 1$. For any sheaf $\mathcal{F}$, the restriction map gives an invertible map $\mathcal{F}(D) \to H^0(\underline{D}; \mathcal{F})$.*

Proof. First consider the second claim. If $f \in C^0(\underline{D}; \mathcal{F})$ and $\partial_0 f = 0$, then for all $1 \leq i < j \leq N$, we have

$$0 = \langle \partial_0 f, e_{\{i,j\}} \rangle = \langle f, e_j \rangle|_{D_{\{i,j\}}} - \langle f, e_i \rangle|_{D_{\{i,j\}}}.$$

Thus there is a unique function $f \in \mathcal{F}(D)$ such that $f|_{D_k} = \langle f, e_k \rangle$. Since $\partial_{-1} = 0$, this proves the statement.

Now let $\mathcal{F}$ be a fine sheaf, $k \geq 1$, and $f \in C^k(\underline{D}; \mathcal{F})$ with $\partial_k f = 0$. Pick a partition of unity $\{\eta_j\}_{j=1}^N$ subordinate to $\underline{D}$, so that $\operatorname{supp} \eta_j \subset D_j$ and $\sum_j \eta_j = 1$ on D. Define a $(k-1)$-cochain

$$\langle g, e_t \rangle := \sum_i \eta_i \langle f, e_i \wedge e_t \rangle, \quad |t| = k.$$

Note that $\langle f, e_i \wedge e_t \rangle$ is defined only on $D_t \cap D_i$, but that after multiplication by η_i, the product can be extended by zero across $(\partial D_i) \cap D_t$ to all D_t. The anticommutation relation from Theorem 2.8.1 yields

$$\langle \partial_{k-1} g, e_s \rangle = \sum_j \langle g, e_j \lrcorner e_s \rangle|_{D_s} = \sum_j \left(\sum_i \eta_i \langle f, e_i \wedge (e_j \lrcorner e_s) \rangle \right) \Big|_{D_s}$$

$$= \sum_j \left(\sum_i \eta_i (\delta_{i,j} \langle f, e_s \rangle - \langle f, e_j \lrcorner (e_i \wedge e_s) \rangle) \right) \Big|_{D_s}$$

$$= \left(\sum_i \eta_i \right) \langle f, e_s \rangle - \sum_i \eta_i \left(\sum_j \langle f, e_j \lrcorner (e_i \wedge e_s) \rangle|_{D_i \cap D_s} \right) \Big|_{D_s}$$

$$= \langle f, e_s \rangle - 0 = \langle f, e_s \rangle,$$

where $\delta_{i,j} = 1$ if $i = j$ and otherwise 0. This shows that $\mathsf{N}(\partial_k) = \mathsf{R}(\partial_{k-1})$, as desired. $\qquad \square$

We finish this section with two algebraic techniques that are useful in studying complexes. By a diagram of maps being *commutative*, we mean that whenever we have two different compositions of maps $A \to D$,

$$\begin{array}{ccc} A & \xrightarrow{f_1} & B \\ {\scriptstyle f_2}\downarrow & & \downarrow{\scriptstyle f_3} \\ C & \xrightarrow{f_4} & D \end{array}$$

then we have $f_3 \circ f_1 = f_4 \circ f_2$.

Lemma 10.5.6 (Snake lemma). *Let (U_j, ∂_j^u), (V_j, ∂_j^v) and (W_j, ∂_j^w) be complexes of linear spaces, and for each j, let*

$$0 \to U_j \xrightarrow{f_j} V_j \xrightarrow{g_j} W_j \to 0$$

be a short exact sequence such that $\partial_j^v f_j = f_{j+1} \partial_j^u$ and $\partial_j^w g_j = g_{j+1} \partial_j^v$ for all j. This hypothesis is summarized in the following commutative diagram with exact columns:

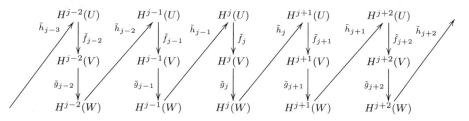

Then there are connecting linear maps $\tilde{h}_j : H^j(W) \to H^{j+1}(U)$ and induced linear maps $\tilde{f}_j : H^j(U) \to H^j(V)$ and $\tilde{g}_j : H^j(V) \to H^j(W)$, for all j, such that the cohomology sequence

$$
\begin{array}{ccccccccc}
H^{j-2}(U) & & H^{j-1}(U) & & H^j(U) & & H^{j+1}(U) & & H^{j+2}(U) \\
\tilde{h}_{j-3} \nearrow \; \downarrow \tilde{f}_{j-2} & & \tilde{h}_{j-2} \nearrow \; \downarrow \tilde{f}_{j-1} & & \tilde{h}_{j-1} \nearrow \; \downarrow \tilde{f}_j & & \tilde{h}_j \nearrow \; \downarrow \tilde{f}_{j+1} & & \tilde{h}_{j+1} \nearrow \; \downarrow \tilde{f}_{j+2} & \tilde{h}_{j+2} \nearrow \\
H^{j-2}(V) & & H^{j-1}(V) & & H^j(V) & & H^{j+1}(V) & & H^{j+2}(V) \\
\tilde{g}_{j-2} \downarrow & & \tilde{g}_{j-1} \downarrow & & \tilde{g}_j \downarrow & & \tilde{g}_{j+1} \downarrow & & \tilde{g}_{j+2} \downarrow \\
H^{j-2}(W) & & H^{j-1}(W) & & H^j(W) & & H^{j+1}(W) & & H^{j+2}(W)
\end{array}
$$

is an exact complex.

Exercise 10.5.7 (Diagram chasing). Prove the snake lemma through *diagram chasing*. To see an example of this, consider the definition of the connecting map $\tilde{h}_j : H^j(W) = \mathsf{N}(\partial_j^w)/\mathsf{R}(\partial_{j-1}^w) \to H^{j+1}(U) = \mathsf{N}(\partial_{j+1}^u)/\mathsf{R}(\partial_j^u)$. Take $w \in \mathsf{N}(\partial_j^w)$. Surjectivity of g_j gives $v \in V_j$, which maps to a $v' \in V_{j+1}$. Commutativity $\partial_j^w g_j = g_{j+1} \partial_j^v$ shows that v' maps to 0 in W_{j+1}, and so exactness at V_{j+1} gives $u \in U_{j+1}$ such that $f_{j+1}(u) = v'$. Since v' maps to 0 in V_{j+2}, commutativity

$\partial^v_{j+1} f_{j+1} = f_{j+2} \partial^u_{j+1}$ shows that $u \in \mathsf{N}(\partial_{j+1})$, since f_{j+2} is injective.

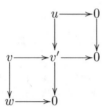

The connecting map is defined as $\tilde{h}_j([w]) := [u]$. Further diagram chasing shows that $u \in \mathsf{R}(\partial^u_j)$ if $w \in \mathsf{R}(\partial^w_{j-1})$, so that $\tilde{h}_j$ is a well-defined map between cohomology spaces.

Through diagram chasing, one can similarly prove the following useful result.

Lemma 10.5.8 (Five lemma). *Consider the following commutative diagram of linear spaces and linear maps, where the two rows form complexes:*

$$
\begin{array}{ccccccccc}
U_1 & \xrightarrow{f_1} & U_2 & \xrightarrow{f_2} & U_3 & \xrightarrow{f_3} & U_4 & \xrightarrow{f_4} & U_5 \\
\downarrow{\scriptstyle h_1} & & \downarrow{\scriptstyle h_2} & & \downarrow{\scriptstyle h_3} & & \downarrow{\scriptstyle h_4} & & \downarrow{\scriptstyle h_5} \\
V_1 & \xrightarrow{g_1} & V_2 & \xrightarrow{g_2} & V_3 & \xrightarrow{g_3} & V_4 & \xrightarrow{g_4} & V_5
\end{array}
$$

(i) *If the row complexes are exact at U_3 and V_2, and if h_2, h_4 are injective and h_1 is surjective, then the middle map h_3 is injective.*

(ii) *If the row complexes are exact at V_3 and U_4, and if h_5 is injective and h_2, h_4 are surjective, then the middle map h_3 is surjective.*

In particular, if the row complexes are exact, and if h_1, h_2, h_4, and h_5 are invertible, then h_3 is also invertible.

10.6   De Rham Cohomology

Let D be a bounded Lipschitz domain in an n-dimensional Euclidean space X, and consider the operators d and $\underline{\delta} = -d^*$ in $L_2(D)$. In this section, we study the finite-dimensional subspace $\mathcal{C}_\|(D)$ in the Hodge decomposition

$$L_2(D) = \mathsf{R}(d) \oplus \mathcal{C}_\|(D) \oplus \mathsf{R}(\underline{\delta}).$$

Definition 10.6.1 (De Rham cohomology spaces). Let D be a bounded Lipschitz domain. The *De Rham cohomology spaces* are the quotient spaces

$$H^k(D) := \mathsf{N}(d; \wedge^k)/\mathsf{R}(d; \wedge^k).$$

We identify their direct sum with $H(D) := \mathsf{N}(d)/\mathsf{R}(d)$.

These cohomology spaces $H^k(D)$ should not be confused with the Čech cohomology spaces $H^k(\underline{D}; \mathcal{F})$ from Section 10.5. We shall, however, show in this section that for the sheaf $\mathcal{F} = \mathbf{R}$ they are indeed closely related.

We note that the following spaces essentially are the same. The last three are indeed equal.

- The De Rham space $H(D)$.

- The dual quotient space $\mathsf{N}(\underline{\delta})/\mathsf{R}(\underline{\delta})$.

- The intersection of the Hodge subspaces $\mathcal{C}_{\parallel}(D) = \mathsf{N}(d) \cap \mathsf{N}(\underline{\delta})$.

- The null space of the Hodge–Dirac operator $d + \underline{\delta}$ from Proposition 10.1.3.

- The null space of the Hodge–Laplace operator $(d + \underline{\delta})^2 = d\underline{\delta} + \underline{\delta}d$.

Note that the orthogonal complement $\mathcal{C}_{\parallel}(D)$ of $\mathsf{R}(d)$ in $\mathsf{N}(d)$ is different from, but can be identified with the quotient space $H(D)$. One can show that for smooth domains, $\mathcal{C}_{\parallel}(D)$ is a subspace of $C^{\infty}(\overline{D})$. For Lipschitz domains this is not true, but using the potential maps from Section 10.4, one can show, at least for strongly Lipschitz domains, that there is another complement of $\mathsf{R}(d)$ in $\mathsf{N}(d)$ that is contained in $C^{\infty}(\overline{D})$. This means that the de Rham cohomology space $H(D)$ can be represented by $C^{\infty}(\overline{D})$ fields.

The cohomology space $\mathcal{C}_{\parallel}(D)$ splits into its homogeneous k-vector parts $\mathcal{C}_{\parallel}(D; \wedge^k)$, and it is our aim in this section to calculate the Betti numbers

$$b_k(D) = \dim \mathcal{C}_{\parallel}(D; \wedge^k) = \dim H^k(D)$$

from Definition 7.6.3 for a given domain D.

On a domain D with boundary, we can similarly consider the cohomology space $\mathcal{C}_{\perp}(D)$ with normal boundary conditions. But by Hodge star duality as in Proposition 10.2.2, we have

$$\dim \mathcal{C}_{\perp}(D; \wedge^k) = \dim \mathcal{C}_{\parallel}(D; \wedge^{n-k}) = b_{n-k}(D).$$

For the remainder of this section, we therefore consider only tangential boundary conditions. The following observation shows that the Betti numbers do not depend on the geometry, but only on the topology of the domain.

Proposition 10.6.2 (Topological invariance). *Let $\rho : D_1 \to D_2$ be a Lipschitz diffeomorphism between bounded Lipschitz domains in Euclidean space. Then the pullback induces an invertible linear map $\rho^* : H(D_2) \to H(D_1)$. In particular,*

$$b_k(D_1) = b_k(D_2)$$

for all $k = 0, 1, 2, \ldots, n$.

Note that the pullback will not in general map between $\mathsf{R}(d)^{\perp} = \mathsf{N}(\underline{\delta})$ for the two domains, and that consequently the spaces $\mathcal{C}_{\parallel}(D, \wedge^k)$ depend on the geometry of D. It is only the dimensions that are topological invariants.

Proof. The result is immediate from Lemma 10.2.6, which shows that ρ^* yields invertible maps between the ranges $\mathsf{R}(d)$ as well as between the null spaces $\mathsf{N}(d)$ for the two domains.   □

The Betti numbers give information about the topology of the domain, the simplest example being

$$b_0(D) = \text{ number of connected components of } D,$$

which is clear, since $dF = 0$ for a scalar function means that F locally constant. Note also that by imposing Dirichlet boundary conditions $\underline{d}F = 0$ for a scalar function forces $F = 0$. By Hodge star duality, this translates to

$$b_n(D) = 0$$

for every bounded Lipschitz domain in Euclidean space. The heuristic is that for general k, the Betti number $b_k(D)$ measures the number of k-dimensional obstructions in D. For the topologically trivial domains, those that are Lipschitz diffeomorphic to the ball, the following result is clear from Poincaré's theorem (Theorem 7.5.2) and its L_2 extension in Section 10.4.

Proposition 10.6.3 (Cohomology for balls). *If D is Lipschitz diffeomorphic to a ball, then $b_0(D) = 1$ and $b_k(D) = 0$ for $k = 1, 2, \ldots, n$.*

To calculate the Betti numbers for a general domain, we cover D by subsets D_j, all diffeomorphic to balls, such that

$$D = D_1 \cup \cdots \cup D_N.$$

We also require that all intersections be topologically trivial as follows.

Definition 10.6.4 (Good cover). Let D be a bounded Lipschitz domain, and assume that $D = D_1 \cup \cdots \cup D_N$ is a finite cover of D by open subsets. We say that $\underline{D} = \{D_j\}_{j=1}^N$ is a *good cover* of D if all nonempty intersections $D_s = D_{s_1} \cap \cdots \cap D_{s_k}$ are Lipschitz diffeomorphic to balls.

We use the algebra from Section 10.5 and three simple examples of sheaves $\mathcal{F}$ on D.

- The sheaf $\mathsf{D}(d; \wedge^k)$, where the Hilbert space associated with an open set $D' \subset D$ consists of k-vector fields $F \in L_2(D')$ such that $dF \in L_2(D')$.

- The sheaf $\mathsf{N}(d; \wedge^k)$, where the Hilbert space associated with an open set $D' \subset D$ consists of k-vector fields $F \in L_2(D')$ such that $dF = 0$.

- The sheaf $\mathbf{R}$, where a real line is associated with an open set $D' \subset D$, and restriction is the identity map.

Note that $\mathsf{D}(d;\wedge^k)$ is a fine sheaf, but not $\mathsf{N}(d;\wedge^k)$ or $\mathbf{R}$. Note also that we cannot use $\underline{\delta}$ here to define sheaves, since restriction does not preserve the boundary conditions. The main result of this section is the following characterization of the Betti numbers.

Theorem 10.6.5 (De Rham = Čech). *Let D be a bounded Lipschitz domain in Euclidean space, and let $\underline{D}$ be a good cover as in Definition 10.6.4. Then*

$$b_k(D) = \dim H^k(\underline{D};\mathbf{R}), \quad k = 0, 1, \ldots, n, \tag{10.13}$$

where $H^k(\underline{D};\mathbf{R})$ is the Čech cohomology space for the constant sheaf $\mathbf{R}$.

This shows in particular that the Betti numbers do not depend on the exterior and interior derivative operators, since the right-hand side in (10.13) does not. Conversely, $\dim H^k(\underline{D};\mathbf{R})$ does not depend on the choice of good cover, since the left-hand side in (10.13) does not.

Proof. Consider the following sequence of maps on D_s:

$$0 \to \mathsf{N}(d_{D_s};\wedge^k) \xrightarrow{i} \mathsf{D}(d_{D_s};\wedge^k) \xrightarrow{d} \mathsf{N}(d_{D_s};\wedge^{k+1}) \to 0,$$

where i denotes inclusion. Assuming that the intersection D_s is Lipschitz diffeomorphic to a ball, we know that this is an exact complex for $k \geq 0$. Acting componentwise, this induces an exact complex

$$0 \to C^j(\underline{D};\mathsf{N}(d;\wedge^k)) \xrightarrow{i} C^j(\underline{D};\mathsf{D}(d;\wedge^k)) \xrightarrow{d} C^j(\underline{D};\mathsf{N}(d;\wedge^{k+1})) \to 0.$$

Consider the following commutative diagram:

According to Proposition 10.5.5, the cohomology spaces $H^j(\underline{D};\mathsf{D}(d;\wedge^k))$ for the second row vanish when $j \geq 1$. The exact cohomology complex provided by the snake lemma (Lemma 10.5.6), thus splits into exact sequences

$$0 \to H^j(\underline{D};\mathsf{N}(d;\wedge^{k+1})) \to H^{j+1}(\underline{D};\mathsf{N}(d;\wedge^k)) \to 0$$

for $k \geq 0$, $j \geq 1$, and for $j = 0$ the exact complex

$$0 \to \mathsf{N}(d; \wedge^k) \to \mathsf{D}(d; \wedge^k) \to \mathsf{N}(d; \wedge^{k+1}) \to H^1(\mathsf{N}(d; \wedge^k)) \to 0.$$

This shows that $\dim H^{j+1}(\underline{D}; \mathsf{N}(d; \wedge^k)) = \dim H^j(\underline{D}; \mathsf{N}(d; \wedge^{k+1}))$ for $j \geq 1$, and $\dim H^1(\mathsf{N}(d; \wedge^k)) = \dim(\mathsf{N}(d; \wedge^{k+1})/d\mathsf{D}(d; \wedge^k)) = b_{k+1}(D)$. Thus for $k \geq 1$, we get

$$\begin{aligned}
\dim H^k(\underline{D}; \mathbf{R}) &= \dim H^k(\underline{D}; \mathsf{N}(d; \wedge^0)) = \dim H^{k-1}(\underline{D}; \mathsf{N}(d; \wedge^1)) \\
&= \dim H^{k-2}(\underline{D}; \mathsf{N}(d; \wedge^2)) = \cdots = \dim H^1(\underline{D}; \mathsf{N}(d; \wedge^{k-1})) = b_k(D),
\end{aligned}$$

since the sheaves $\mathsf{N}(d; \wedge^0)$ and $\mathbf{R}$ coincide. Proposition 10.5.5 shows that $H^0(\underline{D}; \mathbf{R}) = \mathbf{R}(D)$, which equals the space $\mathcal{C}_{\parallel}(D; \wedge^0)$ of locally constant functions on D. Thus $\dim H^0(\underline{D}; \mathbf{R}) = b_0(D)$, which completes the proof. $\square$

Theorem 10.6.5 reduces the computation of the Betti numbers to a finite problem, although the construction of a good cover can be nontrivial. Note that we started by defining the Betti numbers as the dimension of the finite-dimensional space $\mathsf{N}(d; \wedge^k)/\mathsf{R}(d; \wedge^k)$. However, note that both the numerator and denominator are infinite-dimensional Hilbert spaces in general. On the other hand, we have now characterized the Betti numbers as the dimensions of the spaces

$$\mathsf{N}(\partial_k; C^k(\underline{D}; \mathbf{R}))/\mathsf{R}(\partial_{k-1}; C^k(\underline{D}; \mathbf{R})),$$

where all spaces involved are finite-dimensional.

Example 10.6.6 (Annulus). The simplest domain with nontrivial topology is the two-dimensional annulus $D = \{x = (x_1, x_2) \; ; \; r < |x| < R\}$. We see that a good cover of D requires three subsets D_1, D_2, D_3. For example, $D_1 := \{x \in D \; ; \; x_2 > 0\}$, $D_2 := \{x \in D \; ; \; x_1 > x_2\}$, and $D_3 := \{x \in D \; ; \; x_1 + x_2 < 0\}$ give a good cover. The nonempty intersections are $D_1, D_2, D_3, D_{12}, D_{13}$, and D_{23}. We see that $C^0(\underline{D}; \mathbf{R})$ is a three-dimensional space, with a basis $(\omega_1, \omega_2, \omega_3)$, where $\langle \omega_i, e_i \rangle = 1$, and 0 on the other subsets. Similarly, $C^1(\underline{D}; \mathbf{R})$ is a three-dimensional space, with a basis $(\omega_{12}, \omega_{13}, \omega_{23})$, where $\langle \omega_{ij}, e_{ij} \rangle = 1$, and 0 on the other subsets. By Definition 10.5.1, the matrix for ∂_0 is

$$\begin{bmatrix} -1 & 1 & 0 \\ -1 & 0 & 1 \\ 0 & -1 & 1 \end{bmatrix}.$$

This has a one-dimensional null space, so that $b_0(D) = \dim \mathsf{N}(\partial_0) = 1$. Since ∂_0 has two-dimensional range and $\partial_1 = 0$, since $D_{123} = \emptyset$, we get $b_1(D) = \dim \mathsf{N}(\partial_1) - \dim \mathsf{R}(\partial_0) = 3 - 2 = 1$. We have shown that the Betti numbers for D are

$$b_i(D) = (1, 1, 0).$$

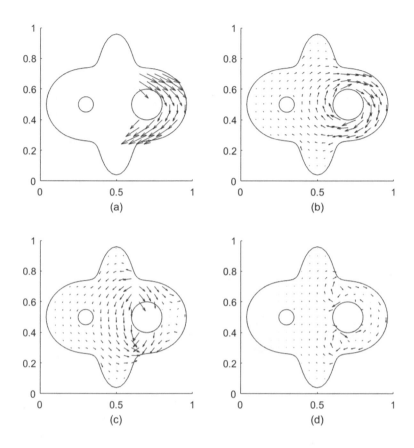

Figure 10.2: (a) Vector field F to be decomposed. (b) The cohomology part H_1 of F in the two-dimensional space $\mathcal{C}_\|(D)$. By Exercise 10.6.8, the cohomology part of any F is a linear combination of this H_1 and H_2 appearing in Figure 10.3. (c) The curl-free part ∇U of F. (d) The divergence-free part $\nabla \lrcorner (V j)$ of F, with tangential boundary conditions.

Exercise 10.6.7 (3D spherical shell). Show by constructing a good cover of the three dimensional spherical shell $D := \{x = (x_1, x_2, x_3) \; ; \; r < |x| < R\}$ with four subsets that
$$b_i(D) = (1, 0, 1, 0).$$

Exercise 10.6.8 (General plane domain). For a plane domain consisting of m disks with n_i smaller interior disks removed in their respective disks, $i = 1, \dots, m$,

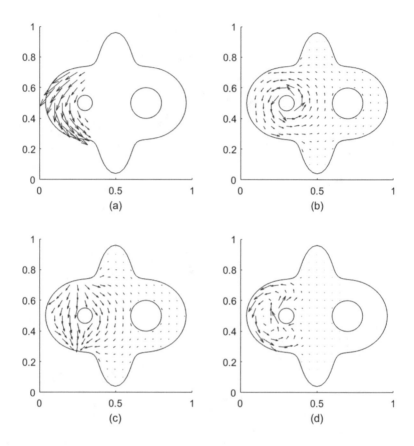

Figure 10.3: (a) Vector field F to be decomposed. (b) The cohomology part H_2 of F in the two-dimensional space $\mathcal{C}_{\parallel}(D)$. By Exercise 10.6.8, the cohomology part of any F is a linear combination of this H_2 and H_1 appearing in Figure 10.2. (c) The curl-free part ∇U of F. (d) The divergence-free part $\nabla \lrcorner (Vj)$ of F, with tangential boundary conditions.

construct a good cover and show that the Betti numbers are

$$b_i(D) = \left(m, \sum_{i=1}^{m} n_i, 0\right).$$

The case $m = 1$, $n = n_1 = 2$ is illustrated in Figures 10.2 and 10.3.

The disadvantage with Theorem 10.6.5 is that the construction of good covers soon gets complicated in higher dimensions. We therefore discuss two complementary techniques for computing Betti numbers: The *Mayer–Vietoris sequence* and the *Künneth formula*. These enable us to calculate Betti numbers for unions and Cartesian products of domains.

Theorem 10.6.9 (Mayer–Vietoris sequence). *Let D_1, D_2 be bounded Lipschitz domains such that $D_1 \cup D_2$ and $D_1 \cap D_2$ also are Lipschitz domains. Then we have the following exact complex, the Mayer–Vietoris sequences:*

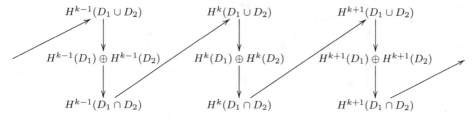

Proof. Consider the complex

$$0 \to \mathsf{D}(d_{D_1 \cup D_2}; \wedge^k) \xrightarrow{i} \mathsf{D}(d_{D_1}; \wedge^k) \oplus \mathsf{D}(d_{D_2}; \wedge^k) \xrightarrow{j} \mathsf{D}(d_{D_1 \cap D_2}; \wedge^k) \to 0,$$

where $i(f) := (f|_{D_1}, f|_{D_2})$ is restriction, and j is the map $j(g_1, g_2) := g_1|_{D_1 \cap D_2} - g_2|_{D_1 \cap D_2}$. We see that i is injective and that the sequence is exact at the middle space. To verify that j is surjective, take $h \in \mathsf{D}(d_{D_1 \cap D_2}; \wedge^k)$. Proposition 10.2.7 shows that we can extend h to $g \in \mathsf{D}(d_{D_1}; \wedge^k)$ such that $g|_{D_1 \cap D_2} = h$. Then $j(g, 0) = h$. The stated Mayer–Vietoris sequence is obtained by applying the snake lemma (Lemma 10.5.6) with $\partial = d$ to this complex. $\square$

Theorem 10.6.10 (Künneth formula). *Let D_1 and D_2 be bounded Lipschitz domains in Euclidean spaces X_1 and X_2 respectively, not necessarily of the same dimension. Then the Betti numbers of the Cartesian product $D_1 \times D_2 \subset X_1 \times X_2$ are given by the Künneth formula*

$$b_k(D_1 \times D_2) = \sum_{j=0}^{k} b_j(D_1)\, b_{k-j}(D_2).$$

Proof of Theorem 10.6.10. Let $p_i : D_1 \times D_2 \to D_i$, $i = 1, 2$, denote the coordinate projections and use pullbacks to form the bilinear map

$$(f_1(x_1), f_2(x_2)) \mapsto (p_1^*(f_1) \wedge p_2^*(f_2))(x_1, x_2) = f_1(x_1) \wedge f_2(x_2),$$

where $f_i : D_i \to \wedge^{k_i} V_i$. Note that $d(p_1^* f_1 \wedge p_2^* f_2) = p_1^*(df_1) \wedge p_2^* f_2 + (-1)^{k_1} p_1^* f_1 \wedge p_2^*(df_2)$. This shows that $p_1^* f_1 \wedge p_2^* f_2$ is closed if f_1 and f_2 are closed, and that it is exact if one of the factors is exact and the other factor is closed. Using the universal property for tensor products, we obtain a well-defined linear map

$$H^{k_1}(D_1) \otimes H^{k_2}(D_2) \to H^{k_1 + k_2}(D_1 \times D_2) : [f_1(x_1)] \otimes [f_2(x_2)] \mapsto [f_1(x_1) \wedge f_2(x_2)].$$

Assume that D_i have a good cover by N_i subsets, $i = 1, 2$. We prove the Künneth formula by induction on $N_1 + N_2$, the induction hypothesis being that the map

$$\bigoplus_i (H^i(D_1) \otimes H^{k-i}(D_2)) \to H^k(D_1 \times D_2),$$

defined as the direct sum of the maps above, is invertible for all fixed k. Evaluating the dimensions will then give the Künneth formula.

If $N_1 + N_2 = 2$, then D_1, D_2, and $D_1 \times D_2$ are all diffeomorphic to balls in the respective spaces, and the induction hypothesis is readily verified. For the induction step, write, for example, $D_1 = U \cup V$, where U is one of the sets in a good cover of D_1 and V is the union of the rest. The induction hypothesis applies to $U \times D_2$, $V \times D_2$, and $(U \cap V) \times D_2$, since both V and $U \cap V$ have good covers by at most $N_1 - 1$ sets. Consider the diagram

$$(H^{i-1}(U) \otimes H^{k-i}(D_2)) \oplus (H^{i-1}(V) \otimes H^{k-i}(D_2)) \longrightarrow H^{k-1}(U \times D_2) \oplus H^{k-1}(V \times D_2)$$

$$H^{i-1}(U \cap V) \otimes H^{k-i}(D_2) \longrightarrow H^{k-1}((U \cap V) \times D_2)$$

$$H^i(U \cup V) \otimes H^{k-i}(D_2) \longrightarrow H^k((U \cup V) \times D_2)$$

$$(H^i(U) \otimes H^{k-i}(D_2)) \oplus (H^i(V) \otimes H^{k-i}(D_2)) \longrightarrow H^k(U \times D_2) \oplus H^k(V \times D_2)$$

$$H^i(U \cap V) \otimes H^{k-i}(D_2) \longrightarrow H^k((U \cap V) \times D_2).$$

The horizontal maps are defined as above in the natural way, and the vertical maps come from the Mayer–Vietoris sequence. On the left, we have taken the tensor product by $H^{k-i}(D_2)$, and the maps act trivially in this factor. For the first and fourth rows we have a distributional rule for tensor products and direct sums. It is straightforward to verify that this diagram commutes. Taking the direct sum over i, we can apply the induction hypothesis and the five lemma (Lemma 10.5.8), which proves the theorem. $\qquad\square$

Example 10.6.11 ($\mathbf{R}^n$ annulus). Consider an n-dimensional annulus $A^n = \{x = (x_1, \ldots, x_n) \; ; \; 2 < |x| < 3\}$, and let $A^n_+ := A^n \cap \{x_n > -1\}$ and $A^n_- := A^n \cap \{x_n < 1\}$. Then $A^n_\pm$ are both diffeomorphic to the n-dimensional unit ball B^n, and $A^n_+ \cap A^n_-$ is diffeomorphic to $A^{n-1} \times (0, 1)$. We prove by induction that

$$b_i(A^n) = (1, 0, 0, \ldots, 0, 0, 1, 0),$$

so that $b_{n-1}(A^n) = 1$. The cases $n = 2, 3$ are known from Example 10.6.6 and Exercise 10.6.7. For $n > 3$, the Mayer–Vietoris sequence and Künneth formula give exact sequences

$$0 \to H^{k-1}(A^{n-1}) \to H^k(A^n) \to 0$$

for $k \geq 2$ and using that $b_1(A^n) = 1$, we have for $k = 1$ that

$$0 \to \mathbf{R} \to \mathbf{R}^2 \to \mathbf{R} \to \mathcal{H}^1(A^n) \to 0$$

is exact. This proves the stated formula for the Betti numbers. Note that $x/|x|^n$ is a divergence and curl-free vector field that is normal on the boundary of the annulus. Applying the Hodge star map, it follows that $\mathcal{C}_{\parallel}(A^n; \wedge^{n-1})$ is spanned by the tangential $(n-1)$-vector field $*x/|x|^n$.

Exercise 10.6.12 (3D cohomology). Let D be a three-dimensional ball with m smaller balls and n cylinders removed, all of these disjoint. Show by inductively applying a Mayer–Vietoris sequence that

$$b_i(D) = (1, n, m, 0).$$

10.7 Comments and References

10.1 The formulation of Hodge decompositions presented here is based on the survey paper by Axelsson and McIntosh [14]. The notation Γ for an abstract nilpotent operator, generalizing d, was introduced there, along with $\Pi = \Gamma \pm \Gamma^*$ for an abstract Hodge–Dirac operator. Playing with symbols, Γ^* can be viewed as a backward Γ, and together they combine to a Π.

   As in Sections 9.3–9.6, a main philosophy in this chapter is to handle Hodge decomposition by first-order operators. Concretely, this means that we study Hodge decompositions as far as possible using Γ, Γ^*, and Π, which in applications are always first-order differential operators, rather than involving the abstract Laplace operator Π^2. The latter is standard in the literature, but may sometimes complicate the problems. A concrete example is the proof of Theorem 10.3.1. The reason that this works is that we work with the first-order operator $\Gamma = d$ for which the commutation theorem is available.

   A reference for the classical variational second-order approach to Hodge decompositions is Morrey [72].

10.2 We have two natural choices of boundary conditions leading to skew-adjoint Hodge–Dirac operators $d + \underline{\delta}$ and $\underline{d} + \delta$ respectively. It is standard to consider the second-order Laplace operators. Here the generalized Dirichlet Laplacian $(\underline{d} + \delta)^2 = \underline{d}\delta + \delta\underline{d}$ is said to have relative boundary conditions, and the generalized Neumann Laplacian $(d + \underline{\delta})^2 = d\underline{\delta} + \underline{\delta}d$ is said to have absolute boundary conditions. See, for example, [91]. For geometrical reasons, we use the terminology normal and tangential, rather than relative and absolute, in this book.

10.3 The beautiful and simple reduction from Lipschitz domains to smooth domains using the commutation theorem in Theorem 10.3.1 appears in the work by R. Picard [74].

For strongly Lipschitz domains there is a singular integral proof, based on Theorems 8.3.2 and 9.5.1, showing that $\mathsf{D}(d+\underline{\delta})$ and $\mathsf{D}(\underline{d}+\delta)$ are contained in $H^{1/2}(D)$. This exponent $1/2$ is sharp for the class of strongly Lipschitz domains. See [14].

10.4 These regularity and support properties of potentials where proved by Mitrea, Mitrea, and Monniaux [70] and Costabel and McIntosh [31]. In fact, [31] shows the stronger result that the constructed operators T_D and T_D^* are pseudodifferential operators of order -1. By well-known estimates for such operators, bounds on a large number of scales of function spaces follow immediately. Rather than relying on the theory of pseudodifferential operators, however, we follow the proof from [70].

10.5-10.6 Two references for Čech cohomology theory are Bott and Tu [21] and Fulton [40]. The author's original inspiration for this section comes from lecture notes by Ben Andrews at the Australian National University.

The snake and five lemmas are usually used in an algebraic context. However, Pryde [76] shows that such techniques also are useful in the context of Fredholm operators on Banach spaces in analysis.

Two references for de Rham cohomology are [21] and Madsen and Tornehave [65].

Chapter 11

Multivector and Spinor Bundles

Prerequisites:

The reader should be familiar with the basic ideas of differential geometry. Section 11.1 gives a short survey of the required material and fixes notation. Section 11.2 builds on Chapter 7, and Section 11.6 builds on Chapter 5. Section 11.4 uses Section 4.5. The material from Section 10.5 is used in Section 11.6 and some in Section 11.2.

Road map:

In many situations the appropriate model of space is not that of an affine space, but rather that of a manifold as defined in Section 6.1. A manifold behaves only locally like affine space, whereas the global behavior can be quite different. So far we have considered only manifolds embedded as k-dimensional surfaces in some surrounding affine space. A general abstract C^∞ manifold M can always be embedded in some finite-dimensional affine space X. However, when the geometry of the manifold is prescribed by a Riemannian metric, most manifolds do not appear as k-surfaces embedded in some affine space.

In the remaining chapters we consider general compact Riemannian manifolds M, from an intrinsic point of view. We also change focus: instead of being mainly concerned with local nonsmooth analysis as in the previous two chapters, we now focus on global smooth analysis in the remaining two chapters. Extending the concepts of multivectors and spinors to such manifolds presents new problems, although their relation is fairly straightforward after our having developed the affine theory thoroughly in the previous chapters. The first step is to construct the space T_pM of tangent vectors at a point $p \in M$. These form a family of vector spaces

$$\{T_pM\}_{p\in M}$$

indexed by M, that is, a vector bundle. In contrast to the case in which M is

© Springer Nature Switzerland AG 2019

A. Rosén, *Geometric Multivector Analysis*, Birkhäuser Advanced Texts Basler Lehrbücher,
https://doi.org/10.1007/978-3-030-31411-8_11

embedded as a k-surface in an affine space (X, V), in which case all tangent spaces T_pM are subspaces of V, in general two tangent spaces T_pM and T_qM, $p \neq q$, are incomparable. As a consequence, some care is needed to define directional derivatives, referred to as covariant derivatives in this context, of tangential vector fields along M.

It is straightforward to construct a bundle of multivectors $\{\wedge(T_pM)\}_{p \in M}$ and to extend the multivector calculus from Chapter 7, which we do in Section 11.2. Particularly useful are the pullback and pushforward operations from Section 7.2, which allow us to pass between different charts on the manifold and give an invariant meaning to multivectors on the manifold. For spinors we also have induced maps from Proposition 5.3.5. However, these are defined only up to a sign, which makes the definition and even the existence of spinor bundles $\{\triangle(T_pM)\}_{p \in M}$ a delicate matter. We use Čech $\mathbf{Z}_2$ cohomology in Section 11.6 to investigate when there are topological obstructions for spinor bundles to exist, and if they exist, how many different such spinor bundles there are globally over M.

The most central concept for Riemannian manifolds is curvature, which in some sense measures how much the manifold locally differs from flat Euclidean space. The curvature operator in vector notation can be a confusing beast. Having access to multivectors, though, we show in Section 11.3 that this is a symmetric operator on bivectors

$$\wedge^2(T_pM) \to \wedge^2(T_pM) : b \mapsto R(b).$$

The input bivector b represents the oriented measure of an infinitesimal 2-surface Σ at p, and the output bivector $R(b)$ represents, as in Proposition 4.2.3, an infinitesimal rotation determined by the total variation that a vector undergoes as we move it around $\partial\Sigma$, keeping it constant as much as the curved manifold allows.

In Sections 11.5 and 11.6, we define Dirac operators $\mathbf{D}$ and $\rlap{/}{D}$ acting on multivector and spinor fields respectively. In flat Euclidean space we saw in Section 9.1 that both these Dirac operators are first-order partial differential operators that are square roots of the Laplace operator Δ. This is no longer true on a curved manifold, but the squares $\mathbf{D}^2$ and $\rlap{/}{D}^2$ differ from Δ by a zeroth-order term determined by the curvature R of the manifold. We prove such Weitzenböck identities

$$\mathbf{D}^2 = \Delta + R$$

in preparation for Chapter 12, although there they have many other applications to geometry. We have seen an analogue for bounded domains in Euclidean space in Theorem 10.3.6, where the curvature of the boundary appears rather than the curvature of space itself.

Section 11.4 contains a proof of Liouville's theorem on conformal maps in dimension $n \geq 3$. This completes Theorem 4.5.12, which is otherwise not used elsewhere.

Highlights:

- Curvature as infinitesimal rotations around infinitesimal bivectors:11.3.2

- Liouville's theorem on conformal maps: 11.4.2

- $\triangle M$ and $\not\triangle M$ Weitzenböck identities: 11.5.9 and 11.6.10

- Čech cohomology of spinor bundles: 11.6.3 and 11.6.5

11.1   Tangent Vectors and Derivatives

We define general vector bundles E following standard terminology, although from the discussion in Section 1.1 we keep in mind that we in general are considering bundles of linear spaces. Let M be a manifold, and let L be an N-dimensional linear space, over $\mathbf{R}$ or possibly $\mathbf{C}$. When considering nonembedded manifolds and vector bundles over them, we shall assume C^∞ regularity.

- A *vector bundle* E over M, with *fiber space L*, is a collection of linear spaces $\{E_p\}_{p\in M}$ together with a cover $M = \bigcup_{\alpha\in\mathcal{I}} U_\alpha$ of M by open sets $U_\alpha \subset M$, and linear invertible maps $\underline{\mu}_\alpha(p) \in \mathcal{L}(L; E_p)$, $p \in U_\alpha$, $\alpha \in \mathcal{I}$, such that each *bundle transition map*

$$\underline{\mu}_{\beta\alpha}(p) := (\underline{\mu}_\beta(p))^{-1}\underline{\mu}_\alpha(p) \in \mathcal{L}(L)$$

  is a C^∞ function of $p \in U_{\beta\alpha} := U_\beta \cap U_\alpha$ for all $\alpha, \beta \in \mathcal{I}$. The linear space E_p is referred to as the *fiber* of E over $p \in M$.

  We refer to $\underline{\mu}_\alpha$ as *bundle charts* and to $\{\underline{\mu}_\alpha\}_{\alpha\in\mathcal{I}}$ as a *bundle atlas*. More generally, a *bundle chart* is a family $\underline{\mu}(p) \in \mathcal{L}(L; E_p)$ of linear invertible maps defined in open sets U such that the bundle transition maps $(\underline{\mu}(p))^{-1}\underline{\mu}_\alpha(p) \in \mathcal{L}(L)$ are C^∞ functions of $p \in U \cap U_\alpha$ for all $\alpha \in \mathcal{I}$.

- A *section w* of E is a map that, to each $p \in M$ associates a vector

$$w(p) \in E_p.$$

  We let $C^\infty(M; E)$ be the space of all sections w such that $\mu(p)^{-1}(w(p)) \in L$ is a C^∞-regular function of $p \in U$ for every bundle chart $\underline{\mu}$. Spaces of sections like $C^k(M; E)$ and $L_2(M; E)$ are defined similarly.

- A (local) *frame* for E in an open set $U \subset M$ is a set of sections $\mathbf{e}_i \in C^\infty(U; E)$, $i = 1, \dots, N$, such that $\{\mathbf{e}_i(p)\}$ is a basis for E_p for each $p \in U$. Thus, if w is a section of E in U, then $w(p) = \sum_{i=1}^N w_i(p)\mathbf{e}_i(p)$, where w_i are the scalar coordinate functions for w in the frame $\{\mathbf{e}_i\}$.

  Note that in general there may not exist a frame defined on all M.

- A *metric* g on E is a family of inner products $g(\cdot,\cdot) = \langle \cdot,\cdot \rangle_p$, one on each fiber E_p, $p \in M$. It is assumed that the symmetric nondegenerate bilinear forms

$$L \times L \to \mathbf{R} : (u,v) \mapsto \langle \underline{\mu}(p)u, \underline{\mu}(p)v \rangle_p$$

are C^∞ functions of $p \in U$ for all bundle charts $\underline{\mu}$.

Let $\{\mathbf{e}_i\}_{i=1}^N$ be a frame in $U \subset M$. The *metric coordinates* in this frame are the functions

$$g_{ij}(p) := \langle \mathbf{e}_i(p), \mathbf{e}_j(p) \rangle_p, \quad p \in U, i,j = 1,\ldots,N.$$

A vector bundle equipped with a metric is called a *metric vector bundle*. We use only Euclidean metrics on real vector bundles and only Hermitian metrics on complex vector bundles.

The metric coordinates for the constant standard inner product on $\mathbf{R}^n$ are the *Kronecker delta*

$$\delta_{ij} := \begin{cases} 1, & i = j, \\ 0, & i \neq j. \end{cases}$$

Example 11.1.1 ($E^* = E$ identification)**.** If E is a vector bundle over M, with bundle charts $\underline{\mu}(p) : L \to E_p$, then the dual vector bundle E^* is defined to be the bundle $\{E_p^*\}_{p \in M}$ of dual spaces, with fiber space L^* and bundle charts

$$(\underline{\mu}(p)^{-1})^* : L^* \to E_p^*.$$

When a vector bundle E is equipped with a metric, the dual bundle E^* and E can be identified by applying Proposition 1.2.3 to each fiber. However, some care has to be taken with regard to charts. We now have twins of bundle charts

$$\underline{\mu}(p) : L \to E_p, \quad p \in U,$$

and

$$(\underline{\mu}(p)^{-1})^* : L^* \to E_p^*, \quad p \in U,$$

for $E_p^* = E_p$. That $L^* \neq L$ is a technicality; what is important is that these spaces do not depend on p. Fixing an auxiliary inner product on L, we have $L^* = L$ as the fiber space for $E^* = E$. This gives a transition map $\underline{\mu}(p)^* \underline{\mu}(p) : L \to L^*$ between the twin bundle charts.

The most fundamental vector bundle is the *tangent vector bundle*. To define the *tangent space* $T_p M$ to a nonembedded manifold M at a point $p \in M$, we proceed as follows. We use a C^1 curve $\gamma : (-\epsilon, \epsilon) \to M$ with $\gamma(0) = p$ and $\epsilon > 0$ to define a tangent vector v. If μ_α is any chart for M around p, then $t \mapsto \mu_\alpha^{-1}(\gamma(t))$ yields a curve through x in $\mathbf{R}^n$, where $\mu_\alpha(x) = p$. We define two curves γ_1 and γ_2 through p to be equivalent, $\gamma_1 \sim \gamma_2$, if $(\mu_\alpha^{-1} \circ \gamma_1)'(0) = (\mu_\alpha^{-1} \circ \gamma_2)'(0)$ in $\mathbf{R}^n$. From the chain rule, we see that this equivalence relation is independent of the choice of chart μ_α. A *tangent vector* v at $p \in M$ is defined to be an equivalence class $[\gamma]$ of curves γ through p on M, under the equivalence relation $\sim$.

Definition 11.1.2 (Tangent bundle). Let $(M, \{\mu_\alpha\}_{\alpha \in \mathcal{I}})$ be an n-dimensional manifold, and let $p \in M$. We define the *tangent space* $T_p M$ to be the set of equivalence classes $v = [\gamma]$ of C^1 curves γ through p on M as above. Defining the bijective map

$$T_p M \to \mathbf{R}^n : v = [\gamma] \mapsto (\mu_\alpha^{-1} \circ \gamma)'(0) \tag{11.1}$$

to be linear gives $T_p M$ the structure of an n-dimensional linear space.

The *tangent bundle* TM is the vector bundle $\{T_p M\}_{p \in M}$, with fiber space $\mathbf{R}^n$. Each manifold chart $\mu_\alpha : D_\alpha \to M_\alpha$ induces a bundle chart $\underline{\mu}_\alpha(p) \in \mathcal{L}(\mathbf{R}^n; T_p M)$, $p \in M_\alpha$, being the inverse of the map (11.1). We say that M is a *Riemannian manifold* if TM is equipped with a metric. We refer to sections of TM as *tangent vector fields*. From the standard basis $\{e_i\}$ for $\mathbf{R}^n$, we define the *coordinate frame* $\{\underline{\mu}_\alpha(e_i)\}$ in M_α.

Note that for TM, the bundle transition functions $\underline{\mu}_{\beta\alpha}(p) : \mathbf{R}^n \to \mathbf{R}^n$ equal the total derivatives $\underline{\mu_{\beta\alpha}}_x : \mathbf{R}^n \to \mathbf{R}^n$, $p = \mu_\alpha(x)$, of the manifold transition functions $\mu_{\beta\alpha}$.

Definition 11.1.3 (Directional derivatives). Let $f \in C^1(M; \mathbf{R})$ be a scalar-valued function on a manifold M, and let $v \in T_p M$ be a tangent vector at $p \in M$. We define the *tangential directional derivative*

$$(\partial_v f)(p) := (f \circ \gamma)'(0)$$

if $v = [\gamma]$.

In a chart μ_α around $p = \mu_\alpha(x)$, let $v = \underline{\mu}_\alpha(p) u$, $u \in \mathbf{R}^n$. Then the chain rule shows that $\partial_v f(p) = \partial_u f_\alpha(x)$, so $\partial_v f(p)$ is independent of the choice of curve γ representing v.

Next consider a vector bundle E over M and a section $w \in C^\infty(M; E)$. We ask whether there is a well-defined tangential directional derivative

$$\partial_v w(p).$$

Proceeding similarly to the above, we consider $w_\alpha(p) := \underline{\mu}_\alpha(p)^{-1} w(p) \in L$ and $w_\beta(p) := \underline{\mu}_\beta(p)^{-1} w(p) \in L$. In this case, $w_\beta(p) = \underline{\mu}_{\beta\alpha}(p)(w_\alpha(p))$. However, because of the p-dependence of the linear maps $\underline{\mu}_{\beta\alpha}(p)$, the derivatives $\partial_v w_\alpha(p)$ and $\partial_v w_\beta(p)$ will in general differ by a zeroth-order term according to the product rule. More precisely,

$$(\partial_v w_\beta)(p) = \underline{\mu}_{\beta\alpha}(p)(\partial_v w_\alpha)(p) + (\partial_v \underline{\mu}_{\beta\alpha}(p)) w_\alpha(p),$$

where the directional derivative is taken componentwise. Directional derivatives of sections of vector bundles in differential geometry are traditionally called *covariant derivatives*, denoted by ∇_v rather than ∂_v, and amount to a choice of zeroth-order terms, consistent with the transition maps as above.

Definition 11.1.4 (Covariant derivative). Let E be a vector bundle over a manifold M. A *covariant derivative* on E is a map $C^\infty(U;TM) \times C^\infty(U;E) \mapsto C^\infty(U;E)$: $(v,w) \mapsto \nabla_v w$ such that

- $\nabla_{v_1+v_2} w = \nabla_{v_1} w + \nabla_{v_2} w$, for all $v_1, v_2 \in C^\infty(U;TM)$ and $w \in C^\infty(U;E)$,

- $\nabla_{fv} w = f(\nabla_v w)$, for all $f \in C^\infty(U;\mathbf{R})$, $v \in C^\infty(U;TM)$ and $w \in C^\infty(U;E)$,

- $\nabla_v(w_1 + w_2) = \nabla_v w_1 + \nabla_v w_2$, for all $v \in C^\infty(U;TM)$ and $w_1, w_2 \in C^\infty(U;E)$,

- $\nabla_v(fw) = f(\nabla_v w) + (\partial_v f)w$, for all $f \in C^\infty(U;\mathbf{R})$, $v \in C^\infty(U;TM)$, and $w \in C^\infty(U;E)$.

Let E be a vector bundle over a Riemannian manifold M, and let $\{\mathbf{e}_i(p)\}_{i=1}^N$ be a frame for E in an open subset $U \subset M$. By the Leibniz rule, a section $w(p) = \sum_i f_i(p)\mathbf{e}_i(p)$ of E in U has covariant derivative

$$\nabla_v w = \sum_i (\partial_v f_i)\mathbf{e}_i + \sum_i f_i(\nabla_v \mathbf{e}_i).$$

Since $v \mapsto \nabla_v w$ is linear, this uniquely defines vector fields $w_{ji} \in C^\infty(U;TM)$ in U such that

$$\nabla_v \mathbf{e}_i = \sum_{j=1}^N \langle \omega_{ji}, v \rangle \mathbf{e}_j, \quad v \in C^\infty(U;TM), \; i = 1, \ldots, N.$$

This shows in particular that $\nabla_v w$ is well defined at $p \in M$ whenever $v \in T_p M$ is given and w is a section differentiable at p.

Definition 11.1.5 (Christoffel symbols). Let E be an N-dimensional vector bundle over a manifold M, with a covariant derivative ∇_v. Let $\mathbf{e} = \{\mathbf{e}_i(p)\}_{i=1}^N$ be a frame for E in an open subset $U \subset M$. Then the N^2 vector fields $w_{ij} \in C^\infty(U;TM)$ specifying ∇_v in the frame $\mathbf{e}$ are called the *Christoffel symbols* of ∇ in $\mathbf{e}$. We write

$$\omega_{\mathbf{e}} \in C^\infty(U;\mathcal{L}(TM, \mathcal{L}(E)))$$

for the section such that the matrix for the map $\omega_{\mathbf{e}}(v) \in \mathcal{L}(E)$ in the basis $\{\mathbf{e}_i\}$ is $\{\langle \omega_{ij}, v \rangle\}_{ij}$.

With this notation, we have

$$\nabla_v \left(\sum_i f_i \mathbf{e}_i \right) = \sum_i (\partial_v f_i)\mathbf{e}_i + \omega_{\mathbf{e}}(v)\left(\sum_i f_i \mathbf{e}_i \right)$$

in the frame $\mathbf{e} = \{\mathbf{e}_i(p)\}_{i=1}^N$. To single out a canonical covariant derivative, we demand that it be compatible with natural structures of the vector bundle. In particular, if E is equipped with a metric $\{\langle \cdot, \cdot \rangle_p\}_{p \in M}$, then we demand that the product rule

$$\partial_v \langle w_1, w_2 \rangle = \langle \nabla_v w_1, w_2 \rangle + \langle w_1, \nabla_v w_2 \rangle$$

hold for all $w_1, w_2 \in C^\infty(U; E)$ and $v \in C^\infty(U; TM)$ and say that ∇_v is a *metric covariant derivative*. If ω_{ij} are the Christoffel symbols in an ON-basis, this happens if and only if

$$\omega_{ji} = -\omega_{ij}$$

holds for all $i, j = 1, \ldots, N$.

To obtain the unique existence of a covariant derivative on the tangent bundle TM, one imposes, besides it being metric, a condition in terms of Lie brackets of vector fields.

Exercise 11.1.6 (Lie brackets on manifolds). Let $\mu_\alpha : D_\alpha \to M$ and $\mu_\alpha : D_\beta \to M$ be two charts on a manifold M. Push forward two vector fields u and v in $D_\alpha \subset \mathbf{R}^n$, by the transition map $\rho := \mu_{\beta\alpha}$, to $u' := \rho_* u$ and $v' := \rho_* v$ in D_β. Consider the Lie brackets $[u, v] := \partial_u v - \partial_v u$ and $[u', v'] := \partial_{u'} v' - \partial_{v'} u'$. Show that

$$[u', v'] = \rho_*[u, v],$$

and deduce that the Lie bracket of tangent vector fields on M is well defined.

Proposition 11.1.7 (Levi-Civita covariant derivative). *Let M be a Riemannian manifold. Then there exists a unique covariant derivative on TM, the Levi-Civita covariant derivative, which is both metric and torsion-free, in the sense that*

$$\partial_v \langle v_1, v_2 \rangle = \langle \nabla_v v_1, v_2 \rangle + \langle v_1, \nabla_v v_2 \rangle,$$
$$\nabla_{v_1} v_2 - \nabla_{v_2} v_1 = [v_1, v_2],$$

for all $v, v_1, v_2 \in C^\infty(M; TM)$.

Proof. We use an ON-frame $\{\mathbf{e}_i(p)\}_{i=1}^N$ in some open set $U \subset M$. If a covariant derivative is metric and torsion-free, then its Christoffel symbols satisfy $\omega_{ij} = -\omega_{ji}$ and

$$\langle \omega_{kj}, \mathbf{e}_i \rangle - \langle \omega_{ki}, \mathbf{e}_j \rangle = \langle [\mathbf{e}_i, \mathbf{e}_j], \mathbf{e}_k \rangle,$$

for all i, j, k. Permuting (i, j, k) to (j, k, i), (k, i, j) and subtracting and adding these two obtained equations gives the formula

$$2\langle \omega_{kj}, \mathbf{e}_i \rangle = \langle [\mathbf{e}_i, \mathbf{e}_j], \mathbf{e}_k \rangle + \langle [\mathbf{e}_k, \mathbf{e}_i], \mathbf{e}_j \rangle - \langle [\mathbf{e}_j, \mathbf{e}_k], \mathbf{e}_i \rangle.$$

Conversely, the covariant derivative defined through this formula is seen to be metric and torsion-free. $\qquad\square$

Exercise 11.1.8. Generalize the above argument, and show that in a general frame $\{\mathbf{e}_i\}$ for TM over a Riemannian manifold, the Levi-Civita covariant derivative is given by

$$2\langle \nabla_{\mathbf{e}_i}\mathbf{e}_j, \mathbf{e}_k \rangle = \partial_{\mathbf{e}_i}\langle \mathbf{e}_j, \mathbf{e}_k \rangle + \partial_{\mathbf{e}_j}\langle \mathbf{e}_k, \mathbf{e}_i \rangle - \partial_{\mathbf{e}_k}\langle \mathbf{e}_i, \mathbf{e}_j \rangle$$
$$+ \langle [\mathbf{e}_i, \mathbf{e}_j], \mathbf{e}_k \rangle + \langle [\mathbf{e}_k, \mathbf{e}_i], \mathbf{e}_j \rangle - \langle [\mathbf{e}_j, \mathbf{e}_k], \mathbf{e}_i \rangle.$$

Note that in a coordinate frame the last three terms vanish, whereas in an ON-frame the first three terms vanish.

The Christoffel symbols for a metric covariant derivative in an ON-frame give a skew-symmetric map on the fibers E_p. This is true in particular for the Levi-Civita covariant derivative on TM, and in this case we will identify these skew-symmetric maps and the corresponding bivector using Proposition 4.2.3.

Definition 11.1.9 (*TM* Christoffel bivectors). Let M be a Riemannian manifold and let $\mathbf{e} = \{\mathbf{e}_i(p)\}$ be an ON-frame for the tangent bundle TM in an open set $U \subset M$. Denote the Christoffel symbols for the Levi-Civita covariant derivative by

$$\Gamma_{\mathbf{e}} \in C^\infty(U; \mathcal{L}(TM; \wedge^2 M)),$$

so that $\nabla_v \mathbf{e}_i = \Gamma_{\mathbf{e}}(v) \llcorner \mathbf{e}_i$.

For the definition of $\wedge^2 M$, see Section 11.2.

11.2   Multivector Calculus on Manifolds

In this section we show how the basic affine multivector calculus from Chapter 7 generalizes to manifolds. To simplify the presentation, we consider only compact Riemannian manifolds.

Definition 11.2.1 (Multivector bundle). Let M be an n-dimensional Riemannian manifold with atlas $\{\mu_\alpha\}_{\alpha \in \mathcal{I}}$. The *multivector bundle* $\wedge M$ over M is the vector bundle $\{\wedge(T_pM)\}_{p \in M}$, with fiber space $L = \wedge \mathbf{R}^n$ and a bundle atlas of bundle charts $\underline{\mu}_\alpha : \wedge \mathbf{R}^n \to \wedge(T_pM)$ comprising the linear maps induced by the bundle charts of TM as in Definition 2.3.1. We refer to sections of $\wedge M$ as (tangent) *multivector fields* on M. Given a frame $\{\mathbf{e}_i\}_{i=1}^n$ for TM, we obtain an induced frame $\{\mathbf{e}_s\}_{s \subset \overline{n}}$ for $\wedge M$.

The bundle $\wedge M$ is a metric bundle equipped with the metric induced from TM, as in Definition 2.5.2, on each fiber. Using the standard inner product on the fiber space $\wedge \mathbf{R}^n$, in $M_\alpha = \mu_\alpha(D_\alpha)$ the manifold chart μ_α gives rise to the following three bundle charts for $\wedge M$.

- The *pushforward chart* $\underline{\mu}_\alpha(p) : \wedge \mathbf{R}^n \to \wedge T_pM$.

- The *pullback chart* $(\underline{\mu}_\alpha^*)^{-1}(p) : \wedge \mathbf{R}^n \to \wedge T_pM$.

- *The normalized pushforward chart* $|J_{\mu_\alpha}|^{-1}\underline{\mu}_\alpha(p) : \wedge\mathbf{R}^n \to \wedge T_p M$.

Denote by $\wedge^k M$ the subbundle of tangential k-vectors, $k = 0, 1 \ldots, n$, so that

$$\wedge M = \wedge^0 M \oplus \wedge^1 M \oplus \wedge^2 M \oplus \cdots \oplus \wedge^n M,$$

where $\wedge^1 M = TM$ and $\wedge^0 M = M \times \mathbf{R}$.

Note that we shall not use the dual multicovector bundle $(\wedge M)^*$, since we identify it with $\wedge M$ as in Example 11.1.1.

Exercise 11.2.2 (Bundle transition maps). In a fixed manifold chart $M_\alpha = \mu_\alpha(D_\alpha)$, show that the transition map for $\wedge M$ from the pushforward chart to the pullback chart is $G(p) : \wedge\mathbf{R}^n \to \wedge\mathbf{R}^n$, the $\wedge$-extension of the linear map of $\mathbf{R}^n$ corresponding to the metric

$$g_{ij}(x) = \langle \underline{\mu}_\alpha(e_i), \underline{\mu}_\alpha(e_j) \rangle_p, \quad p = \mu_\alpha(x).$$

Also show that the transition map from the normalized pushforward chart to the pullback chart is $G/\sqrt{\det g}$.

If $\mu_\beta : D_\beta \to M_\beta$ is a second manifold chart and the manifold transition map from D_α to D_β is $\mu_{\beta\alpha}$, show that the bundle transition maps between the pushforward, pullback, and normalized pushforward charts are $\mu_{\beta\alpha*}$, $\mu_{\beta\alpha}^*$, and $\tilde{\mu}_{\beta\alpha*}$ respectively.

Note for $\wedge M$ that the pushforward charts $\underline{\mu}_\alpha$ are not only linear, but also $\wedge$-homomorphisms, as are the transition maps $\underline{\mu}_{\beta\alpha}$. The same is true for the pullback charts, but not for the normalized pushforward charts. We have more precisely constructed not only a bundle of linear spaces but a bundle of associative algebras equipped with the exterior product. Applying the affine theory on each fiber, we also have interior products and Hodge star operations defined on $\wedge M$. The latter require a choice of orientation of M, that is, a choice of frame $e_{\overline{n}}$ for the line bundle, that is, a vector bundle with one-dimensional fibers, $\wedge^n M$ with $|e_{\overline{n}}| = 1$. The local existence of such $e_{\overline{n}}$ in a chart M_α presents no problem. The global existence is discussed below.

A C^1 map $\rho : M \to M'$ between two different manifolds M and M' induces, by working with appropriate types of charts, maps between the respective multivector bundles $\wedge M$ and $\wedge M'$. Let $\mu_\alpha : D_\alpha \to M$ be a chart around $p \in M$, let $\mu'_\beta : D'_\beta \to M'$ be a chart around $q = \rho(p) \in M'$, and define $\rho_{\beta\alpha}(x) := (\mu'_\beta)^{-1}(\rho(\mu_\alpha(x)))$. By Exercise 11.2.2, the following constructions are independent of choices of charts.

- We define the pullback of a multivector field F' on M' to be the multivector field $\rho^* F'$ on M such that $\mu_\alpha^*(\rho^* F'(p)) = \rho_{\beta\alpha}^*(\mu'_\beta{}^* F'(q))$.

- If ρ is a diffeomorphism, then we define the pushforward of a multivector field F on M to be the multivector field $\rho_* F$ on M' such that $(\mu'_{\beta*})^{-1}(\rho_* F(q)) = \rho_{\beta\alpha*}(\mu_{\alpha*}^{-1} F(p))$. The normalized pushforward $\tilde{\rho}_* F$ on M' is defined similarly by demanding that $(\tilde{\mu}'_{\beta*})^{-1}(\tilde{\rho}_* F(q)) = \tilde{\rho}_{\beta\alpha*}(\tilde{\mu}_{\alpha*}^{-1} F(p))$.

Exercise 11.2.3 (Tangential restriction). Let M be a k-surface in a Riemannian manifold N with embedding $\rho : M \to N$, and let $p \in M \subset N$. This means in particular that we can regard the tangent space $T_p M$ as a subspace of $T_p N$. Generalize Exercise 7.2.2 and show that $\rho^* F(p)$, where F is a multivector field on N, equals the part of $F(p) \in \wedge T_p N$ tangential to $T_p M$.

We next consider how directional, exterior, and interior derivatives generalize to manifolds.

Proposition 11.2.4 ($\wedge M$ covariant derivative). *Let M be a Riemannian manifold. Then there exists a unique covariant derivative ∇_v on $\wedge M$ that*

- *equals the Levi-Civita covariant derivative from Proposition 11.1.7 on $\wedge^1 M = TM$,*

- *equals the tangential directional derivative on $\wedge^0 M = M \times \mathbf{R}$, and*

- *satisfies the product rule $\nabla_v (w_1 \wedge w_2) = (\nabla_v w_1) \wedge w_2 + w_1 \wedge (\nabla_v w_2)$ for all C^1 multivector fields w_1 and w_2 and vectors v.*

Note that this induced covariant derivative, which we refer to as the Levi-Civita covariant derivative on $\wedge M$, preserves the subbundles $\wedge^k M$, $k = 0, 1, \ldots, n$.

Proof. Consider an induced ON-frame $\{e_s\}_{s \subset \overline{n}}$ for $\wedge M$. It is clear that a covariant derivative with the stated properties is unique, since

$$\nabla_v (e_{s_1} \wedge \cdots \wedge e_{s_k}) = \sum_{i=1}^{k} e_{s_1} \wedge \cdots \wedge \nabla_v e_{s_i} \wedge \cdots \wedge e_{s_k} \qquad (11.2)$$

and $\nabla_v 1 = 0$. To show existence, we define a covariant derivative by these identities. It suffices to verify the product rule for $w_1 = e_{s_1} \wedge \cdots \wedge e_{s_k}$ and $w_2 = e_{t_1} \wedge \cdots \wedge e_{t_l}$. We note that (11.2) continues to hold for arbitrary indices $1 \leq s_i \leq n$, not necessarily distinct or in increasing order, as a consequence of

$$\nabla_v e_{s_i} \wedge e_{s_i} + e_{s_i} \wedge \nabla_v e_{s_i} = 0.$$

This shows that

$$\nabla_v (e_s \wedge e_t) = (\nabla_v e_s) \wedge e_t + e_s \wedge (\nabla e_t). \qquad \square$$

By its definition above, the Levi-Civita covariant derivative on $\wedge M$ satisfies the product rule with respect to exterior multiplication. The following shows that it also satisfies the natural product rules with respect to other products. This is also true for the Clifford product, but we postpone this discussion to Section 11.5.

Proposition 11.2.5 (Covariant product rules). *Let M be a Riemannian manifold, and let ∇_v denote the Levi-Civita covariant derivative on $\wedge M$. This is a metric covariant derivative, that is, $\partial_v \langle F, G \rangle = \langle \nabla_v F, G \rangle + \langle F, \nabla_v G \rangle$. Moreover, we have*

$$\nabla_v (F \lrcorner G) = (\nabla_v F) \lrcorner G + F \lrcorner (\nabla_v G),$$

for all $v \in C(M; TM)$ and $F, G \in C^1(M; \wedge M)$. In particular, $\nabla_v (F*) = (\nabla_v F)*$. The analogous product rules for the right interior product also hold.

Proof. It is clear from the definition that ∇_v is metric on $\wedge^0 M$ and $\wedge^1 M$. By bilinearity, it suffices to consider simple k-vector fields $F = F_1 \wedge \cdots \wedge F_k$ and $G = G_1 \wedge \cdots \wedge G_k$. In this case, we have

$$\partial_v \langle F, G \rangle = \partial_v \begin{vmatrix} \langle F_1, G_1 \rangle & \cdots & \langle F_1, G_k \rangle \\ \vdots & \ddots & \vdots \\ \langle F_k, G_1 \rangle & \cdots & \langle F_k, G_k \rangle \end{vmatrix} = \partial_v \sum (\pm) \prod \langle F_i, G_j \rangle$$

$$= \sum (\pm) \partial_v \prod \langle F_i, G_j \rangle = \cdots = \langle \nabla_v F, G \rangle + \langle F, \nabla_v G \rangle,$$

since each vector field F_i and G_j appears exactly once in each product $\prod \langle F_i, G_j \rangle$.

The covariant derivative being metric, the identity for the left interior product follows by dualilty from the product rule for the exterior derivative, since

$$\langle \nabla_v (F \lrcorner G), H \rangle = \partial_v \langle F \lrcorner G, H \rangle - \langle F \lrcorner G, \nabla_v H \rangle = \partial_v \langle G, F \wedge H \rangle - \langle G, F \wedge \nabla_v H \rangle$$

$$= \langle \nabla_v G, F \wedge H \rangle + \langle G, \nabla_v (F \wedge H) - F \wedge \nabla_v H \rangle = \langle F \lrcorner \nabla_v G, H \rangle + \langle \nabla_v F \lrcorner G, H \rangle$$

for all $H \in C^1(M; \wedge M)$. From this, the identity for $\nabla_v (F*)$ will follow if we prove that $\nabla_v \mathbf{e}_{\overline{n}} = 0$ for $\mathbf{e}_{\overline{n}} \in C^\infty(M; \wedge^n M)$ such that $|\mathbf{e}_{\overline{n}}| = 1$. This follows from the fact that $\wedge^n M$ has one-dimensional fibers and

$$0 = \partial_v |\mathbf{e}_{\overline{n}}|^2 = 2 \langle \nabla_v \mathbf{e}_{\overline{n}}, \mathbf{e}_{\overline{n}} \rangle.$$

The proof of the identities for the right interior product and Hodge star are similar.   □

Using the pullback and normalized pushforward charts respectively, Exercise 11.2.2 and the commutation theorem (Theorem 7.2.9) show that the following gives well-defined exterior and interior derivatives of multivector fields on a manifold.

Definition 11.2.6 (d and δ on manifolds). Consider the exterior bundle $\wedge M$ over a Riemannian manifold M. We define the *exterior and interior derivatives* dF and δF of a multivector field $F \in C^1(M; \wedge M)$ as follows. In a chart $\mu_\alpha : D_\alpha \to M_\alpha$, we define

$$dF(p) := (\underline{\mu}_\alpha^*)^{-1} d(\underline{\mu}_\alpha^* F(x)),$$

$$\delta F(p) := |J_{\mu_\alpha}|^{-1} \underline{\mu}_\alpha \, \delta \, (|J_{\mu_\alpha}| (\underline{\mu}_\alpha^{-1}) F(x)),$$

at $p = \mu_\alpha(x)$, $x \in D_\alpha$, where d and δ on the right-hand sides denote the $\mathbf{R}^n$ derivatives from Definition 7.1.5.

Exercise 11.2.7 (d and δ duality). Let M be a Riemannian manifold, with measure dp. Generalize Proposition 7.1.7 to Riemannian manifolds and prove the following duality relations.

For all $F \in C^1(M; \wedge^k M)$, we have

$$\delta(F*) = (-1)^k (dF)*$$

at $p \in M$, where the Hodge star uses any fixed orientation of the manifold locally around p.

For all $F \in C^1(M; \wedge M)$ and $G \in C^1(M; \wedge M)$ that vanish on ∂M, we have

$$\int_M \langle dF, G \rangle \, dp = - \int_M \langle F, \delta G \rangle \, dp.$$

Hint: Use a partition of unity to localize to charts.

The following main result in this section shows that the exterior and interior derivatives on a Riemannian manifold are nabla operators in the natural covariant sense.

Proposition 11.2.8 (Covariant nabla). *Let M be a Riemannian manifold and let ∇_v be the Levi-Civita covariant derivative on $\wedge M$. If $\{\mathbf{e}_i\}$ is a frame for TM, with dual frame $\{\mathbf{e}_i^*\}$, then*

$$dF = \sum_i \mathbf{e}_i^* \wedge \nabla_{\mathbf{e}_i} F, \quad F \in C^1(M; \wedge M),$$

$$\delta F = \sum_i \mathbf{e}_i^* \lrcorner \nabla_{\mathbf{e}_i} F, \quad F \in C^1(M; \wedge M).$$

Proof. The proof for the exterior derivative uses the expansion rule for d from Exercise 7.4.9. Applying pullbacks, pushforwards, and Exercise 11.1.6, this is seen to hold on M and shows that

$$\langle dF, v_0 \wedge \cdots \wedge v_k \rangle = \sum_{i=0}^k (-1)^i \partial_{v_i} \langle F, v_0 \wedge \cdots \check{v}_i \cdots \wedge v_k \rangle$$

$$+ \sum_{0 \leq i < j \leq k} (-1)^{i+j} \langle F, [v_i, v_j] \wedge v_0 \wedge \cdots \check{v}_i \cdots \check{v}_j \cdots \wedge v_k \rangle,$$

for $F \in C^1(M; \wedge^k M)$ and $v_j \in C^1(M; \wedge^1 M)$. Using that the Levi-Civita covariant derivative is metric, the first sum on the right-hand side is seen to equal

$$\sum_{i=0}^k (-1)^i \langle \nabla_{v_i} F, v_0 \wedge \cdots \check{v}_i \cdots \wedge v_k \rangle$$

$$- \sum_{0 \leq i < j \leq k} (-1)^{i+j} \langle F, (\nabla_{v_i} v_j - \nabla_{v_j} v_i) \wedge v_0 \wedge \cdots \check{v}_i \cdots \check{v}_j \cdots \wedge v_k \rangle$$

Since ∇_v is also torsion-free, it follows that $\langle dF, v_0 \wedge \cdots \wedge v_k \rangle = \sum_{i=0}^k \langle \nabla_{v_i} F, v_i \lrcorner (v_0 \wedge \cdots \wedge v_k) \rangle$. This proves the stated formula, since both sides equal the gradient in the case $k = 0$ of scalar functions.

To transfer this result to δ, we use Hodge duality. Write $F = G*$, where we may assume that G is a k-vector section. Then

$$\delta F = (-1)^k (dG)* = (-1)^k \Big(\sum_i \mathbf{e}_i^* \wedge \nabla_{\mathbf{e}_i} G \Big)*$$

$$= \Big(\sum_i \nabla_{\mathbf{e}_i} G \wedge \mathbf{e}_i^* \Big)* = \sum_i \mathbf{e}_i^* \lrcorner (\nabla_{\mathbf{e}_i} G)* = \sum_i \mathbf{e}_i^* \lrcorner \nabla_{\mathbf{e}_i} F,$$

where we have used Exercise 11.2.7 in the first step and Proposition 11.2.5 in the last step. $\qquad \square$

In the case of an affine space (X, V), the two one-dimensional linear spaces $\wedge^0 V = \mathbf{R}$ and $\wedge^n V$ are isomorphic. For manifolds, the vector bundles $\wedge^0 M$ and $\wedge^n M$ need not be globally isomorphic: this happens precisely when M is orientable, that is, when there exists a globally defined section $e_{\overline{n}} \in C^\infty(M; \wedge^n M)$ with $|e_{\overline{n}}| = 1$. We can characterize orientability of the manifold M using Čech cohomology.

Example 11.2.9 (The sheaf $\mathbf{Z}_2$). Let M be an n-dimensional manifold, with a cover $M = \bigcup_{\alpha \in \mathcal{I}} U_\alpha$ by open sets. For simplicity, we shall always assume that this is a good cover, although we will not always need all of this hypothesis. Analogous to Definition 10.6.4, this means that each nonempty intersection

$$U_s := U_{\alpha_1} \cap U_{\alpha_2} \cap \cdots \cap U_{\alpha_k}, s = \{\alpha_1, \ldots, \alpha_k\}$$

is diffeomorphic to a ball in $\mathbf{R}^n$. Consider the sheaf $\mathbf{Z}_2$, which to each U_s associates the additive group $\mathbf{Z}_2 = \{0, 1\}$. The Čech algebra from Section 10.5 generalizes to $\mathbf{Z}_2$, since it really only requires the spaces to be abelian groups and not vector spaces. In this simplest group $\mathbf{Z}_2$ we have $-1 = 1$, which means that we do not need an alternating algebra. Therefore we write

$$f(s) := \langle f, e_s \rangle = -\langle f, e_s \rangle$$

for the value of the Čech cochain at the intersection U_s. Also the coboundary operator simplifies to

$$\partial_k f(s) = \sum_{\alpha \in s} f(s \setminus \{\alpha\}), \quad |s| = k + 2, f \in C^k(\underline{U}; \mathbf{Z}_2).$$

Proposition 11.2.10 (First Stiefel–Whitney class). *Let M be a manifold with a bundle atlas of bundle charts $\underline{\mu}_\alpha(p) \in \mathcal{L}(\mathbf{R}^n; T_p M)$, $p \in U_\alpha$, $\alpha \in \mathcal{I}$, for TM. Consider the sheaf $\mathbf{Z}_2$ on the good cover $\{U_\alpha\}$ from Example 11.2.9. Let $f \in C^1(\underline{U}; \mathbf{Z}_2)$ be the 1-cochain, for which the value on the intersection $U_\alpha \cap U_\beta$ is*

$$f(\{\alpha, \beta\}) := \begin{cases} 0, & \text{if } \det \underline{\mu}_{\beta\alpha} > 0, \\ 1, & \text{if } \det \underline{\mu}_{\beta\alpha} < 0. \end{cases}$$

Then $\partial_1 f = 0$, and the Čech cohomology class $[f]$ does not depend on the choice of local orientation in each U_α specified by the bundle charts μ_α. The manifold M is orientable if and only if there exists $g \in C^0(\underline{U}; \mathbf{Z}_2)$ such that $\partial_0 g = f$, that is, if $[f] = [0]$.

The Čech cohomology class $[f] \in H^1(\underline{U}; \mathbf{Z}_2)$ is called the *first Stiefel–Whitney class* $w_1(M)$ of M. The two orientations of M correspond to $H^0(\underline{U}; \mathbf{Z}_2) = \mathbf{Z}_2$.

Proof. That

$$\partial_1 f(\{\alpha, \beta, \gamma\}) = f(\{\alpha, \beta\}) + f(\{\alpha, \gamma\}) + f(\{\beta, \gamma\}) = 0 \quad \mathrm{mod}\ 2$$

follows from the transitivity $\mu_{\gamma\beta}\mu_{\beta\alpha} = \mu_{\gamma\alpha}$ of the transition maps.

Another choice of local orientation in each U_α corresponds to $g \in C^0(\underline{U}; \mathbf{Z}_2)$, where $g(\{\alpha\}) = 1$ if we change the orientation in U_α, and $g(\{\alpha\}) = 0$ otherwise. If $f' \in C^1(\underline{U}; \mathbf{Z}_2)$ is the 1-cochain for this new choice, then

$$f'(\{\alpha, \beta\}) = f(\{\alpha, \beta\}) + g(\{\alpha\}) + g(\{\beta\}) \quad \mathrm{mod}\ 2,$$

that is, $f' = f + \partial_0 g$. Since M is orientable, we can choose $f' = 0$; this completes the proof. $\qquad\square$

The integral calculus from Section 7.3 carries over with minor changes from affine spaces to general manifolds M. In the absence of an affine space X in which M is embedded, the main change is that we no longer can consider integrals like oriented measure $\wedge^k M$ or integrals $\int_M F(p)dp$ of tangent vector fields F. Such notions are not defined for general manifolds, for the simple reason that if E is a general vector bundle over a manifold M, then two fibers E_p and E_q, $p \neq q$, are incomparable. Two consequences are the following.

- The vector sum
$$v_p + v_q, \qquad v_p \in E_p, v_q \in E_q,$$
  is not defined, except as a direct sum $v_p \oplus v_q = (v_p, v_q)$, which is not what we want. The continuous analogue of this, the integral, is therefore not possible either.

- It is not well defined what is meant by a constant section of a general vector bundle.

When the form we integrate takes values in a fixed linear space L for all $p \in M$, then we can proceed as before. We outline the main steps of the extension of Section 7.3 to an integral calculus of forms on k-surfaces M embedded in an n-dimensional manifold N.

- Extending Definition 7.3.1, a general k-form on N is now a map that to each $p \in N$, associates a homogeneous function
$$\Theta : \wedge^k T_p N \to L : w \mapsto \Theta(p, w).$$

If these functions are linear, we refer to Θ as a linear k-form.

- Using charts and a partition of unity, we define the integral

$$\int_M \Theta(p, \underline{dp})$$

of a k-form Θ in N over an oriented C^1-regular k-surface M. When Θ is an even form, for example when we are integrating a scalar function $\int_M f(p)dp$ in a Riemannian manifold, where $dp = |\underline{dp}|$, we do not need M to be oriented.

- A linear k-form Θ on N can be uniquely written

$$\Theta(p, w) = \langle F_1(p), w \rangle v_1 + \cdots + \langle F_m(p), w \rangle v_m,$$

if $\{v_j\}_{j=1}^m$ is a basis for L, where F_j and w are k-vector fields on N. By Proposition 11.2.8, its exterior derivative is

$$\Theta(\dot{p}, \nabla \lrcorner w) = \sum_{j=1}^m \langle dF_j, w \rangle = \sum_{j=1}^m \sum_{i=1}^n \langle \nabla_{\mathbf{e}_i} F_j, \mathbf{e}_i^* \lrcorner w \rangle v_j$$

in a frame $\{\mathbf{e}_i\}$ for TM. Note that we cannot regard w as a constant section of $\wedge^k M$, and the product rule formula in Definition 7.3.7 needs to be adjusted by a term involving the covariant derivative of $\mathbf{e}_i^* \lrcorner w$ if we want to use this on a manifold N.

- If the k-surface M in N is oriented and C^1-regular, and if ∂M has the orientation induced by M, then for every C^1-regular k-vector field F in a neighborhood of M in N, we have the Stokes formula

$$\int_M \langle dF(p), \underline{dp} \rangle = \int_{\partial M} \langle F(q), \underline{dq} \rangle.$$

Here dF denotes the exterior derivative. Extending the nabla notation to manifolds, we may write $\nabla \wedge F$ to avoid notational confusion between the exterior derivative dF and the oriented measure $\underline{dp}$. For L-valued linear k-forms, the Stokes formula reads

$$\int_M \Theta(\dot{p}, \nabla \lrcorner \underline{dp}) = \int_{\partial M} \Theta(q, \underline{dq}).$$

Also the Hodge dual Stokes formula (7.4) extends to manifolds, by inspection of the affine proof, using Proposition 11.2.5. In particular, the following special case is useful.

Exercise 11.2.11 (Divergence theorem). Let D be a domain in a Riemannian manifold N, with outward pointing unit normal vector field ν on ∂D. Show that

$$\int_D \operatorname{div} F(p)\, dp = \int_{\partial D} \langle F(q), \nu(q) \rangle\, dq,$$

for a C^1 vector field F on $\overline{D}$. Here $\operatorname{div} F = \delta F = \sum_i \langle \mathbf{e}_i, \nabla_{\mathbf{e}_i} F \rangle$, using Proposition 11.2.8.

11.3   Curvature and Bivectors

Consider a vector bundle E over a manifold M, with a given covariant derivative ∇_v as in Definition 11.1.4 that allows us to differentiate sections of E. Given a frame $\{\mathbf{e}_i\}_{i=1}^n$ for E, we have Christoffel symbols $\omega_\mathbf{e}$ defined by

$$\nabla_v \mathbf{e}_i = \sum_j \langle \omega_{ji}, v \rangle \mathbf{e}_j$$

as in Definition 11.1.5. Although in general we cannot give a definition of what it means for a section to be constant, using the covariant derivative, we can define what is meant by a section being constant along a given curve on M as follows.

Definition 11.3.1 (Parallel sections). Let E be a vector bundle over a manifold M, with covariant derivative ∇_v, and consider a curve γ, that is, a 1-surface, in M. A section F of E over γ is said to be *parallel* if

$$\nabla_v F = 0$$

at each $p \in \gamma$ and tangent vector v to γ at p.

The following calculation shows that in general we cannot even extend this notion of parallel sections to 2-surfaces, which leads us to the fundamental notion of curvature.

Example 11.3.2 (Parallel transport around triangles). Let E be a vector bundle over a manifold M, with covariant derivative ∇_v. Consider a 2-surface Σ in M, at a given point $p \in M$, defined by a chart

$$\mu : D \to \Sigma \subset M.$$

Here D is the triangle $D := \{(x,y) \; ; \; x,y \geq 0, x+y \leq 1\} \subset \mathbf{R}^2$ and $\mu(0,0) = p$. For $0 < \epsilon < 1$ we also consider the subsurfaces $\Sigma_\epsilon = \mu(D_\epsilon)$, where $D_\epsilon := \{(x,y) \; ; \; x,y \geq 0, x+y \leq \epsilon\}$.

Assume that F is a section of E over the curve $\partial\Sigma_\epsilon$ that is parallel along the three sides and continuous at $\mu(\epsilon,0)$ and $\mu(0,\epsilon)$. We do not assume continuity at p. This defines a map

$$E_p \to E_p : \lim_{x \to 0^+} F(\mu(x,0)) \mapsto \lim_{y \to 0^+} F(\mu(0,y)),$$

which is referred to as the *parallel transport* around $\partial\Sigma_\epsilon$. We want to calculate this map to order ϵ^2. To this end, fix a frame $\{\mathbf{e}_i\}_{i=1}^N$ for E in a neighborhood of Σ and write $F = \sum_i F_i \mathbf{e}_i$. The equation $\nabla_v F = 0$ becomes

$$\sum_i \left(\partial_v F_i + \sum_j \langle \omega_{ij}, v \rangle F_j \right) \mathbf{e}_i = 0.$$

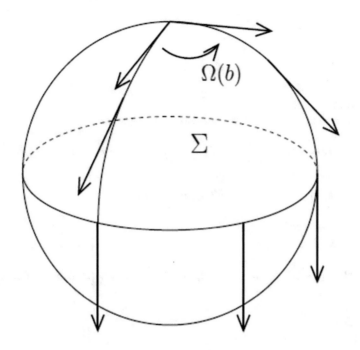

Figure 11.1: Parallel transport of a vector around a triangle Σ. Since the sphere has positive curvature, the resulting rotation will be in the direction corresponding to the orientation of Σ.

Write $\omega_{ij}^0 := \mu^*(\omega_{ij})$ for the pulled back vector field on D_ϵ. Then the equation becomes N scalar differential equations

$$\partial_u f_i + \sum_j \langle \omega_{ij}^0, u \rangle f_j = 0, \quad i = 1, \ldots, N,$$

along the triangle curve ∂D_ϵ, where $f_i := \mu^* F_i = F_i \circ \mu$ and u is a tangent vector to ∂D_ϵ.

Let $z_i := f_i(0^+, 0)$ be given, and write $a_{ij}(x, y) := \langle \omega_{ij}^0, e_1 \rangle$ and $b_{ij}(x, y) := \langle \omega_{ij}^0, e_2 \rangle$. Also write $a_{ij} := a_{ij}(0, 0)$, $\partial_x a_{ij} := \partial_x a_{ij}(0, 0)$, $\partial_y a_{ij} := \partial_y a_{ij}(0, 0)$, and similarly for b_{ij}.

(i) We first solve the parallel transport equations

$$\partial_x f_i(x, 0) + \sum_j a_{ij}(x, 0) f_j(x, 0) = 0$$

along $\{(x, 0) \ ; \ 0 < x < \epsilon\}$. With the ansatzes $f_i(x, 0) = \phi_i^0 + \phi_i^1 x + \phi_i^2 x^2 + O(x^3)$

and $a_{ij}(x,0) = \alpha_{ij}^0 + \alpha_{ij}^1 x + O(x^2)$, we obtain

$$f_i(\epsilon,0) = z_i - \epsilon \sum_j a_{ij} z_j + \frac{\epsilon^2}{2} \sum_j \left(\sum_k a_{ik} a_{kj} - (\partial_x a_{ij}) \right) z_j + O(\epsilon^3)$$

under the initial condition $\phi_i^0 = z_i$.

(ii) We next solve the parallel transport equations

$$\partial_x f_i(\epsilon - x, x) + \sum_j (-a_{ij}(\epsilon - x, x) + b_{ij}(\epsilon - x, x)) f_j(\epsilon - x, x) = 0$$

along $\{(\epsilon - x, x) \; ; \; 0 < x < \epsilon\}$. We approximate

$$- a_{ij}(\epsilon - x, x) + b_{ij}(\epsilon - x, x)$$
$$= (-a_{ij} + b_{ij} - \epsilon \partial_x a_{ij} + \epsilon \partial_x b_{ij}) + (\partial_x a_{ij} - \partial_y a_{ij} - \partial_x b_{ij} + \partial_y b_{ij}) x + O(\epsilon^2).$$

Letting $\phi_i^0 = f_i(\epsilon,0)$, $\alpha_{ij}^0 = -a_{ij} + b_{ij} - \epsilon \partial_x a_{ij} + \epsilon \partial_x b_{ij}$, and $\alpha_{ij}^1 = \partial_x a_{ij} - \partial_y a_{ij} - \partial_x b_{ij} + \partial_y b_{ij}$ in (i), we get

$$f_i(0,\epsilon) = z_i - \epsilon \sum_j b_{ij} z_j$$

$$+ \frac{\epsilon^2}{2} \sum_j \left(\sum_k (-a_{ik} b_{kj} + b_{ik} a_{kj} + b_{ik} b_{kj}) - (\partial_y a_{ij} - \partial_x b_{ij} - \partial_y b_{ij}) \right) z_j + O(\epsilon^3).$$

(iii) We finally solve the parallel transport equations

$$\partial_x f_i(0, \epsilon - x) - \sum_j b_{ij}(0, \epsilon - x) f_j(0, \epsilon - x) = 0$$

along $\{(0, \epsilon - x) \; ; \; 0 < x < \epsilon\}$. We approximate

$$-b_{ij}(0, \epsilon - x) = (-b_{ij} - \epsilon \partial_y b_{ij}) + (\partial_y b_{ij}) x + O(\epsilon^2).$$

Letting $\phi_i^0 = f_i(0,\epsilon)$, $\alpha_{ij}^0 = -b_{ij} - \epsilon \partial_y b_{ij}$, and $\alpha_{ij}^1 = \partial_y b_{ij}$ in (i), we get

$$f_i(0, 0^+) = z_i + \frac{\epsilon^2}{2} \sum_j \left(\sum_k (-a_{ik} b_{kj} + b_{ik} a_{kj}) + (\partial_y a_{ij} - \partial_x b_{ij}) \right) z_j + O(\epsilon^3).$$

To interpret this result, we push forward the oriented measure $\frac{\epsilon^2}{2} e_1 \wedge e_2$ of D_ϵ to

$$\wedge^2(\Sigma_\epsilon) := \underline{\mu}_{(0,0)} \left(\frac{\epsilon^2}{2} e_1 \wedge e_2 \right) \in \wedge^2(T_p M).$$

Although there is no notion of oriented measure of k-surfaces in manifolds in general, for $\epsilon \approx 0$ the bivector $\wedge^2(\Sigma_\epsilon)$ approximately describes an oriented measure of the small 2-surface Σ_ϵ. For the vector fields $\omega_{ij}^0 = a_{ij}e_1 + b_{ij}e_2$ in $\mathbf{R}^2$, we note that

$$d\omega_{ij}^0 = (\partial_x b_{ij} - \partial_y a_{ij})e_{12},$$
$$\omega_{ik}^0 \wedge \omega_{kj}^0 = (a_{ik}b_{kj} - b_{ik}a_{kj})e_{12}.$$

Therefore our calculation shows that

$$f_i(0,0^+) = z_i - \sum_j \langle d\omega_{ij}^0 + \sum_k \omega_{ik}^0 \wedge \omega_{kj}^0, \tfrac{\epsilon^2}{2}e_{12}\rangle z_j + O(\epsilon^3).$$

Back on M we conclude that parallel transport around Σ_ϵ equals

$$E_p \to E_p : F \mapsto F - \sum_j \langle \Omega_{ij}, \wedge^2(\Sigma_\epsilon)\rangle F_j \mathbf{e}_i + O(\epsilon^3),$$

where Ω is the curvature operator defined below.

The following encodes the curvature of a general manifold, as compared to Definition 10.3.5 which encodes the curvature of an $n - 1$ surface embedded in n-dimensional Euclidean space.

Definition 11.3.3 (Curvature operator). Consider a vector bundle E with a covariant derivative ∇_v, over a manifold M. Let $\{\mathbf{e}_i\}_{i=1}^N$ be a frame for E, and let ω_{ij} denote the Christoffel symbols of ∇_v in this frame. Assuming that M is a Riemannian manifold, these are vector fields on M, defined in the domain of the frame. The *curvature operator* Ω for ∇_v is the linear map that, at each p, to each bivector $b \in \wedge^2(T_pM)$ associates the linear map $\Omega(b) \in \mathcal{L}(E_p)$, which in the given frame has matrix $\{\langle \Omega_{ij}, b\rangle\}_{i,j=1}^N$, where the bivector fields are

$$\Omega_{ij} := d\omega_{ij} + \sum_{k=1}^N \omega_{ik} \wedge \omega_{kj}, \quad i,j = 1,\ldots,N.$$

Note that we do not write $\Omega_{\mathbf{e}}$ as for $\omega_{\mathbf{e}}$. As Example 11.3.2 indicates, unlike the Christoffel symbols, the operator $\Omega(b)$ is independent of the choice of frame, as the following shows.

Proposition 11.3.4. *Let E be a vector bundle with a covariant derivative ∇_v, over a manifold M. Consider two frames $\{\mathbf{e}_i\}$ and $\{\tilde{\mathbf{e}}_i\}$, defined on the same open set, with associated Christoffel symbols ω, $\tilde{\omega}$ and curvature operators Ω and $\tilde{\Omega}$ respectively. If the relation between frames is $\tilde{\mathbf{e}}_i = \sum_j \mathbf{e}_j \alpha_{ji}$, then*

$$\tilde{\omega}_{ij} = \sum_{kl} \alpha^{ik}\omega_{kl}\alpha_{lj} + \sum_k \alpha^{ik}(d\alpha_{kj}),$$
$$\tilde{\Omega}_{ij} = \sum_{kl} \alpha^{ik}\Omega_{kl}\alpha_{lj},$$

for all $i,j = 1,\ldots,N$, where $\{\alpha^{ij}\}_{ij}$ denotes the inverse matrix to $A = \{\alpha_{ij}\}_{ij}$.

These transformation properties of Ω_{ij} mean that the curvature operator is a well-defined section

$$\Omega \in C^\infty(M; \mathcal{L}(\wedge^2 M; \mathcal{L}(E))),$$

globally on all M, that does not depend on the choice of frame. Note that the corresponding statement is not true for the Christoffel symbols, because of term $A^{-1}dA$. It is only when the change of frame matrix A is constant that the Christoffel symbols transform as an operator.

Proof. On the one hand, $\nabla_v \tilde{e}_i = \sum_j \langle \tilde{\omega}_{ji}, v \rangle \tilde{e}_j$. On the other hand,

$$\nabla_v \tilde{e}_i = \sum_j \nabla_v(e_j \alpha_{ji}) = \sum_{jk} \langle \omega_{kj}, v \rangle e_k \alpha_{ji} + \sum_j e_j (\partial_v \alpha_{ji}),$$

and $\partial_v \alpha_{ji} = \langle d\alpha_{ji}, v \rangle$. Comparing these expressions, we obtain

$$\sum_{jk} \langle \tilde{\omega}_{ji}, v \rangle \alpha_{kj} e_k = \sum_{jk} \langle \omega_{kj}, v \rangle e_k \alpha_{ji} + \sum_k \langle d\alpha_{ki}, v \rangle e_k,$$

from which the relation for the Christoffel symbols follows. Furthermore,

$$\tilde{\Omega}_{ij} = \sum_k d\alpha^{ik} \wedge \left(\sum_l \omega_{kl}\alpha_{lj} + d\alpha_{kj} \right) + \sum_k \alpha^{ik} \left(\sum_l (d\omega_{kl})\alpha_{lj} - \omega_{kl} \wedge d\alpha_{lj} \right)$$

$$+ \sum_{kmk'} \alpha^{ik} \left(\sum_l \omega_{kl}\alpha_{lm} + d\alpha_{km} \right) \wedge \alpha^{mk'} \left(\sum_{l'} \omega_{k'l'}\alpha_{l'j} + d\alpha_{k'j} \right).$$

Here the terms containing $d\omega$ and $\omega \wedge \omega$ combine to the stated relation, and the remaining terms cancel pairwise, using $0 = \sum_k d(\alpha^{ik}\alpha_{kj}) = (d\alpha^{ik})\alpha_{kj} + \alpha^{ik}(d\alpha_{kj})$. $\qquad \square$

A fundamental fact about second-order partial derivatives in affine space is that

$$\partial_i \partial_j = \partial_j \partial_i.$$

The quantity that describes the failure of this for covariant derivatives on vector bundles turns out to be the curvature operator, as the following shows.

Proposition 11.3.5 (Curvature as commutator). *Let E be a vector bundle with a covariant derivative ∇_v, over a manifold M. Then*

$$\nabla_u(\nabla_v F) - \nabla_v(\nabla_u F) = \nabla_{[u,v]} F + \Omega(u \wedge v)F,$$

for every section $F \in C^2(M; E)$.

Proof. Calculating in a frame $\{\mathbf{e}_i\}$, we have

$$\nabla_u(\nabla_v F) = \nabla_u\left(\sum_i (\partial_v F_i)\mathbf{e}_i + \sum_{ki}\langle\omega_{ki}, v\rangle F_i\mathbf{e}_k\right)$$

$$= \sum_i (\partial_u\partial_v F_i)\mathbf{e}_i + \sum_{ik}\left(((\partial_u\langle\omega_{ki}, v\rangle)F_i + \langle\omega_{ki}, v\rangle(\partial_u F_i))\mathbf{e}_k + (\partial_v F_i)\langle\omega_{ki}, u\rangle\mathbf{e}_k\right)$$

$$+ \sum_{ijk}\langle\omega_{ji}, v\rangle F_i\langle\omega_{kj}, u\rangle\mathbf{e}_k.$$

Subtracting the corresponding formula for $\nabla_v(\nabla_u F)$, it suffices to show that

$$\partial_u\partial_v F_i - \partial_v\partial_u F_i = \partial_{[u,v]}F_i,$$
$$\partial_u\langle\omega_{ki}, v\rangle - \partial_v\langle\omega_{ki}, u\rangle = \langle d\omega_{ki}, u \wedge v\rangle + \langle\omega_{ki}, [u, v]\rangle,$$
$$\langle\omega_{ji}, v\rangle\langle\omega_{kj}, u\rangle - \langle\omega_{ji}, u\rangle\langle\omega_{kj}, v\rangle = \langle\omega_{kj} \wedge \omega_{ji}, u \wedge v\rangle.$$

The first and last equations are straightforward to verify, and the second follows from Exercise 7.4.9, generalized to manifolds. This proves the proposition. □

Our interest is in metric vector bundles and covariant derivatives, in which case it is clear from Definition 11.3.3 and the skew-symmetry of the Christoffel symbols and the exterior product of vectors that

$$\Omega_{ji} = -\Omega_{ij}$$

in an ON-frame $\{\mathbf{e}_i\}$. The map $\Omega(b) \in \mathcal{L}(E)$ being skew-symmetric, recalling Proposition 4.2.3 we can equivalently view it as the bivector

$$\Omega(b) = \sum_{i<j}\langle\Omega_{ij}, b\rangle\mathbf{e}_i \wedge \mathbf{e}_j \in \wedge^2 E,$$

so that $\Omega \in C^\infty(M; \mathcal{L}(\wedge^2 M; \wedge^2 E))$. In this setup, the action of $\Omega(b)$ on a section $F \in C^\infty(M; E)$ is

$$\Omega(b)F = \Omega(b) \llcorner F,$$

using the right interior product in exterior bundle $\wedge E$. Although possible for any metric vector bundle, we shall use only this bivector representation of the curvature operator for the tangent bundle $E = TM$.

Definition 11.3.6 (Riemann curvature operator). Let M be a Riemannian manifold. By the *Riemann curvature operator* we mean the curvature operator

$$R \in C^\infty(M; \mathcal{L}(\wedge^2 M))$$

for the Levi-Civita covariant derivative on the tangent bundle TM. Using the relation $\wedge^2 M = \underline{SO}(TM) \subset \mathcal{L}(TM)$ provided by Proposition 4.2.3, we have $R \in C^\infty(M; \mathcal{L}(TM) \otimes \mathcal{L}(TM))$. Using the contractions

$$\mathcal{L}(TM) \otimes \mathcal{L}(TM) \to \mathcal{L}(TM) = TM \otimes TM \to M \times \mathbf{R},$$

obtained by lifting the bilinear products $A_1 A_2^*$ and $\langle v_1, v_2 \rangle$ of matrices and vectors respectively to the tensor products, we define from R the *Ricci curvature* Ric $\in C^\infty(M; \mathcal{L}(TM))$ and the *scalar curvature* $S \in C^\infty(M; \mathbf{R})$.

If $\{\mathbf{e}_i\}$ is an ON-frame for TM, then we define the *Riemann curvature coefficients*

$$R_{ijkl} := \langle \mathbf{e}_i \wedge \mathbf{e}_j, R(\mathbf{e}_k \wedge \mathbf{e}_l) \rangle.$$

Then the matrix for the Ricci curvature operator in this frame is

$$\mathrm{Ric}_{ij} := \langle \mathbf{e}_i, \mathrm{Ric}(\mathbf{e}_j) \rangle = \sum_k R_{ikjk},$$

and the scalar curvature is $S = \sum_i \mathrm{Ric}_{ii} = \sum_{ik} R_{ikik}$.

Proposition 11.3.7 (Symmetries of R). *Let $R \in C^\infty(M; \mathcal{L}(\wedge^2 M))$ be the Riemann curvature operator on a Riemannian manifold M. Then at each $p \in M$, the operator $R : (\wedge^2 M)_p \to (\wedge^2 M)_p$ is symmetric, and the Bianchi identities*

$$R(u \wedge v) \llcorner w + R(v \wedge w) \llcorner u + R(w \wedge u) \llcorner v = 0$$

hold for all vectors $u, v, w \in T_p M$.

Proof. To prove the Bianchi identities, by trilinearity it suffices to consider a coordinate frame $\{\mathbf{e}_i\}$. In this case, $[\mathbf{e}_i, \mathbf{e}_j] = 0$, so by Proposition 11.3.5 the identities follow from the computation

$$\left(\nabla_{\mathbf{e}_i} \nabla_{\mathbf{e}_j} \mathbf{e}_k - \nabla_{\mathbf{e}_j} \nabla_{\mathbf{e}_i} \mathbf{e}_k \right) + \left(\nabla_{\mathbf{e}_j} \nabla_{\mathbf{e}_k} \mathbf{e}_i - \nabla_{\mathbf{e}_k} \nabla_{\mathbf{e}_j} \mathbf{e}_i \right) + \left(\nabla_{\mathbf{e}_k} \nabla_{\mathbf{e}_i} \mathbf{e}_j - \nabla_{\mathbf{e}_i} \nabla_{\mathbf{e}_k} \mathbf{e}_j \right)$$
$$= \nabla_{\mathbf{e}_i}(\nabla_{\mathbf{e}_j} \mathbf{e}_k - \nabla_{\mathbf{e}_k} \mathbf{e}_j) + \nabla_{\mathbf{e}_j}(\nabla_{\mathbf{e}_k} \mathbf{e}_i - \nabla_{\mathbf{e}_i} \mathbf{e}_k) + \nabla_{\mathbf{e}_k}(\nabla_{\mathbf{e}_i} \mathbf{e}_j - \nabla_{\mathbf{e}_j} \mathbf{e}_i) = 0,$$

since $\nabla_u v - \nabla_v u = [u, v]$ for the Levi-civita covariant derivative.

Using the Bianchi identities, we compute

$$\langle u_2 \wedge v_2, R(u_1 \wedge v_1) \rangle = \langle u_2, R(u_1 \wedge v_1) \llcorner v_2 \rangle$$
$$= -\langle u_2, R(v_1 \wedge v_2) \llcorner u_1 \rangle - \langle u_2, R(v_2 \wedge u_1) \llcorner v_1 \rangle$$
$$= \langle u_1, R(v_1 \wedge v_2) \llcorner u_2 \rangle + \langle v_1, R(v_2 \wedge u_1) \llcorner u_2 \rangle$$
$$= -\langle u_1, R(v_2 \wedge u_2) \llcorner v_1 \rangle - \langle u_1, R(u_2 \wedge v_1) \llcorner v_2 \rangle - \langle v_1, R(u_1 \wedge u_2) \llcorner v_2 \rangle$$
$$\quad - \langle v_1, R(u_2 \wedge v_2) \llcorner u_1 \rangle$$
$$= 2\langle u_1 \wedge v_1, R(u_2 \wedge v_2) \rangle + \langle v_2, R(u_2 \wedge v_1) \llcorner u_1 \rangle + \langle v_2, R(u_1 \wedge u_2) \llcorner v_1 \rangle$$
$$= 2\langle u_1 \wedge v_1, R(u_2 \wedge v_2) \rangle - \langle v_2, R(v_1 \wedge u_1) \llcorner u_2 \rangle,$$

from which the stated symmetry follows. $\square$

11.4 Conformal Maps and ON-Frames

The main result in this section, Liouville's theorem on conformal maps, is difficult to place. It concerns the fractional linear maps from Section 4.5 and completes the proof of Theorem 4.5.12. It is also relevant for the hypercomplex analysis from Chapter 8. We have nevertheless placed it here, since the proof makes use of ON-frames and curvature.

We start with the following method of Cartan for calculating the Christoffel symbols in an ON-frame.

Proposition 11.4.1 (Cartan). *Let $\{e_i\}$ be an ON-frame for TM on a Riemannian manifold M. Then the Christoffel symbols for the Levi-Civita covariant derivative in $\{e_i\}$ are the unique vector fields ω_{ij} that satisfy $\omega_{ji} = -\omega_{ij}$ and*

$$d\mathbf{e}_i + \sum_{j=1}^{n} \omega_{ij} \wedge \mathbf{e}_j = 0, \quad j = 1, \ldots, n. \tag{11.3}$$

Note that the uniqueness part of the proposition means that any method of writing the exterior derivatives of the ON-frame vectors as in (11.3) must give the Christoffel symbols.

Proof. Using Proposition 11.2.8, we have

$$d\mathbf{e}_i = \sum_j \mathbf{e}_j \wedge \nabla_{\mathbf{e}_j} \mathbf{e}_i = \sum_j \mathbf{e}_j \wedge \sum_k \langle \omega_{ki}, \mathbf{e}_j \rangle \mathbf{e}_k = \sum_k \sum_j \langle \omega_{ki}, \mathbf{e}_j \rangle \mathbf{e}_j \wedge \mathbf{e}_k = \sum_k \omega_{ki} \wedge \mathbf{e}_k$$

for the Christoffel symbols. For uniqueness, assume that ω'_{ij} is any skew-symmetric family of vector fields that satisfies (11.3), and consider the differences $\tilde{\omega}_{ij} := \omega_{ij} - \omega'_{ij}$. We have

$$\sum_j \tilde{\omega}_{ij} \wedge \mathbf{e}_j = 0, \quad i = 1, \ldots, n.$$

Forming the inner product with frame bivectors gives

$$0 = \sum_j \langle \mathbf{e}_k \wedge \mathbf{e}_l, \tilde{\omega}_{ij} \wedge \mathbf{e}_j \rangle = \sum_j \left(\langle \mathbf{e}_k, \tilde{\omega}_{ij} \rangle \langle \mathbf{e}_l, \mathbf{e}_j \rangle - \langle \mathbf{e}_k, \mathbf{e}_j \rangle \langle \mathbf{e}_l, \tilde{\omega}_{ij} \rangle \right)$$

$$= \langle \mathbf{e}_k, \tilde{\omega}_{il} \rangle - \langle \mathbf{e}_l, \tilde{\omega}_{ik} \rangle.$$

So $\langle \mathbf{e}_k, \tilde{\omega}_{il} \rangle = \langle \mathbf{e}_l, \tilde{\omega}_{ik} \rangle$ for all i, k, l, as well as $\tilde{\omega}_{ij} = -\tilde{\omega}_{ji}$ for all i, j. This yields

$$\langle \mathbf{e}_k, \tilde{\omega}_{ji} \rangle = \langle \mathbf{e}_i, \tilde{\omega}_{jk} \rangle = -\langle \mathbf{e}_i, \tilde{\omega}_{kj} \rangle = -\langle \mathbf{e}_j, \tilde{\omega}_{ki} \rangle = \langle \mathbf{e}_j, \tilde{\omega}_{ik} \rangle = \langle \mathbf{e}_k, \tilde{\omega}_{ij} \rangle = -\langle \mathbf{e}_k, \tilde{\omega}_{ji} \rangle$$

for all k, i, j. This proves that the ω'_{ij} coincide with the Christoffel symbols. $\square$

A two-dimensional Euclidean space abounds with conformal maps: all analytic functions are conformal where the derivative is nonzero. The goal of this section is to prove Liouville's theorem on conformal maps, which shows that in n-dimensional Euclidean space, $n \geq 3$, the situation is drastically different: the only conformal maps are the fractional linear maps / Möbius maps from Theorem 4.5.12. These are very few indeed, since they depend only on as many parameters as the Lorentz isometries in dimension $n+2$, that is, $\binom{n+2}{2}$ real parameters.

Theorem 11.4.2 (Liouville). *Let D be a connected open subset of a Euclidean space (X, V) of dimension $\dim X \geq 3$. Assume that $f : D \to V$ is a conformal map of class C^3. Then there exists a fractional linear map $g : \overline{V} \to \overline{V}$ such that $g|_D = f$. Here $X = V$ with origin fixed, and extended $V \subset \overline{V}$ with notation as in Section 4.5.*

Even though the conclusion is global in the sense that we obtain a conformal diffeomorphism of the extended space $\overline{V}$, the problem is local. Indeed, since fractional linear maps are real analytic, it suffices to prove that f coincides with a fractional linear map in D, when D is a ball. A concrete proof of this is as follows. Assume that we have proved that $f = f_1$ on a ball B_1, and that $f = f_2$ on a ball B_2, where $B_1 \cap B_2 \neq \emptyset$ and f_1 and f_2 are fractional linear maps. To prove that $f_1 = f_2$ on all $\overline{V}$, write $f_0(x) := f_2^{-1}(f_1(x)) = (ax + b)(cx + d)^{-1}$ as in Theorem 4.5.16. Then

$$x = (ax + b)(cx + d)^{-1}, \quad \text{for all } x \in B_1 \cap B_2,$$

or equivalently $xcx + xd - ax - b = 0$. If this second-order polynomial vanishes on an open set, then $c = b = 0$ and $xd = ax$ for all vectors x. This proves that $f_0(x) = axd^{-1} = x$ for all $x \in V$.

Proof of Theorem 11.4.2. Without loss of generality, we assume that D is a ball with small enough radius that $f : D \to f(D)$ is a C^3-diffeomorphism by the inverse function theorem. Since f is conformal, there exists a unique scalar function $\lambda(x) > 0$ in D such that

$$\lambda(x) \underline{f}_x \quad \text{is an isometry at each } x \in D,$$

and by assumption λ is C^2-regular. Fix an ON-basis $\{e_i\}$ for V and define the ON-frame

$$\mathbf{e}_i(y) := \lambda(x) \underline{f}_x(e_i), \quad y = f(x),$$

in the open set $f(D)$. By definition of λ, $f_* \lambda$ is an isometry, so $f^* f_* = \lambda^{-2} I$ when acting on vector fields in D. Therefore, pulling back $\{\mathbf{e}_i(y)\}$ to D gives $f^*(\mathbf{e}_i)(x) = 1/\lambda(x) e_i$, $x \in D$. Applying the exterior derivative and using Theorem 7.2.9, we get

$$f^*(\nabla \wedge \mathbf{e}_i) = -\lambda^{-2}(\nabla \lambda) \wedge e_i = -\lambda^{-2} \sum_j \lambda_j \, e_j \wedge e_i,$$

since e_i is a constant vector field, where $\lambda_i := \partial_i \lambda$ and ∂_i is the partial derivative along e_i. Pushing forward this equation to $f(D)$, noting that $f_* f^* = \mu^{-4} I$

when acting on bivector fields in $f(D)$, where $\mu := \lambda \circ f^{-1}$, gives $\mu^{-4}\nabla \wedge \mathbf{e}_i = -\mu^{-2}\sum_j \mu_i(\mu^{-1}\mathbf{e}_j) \wedge (\mu^{-1}\mathbf{e}_i)$, where $\mu_j := f_*(\lambda_j) = \lambda_j \circ f^{-1}$. We obtain

$$\nabla \wedge \mathbf{e}_i = -\sum_j \mu_j \mathbf{e}_j \wedge \mathbf{e}_i = -\sum_j (\mu_i \mathbf{e}_j - \mu_j \mathbf{e}_i) \wedge \mathbf{e}_j.$$

Uniqueness of the Christoffel symbols in Proposition 11.4.1 shows that $\omega_{ij} = \mu_i \mathbf{e}_j - \mu_j \mathbf{e}_i$. This gives

$$\nabla \wedge \omega_{ij} = (\nabla\mu_i) \wedge \mathbf{e}_j - (\nabla\mu_j) \wedge \mathbf{e}_i + \sum_k \left(-\mu_i \mathbf{e}_k \wedge \omega_{kj} + \mu_j \mathbf{e}_k \wedge \omega_{ki} \right)$$

$$= (\nabla\mu_i) \wedge \mathbf{e}_j - (\nabla\mu_j) \wedge \mathbf{e}_i + \sum_k \left(-\mu_i \mu_k \mathbf{e}_k \wedge \mathbf{e}_j + \mu_j \mu_k \mathbf{e}_k \wedge \mathbf{e}_i \right),$$

$$\sum_k \omega_{ik} \wedge \omega_{kj} = \sum_k (\mu_i \mathbf{e}_k - \mu_k \mathbf{e}_i) \wedge (\mu_k \mathbf{e}_j - \mu_j \mathbf{e}_k)$$

$$= \sum_k (\mu_i \mu_k \mathbf{e}_k \wedge \mathbf{e}_j - \mu_k^2 \mathbf{e}_i \wedge \mathbf{e}_j + \mu_k \mu_j \mathbf{e}_i \wedge \mathbf{e}_k).$$

Since Euclidean space has zero curvature, it follows from Definition 11.3.3 that $(\nabla\mu_i) \wedge \mathbf{e}_j - (\nabla\mu_j) \wedge \mathbf{e}_i = (\sum_k \mu_k^2)\mathbf{e}_i \wedge \mathbf{e}_j$. Pulling back this equation to D gives $(\nabla\lambda_i) \wedge (\lambda^{-1}\mathbf{e}_j) - (\nabla\lambda_j) \wedge (\lambda^{-1}\mathbf{e}_i) = (\sum_k \lambda_k^2)(\lambda^{-1}\mathbf{e}_i) \wedge (\lambda^{-1}\mathbf{e}_j)$. We obtain the following nonlinear system of second-order partial differential equations

$$\lambda \sum_{k=1}^n (\lambda_{ik}\, \mathbf{e}_k \wedge \mathbf{e}_j - \lambda_{jk}\, \mathbf{e}_k \wedge \mathbf{e}_i) = \left(\sum_{k=1}^n \lambda_k^2\right) \mathbf{e}_i \wedge \mathbf{e}_j, \quad i,j = 1,\dots,n, \qquad (11.4)$$

where $\lambda_{ij} := \partial_j \partial_i \lambda$.

If $\dim X \geq 3$, then the system (11.4) is overdetermined, which we exploit as follows. Evaluating the $\mathbf{e}_m \wedge \mathbf{e}_j$ component of the equation (11.4), where i,j,m are distinct, we deduce that $\lambda_{im} = 0$ whenever $i \neq m$. On the other hand, evaluating the $\mathbf{e}_i \wedge \mathbf{e}_j$ component of the same equation, it follows that $\lambda_{ii} + \lambda_{jj} = \lambda^{-1}|\nabla\lambda|^2$. Since this equation holds for all $i \neq j$, we get $\lambda_{ii} = |\nabla\lambda|^2/(2\lambda)$, $i = 1,\dots,n$. Since $\lambda_{im} = 0$ whenever $i \neq m$, λ_i is a function of x_i only, and so is λ_{ii}. This holds for all i, so $|\nabla\lambda|^2/\lambda =: c$ must be a constant independent of all x_i. We have shown that

$$\lambda_{ii}(x) = |\nabla\lambda(x)|^2/(2\lambda(x)) = c, \quad i = 1,\dots,n, \; x \in D.$$

If $c = 0$, then $\nabla\lambda = 0$ and $\lambda(x) =: b$ is constant. Since $\mathbf{e}_i = (bf_*)(e_i) = (f^{-1})^*(b^{-1}e_i)$, it follows from Theorem 7.2.9 that $\nabla \wedge \mathbf{e}_i = 0$, since $b^{-1}e_i$ is constant. Proposition 11.4.1 shows that the Christoffel symbols vanish. Therefore $\mathbf{e}_i$ must be constant in D since the covariant derivatives for the Euclidean space are the standard partial derivatives. Hence the total derivative $f_{\underline{x}}$ is a constant matrix, and we conclude that the conformal map is of the form

$$f\left(\sum_i x_i e_i\right) = b \sum_i x_i \mathbf{e}_i + d, \qquad \sum_i x_i e_i \in D,$$

for some constant vector $d \in V$. Thus f is a restriction of a composition of an isometry, a dilation, and a translation, and thus a fractional linear map.

On the other hand, if $c \neq 0$, then integration gives $\nabla \lambda = c(x - a)$ for some $a \in V$, since λ_i depends only on x_i. This yields

$$\lambda(x) = \frac{1}{2c} |c(x - a)|^2 = \frac{c}{2} |x - a|^2.$$

This can be reduced to the case that $\lambda(x)$ is constant, by composing $y = f(x)$ with the fractional linear map $x = g(z) = 1/z + a$. Since $\lambda > 0$ on $\overline{D}$, it follows that $a \notin \overline{D}$, and therefore g maps a domain $D' \subset V$ bijectively onto D. From Exercise 4.5.18 we have that $\underline{g}_z(h) = |z|^{-2} zhz^{-1}$. Thus $f \circ g : D \to f(D)$ has derivative

$$\underline{f \circ g}_z(h) = \underline{f}_x(|z|^{-2} zhz^{-1}) = |x - a|^2 \underline{f}_x(zhz^{-1}) = \frac{2}{c} \lambda \underline{f}_x(zhz^{-1}),$$

which is a constant times an isometry, so the scale factor λ associated to $f \circ g$ is constant. Thus the calculation above for constant λ applies to $f \circ g$ and shows that $f \circ g$, and hence f, is a fractional linear map. This completes the proof. $\square$

Exercise 11.4.3. In dimension $n = 2$, show that (11.4) reduces to the single equation $\lambda \Delta \lambda = |\nabla \lambda|^2$, that is, $\Delta(\ln \lambda) = 0$, saying that $\ln |f'(z)|$ is harmonic when $f(z)$ is an analytic function.

11.5   Weitzenböck Identities

Proceeding as in Section 3.1, we define on each fiber $\wedge(T_p M)$ of the multivector bundle over a Riemannian manifold M a Clifford product $\vartriangle$.

Definition 11.5.1 (Clifford bundle). Let M be a Riemannian manifold. The Clifford bundle $\triangle M$ over M equals the multivector bundle $\wedge M$ as a bundle of linear spaces, but equipped with the Clifford product $\vartriangle$ on the fibers $\wedge(T_p M)$, making it into a bundle of associative algebras over M.

Just as in the case of a single affine space, we write $\triangle M$ instead of $\wedge M$ when we use the Clifford product rather than the exterior and interior products, but we are not strict in this, since sometimes we use all these products and it may not be clear whether $\triangle M$ or $\wedge M$ is the appropriate notation.

Translating from the multivector bundle, the Clifford bundle is the direct sum of subbundles

$$\triangle M = \triangle^0 M \oplus \triangle^1 M \oplus \triangle^2 M \oplus \cdots \oplus \triangle^n M.$$

Note that since the charts $\underline{\mu}_\alpha : \mathbf{R}^n \to TM$ in general are not isometries, they do not induce isomorphisms between Clifford algebras. However, given an ON-frame $\{\mathbf{e}_i\}_{i=1}^n$ for TM, we obtain an induced ON-frame $\{\mathbf{e}_s\}_{s \subset \overline{n}}$ for $\triangle M$.

Exercise 11.5.2 (Clifford product rule). Show for the Levi-Civita covariant derivative on $\triangle M = \wedge M$ that the product rule

$$\nabla_v(F \vartriangle G) = (\nabla_v F) \vartriangle G + F \vartriangle (\nabla_v G)$$

holds for all multivector fields $F, G \in C^1(M; \triangle M)$ on M.

A first use of the Clifford product on $\triangle M$ is to express the Christoffel symbols and curvature of $\wedge M = \triangle M$ in terms of the bivectors $\Gamma_\mathbf{e}$ and R for TM through Clifford commutators.

Proposition 11.5.3 (Clifford bivector commutators). *Let M be a Riemannian manifold, and let $\mathbf{e} = \{e_i(p)\}$ be an ON-frame for the tangent bundle TM in an open set $U \subset M$. Denote by $\Gamma_\mathbf{e} \in C^\infty(U; \mathcal{L}(\wedge^1 M; \wedge^2 M))$ and $R \in C^\infty(M; \mathcal{L}(\wedge^2 M; \wedge^2 M))$ the Christoffel symbols and curvature operator for TM.*

Let $\tilde{\mathbf{e}} = \{e_s(p)\}$ be the induced ON-frame for $\triangle M$ in U. Denote by $\omega_{\tilde{\mathbf{e}}} \in C^\infty(U; \mathcal{L}(\wedge^1 M; \mathcal{L}(\triangle M)))$ and $\Omega \in C^\infty(M; \mathcal{L}(\wedge^2 M; \mathcal{L}(\triangle M)))$ the Christoffel symbols and curvature operator for $\triangle M$. Then

$$\omega_{\tilde{\mathbf{e}}}(v)F = \tfrac{1}{2}[\Gamma_\mathbf{e}(v), F],$$
$$\Omega(b)F = \tfrac{1}{2}[R(b), F],$$

for $v \in C(M; TM)$, $b \in C(M; \wedge^2 M)$, and $F \in C(M; \triangle M)$, using the Clifford commutator $[w_1, w_2] := w_1 \vartriangle w_2 - w_2 \vartriangle w_1$.

Proof. Consider first the Christoffel symbols. For a basis vector field $\mathbf{e}_i \in C(M; TM)$, we have

$$\omega_{\tilde{\mathbf{e}}}(v)\mathbf{e}_i = \nabla_v \mathbf{e}_i = \Gamma_\mathbf{e}(v) \llcorner \mathbf{e}_i = \tfrac{1}{2}(\Gamma_\mathbf{e}(v) \vartriangle \mathbf{e}_i - \mathbf{e}_i \vartriangle \Gamma_\mathbf{e}(v)) = \tfrac{1}{2}[\Gamma_\mathbf{e}(v), \mathbf{e}_i],$$

using the analogue of (3.3) for the right interior product, and Definitions 11.1.5 and 11.1.9. For scalar functions $F \in C(M; \triangle^0 M)$ both sides of the identity vanish, and for $F \in C(M; \triangle^k M)$, $k \geq 2$, the result follows from the vector case $k = 1$ and the derivation property

$$D(F_1 \vartriangle F_2) = (DF_1) \vartriangle F_2 + F_1 \vartriangle (DF_2), \quad F_1, F_2 \in C^1(M; \triangle M),$$

which holds for $D = \Gamma_\mathbf{e}(v)$, as a consequence of Exercise 11.5.2, as well as for $D : F \mapsto \tfrac{1}{2}[\Gamma_\mathbf{e}(v), F]$, by general properties of commutators.

For the curvature operator, we can argue similarly. Since

$$\Omega(b)\mathbf{e}_i = R(b) \llcorner \mathbf{e}_i = \tfrac{1}{2}(R(b) \vartriangle \mathbf{e}_i - \mathbf{e}_i \vartriangle R(b)) = \tfrac{1}{2}[R(b), \mathbf{e}_i],$$

it suffices to show that

$$\Omega(b)(F_1 \vartriangle F_2) = (\Omega(b)F_1) \vartriangle F_2 + F_1 \vartriangle (\Omega(b)F_2) \tag{11.5}$$

for all $F_1, F_2 \in C^\infty(M; \triangle M)$. By Exercise 11.5.2, we have

$$\nabla_u \nabla_v (F_1 \vartriangle F_2)$$
$$= (\nabla_u \nabla_v F_1) \vartriangle F_2 + (\nabla_v F_1) \vartriangle (\nabla_u F_2) + (\nabla_u F_1) \vartriangle (\nabla_v F_2) + F_1 \vartriangle (\nabla_u \nabla_v F_2),$$

as well as $\nabla_{[u,v]}(F_1 \vartriangle F_2) = (\nabla_{[u,v]} F_1) \vartriangle F_2 + F_1 \vartriangle (\nabla_{[u,v]} F_2)$, so the derivation property (11.5) follows from Proposition 11.3.5.   □

The fundamental first-order differential operator on M acting on sections of $\triangle M$ is the following natural generalization of the Euclidean nabla operator from Definition 9.1.1.

Definition 11.5.4 (The Dirac operator on $\triangle M$). Let M be a Riemannian manifold. The $\triangle$-*Dirac operator* on $\triangle M$ is the operator

$$\mathbf{D}F := dF + \delta F = \sum_i \mathbf{e}_i^* \vartriangle \nabla_{\mathbf{e}_i} F, \quad F \in C^1(M; \triangle M),$$

where $\{\mathbf{e}_i\}$ is a frame for TM, with dual frame $\{\mathbf{e}_i^*\}$.

Note that the Dirac operator $\mathbf{D} = d + \delta$ is not locally similar to the Euclidean Dirac operator. This in contrast to d and δ, where by definition $\mu^* d (\mu^*)^{-1}$ and $(\tilde{\mu}_*)^{-1} \delta \tilde{\mu}_*$ are the Euclidean exterior and interior derivatives, locally in any chart μ for M.

As in Euclidean space, $d^2 = 0$ and $\delta^2 = 0$, and as a consequence,

$$\mathbf{D}^2 = d\delta + \delta d$$

for every Riemannian manifold. We saw in Section 8.1 that $\mathbf{D}^2$ equals the Laplace operator Δ, acting componentwise on multivector fields, in Euclidean space. The situation on a manifold is more subtle, where $\mathbf{D}^2$ differs from the following Laplace operator, on the bundle $E = \triangle M$, by a curvature term.

Proposition 11.5.5. *Let E be a metric vector bundle with a metric covariant derivative, over a Riemannian manifold M. If $\{\mathbf{e}_i\}$ is an ON-frame for TM in an open set $U \subset M$, then*

$$\int_M \sum_i \langle \nabla_{\mathbf{e}_i} F, \nabla_{\mathbf{e}_i} G \rangle dp = - \int_M \sum_i \langle \nabla_{\mathbf{e}_i} \nabla_{\mathbf{e}_i} F - \nabla_{\nabla_{\mathbf{e}_i} \mathbf{e}_i} F, G \rangle dp \qquad (11.6)$$

for all $F, G \in C_0^2(U; E)$. Here $\nabla_{\mathbf{e}_i} \mathbf{e}_i$ denotes the Levi-Civita covariant derivative on TM.

Proof. Using that the covariant derivative on E is metric, we get the pointwise identity

$$\partial_{\mathbf{e}_i} \langle \nabla_{\mathbf{e}_i} F, G \rangle = \langle \nabla_{\mathbf{e}_i} \nabla_{\mathbf{e}_i} F, G \rangle + \langle \nabla_{\mathbf{e}_i} F, \nabla_{\mathbf{e}_i} G \rangle.$$

Define the vector field

$$v := \sum_i \langle \nabla_{\mathbf{e}_i} F, G \rangle \mathbf{e}_i \in C_0^1(U; TM).$$

From Proposition 11.2.8 we have

$$\operatorname{div} v = \sum_{ij} \langle \mathbf{e}_j, \nabla_{\mathbf{e}_j}(\langle \nabla_{\mathbf{e}_i} F, G \rangle \mathbf{e}_i) \rangle = \sum_{ij} \langle \mathbf{e}_j, (\partial_{\mathbf{e}_j} \langle \nabla_{\mathbf{e}_i} F, G \rangle) \mathbf{e}_i + \langle \nabla_{\mathbf{e}_i} F, G \rangle (\nabla_{\mathbf{e}_j} \mathbf{e}_i) \rangle$$

$$= \sum_i \left(\partial_{\mathbf{e}_i} \langle \nabla_{\mathbf{e}_i} F, G \rangle + \langle \nabla_{\mathbf{e}_i} F, G \rangle \operatorname{div} \mathbf{e}_i \right).$$

Using the identity

$$\sum_i \operatorname{div}(\mathbf{e}_i) \mathbf{e}_i = \sum_{ij} \langle \mathbf{e}_j, \nabla_{\mathbf{e}_j} \mathbf{e}_i \rangle \mathbf{e}_i = - \sum_{ij} \langle \nabla_{\mathbf{e}_j} \mathbf{e}_j, \mathbf{e}_i \rangle \mathbf{e}_i = - \sum_j \nabla_{\mathbf{e}_j} \mathbf{e}_j,$$

the stated identity now follows from the divergence theorem in Exercise 11.2.11, which shows that $\int_M v\, dp = 0$. $\qquad\square$

Definition 11.5.6 (Laplace–Beltrami operator). Let E be a metric vector bundle with a metric covariant derivative, over a Riemannian manifold M. The *Laplace–Beltrami operator* is the second-order differential operator that in an ON-frame $\{\mathbf{e}_i\}$ for TM is given by

$$\Delta F := \sum_i (\nabla_{\mathbf{e}_i} \nabla_{\mathbf{e}_i} F - \nabla_{\nabla_{\mathbf{e}_i} \mathbf{e}_i} F).$$

The left-hand side in (11.6), for $G = F$, can be viewed as the H^1 Sobolev (semi-)norm $\|\nabla \otimes F\|^2_{L_2(M;E)}$, and $-\Delta$ is the L_2 operator corresponding to this quadratic form.

Exercise 11.5.7 (Second covariant derivative). Show that

$$\nabla^2_{u,v} F := \nabla_u \nabla_v F - \nabla_{\nabla_u v} F$$

is bilinear in u and v. Deduce that the Laplace–Beltrami operator is well defined, that is, independent of choice of ON-basis.

Exercise 11.5.8 (Scalar Laplace–Beltrami). Show that for the trivial vector bundle $E = \mathbf{R}$, the Laplace–Beltrami operator acting on scalar functions $f \in C^2(M; \mathbf{R})$ is

$$\Delta f = \delta df.$$

Verify also that this follows from Proposition 11.5.9 below. With a chart for M, and inverse metric $(g^{ij}) = (g_{ij})^{-1}$ and determinant $g := \det(g_{ij})$, generalize Example 7.2.12 and show that

$$\Delta f = \sum_{ij} \frac{1}{\sqrt{g}} \partial_i (\sqrt{g}\, g^{ij} \partial_j f).$$

The following main result in this section is the key identity used in the proof of the Chern–Gauss–Bonnet theorem in Section 12.3.

Proposition 11.5.9 (Weitzenböck identity for $\mathbf{D}$). *Let M be a Riemannian manifold M, and let $\{\mathbf{e}_i\}$ be an ON-frame for TM. Then*

$$-\mathbf{D}^2 F = -\Delta F - \sum_{i<j}\mathbf{e}_i\mathbf{e}_j\Omega(\mathbf{e}_i \wedge \mathbf{e}_j)F = -\Delta F + \frac{1}{4}SF + \frac{1}{8}\sum_{ijkl}R_{ijkl}\mathbf{e}_i\mathbf{e}_j F\mathbf{e}_k\mathbf{e}_l,$$

for $F \in C^\infty(M; \triangle M)$, where Δ and Ω denote the Laplace–Beltrami and curvature operators on $\triangle M$, and R denotes the Riemann curvature operator on TM with coefficients R_{ijkl} and scalar curvature S.

Proof. With notation as in Exercise 11.5.7 and suppressing $\triangle$, Proposition 11.3.5 yields

$$\mathbf{D}^2 F = \sum_{ij}\mathbf{e}_i\nabla_{\mathbf{e}_i}(\mathbf{e}_j\nabla_{\mathbf{e}_j}F) = \sum_{ij}\mathbf{e}_i\Big((\nabla_{\mathbf{e}_i}\mathbf{e}_j)\nabla_{\mathbf{e}_j}F + \mathbf{e}_j\nabla_{\mathbf{e}_i}\nabla_{\mathbf{e}_j}F\Big)$$

$$= \sum_{ij}\mathbf{e}_i\Big(\mathbf{e}_j\nabla^2_{\mathbf{e}_i,\mathbf{e}_j}F + \mathbf{e}_j\nabla_{\nabla_{\mathbf{e}_i}\mathbf{e}_j}F + (\nabla_{\mathbf{e}_i}\mathbf{e}_j)\nabla_{\mathbf{e}_j}F\Big)$$

$$= \sum_{i}\mathbf{e}_i^2\nabla^2_{\mathbf{e}_i,\mathbf{e}_i}F + \sum_{i<j}\mathbf{e}_i\mathbf{e}_j(\nabla^2_{\mathbf{e}_i,\mathbf{e}_j}F - \nabla^2_{\mathbf{e}_j,\mathbf{e}_i}F)$$

$$+ \sum_{ij}\mathbf{e}_i\Big(\mathbf{e}_j\nabla_{\nabla_{\mathbf{e}_i}\mathbf{e}_j}F + (\nabla_{\mathbf{e}_i}\mathbf{e}_j)\nabla_{\mathbf{e}_j}F\Big)$$

$$= \Delta F + \sum_{i<j}\mathbf{e}_i\mathbf{e}_j\Omega(\mathbf{e}_i \wedge \mathbf{e}_j)F + \sum_{ijk}\mathbf{e}_i\Big(\langle\nabla_{\mathbf{e}_i}\mathbf{e}_j,\mathbf{e}_k\rangle\mathbf{e}_j\nabla_{\mathbf{e}_k}F + \langle\nabla_{\mathbf{e}_i}\mathbf{e}_j,\mathbf{e}_k\rangle\mathbf{e}_k\nabla_{\mathbf{e}_j}F\Big).$$

Since $\langle\nabla_{\mathbf{e}_i}\mathbf{e}_j,\mathbf{e}_k\rangle + \langle\mathbf{e}_j,\nabla_{\mathbf{e}_i}\mathbf{e}_k\rangle = \partial_{\mathbf{e}_i}\langle\mathbf{e}_j,\mathbf{e}_k\rangle = 0$, this proves the first identity.

To express the identity in terms of the Riemann curvature operator, we use Proposition 11.5.3 to write

$$\sum_{i<j}\mathbf{e}_i\mathbf{e}_j\Omega(\mathbf{e}_i \wedge \mathbf{e}_j)F = \frac{1}{2}\sum_{i<j}\mathbf{e}_i\mathbf{e}_j[R(\mathbf{e}_i \wedge \mathbf{e}_j), F]$$

$$= \frac{1}{4}\sum_{i<j}\sum_{kl}R_{ijkl}\mathbf{e}_i\mathbf{e}_j[\mathbf{e}_k\mathbf{e}_l, F] = \frac{1}{8}\sum_{ijkl}R_{ijkl}(\mathbf{e}_i\mathbf{e}_j\mathbf{e}_k\mathbf{e}_l F - \mathbf{e}_i\mathbf{e}_j F\mathbf{e}_k\mathbf{e}_l).$$

It remains to simplify $\sum_{ijkl}R_{ijkl}\mathbf{e}_i\mathbf{e}_j\mathbf{e}_k\mathbf{e}_l$. From the Bianchi identities in Proposition 11.3.7 and multivector algebra, we obtain

$$\sum_{ijk}R_{ijkl}\mathbf{e}_i\mathbf{e}_j\mathbf{e}_k = -\sum_{ijk}(R_{jkil} + R_{kijl})\mathbf{e}_i\mathbf{e}_j\mathbf{e}_k = -\sum_{ijk}R_{ijkl}(\mathbf{e}_k\mathbf{e}_i\mathbf{e}_j + \mathbf{e}_j\mathbf{e}_k\mathbf{e}_i)$$

$$= -\sum_{ijk}R_{ijkl}(\mathbf{e}_i\mathbf{e}_j\mathbf{e}_k + 2\mathbf{e}_k \lrcorner (\mathbf{e}_i\mathbf{e}_j) + \mathbf{e}_i\mathbf{e}_j\mathbf{e}_k + 2(\mathbf{e}_j\mathbf{e}_k) \llcorner \mathbf{e}_i)$$

$$= -2\sum_{ijk}R_{ijkl}\mathbf{e}_i\mathbf{e}_j\mathbf{e}_k - 4\sum_{ij}R_{ijil}\mathbf{e}_j + 2\sum_{ij}R_{ijjl}\mathbf{e}_i + 2\sum_{ik}R_{iikl}\mathbf{e}_k.$$

Since $R_{iikl} = 0$ and $R_{ijil} = -R_{jiil}$, this yields $-\sum_{ijk} R_{ijkl}\mathbf{e}_i\mathbf{e}_j\mathbf{e}_k = 2\sum_i \mathrm{Ric}_{il}\mathbf{e}_i$, and in particular

$$-\sum_{ijkl} R_{ijkl}\mathbf{e}_i\mathbf{e}_j\mathbf{e}_k\mathbf{e}_l = 2\sum_i \mathrm{Ric}_{il}\mathbf{e}_i\mathbf{e}_l = 2S, \tag{11.7}$$

since the Ricci curvature is symmetric. This proves the second form of the Weitzenböck identity. $\qquad\square$

11.6 Spinor Bundles

Let M be a Riemannian manifold. The multivector bundle $\wedge M = \triangle M$ over M is a globally well-defined vector bundle over M, built fiberwise from the tangent vector bundle TM. In this section we investigate when and how it is possible to construct a bundle of spinor spaces $\slashed{\triangle} M$ globally over M. What we want to achieve is the following.

Definition 11.6.1 (Spinor bundle). Let M be an oriented Riemannian manifold of real dimension n. A (complex) *spinor bundle* over M is a complex vector bundle $\slashed{\triangle} M$, of complex dimension $2^{\lfloor n/2 \rfloor}$, together with linear maps $\rho = \rho_p : T_pM \to \mathcal{L}((\slashed{\triangle} M)_p)$ depending smoothly on $p \in M$ such that

$$\rho(v)^2\psi = |v|^2\psi, \quad v \in C(M; TM), \psi \in C(M; \slashed{\triangle} M).$$

We assume that $\slashed{\triangle} M$ is normed in the sense of Definition 5.3.3, that is, that each fiber $(\slashed{\triangle} M)_p$ is equipped with a spinor inner product $(\cdot, \cdot)$ and a spinor conjugation $\cdot^\dagger$, which vary smoothly with p.

Complexifying each fiber of the multivector bundle as in Section 1.5, we obtain a globally well-defined complex vector bundle $\triangle M_c$. Applying Lemma 5.1.4 fiberwise, we see that for a spinor bundle we have homomorphisms $\triangle M_c \to \mathcal{L}(\slashed{\triangle} M)$ of complex algebras at each $p \in M$. Assuming M to be oriented, we have globally defined main reflectors w_n as in Definition 5.2.1. Since $\rho(w_n) = \pm I$ in odd dimension, we see that orientability is necessary for the existence of spinor bundles in this case. In even dimension, $\rho(w_n)$ will yield globally defined subbundles $\slashed{\triangle}^\pm M$, which are fundamental in Chapter 12.4.

For the construction of spinor bundles over an orientable Riemannian manifold M, we fix a bundle atlas for TM of bundle charts $\underline{\mu}_\alpha(p) \in \mathcal{L}(\mathbf{R}^n; T_pM)$, $p \in U_\alpha, \alpha \in \mathcal{I}$. We assume that each $\underline{\mu}_\alpha(p)$ is an isometry and that $\{U_\alpha\}$ is a good cover of M. We also assume that all the charts $\underline{\mu}_\alpha(p)$ are orientation-preserving. In particular, we have orientation-preserving transition maps $\underline{\mu}_{\beta\alpha}(p) \in \mathrm{SO}(\mathbf{R}^n)$.

Locally over U_α, the existence of a spinor bundle is clear. Indeed, let

$$\rho_0 : \mathbf{R}^n \to \mathcal{L}(\slashed{\triangle}\mathbf{R}^n)$$

be the complex spinor space from Definition 5.2.4. Define $(\not\!\!\triangle M)_p := \not\!\!\triangle \mathbf{R}^n$ and $\rho_\alpha(v) := \rho_0(\underline{\mu}_\alpha^{-1}(p)v)$, for $v \in T_pM$, $p \in U_\alpha$. This clearly yields a spinor bundle locally over U_α. To construct a spinor bundle globally over M, we study the transition maps $\underline{\mu}_{\beta\alpha}(p) = \underline{\mu}_\beta(p)^{-1}\underline{\mu}_\alpha(p) \in \mathrm{SO}(\mathbf{R}^n)$. These clearly satisfy the symmetry condition

$$\underline{\mu}_{\alpha\beta} = \underline{\mu}_{\beta\alpha}^{-1}$$

on $U_{\beta\alpha}$, and the transitivity condition

$$\underline{\mu}_{\gamma\beta}\underline{\mu}_{\beta\alpha} = \underline{\mu}_{\gamma\alpha}$$

on $U_{\gamma\beta\alpha} := U_\gamma \cap U_\beta \cap U_\alpha$.

Definition 11.6.2 (Spin structure). Let M be an oriented Riemannian manifold, with a bundle atlas for TM with transition maps $\underline{\mu}_{\beta\alpha}(p) \in \mathrm{SO}(\mathbf{R}^n)$ as above. A *spin structure* for M is a choice of smooth functions $q_{\beta\alpha}(p) \in \mathrm{Spin}(\mathbf{R}^n) \subset \triangle^{\mathrm{ev}}\mathbf{R}^n$ such that

$$\underline{\mu}_{\beta\alpha}(p)v = q_{\beta\alpha}(p)vq_{\beta\alpha}(p)^{-1}, \quad v \in T_pM, p \in U_{\beta\alpha}, \alpha, \beta \in \mathcal{I},$$

and that satisfy the symmetry condition $q_{\alpha\beta} = q_{\beta\alpha}^{-1}$ on $U_{\beta\alpha}$, and the transitivity condition

$$q_{\gamma\beta}q_{\beta\alpha} = q_{\gamma\alpha}, \quad \text{on } U_{\gamma\beta\alpha}.$$

Since the map

$$\mathrm{Spin}(\mathbf{R}^n) \to \mathrm{SO}(\mathbf{R}^n)$$

from Proposition 4.1.9 is a covering map, there are lifts $q_{\beta\alpha}$ such that $\underline{\mu}_{\beta\alpha}v = q_{\beta\alpha}vq_{\beta\alpha}^{-1}$, and there are two possible such lifts, differing only in sign. Choosing appropriate signs, we can always satisfy the symmetry condition $q_{\alpha\beta} = q_{\beta\alpha}^{-1}$. However, Proposition 4.1.9 shows only that $q_{\gamma\beta}q_{\beta\alpha} = \pm q_{\gamma\alpha}$. To investigate when it is possible to choose signs so that the transitivity condition holds, there is a Čech cohomology criterion for existence of spin structures analogous to that for orientability in Proposition 11.2.10.

Proposition 11.6.3 (Second Stiefel–Whitney class). *Let M be an oriented Riemannian manifold, with a bundle atlas for TM with transition maps $\underline{\mu}_{\beta\alpha}(p) \in \mathrm{SO}(\mathbf{R}^n)$, and let $q_{\beta\alpha}(p) \in \mathrm{Spin}(\mathbf{R}^n)$, $p \in U_{\beta\alpha}$ as above. Consider the sheaf $\mathbf{Z}_2$ on the good cover $\{U_\alpha\}$ as in Example 11.2.9. Define the Čech 2-cochain $f \in C^2(\underline{U}; \mathbf{Z}_2)$ by*

$$f(\{\alpha, \beta, \gamma\}) := \begin{cases} 0, & q_{\alpha\gamma}q_{\gamma\beta}q_{\beta\alpha} = 1, \\ 1, & q_{\alpha\gamma}q_{\gamma\beta}q_{\beta\alpha} = -1. \end{cases}$$

Then $\partial_2 f = 0$, and the Čech cohomology class $[f]$ does not depend on the choice of lifts $q_{\beta\alpha}$. There exists a spin structure for M if and only if there exists $g \in C^1(\underline{U}; \mathbf{Z}_2)$ such that $\partial_1 g = f$, that is, if $[f] = [0]$.

The Čech cohomology class $w_2(M) := [f]$ is called the *second Stiefel–Whitney class* of M.

Proof. A straightforward verification shows that f is a well-defined element of $C^2(\underline{U}; \mathbf{Z}_2)$, that is, symmetric with respect to permutations of α, β, γ. To show that

$$\partial_2 f(\{\alpha, \beta, \gamma, \delta\}) = f(\{\alpha, \beta, \gamma\}) + f(\{\alpha, \beta, \delta\}) + f(\{\alpha, \gamma, \delta\}) + f(\{\beta, \gamma, \delta\})$$
$$= 0 \mod 2,$$

it suffices to see that

$$(q_{\alpha\gamma} q_{\gamma\beta} q_{\beta\alpha})(q_{\alpha\beta} q_{\beta\delta} q_{\delta\alpha})(q_{\alpha\delta} q_{\delta\gamma} q_{\gamma\alpha})(q_{\beta\gamma} q_{\gamma\delta} q_{\delta\beta}) = 1,$$

by the symmetry of f. Since the left-hand side is ± 1, it suffices to show that the scalar part w_0 of this multivector w is 1. Using that $(u \wedge v)_0 = (v \wedge u)_0$, this follows from the assumed symmetry condition $q_{\beta\alpha} = q_{\alpha\beta}^{-1}$.

If $q'_{\beta\alpha}$ are other choices of lifts, define

$$g(\{\alpha, \beta\}) := \begin{cases} 0, & q'_{\alpha\beta} = q_{\alpha\beta}, \\ 1, & q'_{\alpha\beta} = -q_{\alpha\beta}. \end{cases}$$

This is a well-defined element of $C^1(\underline{U}; \mathbf{Z}_2)$ because of the assumed symmetry condition on $q_{\beta\alpha}$ and $q'_{\beta\alpha}$, and we see that

$$f'(\{\alpha, \beta, \gamma\}) = f(\{\alpha, \beta, \gamma\}) + g(\{\alpha, \beta\} + g(\{\alpha, \gamma\}) + g(\{\beta, \gamma\}) \mod 2,$$

where $f' \in C^2(\underline{U}; \mathbf{Z}_2)$ is the 2-cochain for this new choice. Therefore $f' = f + \partial_1 g$, so that $[f'] = [f]$. Since there exists a spin structure for M if and only if lifts $q'_{\beta\alpha}$ can be chosen such that $f' = 0$, this completes the proof. $\qquad\square$

We now define a normed spinor bundle $\not{\triangle} M$, using a given spin structure $\{q_{\beta\alpha}\}$. Locally in each U_α, by fixing an ON-frame, we have a spinor bundle over U_α with fiber $\not{\triangle} \mathbf{R}^n$ and representation

$$\rho_\alpha(v)\psi = \rho_0(\underline{\mu}_\alpha^{-1}(p)v)$$

over each $p \in U_\alpha$, $\alpha \in \underline{\mathcal{I}}$. In terms of the spin structure, the bundle transition maps for TM are

$$\underline{\mu}_{\beta\alpha}(p)v = q_{\beta\alpha}(p)v q_{\beta\alpha}(p)^{-1}, \quad \alpha\beta \in \underline{\mathcal{I}}.$$

To define a fiber of a global spinor bundle at $p \in M$, let

$$\underline{\mathcal{I}}_p := \{\alpha \in \underline{\mathcal{I}} \; ; \; p \in U_\alpha\}$$

and consider the set $\underline{\mathcal{I}}_p \times \boldsymbol{\Lambda\!\!\!\!/}\,\mathbf{R}^n$. The symmetry and transitivity conditions on $q_{\beta\alpha}$ show that the relation $(\alpha, \psi) \sim (\beta, \phi)$, defined to hold if

$$\phi = q_{\beta\alpha}(p).\psi$$

is an equivalence relation on $\underline{\mathcal{I}}_p \times \boldsymbol{\Lambda\!\!\!\!/}\,\mathbf{R}^n$. We define the fiber of the global spinor bundle at $p \in M$ to be the set of equivalence classes

$$(\boldsymbol{\Lambda\!\!\!\!/}\,M)_p := \{[(\alpha, \psi)] \ ; \ \alpha \in \underline{\mathcal{I}}_p, \psi \in \boldsymbol{\Lambda\!\!\!\!/}\,\mathbf{R}^n\}.$$

With fiber space $\boldsymbol{\Lambda\!\!\!\!/}\,\mathbf{R}^n$ and bundle charts

$$q_\alpha(p)\psi := [(\alpha, \psi)], \qquad \psi \in \boldsymbol{\Lambda\!\!\!\!/}\,\mathbf{R}^n, p \in U_\alpha, \alpha \in \underline{\mathcal{I}},$$

we obtain a well-defined vector bundle $\boldsymbol{\Lambda\!\!\!\!/}\,M$ globally over M. Furthermore, defining the T_pM representation on $(\boldsymbol{\Lambda\!\!\!\!/}\,M)_p$ to be

$$\rho(v)[(\alpha, \psi)] := [(\alpha, \rho_\alpha(v)\psi)], \quad v \in T_pM, \psi \in \boldsymbol{\Lambda\!\!\!\!/}\,\mathbf{R}^n, p \in U_\alpha, \alpha \in \underline{\mathcal{I}},$$

we obtain $\boldsymbol{\Lambda\!\!\!\!/}\,M$ as a spinor bundle over M.

To norm $\boldsymbol{\Lambda\!\!\!\!/}\,M$, we fix spinor inner products and conjugations on $\boldsymbol{\Lambda\!\!\!\!/}\,\mathbf{R}^n$ as in Definition 5.3.3. Define a spinor inner product and conjugation on each fiber $(\boldsymbol{\Lambda\!\!\!\!/}\,M)_p$ by

$$\langle[(\alpha, \psi)], [(\alpha, \phi)]\rangle := \langle\psi, \phi\rangle,$$
$$[(\alpha, \psi)]^\dagger := [(\alpha, \psi^\dagger)],$$

for $\psi, \phi \in \boldsymbol{\Lambda\!\!\!\!/}\,\mathbf{R}^n, p \in U_\alpha, \alpha \in \underline{\mathcal{I}}$.

Exercise 11.6.4. Verify the details of the above construction and show that $\boldsymbol{\Lambda\!\!\!\!/}\,M$ indeed is a well-defined normed spinor bundle over M. Concretely, show

- that $\sim$ is an equivalence relation and that $(\boldsymbol{\Lambda\!\!\!\!/}\,M)_p$ is a linear space in a natural way,

- that $q_\alpha(p)$ defines bundle charts for a vector bundle $\boldsymbol{\Lambda\!\!\!\!/}\,M$,

- that ρ is well defined independent of α as a consequence of $q_{\beta\alpha}$ being the rotors representing the rotations $\underline{\mu}_{\beta\alpha}$,

- that the spinor inner product and conjugation are well defined independent of α as a consequence of $\mathbf{R}^n$ rotors $q_{\beta\alpha}$ being real multivectors acting isometrically on $\boldsymbol{\Lambda\!\!\!\!/}\,\mathbf{R}^n$.

Each spin structure on M, represented by a Čech 1-cochain

$$g \in C^1(\underline{U}; \mathbf{Z}_2), \quad \partial_1 g = [f] = [0],$$

yields a normed spinor bundle $\mathbb{A}M$ over M as above. Consider two spinor bundles $\mathbb{A}_1 M$ and $\mathbb{A}_2 M$ corresponding to $g_1, g_2 \in C^1(\underline{U}; \mathbf{Z}_2)$ with $\partial_1 g_1 = \partial_1 g_2 = 0$. We say that $\mathbb{A}_1 M$ and $\mathbb{A}_2 M$ are *isomorphic as normed spinor bundles* over M if there exists a section $F \in C^\infty(M; \mathcal{L}(\mathbb{A}_1 M; \mathbb{A}_2 M))$ such that at each $p \in M$, the linear map

$$F = F(p) : (\mathbb{A}_1 M)_p \to (\mathbb{A}_2 M)_p$$

is an isometry that commutes with spinor conjugation and

$$F(\rho_1(v)\psi) = \rho_2(v)(F\psi), \quad v \in T_p M, \psi \in (\mathbb{A}_1 M)_p.$$

Proposition 11.6.5 (Isomorphic spinor bundles). *Let M be an oriented Riemannian manifold, and consider two spinor bundles $\mathbb{A}_1 M$ and $\mathbb{A}_2 M$ corresponding to $g_1, g_2 \in C^1(\underline{U}; \mathbf{Z}_2)$ with $\partial_1 g_1 = \partial_1 g_2 = 0$ as above. Then $\mathbb{A}_1 M$ and $\mathbb{A}_2 M$ are isomorphic as normed spinor bundles over M if and only if there exists $h \in C^0(\underline{U}; \mathbf{Z}_2)$ such that*

$$g_1 = g_2 + \partial_0 h.$$

This means that there is a one-to-one correspondence between the Čech cohomology classes in $H^1(\underline{U}; \mathbf{Z}_2)$ and isomorphism classes of spinor bundles over M, provided that $w_2(M) = [0]$. Analogously to the remark following Theorem 10.6.5, we note that this shows in particular that the number of elements in $H^1(\underline{U}; \mathbf{Z}_2)$ does not depend on the choice of good cover, and that the number of isomorphism classes of spinor bundles over M does not depend on the choice of Riemannian metric, but only the global topology of M.

Proof. Assume that $g_1 = g_2 + \partial_0 h$. This means that

$$q_{\beta\alpha}^1 = (-1)^{h_\alpha + h_\beta} q_{\beta\alpha}^2,$$

writing $q_{\beta\alpha}^j$ for the two spin structures, $j = 1, 2$, and $h_\alpha := h(\{\alpha\})$. Consider the linear maps

$$\mathbb{A}\mathbf{R}^n \to \mathbb{A}\mathbf{R}^n : \psi \mapsto (-1)^{h_\alpha}\psi$$

in U_α. These are seen to be compatible, that is, give well-defined maps $(\mathbb{A}_1 M)_p \to (\mathbb{A}_2 M)_p$ by inspection of the equivalence relations $\sim_1$ and $\sim_2$, and yield an isomorphism of normed spinor spaces.

Conversely, assume that there exists an isomorphism

$$F \in C^\infty(M; \mathcal{L}(\mathbb{A}_1 M; \mathbb{A}_2 M)).$$

In U_α, consider $F_\alpha : \mathbb{A}\mathbf{R}^n \to \mathbb{A}\mathbf{R}^n$ defined by

$$[(\alpha, F_\alpha(\psi))] = F([(\alpha, \psi)]), \quad \psi \in \mathbb{A}\mathbf{R}^n.$$

We verify that F_α is a spinor map induced by the identity on $\mathbf{R}^n$, as in Proposition 5.3.5. By uniqueness, there exists $h_\alpha \in \mathbf{Z}_2$ such that $F_\alpha = (-1)^{h_\alpha} I$. At

$p \in U_{\beta\alpha}$, we have

$$[(\beta, (-1)^{h_\alpha} q^2_{\beta\alpha} \psi)] = [(\alpha, (-1)^{h_\alpha} \psi)] = [(\alpha, F_\alpha \psi)] = F([(\alpha, \psi)])$$
$$= F([(\beta, q^1_{\beta\alpha} \psi)]) = [(\beta, F_\beta q^1_{\beta\alpha} \psi)] = [(\beta, (-1)^{h_\beta} q^1_{\beta\alpha} \psi)],$$

where the equivalence classes on the first line are with respect to $\sim_2$, and those on the second line are with respect to $\sim_1$. It follows that $q^1_{\beta\alpha} = (-1)^{h_\alpha + h_\beta} q^2_{\beta\alpha}$, so that $g_1 = g_2 + \partial_0 h$. $\qquad\square$

We next proceed to analysis on a fixed normed spinor bundle $\slashed{\triangle} M$ over a spin manifold M.

Definition 11.6.6 (Spinor fields and frames). Sections of a normed spinor bundle $\slashed{\triangle} M$ over an oriented Riemannian manifold M are referred to as *spinor fields* on M. We mainly use Ψ, Φ to denote spinor fields.

We use induced ON-frames for $\slashed{\triangle} M$, constructed as follows. Fix the standard representation of $\mathbf{R}^n$ from Example 5.1.5. This means that we index the standard basis $\{e_i\}$ for $\mathbf{R}^n$ by $-m \le i \le m$, including $i = 0$ if $n = 2m + 1$ and excluding $i = 0$ if $n = 2m$. The spinor space $\slashed{\triangle}\mathbf{R}^n$ equals $\wedge \mathbf{C}^m$ as a complex linear space, with basis $\{e_s\}$ indexed by $s \subset \{1, \ldots, m\}$. When viewing $\wedge \mathbf{C}^m$ as $\slashed{\triangle}\mathbf{R}^n$, we write $\slashed{e}_s := e_s$.

Given an ON-frame $\{e_i\}$ for TM, indexed as the $\mathbf{R}^n$ standard basis above, consider the isometry $\mathbf{R}^n \to T_p M$ that maps the standard basis onto the frame. By Proposition 5.3.5, there is an isometry $\slashed{\triangle}\mathbf{R}^n \to (\slashed{\triangle} M)_p$ compatible with the representation and spinor conjugation, unique up to sign. We refer to either of these two ON-bases $\{\slashed{e}_s\}$ corresponding to $\{\slashed{e}_s\}$ under these isometries as a *spinor frame for $\slashed{\triangle} M$ induced by* $\{e_i\}$.

By construction, frame $\mathbf{e}_s$ multivector fields act on induced frame $\slashed{e}_t$ spinor fields on M, in the same way as the e_s act on $\slashed{e}_t$ in $\mathbf{R}^n$. Note that s and t are subsets of different index sets as above. To avoid technicalities, we shall not write out these index sets $\{-m, \ldots, m\}$, modulo 0, and $\{1, \ldots, m\}$. Note that if $\slashed{\triangle} M$ is constructed from local ON bundle charts $\mu_\alpha(p)$ for TM in U_α as above, then the spinor frame induced by the vector ON-frame $\{\mu_\alpha(p)e_i\}$ equals $\{[(\alpha, \slashed{e}_s)]\}$.

The first step in setting up the calculus on $\slashed{\triangle} M$ is to identify a canonical covariant derivative on $\slashed{\triangle} M$. As in Chapter 5, we shall abbreviate the action of multivectors on spinors by writing

$$w.\psi := \rho(w)\psi, \quad w \in \triangle_p M_c, \ \psi \in \slashed{\triangle}_p M, \ p \in M.$$

Recall from Lemma 5.1.4 that the representation ρ extends to a complex algebra representation of $\triangle M_c$ on $\slashed{\triangle} M$.

Proposition 11.6.7 ($\slashed{\triangle} M$ covariant derivative). *Let $\slashed{\triangle} M$ be a normed spinor bundle over a Riemannian manifold M. Then there exists a unique covariant derivative ∇_v on $\slashed{\triangle} M$ that is*

- *compatible with the representation in the sense that*

$$\nabla_v(F.\Psi) = (\nabla_v F).\Psi + F.(\nabla_v \Psi),$$

  where $\nabla_v F$ denotes the Levi-Civita covariant derivative on $\triangle M_c$,

- *metric in the sense that $\partial_v \langle \Psi, \Phi \rangle = \langle \nabla_v \Psi, \Phi \rangle + \langle \Psi, \nabla_v \Phi \rangle$, and*

- *compatible with spinor conjugation in the sense that $(\nabla_v \Psi)^\dagger = \nabla_v(\Psi^\dagger)$,*

for all vectors $v \in C(M; TM)$, multivector fields $F \in C^1(M; \triangle M_c)$, and all spinor fields $\Psi, \Phi \in C^1(M; \triangle M)$.

Proof. To prove uniqueness, assume that ∇_v and $\widetilde{\nabla}_v$ are two covariant derivatives with the desired properties, and let $L_v := \widetilde{\nabla}_v - \nabla_v$. Since

$$L_v(f\Psi) = ((\partial_v f)\Psi + f\widetilde{\nabla}_v \Psi) - ((\partial_v f)\Psi + f\nabla_v \Psi) = fL_v \Psi,$$

for scalar functions f and spinor fields Ψ, it follows that L_v is a linear map on each fiber $(\triangle M)_p$. A similar subtraction of the identities assumed to hold for ∇ and $\widetilde{\nabla}$ yields the following. From $L_v(F.\Psi) = F.(L_v \Psi)$ and the uniqueness result in Theorem 5.2.3, it follows that $L_v \Psi = \lambda_v \Psi$ for some $\lambda_v \in C^\infty(M; \mathbf{C})$. From $\langle \lambda_v \Psi, \Phi \rangle + \langle \Psi, \lambda_v \Phi \rangle = 0$ it follows that $\mathrm{Re}\, \lambda_v = 0$, and from $(\lambda_v \Psi)^\dagger = \lambda_v(\Psi^\dagger)$ it follows that $\mathrm{Im}\, \lambda = 0$. This proves that $\widetilde{\nabla}_v = \nabla_v$.

To show the existence of a covariant derivative with these properties, it suffices to consider the problem locally in a frame. Fix a vector ON-frame $\mathbf{e}$ in $U \subset M$, with induced spinor frame $\not{e}$. Let $\Gamma_\mathbf{e} \in C^\infty(U; \mathcal{L}(TM; \wedge^2 M))$ be the Christoffel symbols for TM and define a covariant derivative

$$\nabla_v \Psi := \sum_s (\partial_v \Psi_s)\not{e}_s + \tfrac{1}{2}\Gamma_\mathbf{e}(v).\Psi$$

of spinor fields $\Psi = \sum_s \Psi_s \not{e}_s$. To prove $\nabla_v(F.\Psi) = (\nabla_v F).\Psi + F.(\nabla_v \Psi)$, it suffices to consider multivector and spinor frame fields $F = \mathbf{e}_s$ and $\Psi = \not{e}_t$. Since $\{\not{e}_t\}$ is an induced spinor frame, it is clear that $\mathbf{e}_s.\not{e}_t$ is ± 1 times a spinor frame element. Therefore, by Proposition 11.5.3 we need to prove

$$\tfrac{1}{2}\Gamma_\mathbf{e}(v).(\mathbf{e}_s.\not{e}_t) = \tfrac{1}{2}[\Gamma_\mathbf{e}(v), \mathbf{e}_s].\not{e}_t + \mathbf{e}_s.(\tfrac{1}{2}\Gamma_\mathbf{e}(v).\not{e}_t),$$

which is clear. Further we note that ∇_v is a metric covariant derivative, since $\overline{\Gamma_\mathbf{e}(v)} = -\Gamma_\mathbf{e}(v)$, because $\Gamma_\mathbf{e}(v) \in \wedge^2 M$, and it is compatible with spinor conjugation since $\Gamma_\mathbf{e}(v)$ is a real bivector field. This completes the proof. $\square$

Proposition 11.6.8 (Bivector derivations). *Let $\triangle M$ be a normed spinor bundle over a Riemannian manifold M, with the covariant derivative from Proposition 11.6.7, and let $\mathbf{e} = \{\mathbf{e}_i(p)\}$ be an ON-frame for TM in an open set $U \subset M$. Denote by $\Gamma_\mathbf{e} \in C^\infty(U; \mathcal{L}(\wedge^1 M; \wedge^2 M))$ and $R \in C^\infty(M; \mathcal{L}(\wedge^2 M; \wedge^2 M))$ the Christoffel symbols and curvature operator for TM.*

Let $\not\phi = \{\not\phi_s(p)\}$ be the induced ON-frame for $\not\triangle M$ in U. Denote by $\omega_{\not\phi} \in$ $C^\infty(U; \mathcal{L}(\wedge^1 M; \mathcal{L}(\not\triangle M)))$ and $\Omega \in C^\infty(M; \mathcal{L}(\wedge^2 M; \mathcal{L}(\not\triangle M)))$ the Christoffel symbols and curvature operator for $\not\triangle M$. Then

$$\omega_{\not\phi}(v)\Psi = \tfrac{1}{2}\Gamma_{\mathbf{e}}(v).\Psi,$$
$$\Omega(b)\Psi = \tfrac{1}{2}R(b).\Psi,$$

for $v \in C(M; TM)$, $b \in C(M; \wedge^2 M)$, and $\Psi \in C(M; \not\triangle M)$.

Proof. The result for the Christoffel symbols is contained in the proof of Proposition 11.6.7. Consider therefore the curvature operator. Defining

$$\widetilde{\Omega}(b)\Psi := \Omega(b)\Psi - \tfrac{1}{2}R(b).\Psi,$$

we have

$$\widetilde{\Omega}(b)(F.\Psi) = \tfrac{1}{2}[R(b), F].\Psi + F.(\Omega(b)\Psi) - \tfrac{1}{2}R(b).(F.\Psi) = F.(\widetilde{\Omega}(b)\Psi),$$

by a computation similar to the proof of (11.5). The uniqueness result in Theorem 5.2.3 shows that $\widetilde{\Omega}(b)\Psi = \lambda_b \Psi$. Since $\widetilde{\Omega}(b)$ is skew-symmetric, we have $\operatorname{Re}\lambda_b = 0$, and since it commutes with spinor conjugation, we have $\operatorname{Im}\lambda_b = 0$. $\square$

The fundamental first-order differential operator on M acting on sections of $\not\triangle M$ is the following natural generalization of the Euclidean nabla operator from Definition 9.1.3.

Definition 11.6.9 (The Atiyah–Singer Dirac operator). Let $\not\triangle M$ be a normed spinor bundle over a Riemannian manifold M. The $\not\triangle$-*Dirac operator* on $\not\triangle M$ is the operator

$$\not{D}\Psi := \sum_i \mathbf{e}_i^*.(\nabla_{\mathbf{e}_i}\Psi), \quad \Psi \in C^1(M; \not\triangle M),$$

where $\{\mathbf{e}_i\}$ is a frame for TM, with dual frame $\{\mathbf{e}_i^*\}$.

The analogue of Proposition 11.5.9 is the following result relating $\not{D}^2$ to the Laplace–Beltrami operator Δ on $\not\triangle M$. This is the key identity used in the proof of the Atiyah–Singer index theorem in Section 12.4.

Proposition 11.6.10 (Weitzenböck identity for $\not{D}$). *Let $\not\triangle M$ be a normed spinor bundle over a Riemannian manifold M, and let $\{\mathbf{e}_i\}$ be an ON-frame for TM. Then*

$$-\not{D}^2\Psi = -\Delta\Psi - \sum_{i<j} \mathbf{e}_i\mathbf{e}_j.\Omega(\mathbf{e}_i \wedge \mathbf{e}_j)\Psi = -\Delta\Psi + \frac{1}{4}S\Psi,$$

for $\Psi \in C^2(M; \not\triangle M)$, where Δ is the Laplace–Beltrami operator on $\not\triangle M$ as in Definition 11.5.6, Ω is the curvature operator on $\not\triangle M$, and S is the scalar curvature S of TM as in Definition 11.3.6.

Proof. Calculations identical to those in Proposition 11.5.9 lead to the first identity. From Propositions 11.6.8, and (11.7) in the proof of Proposition 11.5.9, we then obtain

$$\sum_{i<j} \mathbf{e}_i \mathbf{e}_j . \Omega(\mathbf{e}_i \wedge \mathbf{e}_j) \Psi = \frac{1}{8} \sum_{ijkl} R_{ijkl} \mathbf{e}_i \mathbf{e}_j \mathbf{e}_k \mathbf{e}_l . \Psi = -\tfrac{1}{4} S \Psi. \qquad \square$$

11.7 Comments and References

1.1–11.3 A reference for the differential geometry needed in this book, and more, is Taubes [90].

It is standard in differential geometry to identify tangent vectors v and directional derivatives ∂_v. In coordinates $\{x_1, \ldots, x_n\}$, it is standard to denote the coordinate basis vectors $\{e_1, \ldots, e_n\}$ by $\{\partial/\partial x_1, \ldots, \partial/\partial x_n\}$, and to denote the dual basis $\{e_1^*, \ldots, e_n^*\}$ by $\{dx_1, \ldots, dx_n\}$. There is of course a canonical one-to-one correspondence $v \leftrightarrow \partial_v$, but to identify v and ∂_v as objects is not natural, and leads to serious notational problems. We therefore refrain from doing so. A main example is the nabla operators, where we define exterior and interior derivatives, and Dirac operators using $\nabla = \sum_j e_i \partial_i$ in an ON-basis. It leads to obvious notational problems if we write the basis vector e_i as a derivative ∂_i, and also writing the dual basis covector e_i^* as dx_i causes problems for the inexperienced reader.

We use the terminology Christoffel symbols in a more general sense than the standard usage in the literature. Normally, Christoffel symbols Γ^i_{jk} refer to the tangent bundle and a coordinate frame. We more generally refer to the zero-order part of the covariant derivative, in a fixed frame for a general vector bundle, as Christoffel symbols.

A reference for the use of differential forms, or multicovector fields in our terminology, in differential geometry is Darling [32]. For the curvature operator, one should note the two different formalisms. Following the Cartan tradition, using differential form, we have Definition 11.3.3. Without using differential forms, the standard definition of the curvature operator is by Proposition 11.3.5.

11.4 Using an ON-frame, equations (11.3), and the equations defining curvature in Definition 11.3.3 provide a way to compute curvature. This is called the orthonormal moving frame method, and the equations are referred to as the structure equations. Some references are [96], [32], and [57].

The proof of Liouville's theorem presented in Section 11.4 follows Flanders [39]. The methods we use require the C^3 hypothesis. However, it has been shown by Rešetnjak [77] that this can be weakened to C^1 only, or even local integrability of $|\partial_i f|^n$, $i = 1, \ldots, n$.

The conformal maps in dimension $n \geq 3$ are indeed very scarce. However, by relaxing the conformal requirement and only requiring the ratio between the largest and smallest singular value of the total derivative to be uniformly bounded, one obtains a much richer class of maps: the quasiconformal maps.

11.5–11.6 Identities like those in Propositions 11.5.9 and 11.6.10, with integral versions as formulated in Section 12.1, or versions for domains with a boundary as in Theorem 10.3.6, are named after Roland Weitzenböck. In the literature they are sometimes referred to as Lichnerowicz identities, in particular that for $\not{D}$ which André Lichnerowicz derived. Also related are Bochner identities, which are second-order analogues of these identities.

In the literature, the standard approach to defining spinor bundles and the $\not{\triangle}$-Dirac operator is via principal bundles. These are similar to vector bundles, but instead of having a family of vector spaces indexed by M, we now have a family of copies $\{P_x\}_{x \in M}$ of a given group G. But in contrast to vector bundles, we do not have a fixed identity element in the fibers P_x, but the group G acts freely and transitively from the right on each P_x. The typical way that this construction is used in connection to vector bundles E is that the fiber P_x collects all bases for E_x of a certain type, described by G, and the coordinates for vectors in E_x are described by a representation of G. For example, to construct spinor bundles starting from TM, we collect all positively oriented bases into a principal bundle P with $G = SO(\mathbf{R}^n)$, then lift this to a principal $\mathrm{Spin}(\mathbf{R}^n)$ bundle $\not{P}$ similar to Proposition 4.1.9, and finally obtain a spinor bundle $\not{\triangle}M$ by combining the frames implicit in $\not{P}$ with coordinates coming from the representation of $\mathrm{Spin}(\mathbf{R}^n)$ on $\not{\triangle}\mathbf{R}^n$.

In this book, we have avoided principal bundles to minimize technicalities. However, one can show that the spinor bundles obtained from a principal spin bundle $\not{P}$ as above are precisely the normed spinor bundles considered in Section 11.6. In particular, they come equipped with a spinor inner product and a spinor conjugation, and the induced spinor frames from Definition 11.6.6 correspond to $\not{P}$.

Our discussion of spin structures follows Gilkey [43].

Chapter 12

Local Index Theorems

Prerequisites:

Chapter 11 should contain the material from differential geometry needed to read the present chapter. Section 12.1 builds on part of Chapter 10.

Road map:

Let M be a two-dimensional closed Riemannian manifold. The famous Gauss–Bonnet theorem states that

$$\chi(M) = \frac{1}{4\pi} \int_M S(p)dp,$$

where $S = 2R_{1212}$ denotes the scalar curvature at $p \in M$ and

$$\chi(M) = b_0(M) - b_1(M) + b_2(M)$$

is the *Euler characteristic* for M. Here $b_j(M)$ are the Betti numbers for M, with the obvious generalization of Definition 7.6.3 to the compact manifold M without boundary. Hiding behind this result is the $\triangle$-Dirac operator $\mathbf{D}$ and the splitting

$$L_2(M; \triangle M) = L_2(M; \triangle^{\mathrm{ev}} M) \oplus L_2(M; \triangle^{\mathrm{od}} M).$$

Indeed, $\mathbf{D}$ is a skew-adjoint operator on $L_2(M; \triangle M)$ that swaps these two subspaces, and $\chi(M)$ equals the index of the restriction

$$\mathbf{D} : L_2(M; \triangle^{\mathrm{ev}} M) \to L_2(M; \triangle^{\mathrm{od}} M).$$

To appreciate the power of the Gauss–Bonnet theorem, one must note that it relates three fundamentally different quantities, S, $\chi(M)$, and $\mathbf{D}$, where S is a local geometric quantity, χ is a global topological quantity, and $\mathbf{D}$ is an analytic object.

© Springer Nature Switzerland AG 2019
A. Rosén, *Geometric Multivector Analysis*, Birkhäuser Advanced Texts Basler Lehrbücher,
https://doi.org/10.1007/978-3-030-31411-8_12

In this chapter, we study two generalizations of the Gauss–Bonnet theorem. The first is the Chern–Gauss–Bonnet theorem, which is the direct generalization to higher-dimensional manifolds. In odd dimension the Euler characteristic vanishes due to Poincaré/Hodge duality. In the interesting case of even dimension, the scalar curvature in the integrand is replaced by the Pfaffian $\mathrm{Pf}(R)$, a quantity pointwise derived from the Riemann curvature operator R.

The second generalization is the Atiyah–Singer index theorem for the $\not\triangle$-Dirac operator $\not{D}$. For this, we consider a spinor bundle $\not\triangle M$ over a closed oriented even-dimensional Riemannian manifold M, and consider the splitting

$$L_2(M; \not\triangle M) = L_2(M; \not\triangle^+ M) \oplus L_2(M; \not\triangle^+ M)$$

into the chiral subspaces of right- and left-handed spinor fields, which are swapped by $\not{D}$. The Atiyah–Singer index theorem, which ranks among the very top achievements in twentieth-century mathematics, states in particular that the index of the restriction

$$\not{D} : L_2(M; \not\triangle^+ M) \to L_2(M; \not\triangle^- M)$$

equals an integral over M of a quantity $\hat{A}(R)$ obtained pointwise from the Riemann curvature operator. It turns out that this index is nonzero only for manifolds of dimension divisible by four, and the integral of $\hat{A}(R)$ does not depend on the choice of spinor bundle. It may happen that the integral over some manifolds M is not be integer-valued, which is for example the case for complex projective plane $\mathbf{C}P^2$, which shows that this four-dimensional real manifold does not possess a spin structure. It is not, however, the purpose of this chapter to pursue the large amount of interesting applications of these results, which can be found in the extensive existing literature. Rather our goal is to demonstrate how our systematic buildup of the multivector and spinor theory makes advanced results in modern mathematics rather easily accessible.

For both proofs, we use the well known heat equation method. For a self-adjoint elliptic differential operator D, with applications $D = i\mathbf{D}$ and $D = i\not{D}$ in mind, the strategy is as follows. Write $L_2 = L_2^+ \oplus L_2^-$ for the splitting in this abstract formulation. Then by definition, the index we consider is

$$\dim \mathsf{N}(D|_{L_2^+}) - \dim \mathsf{N}(D|_{L_2^-}).$$

Aiming at curvature and having the Weitzenböck identities in mind, we note that $\mathsf{N}(D|_{L_2^\pm}) = \mathsf{N}(D^2|_{L_2^\pm})$ and consider the eigenvalues of the nonnegative self-adjoint operator D^2. Write $\lambda_j^\pm$ for the eigenvalues of D^2 on the invariant subspaces $L_2^\pm$. Here the nonzero eigenvalues λ_j^+ are the same as the nonzero eigenvalues λ_j^-, as a consequence of $D_+ : L_2^+ \to L_2^-$ and $D_- : L_2^- \to L_2^+$ being adjoint operators.

We now apply a suitable function $f : \mathbf{R} \to \mathbf{R}$ by Borel functional calculus to the self-adjoint operator $D^2 = D_- D_+ \oplus D_+ D_-$. Choosing $f(0) = 1$ and f to decay fast enough toward ∞, we obtain a trace-class operator $f(D^2)$, and the index of

D_+ equals the difference of traces

$$\sum_j f(\lambda_j^+) - \sum_j f(\lambda_j^-) = \mathrm{Tr} f(D_- D_+) - \mathrm{Tr} f(D_+ D_-).$$

In the heat equation method, we choose $f(\lambda) = e^{-t\lambda^2}$ with a parameter $t > 0$, which yields trace-class operators by the Weyl asymptotics $\lambda_j^\pm \approx j^{2/n}$ on an n-dimensional manifold. The solution operators e^{-tD^2} to the heat equation $\partial_t f + D^2 f = 0$ become local and converge to the identity in a suitable sense as $t \to 0^+$. To prove the index theorems, we need to work with multivector and spinor calculus on M to identify the limit of trace differences

$$\lim_{t \to 0^+} \left(\mathrm{Tr}(e^{-tD_- D_+}) - \mathrm{Tr}(e^{-tD_+ D_-}) \right).$$

Note that the existence of the limit is trivial, since the trace difference is independent of t. Sections 12.1 and 12.2 contain material on L_2 Dirac operators and charts, preliminary to the proofs of the index theorems.

Highlights:

- The Chern–Gauss–Bonnet theorem: 12.3.1

- The Atiyah–Singer index theorem: 12.4.1

12.1 Fredholm Dirac Operators

Throughout this chapter, we consider a closed compact Riemannian manifold M without boundary. In this section we study the L_2 properties of the Dirac operators $\mathbf{D}$ and $\slashed{D}$ on M, analogously to our study of d and δ on affine domains in Chapter 10. However, now our main work concerns the geometry of the manifold M rather than the boundary $\partial M = \emptyset$. In particular, no boundary conditions are needed for the operators.

Consider first the multivector operators $\mathbf{D}$, d, and δ. By transferring the Euclidean result to M using pullbacks and pushforwards as in Exercise 11.2.7, it follows that d and $-\delta$ defined on $C^\infty(M; \wedge M)$ are formally adjoint operators. From this it follows that

$$\mathbf{D} = d + \delta,$$

defined on $C^\infty(M; \triangle M)$, is formally skew-adjoint. Using the covariant nabla expression for $\mathbf{D}$ from Definition 11.5.4, we can also verify this directly on M as follows. Localizing with a partition of unity, we may assume that $F, G \in C_0^\infty(U; \triangle M)$ in an open set $U \subset M$, in which we have an ON-frame $\{\mathbf{e}_i\}$ for TM. Define the vector field

$$v := \sum_i \langle \mathbf{e}_i \mathbin{\triangle} F, G \rangle \mathbf{e}_i.$$

We compute

$$\text{div}\, v = \sum_i \Big(\langle (\nabla_{\mathbf{e}_i} \mathbf{e}_i) \vartriangle F, G \rangle + \langle \mathbf{e}_i \vartriangle \nabla_{\mathbf{e}_i} F, G \rangle + \langle \mathbf{e}_i \vartriangle F, \nabla_{\mathbf{e}_i} G \rangle \Big)$$

$$+ \sum_{ij} \langle \mathbf{e}_i \vartriangle F, G \rangle \langle \mathbf{e}_j, \nabla_{\mathbf{e}_j} \mathbf{e}_i \rangle. \tag{12.1}$$

Since $\sum_i \mathbf{e}_i \langle \mathbf{e}_j, \nabla_{\mathbf{e}_j} \mathbf{e}_i \rangle = -\sum_i \mathbf{e}_i \langle \mathbf{e}_i, \nabla_{\mathbf{e}_j} \mathbf{e}_j \rangle = -\nabla_{\mathbf{e}_j} \mathbf{e}_j$ in the last term, this cancels the first term. The skew-adjointness now follows from the divergence theorem $\int_M \text{div}\, v\, dp = 0$ from Exercise 11.2.11.

A similar calculation, replacing $\vartriangle$ by $\wedge$, gives a second proof of the fact that $-\delta$ is a formal adjoint of d on M. The following extension of the domains of operators in the natural distributional sense yields closed and densely defined linear operators in $L_2(M; \triangle M)$.

Definition 12.1.1 (L_2 operators). Let M be a closed Riemannian manifold. Consider the equation

$$\int_M \Big(\langle F', G \rangle + \langle F, DG \rangle \Big) dp = 0. \tag{12.2}$$

The domain $\mathsf{D}(d)$ of the exterior derivative in $L_2(M; \wedge M)$ is the set of $F \in L_2(M; \wedge M)$ for which there exists $F' \in L_2(M; \wedge M)$ such that (12.2), with $D = \delta$, holds for all $G \in C^\infty(M; \wedge M)$. For $F \in \mathsf{D}(d)$, we define $dF := F'$.

The domain $\mathsf{D}(\delta)$ of the interior derivative in $L_2(M; \wedge M)$ is the set of $F \in L_2(M; \wedge M)$ for which there exists $F' \in L_2(M; \wedge M)$ such that (12.2), with $D = d$, holds for all $G \in C^\infty(M; \wedge M)$. For $F \in \mathsf{D}(\delta)$, we define $\delta F := F'$.

The domain $\mathsf{D}(\mathbf{D})$ of the Dirac operator in $L_2(M; \triangle M)$ is the set of $F \in L_2(M; \triangle M)$ for which there exists $F' \in L_2(M; \triangle M)$ such that (12.2), with $D = \mathbf{D}$, holds for all $G \in C^\infty(M; \triangle M)$. For $F \in \mathsf{D}(\mathbf{D})$, we define $\mathbf{D}F := F'$.

In the absence of boundary ∂M, we can prove full Sobolev H^1 regularity of fields $F \in \mathsf{D}(\mathbf{D})$ following the same route as in Section 10.3. We start with the integral form of the Weitzenböck identity.

Proposition 12.1.2 (Integral Weitzenböck identity). *Let M be a closed Riemannian manifold. Then*

$$\int_M (|dF|^2 + |\delta F|^2) dp = \int_M |\mathbf{D}F|^2 dp$$

$$= \int_M \sum_i |\nabla_{\mathbf{e}_i} F|^2 dp + \frac{1}{4} \int_M S|F|^2 dp + \frac{1}{8} \int_M \sum_{ijkl} \langle R_{ijkl} \mathbf{e}_i \mathbf{e}_j F \mathbf{e}_k \mathbf{e}_l, F \rangle dp, \tag{12.3}$$

for all $F \in H^1(M; \wedge M)$. Here $\{\mathbf{e}_i\}$ is an ON-frame for TM, and R and S denote the Riemann curvature operator and scalar curvature for TM.

Proof. Assume first that $F \in C^2(M; \triangle M)$. Then the result follows from the nilpotence and formal adjointness of d and $-\delta$, and duality from Propositions 11.5.5 and 11.5.9. Since all terms are continuous in H^1 norm, a limiting argument finishes the proof. $\qquad \square$

Proposition 12.1.3 (Regularity). *Let M be a closed Riemannian manifold. Then* $\mathbf{D} = d + \delta$ *with*

$$D(\mathbf{D}) = D(d) \cap D(\delta) = H^1(M; \triangle M)$$

and equivalences of norms.

Proof. Clearly $H^1(M; \triangle M) \subset D(d) \cap D(\delta) \subset D(\mathbf{D})$, so it suffices to show that $D(\mathbf{D}) \subset H^1(M; \triangle M)$. To this end, we argue as in the proof of Theorem 10.3.3, but replacing the ball B by the n-torus T^n.

(i) Consider first the case $M = T^n$ with the flat metric given by the constant standard inner product on $\mathbf{R}^n$. This is the manifold $T^n = \mathbf{R}^n / \mathbf{Z}^n$, with charts obtained from quotient maps $\mathbf{R}^n \to T^n$ in the natural way. A multivector field on T^n corresponds to a $\mathbf{Z}^n$-periodic field on $\mathbf{R}^n$, and $D(\mathbf{D}) = H^1(T^n; \wedge T^n)$ follows from Plancherel's theorem and the Fourier series analogue of Example 10.1.8.

(ii) Next consider a general closed Riemannian manifold M, and $F \in D(\mathbf{D})$. Using a partition of unity, we may assume that $\operatorname{supp} F$ is contained in the range $M_\alpha \subset M$ of a coordinate chart $\mu_\alpha : D_\alpha \to M_\alpha$. Assuming that $D_\alpha \subset \mathbf{R}^n$ is small, we identify D_α with a subset of T^n, and define a C^∞ metric $g_{ij}^1(x)$ on T^n such that

$$\mu_\alpha : D_\alpha \to M_\alpha$$

is an isometry. In this way, we may regard M_α as an open subset of the n-torus T^n with geometry determined by the metric g_{ij}^1.

To show regularity of F, we perturb the metric g_{ij}^1 on T^n continuously to the flat Euclidean metric δ_{ij}, by letting

$$g_{ij}^t(x) := (1 - t)\delta_{ij} + t g_{ij}^1(x), \quad 0 \le t \le 1.$$

It is clear that g_{ij}^t defines a Riemannian metric on T^n for each $0 \le t \le 1$. Write T_t^n for this Riemannian manifold, and note that all T_t^n are the same as C^∞ manifolds, but their Riemannian geometries are distinct. Consider the family of bounded linear operators

$$I + \mathbf{D}_t : H^1(T_t^n; \triangle T_t^n) \to L_2(T_t^n; \triangle T_t^n), \tag{12.4}$$

where $\mathbf{D}_t$ is the Dirac operator on $\triangle T_t^n$. By Proposition 12.1.2, these are all injective semi-Fredholm maps with

$$\|\nabla \otimes F\|^2 + \|F\|^2 \approx \|\mathbf{D}F\|^2 + \|F\|^2 = \|F + \mathbf{D}F\|^2,$$

using the formal skew-adjointness of $\mathbf{D}_t$ for the last equality. We want to show that these are all invertible maps. This is clear for $I + \mathbf{D}_0$ from (i). To apply the method of continuity for semi-Fredholm operators, as explained in Section 6.4, we define auxiliary maps of multivector fields

$$A_t : C(T_0^n; \triangle T_0^n) \to C(T_t^n; \triangle T_t^n).$$

Since the T_t^n are all equal to T^n as C^∞ manifolds, we can realize the multi-vector bundle $\triangle T_t^n$ as $T^n \times \wedge \mathbf{R}^n$, where $\wedge \mathbf{R}^n$ is defined independently of any metric as in Section 2.1, before equipping each fiber with the inner product in-duced by g_{ij}^t. We therefore let A_t be the identity map on $T^n \times \wedge \mathbf{R}^n$, but with different but equivalent metrics in the domain and range. In particular we ob-tain bounded and invertible linear maps $A_t : L_2(T_0^n; \triangle T_0^n) \to L_2(T_t^n; \triangle T_t^n)$ and $A_t : H^1(T_0^n; \triangle T_0^n) \to H^1(T_t^n; \triangle T_t^n)$, and the method of continuity applies to

$$A_t^{-1}(I + \mathbf{D}_t)A_t : H^1(T_0^n; \triangle T_0^n) \to L_2(T_0^n; \triangle T_0^n).$$

It follows that $I + \mathbf{D}_1 : H^1(T_1^n; \triangle T_1^n) \to L_2(T_1^n; \triangle T_1^n)$ is invertible. Since $H^1 \subset \mathsf{D}(\mathbf{D})$ on T_1^n and $I + \mathbf{D}$ is injective on $\mathsf{D}(\mathbf{D})$ by formal skew-adjointness, this shows that $H^1 = \mathsf{D}(\mathbf{D})$. This completes the proof.   □

Recall the definition (6.4) of adjointness in the sense of unbounded operators.

Proposition 12.1.4 (Duality). *Let M be a closed Riemannian manifold. Then d and $-\delta$, with domains as in Definition 12.1.1, are adjoint operators in $L_2(M; \wedge M)$ in the sense of unbounded operators. The Dirac operator $\mathbf{D}$, with domain as in Defi-nition 12.1.1, is a skew-adjoint operator in $L_2(M; \triangle M)$ in the sense of unbounded operators.*

Proof. Consider first $d^* = -\delta$. As in Proposition 10.2.3, it is clear from Defini-tion 12.1.1 that it suffices to show that for every $F \in \mathsf{D}(d)$, there exists a family of fields $F_t \in C^\infty(M; \wedge M)$ such that $F_t \to F$ and $dF_t \to dF$ in $L_2(M; \wedge M)$ as $t \to 0^+$. By localizing with a partition of unity, we may assume that F is supported in a chart M_α. Since d commutes with the pullback μ_α^*, the result follows from the Euclidean case in Section 10.2.

To prove $\mathbf{D}^* = -\mathbf{D}$, we similarly note that it suffices to show that for every $F \in \mathsf{D}(\mathbf{D})$, there exists a family of fields $F_t \in C^\infty(M; \triangle M)$ such that $F_t \to F$ and $\mathbf{D}F_t \to \mathbf{D}F$ in $L_2(M; \triangle M)$ as $t \to 0^+$. In this case, we obtain from Proposition 12.1.3 that $F \in H^1(M; \triangle M)$. This completes the proof, since C^∞ is dense in H^1.   □

From Propositions 10.1.2 and 10.1.6, we now obtain the Hodge decomposition

$$L_2(M; \wedge M) = \mathsf{R}(d) \oplus \mathcal{C}(D) \oplus \mathsf{R}(\delta)$$

of L_2 multivector fields on M, with finite-dimensional cohomology space $\mathcal{C}(D) = \mathsf{N}(d) \cap \mathsf{N}(\delta)$, closed subspaces $\mathsf{R}(d)$ and $\mathsf{R}(\delta)$ of exact and coexact fields, and compact potential maps.

In terms of the Dirac operator, this means that

$$\mathbf{D} : H^1(M; \triangle M) \to L_2(M; \triangle M)$$

is a Fredholm operator. Since it is a skew-adjoint L_2 operator, its index is zero. Concretely,

$$\mathsf{N}(\mathbf{D}) = \mathsf{R}(\mathbf{D})^\perp = \mathcal{C}(D).$$

Splitting the fields further into homogeneous k-vector fields, we write $\mathcal{C}(D; \wedge^k) := \mathcal{C}(D) \cap L_2(M; \wedge^k M)$ and define the *Betti numbers*

$$b_k(D) := \dim \mathcal{C}(D; \wedge^k), \quad k = 0, 1, 2, \ldots, n.$$

As explained in the introduction, we are particularly interested in the following integer.

Definition 12.1.5 (Euler characteristic). Let M be a closed Riemannian manifold. The *Euler characteristic* of M is the alternating sum

$$\chi(M) := \sum_k (-1)^k b_k(M)$$

of Betti numbers, or equivalently the index of the restricted Dirac operator $\mathbf{D} : H^1(M; \triangle^{\mathrm{ev}} M) \mapsto L_2(M; \triangle^{\mathrm{od}} M)$.

Exercise 12.1.6. Compute the three Betti numbers $b_0(M)$, $b_1(M)$, and $b_2(M)$ for the two-dimensional sphere S^2 as well as the two-dimensional torus $T^2 = S^1 \times S^1$, using Hodge star maps and the Gauss–Bonnet theorem stated in the introduction. Note that there exists a flat metric on T^2, that is, a metric for which the curvature operator vanishes. Show also that there exists no flat metric on S^2.

Next consider the spinor Dirac operator $\slashed{D}$ acting on sections of a given normed spinor bundle $\slashed{\triangle} M$ over an oriented closed Riemannian manifold M. For a general real Riemannian manifold, without any further complex structure, we cannot write $\slashed{D}$ in terms of some invariantly defined nilpotent operators Γ and Γ^*, analogously to $\mathbf{D} = d + \delta$ for the $\triangle$-Dirac operator. But besides this, we have an L_2 operator $\slashed{D}$ with similar properties to those of $\mathbf{D}$ above. Some details are as follows.

- A calculation like (12.1), replacing $\mathbf{e}_i \wedge F$ by $\mathbf{e}_i.\Psi$, shows that $\slashed{D}$ is formally skew-adjoint on $C^\infty(M; \slashed{\triangle} M)$.

- Similar to Definition 12.1.1, we extend the domain of $\slashed{D}$ from C^∞ to $\mathsf{D}(\slashed{D})$ consisting of spinor fields $\Psi \in L_2(M; \slashed{\triangle} M)$ for which there exists $\Psi' \in L_2(M; \wedge M)$ such that

$$\int_M \Big(\langle \Psi', \Phi \rangle + \langle \Psi, \slashed{D}\Phi \rangle \Big) dp = 0$$

holds for all $\Phi \in C^\infty(M; \slashed{\triangle} M)$. For $\Psi \in \mathsf{D}(\slashed{D})$, we define $\slashed{D}\Psi := \Psi'$.

- From Propositions 11.5.5 and 11.6.10, we readily obtain an integral Weitzen-böck identity

$$\int_M |\mathbf{D}\Psi|^2 dp = \int_M \sum_i |\nabla_{\mathbf{e}_i}\Psi|^2 dp + \frac{1}{4}\int_M S|\Psi|^2 dp,$$

valid for $\Psi \in H^1(M; \not\!\!\triangle M)$. Here $\{\mathbf{e}_i\}$ is an ON-frame and S is the scalar curvature for TM.

- Also for the $\not\!\!\triangle$-Dirac operator, we have $\mathsf{D}(\not\!\!D) = H^1(M; \not\!\!\triangle M)$. However, some care concerning the map A_t is needed in adapting the perturbation argument in the proof of Proposition 12.1.3. On each n-torus T_t^n we have an ON-frame $\{\mathbf{e}_i\}$ for the tangent bundle globally defined on T_t^n, for example by polar decomposition as in Definition 12.2.4. Upon mapping this onto the standard basis for $\mathbf{R}^n$, we use the trivial spinor bundle

$$\not\!\!\triangle T_t^n = T_t^n \times \not\!\!\triangle\mathbf{R}^n$$

over T_t^n. Note that there is no problem with topological obstructions here, since the problem is local. Defining A_t as the identity map on $T^n \times \not\!\!\triangle\mathbf{R}^n$, but with different metrics on the domain and range, the proof proceeds as for $\mathbf{D}$.

- From H^1 regularity it follows that $\not\!\!D$ is a skew-adjoint operator in $L_2(M; \not\!\!\triangle M)$ in the sense of unbounded operators, and that

$$\not\!\!D : H^1(M; \not\!\!\triangle M) \to L_2(M; \not\!\!\triangle M)$$

is a Fredholm operator. Since it is skew-adjoint, we have $\mathsf{N}(\not\!\!D) = \mathsf{R}(\not\!\!D)^\perp$, and in particular, its index is zero.

- In contrast to the $\triangle$-Dirac operator, here we do not have access to a finer splitting into subspaces like $L_2(M; \triangle^k M)$ and a notion of Betti numbers, but assuming that the dimension $n = 2m$ of M is even, we have the pointwise splitting into chiral subspaces. Let $e_{\overline{n}}$ be the unit n-vector field on M describing the orientation of M, and define the main reflector $w_n := i^{-m}e_{\overline{n}}$ as in Definition 5.2.1. Consider the pointwise orthogonal splitting

$$L_2(M; \not\!\!\triangle M) = L_2(M; \not\!\!\triangle^+ M) \oplus L_2(M; \not\!\!\triangle^- M),$$

where $(\not\!\!\triangle^\pm M)_p$ are the ranges of the projections $w_{\overline{n}}^\pm = \frac{1}{2}(1 \pm w_n)$, at each $p \in M$.

- It follows from Definition 11.6.9 that $\not\!\!D$ swaps the subspaces $L_2(M; \not\!\!\triangle^\pm M)$, since the $\mathbf{e}_i$ swap them, while the ∇_v preserve them, since the $\Gamma_{\mathbf{e}}(v)$ do so. This is so because the vectors $\mathbf{e}_i$ anticommute with w_n, while the bivectors $\Gamma_{\mathbf{e}}(v)$ commute with w_n. Our goal in Section 12.4 is to calculate the index of the restricted $\not\!\!\triangle$-Dirac operator

$$\not\!\!D : H^1(M; \not\!\!\triangle^+ M) \to L_2(M; \not\!\!\triangle^- M).$$

12.2   Normal Coordinates

Let M be a closed Riemannian manifold, with tangent bundle TM, Levi-Civita covariant derivative ∇_v, and Riemann curvature operator R. Fix a base point $q \in M$. To do computations on M near q it is useful to choose a chart for M around that q is as good as possible. We use the exponential map $\exp_q : T_q M \to M$, which is the map taking a tangent vector $v \in T_q M$ to the point $p \in M$ at distance

$$d(p, q) := |v|$$

from q along the geodesic, that is, the length-minimizing curve γ, starting at $\gamma(0) = q$ with tangent vector $\dot\gamma'(0) = v/|v|$ and parametrized by arc length.

Definition 12.2.1. A *normal chart* for M around q is a chart

$$\mu : D \to M,$$

with $D = B(0, r) \subset \mathbf{R}^n$, obtained by fixing an ON-basis for $T_q M$ and identifying it with $\mathbf{R}^n$, and applying the exponential map $T_q M \to M$. The supremum over δ such that $r \geq \delta$ can be chosen at each $q \in M$ is called the *injectivity radius* of M.

We start by formulating a condition for a chart to be normal, in terms of the metric

$$g_{ij} := \langle e_i, e_j \rangle,$$

where we write e_i for the coordinate frame vector fields $\underline{\mu}(e_i)$, by slight abuse of notation. In such a normal chart, we write ω_{ij} for the Christoffel symbols.

Proposition 12.2.2 (Normal chart equations). *Let $\mu : D \to M$ be a chart such that $\mu(0) = q$, with metric coordinates $g_{ij}(x)$. Then μ is a normal chart if and only if*

$$x_i = \sum_j g_{ij}(x) x_j, \quad x \in D, \ i = 1, \ldots, n.$$

Proof. That radial lines are geodesic is equivalent to

$$\nabla_{\underline{\mu}_x(x)}(\underline{\mu}_x(x/|x|)) = 0, \quad x \in D,$$

and that they are parametrized by arc length means that $\langle \underline{\mu}_x(x/|x|), \underline{\mu}_x(x/|x|) \rangle_{\mu(x)} = 1$. In coordinates, the second equation reads

$$\sum_{ij} g_{ij}(x) x_i x_j = \sum_i x_i^2, \tag{12.5}$$

whereas the first equation becomes $\partial_x(x/|x|) + \sum_{ij} e_i \langle \omega_{ij}, x \rangle x_j/|x| = 0$, or equivalently $\sum_j \langle \omega_{ij}, x \rangle x_j = 0$ for all i. Using Exercise 11.1.8, this reads

$$\sum_{ij} (2\partial_i g_{jk}(x) - \partial_k g_{ij}(x)) x_i x_j = 0. \tag{12.6}$$

Under (12.5), equation (12.6) is seen to be equivalent to $\sum_{ij} x_i \partial_i(g_{jk}(x)x_j - x_k) = 0$. Interpreting this as the radial derivatives of $g_{jk}(x)x_j - x_k$ vanishing, the result follows. $\square$

We next show that for normal coordinates, all second derivatives of the metric at q are given by the curvature coefficients.

Proposition 12.2.3 (Metric Taylor expansion). *Let $\mu : D \to M$ be a normal chart for M at q, with metric $g_{ij}(x)$. Then*

$$g_{ij}(x) = \delta_{ij} - \frac{1}{3}\sum_{kl} R_{ikjl}(q)x^k x^l + O(|x|^3), \quad i,j = 1,\ldots,n,$$

where $R_{ijkl}(q)$ are the Riemann curvature coefficients in the coordinate basis $\{e_i\}$, which is ON at q. In particular,

$$g_{ij} = \delta_{ij}, \qquad \partial_i g_{jk} = 0, \qquad \partial_k\langle\omega_{ij}, e_l\rangle = -\partial_j\partial_l g_{ik},$$
$$\partial_i\partial_j g_{kl} + \partial_i\partial_k g_{jl} + \partial_j\partial_k g_{il} = 0, \qquad \partial_i\partial_j g_{kl} = \partial_k\partial_l g_{ij},$$
$$R_{ijkl} = \partial_i\partial_l g_{jk} - \partial_j\partial_l g_{ik}, \qquad \partial_k\partial_l g_{ij} = -\tfrac{1}{3}(R_{ikjl} + R_{iljk}),$$

hold at q, for all i,j,k,l. Here the ω_{ij} denote the Christoffel symbols in the coordinate frame $\{e_i\}$.

Proof. We prove the identities at q, from which the stated Taylor expansion follows. Differentiating $x_i = g_{im}(x)x_m$ three times gives

$$\delta_{ij} = (\partial_j g_{im})x_m + g_{ij},$$
$$0 = (\partial_k\partial_j g_{im})x_m + \partial_j g_{ik} + \partial_k g_{ij},$$
$$0 = (\partial_l\partial_k\partial_j g_{im})x_m + \partial_k\partial_j g_{il} + \partial_l\partial_j g_{ik} + \partial_l\partial_k g_{ij}.$$

At q, the first equation gives $g_{ij} = \delta_{ij}$, using the second equation three times gives

$$\partial_i g_{jk} = -\partial_j g_{ik} = \partial_k g_{ij} = -\partial_i g_{kj},$$

thus $\partial_i g_{jk} = 0$, and the third equation gives $\partial_k\partial_j g_{il} + \partial_l\partial_j g_{ik} + \partial_l\partial_k g_{ij} = 0$. Using this last equation three times gives

$$\partial_i\partial_j g_{kl} = -\partial_i\partial_k g_{jl} - \partial_k\partial_j g_{il} = (\partial_k\partial_l g_{ji} + \partial_i\partial_l g_{jk}) + (\partial_j\partial_l g_{ik} + \partial_k\partial_l g_{ij})$$
$$= 2\partial_k\partial_l g_{ij} - \partial_i\partial_j g_{kl},$$

thus $\partial_i\partial_j g_{kl} = \partial_k\partial_l g_{ij}$. For the Christoffel symbols, we get from Exercise 11.1.8 that

$$2\langle\nabla_{e_i}e_j, e_k\rangle = 2\sum_\alpha g_{\alpha k}\langle\omega_{\alpha j}, e_i\rangle = \partial_i g_{jk} + \partial_j g_{ik} - \partial_k g_{ij}.$$

At q, this gives $\partial_k\langle\omega_{ij}, e_l\rangle = \tfrac{1}{2}\partial_k(\partial_l g_{ji} + \partial_j g_{li} - \partial_i g_{lj}) = \tfrac{1}{2}(-\partial_j\partial_l g_{ki} - \partial_k\partial_l g_{lj}) = -\partial_j\partial_l g_{ik}$. This gives curvature coefficients

$$R_{ijkl} = \langle d\omega_{ij}, e_k \wedge e_l\rangle = \partial_k\langle\omega_{ij}, e_l\rangle - \partial_l\langle\omega_{ij}, e_k\rangle = -\partial_j\partial_l g_{ik} + \partial_i\partial_l g_{jk},$$

using Exercise 7.4.9. Finally, we have $R_{ikjl} + R_{iljk} = (\partial_i \partial_l g_{kj} - \partial_k \partial_l g_{ij}) + (\partial_i \partial_k g_{lj} - \partial_l \partial_k g_{ij}) = -\partial_k \partial_l g_{ij} - 2\partial_k \partial_l g_{ij} = -3\partial_k \partial_l g_{ij}$. This completes the proof. $\qquad \square$

Besides the coordinate frame $\{e_i\}$, we require an ON-frame $\{\mathbf{e}_i\}$ in which to do multivector calculus. We shall use the following construction.

Definition 12.2.4 (Polar ON-frame). Let μ be a normal chart for M around q, with coordinate frame $\{e_i\}$. Let $G = (g_{ij})$ be the metric, with $g_{ij} = \langle e_i, e_j \rangle$, and consider its positive inverse square root $G^{-1/2} = (\alpha_{ij})_{ij}$. The *polar ON-frame* for μ is the ON-frame $\{\mathbf{e}_i\}$, where

$$\mathbf{e}_i := \sum_k \alpha_{ki} e_k = \underline{\mu}(G^{-1/2} e_i).$$

This is a pointwise construction of $\{\mathbf{e}_i\}$ from $\{e_i\}$ based on polar factorization as in Proposition 1.4.4. Note that $\{e_i\}$ is ON at q, whereas $\{\mathbf{e}_i\}$ is an ON-frame in all of the chart. Indeed,

$$\langle \mathbf{e}_i, \mathbf{e}_j \rangle = \langle \underline{\mu}(G^{-1/2} e_i), \underline{\mu}(G^{-1/2} e_j) \rangle = \langle G^{-1/2} e_i, G G^{-1/2} e_j \rangle_{\mathbf{R}^n} = \delta_{ij}.$$

Proposition 12.2.5. *Let $\mu : D \to M$ be a normal chart for M at q, with metric $g_{ij}(x)$. Denote the Christoffel symbols and Riemann curvature coefficients in the associated polar ON-frame $\{\mathbf{e}_i\}$ by Γ_{ij} and R_{ijkl} respectively. Then the identities*

$$\Gamma_{ij} = 0, \quad \partial_k \langle \Gamma_{ij}, \mathbf{e}_l \rangle = \tfrac{1}{2} R_{ijkl},$$

hold at q for all i, j, k, l.

Proof. It follows from Proposition 12.2.3 that $G = (g_{ij}) = I + O(|x|^2)$. Thus the change-of-basis matrix from $\{e_i\}$ to $\{\mathbf{e}_i\}$ is

$$A := G^{-1/2} = I - \tfrac{1}{2}(G - I) + O(|x|^4).$$

By Proposition 11.3.4,

$$\Gamma_{ij} = (A^{-1})_{ik}(\omega_{kl} A_{lj} + dA_{kj}) = \omega_{ij} - \tfrac{1}{2} dg_{ij} + O(|x|^2),$$

where the $\{\omega_{ij}\}$ denote the Christoffel symbols in the coordinate frame $\{e_i\}$. Using Proposition 12.2.3, we get at q that

$$\begin{aligned} \partial_k \langle \Gamma_{ij}, \mathbf{e}_l \rangle = \partial_k \langle \Gamma_{ij}, e_l \rangle &= -\partial_j \partial_l g_{ik} - \tfrac{1}{2} \partial_k \partial_l g_{ij} \\ &= -\partial_j \partial_l g_{ik} - \tfrac{1}{2}(-\partial_j \partial_l g_{ik} - \partial_j \partial_k g_{il}) \\ &= \tfrac{1}{2}(-\partial_j \partial_l g_{ik} + \partial_j \partial_k g_{il}) = \tfrac{1}{2} R_{ijkl}. \end{aligned}$$
$\qquad \square$

Note that at q, the curvature coefficients R_{ijkl} are the same in the frames $\{e_i\}$ and $\{\mathbf{e}_i\}$, since these coincide there.

12.3   The Chern–Gauss–Bonnet Theorem

In this section, we prove the following local index theorem for the $\triangle$-Dirac operator. Recall from Definition 12.1.5 the relation between the Euler characteristic $\chi(M)$, the Betti numbers $b^k(M)$, and the index of **D** restricted to even multivector fields.

Theorem 12.3.1 (Chern–Gauss–Bonnet). *Let M be an n-dimensional compact and closed Riemannian manifold. If $n = 2m$ is even, then*

$$\chi(M) = \left(\frac{1}{2\pi}\right)^m \int_M \langle \mathrm{Pf}(R), \underline{dp} \rangle,$$

where $\mathrm{Pf}(R)$ *denotes the Pfaffian of the Riemann curvature operator R. If n is odd, then* $\chi(M) = 0$.

We begin by explaining $\mathrm{Pf}(R)$. First we replace the antisymmetric matrix $R = (R_{ij})$ of bivectors by a scalar antisymmetric matrix $A = (A_{ij})$, where $A_{ij} \in \mathbf{R}$. Here

$$R_{ij} := R(\mathbf{e}_i \wedge \mathbf{e}_j) \in \wedge^2 M$$

in an ON-frame $\{\mathbf{e}_i\}$, or equivalently, by symmetry of R, the R_{ij} are the bivectors from Definition 11.3.3.

Definition 12.3.2 (Pfaffian). Let the dimension $n = 2m$ be even, let $A \in \underline{\mathrm{SO}}(\mathbf{R}^n)$ be an antisymmetric matrix, and let $b \in \wedge^2 \mathbf{R}^n$ be the bivector that represents A as in Proposition 4.2.3. Then the *Pfaffian* of A is

$$\mathrm{Pf}(A) := \langle b \wedge \cdots \wedge b, e_1 \wedge \cdots \wedge e_n \rangle / (n/2)!,$$

where the first exterior product for b is m-fold.

The Pfaffian behaves like a square root of the determinant for skew-symmetric matrices, as the following shows.

Proposition 12.3.3 (Pfaffian algebra). *Let the dimension $n = 2m$ be even. We have the following properties of the Pfaffian functional of a skew-symmetric matrix A.*

(i) *For A in standard form as in Proposition 4.3.6(ii), we have*

$$\mathrm{Pf}\left(\begin{bmatrix} 0 & x_1 & \cdots & 0 & 0 \\ -x_1 & 0 & \cdots & 0 & 0 \\ \vdots & \vdots & \ddots & \vdots & \vdots \\ 0 & 0 & \cdots & 0 & x_m \\ 0 & 0 & \cdots & -x_m & 0 \end{bmatrix}\right) = x_1 \cdots x_m.$$

(ii) *For general skew-symmetric $A = (A_{ij})$, we have the formula*

$$\mathrm{Pf}(A) = \frac{1}{2^m m!} \sum_{i_1,j_1,\ldots,i_m,j_m} \epsilon(i_1,j_1,\ldots,i_m,j_m) A_{i_1 j_1} \cdots A_{i_m j_m},$$

where $\epsilon(i_1,j_1,\ldots,i_m,j_m) = \langle e_{i_1} \wedge e_{j_1} \wedge \cdots \wedge e_{i_m} \wedge e_{j_m}, e_1 \wedge \cdots \wedge e_n \rangle$ is the sign of the permutation $(i_1,j_1,\ldots,i_m,j_m) \to (1,\ldots,n)$.

(iii) *If $A \in \underline{\mathrm{SO}}(\mathbf{R}^n)$ and $T \in \mathcal{L}(\mathbf{R}^n)$, then $\mathrm{Pf}(TAT^*) = \det(T)\mathrm{Pf}(A)$.*

Proof. (i) and (ii) follow from the facts that $b = x_1 e_{12} + \cdots + x_m e_{2m-1,2m}$ and $b = \frac{1}{2} \sum_{ij} A_{ij} e_{ij}$ in these cases respectively. To prove (iii), we use Proposition 2.7.1 to obtain

$$TAT^* v = T(b \llcorner (T^* v)) = (Tb) \llcorner v.$$

Thus Tb represents TAT^*, and

$$(Tb) \wedge \cdots \wedge (Tb) = T(b \wedge \cdots \wedge b) = \det(T)(n/2)!\mathrm{Pf}(A)e_1 \wedge \cdots \wedge e_n,$$

which proves (iii). $\qquad\square$

Definition 12.3.4 (Exterior Pfaffian). Let M be a Riemannian manifold of dimension $n = 2m$, with Riemann curvature operator R. Define the *Pfaffian*

$$\mathrm{Pf}(R) = \frac{1}{2^m m!} \sum_{i_1,j_1,\ldots,i_m,j_m} \epsilon(i_1,j_1,\ldots,i_m,j_m) R_{i_1 j_1} \wedge \cdots \wedge R_{i_m j_m} \in \wedge^n M.$$

The generalization from A to R amounts to replacing the real field $\mathbf{R}$ by the algebra $(\wedge^{\mathrm{ev}} \mathbf{R}^n, \wedge)$, which is commutative by Proposition 2.1.14. Proposition 12.3.3(iii) generalizes to show that $\mathrm{Pf}(R)$ is independent of the choice of positively oriented ON-frame.

Lemma 12.3.5. *The integrand $\langle \mathrm{Pf}(R), dp \rangle$ in Theorem 12.3.1 does not depend on the choice of ON-frame or orientation.*

Proof. Consider two ON-frames $\{e_i\}$ and $\{\tilde{e}_i\}$, related as $\tilde{e}_i = \sum_j e_j \alpha_{ji}$. By Proposition 11.3.4, the corresponding curvature coefficients are related as $\tilde{R}_{ij} = \sum_{k,l} \alpha_{ki} R_{kl} \alpha_{lj}$. This gives

$$\sum_{i_1,j_1,\ldots,i_m,j_m} \epsilon(i_1,j_1,\ldots,i_m,j_m) \tilde{R}_{i_1 j_1} \wedge \cdots \wedge \tilde{R}_{i_m j_m}$$

$$= \sum_{k_1,l_1,\ldots,k_m,l_m} \Big(\sum_{i_1,j_1,\ldots,i_m,j_m} \langle e_{i_1} \alpha_{k_1 i_1} \wedge e_{j_1} \alpha_{l_1 j_1} \wedge \cdots \wedge e_{i_m} \alpha_{k_m i_m} \wedge e_{j_k} \alpha_{l_m j_m},$$

$$e_1 \wedge \cdots \wedge e_n \rangle \Big) R_{k_1 l_1} \wedge \cdots \wedge R_{k_m l_m}$$

$$= \sum_{k_1,l_1,\ldots,k_m,l_m} \Big(\langle A^*(e_{k_1} \wedge e_{l_1} \wedge \cdots \wedge e_{k_m} \wedge e_{l_m}), e_1 \wedge \cdots \wedge e_n \rangle \Big) R_{k_1 l_1} \wedge \cdots \wedge R_{k_m l_m}$$

$$= \det(A) \sum_{k_1,l_1,\ldots,k_m,l_m} R_{k_1 l_1} \wedge \cdots \wedge R_{k_m l_m},$$

where $A = (\alpha_{ij})$. This, together with the observation that the oriented measure dp also changes sign if $\det(A) = -1$, proves the proposition.   □

Example 12.3.6 (Gauss–Bonnet). If $n = 2$, then $R = \begin{bmatrix} 0 & R_{12} \\ -R_{12} & 0 \end{bmatrix}$, so that $\langle \mathrm{Pf}(R), \mathbf{e}_{12} \rangle = R_{1212} = \frac{1}{2}S$. Thus the Gauss–Bonnet theorem stated in the introduction to this chapter is the two-dimensional case of the Chern–Gauss–Bonnet theorem (Theorem 12.3.1).

Exercise 12.3.7. Write down explicitly the Chern–Gauss–Bonnet integrand in dimension $n = 4$.

We now embark on the proof of the Chern–Gauss–Bonnet theorem, which covers the remainder of this section. When n is odd, consider the Hodge star map

$$* : L_2(M; \triangle^k M) \to L_2(M; \triangle^{n-k} M).$$

This gives an isomorphism between $\mathcal{H}^k(M)$ and $\mathcal{H}^{n-k}(M)$ by Exercise 11.2.7. When n is odd, this Poincaré duality implies that $\chi(M) = 0$.

We next consider the nontrivial case of even dimension $n = 2m$. Following the heat equation method described in the introduction, we calculate

$$\begin{aligned}
\chi(M) &= \dim \mathsf{N}(\mathbf{D}|_{L_2(M; \triangle^{\mathrm{ev}} M)}) - \dim \mathsf{N}(\mathbf{D}|_{L_2(M; \triangle^{\mathrm{od}} M)}) \\
&= \dim \mathsf{N}(\mathbf{D}^2|_{L_2(M; \triangle^{\mathrm{ev}} M)}) - \dim \mathsf{N}(\mathbf{D}^2|_{L_2(M; \triangle^{\mathrm{od}} M)}) \\
&= \mathrm{Tr}(e^{t\mathbf{D}^2}|_{L_2(M; \triangle^{\mathrm{ev}} M)}) - \mathrm{Tr}(e^{t\mathbf{D}^2}|_{L_2(M; \triangle^{\mathrm{od}} M)}),
\end{aligned}$$

for all $t > 0$. The second identity is valid because $\mathbf{D}$ is a normal operator, and the last identity follows from the general fact that the nonzero eigenvalues of operators A^*A and AA^* are the same, in particular for $A = \mathbf{D} : L_2(M; \triangle^{\mathrm{ev}} M) \to L_2(M; \triangle^{\mathrm{od}} M)$. The idea of the proof is to compute this trace difference in the limit as $t \to 0^+$. According to Proposition 11.5.9, the square $\mathbf{D}^2$ of the Dirac operator differs from the Laplace–Beltrami operator only by zero-order terms. Thus it is reasonable to expect the operator $e^{t\mathbf{D}^2}$ to be an integral operator resembling the solution operator

$$e^{t\triangle} f(x) = \int_{\mathbf{R}^n} (4\pi t)^{-n/2} e^{-|x-y|^2/(4t)} f(y) \, dy$$

for the heat equation on $\mathbf{R}^n$ from Example 6.3.3. With this in mind, we make an ansatz of the form

$$(H_t f)(p) := \int_M (4\pi t)^{-m} e^{-d(p,q)^2/(4t)} \sum_{k=0}^{N} t^k H^k(p, q) f(q) \, dq, \qquad (12.7)$$

where $N < \infty$ is to be chosen. Here $d(p, q)$ denotes the shortest distance between points p and q on M, and below, we shall choose linear maps $H^k(p, q) \in$

$\mathcal{L}(\triangle(T_qM); \triangle(T_pM))$ depending smoothly on $p, q \in M$. We want to choose H^k such that $H_t f$ approximates $e^{t\mathbf{D}^2} f$ well for small t, and in particular,

$$H^0(q, q) = I,$$

for all $q \in M$. This will ensure that $\lim_{t\to 0+} H_t f = f = \lim_{t\to 0+} e^{t\mathbf{D}^2} f$ for all f. Secondly, since $\partial_t e^{t\mathbf{D}^2} f = \mathbf{D}^2 e^{t\mathbf{D}^2} f$, we want $(\partial_t - \mathbf{D}^2) H_t f$ to be as small as possible.

Lemma 12.3.8. *Let $\mu : D \to M$ be a normal chart for M around $q = \mu(0)$. Push forward the radial vector field x in D to the vector field $r_q(p) := \underset{-x}{\mu_*}(x)$, $p = \mu(x)$, and let $d_q(p) := |x|$ denote the shortest distance from p to q on M. Then*

$$\nabla(d_q^2) = 2r_q \quad and \quad \Delta(d_q^2) = 2n + \partial_{r_q} \ln g,$$

where $g = \det(g_{ij})$ and Δ is the scalar Laplace–Beltrami operator on M, and

$$(\partial_t - \mathbf{D}^2)(t^{k-m} e^{-d_q^2/(4t)} h) = t^{k-m} e^{-d_q^2/(4t)} \left(t^{-1}\left(\nabla_{r_q} + \tfrac{1}{4}\partial_{r_q}(\ln g) + k\right) h - \mathbf{D}^2 h \right),$$

for all $h \in C^2(M; \triangle M)$.

Proof. Define the frame $e_i^* = \sum_j g^{ij} e_j$ dual to the coodinate frame $\{e_i\}$, where (g^{ij}) denotes the inverse of the metric (g_{ij}). Proposition 12.2.2 shows that

$$\nabla(d_q^2) = \sum_i e_i^* \partial_i |x|^2 = \sum_{ij} g^{ij} e_j 2x_i = 2 \sum_j x_j e_j = 2r_q.$$

To compute $\Delta(d_q^2)$, we use Exercise 11.5.8 to get

$$\Delta(d_q^2) = \tfrac{1}{\sqrt{g}} \sum_{ij} \partial_i(\sqrt{g} g^{ij} 2x_j) = \tfrac{2}{\sqrt{g}} \sum_i \partial_i(\sqrt{g} x_i) = 2n + \partial_{r_q} \ln g.$$

For the last formula, clearly

$$\partial_t(t^{k-m} e^{-d_q^2/(4t)}) = (d_q^2/(4t^2) + (k - m)/t) t^{k-m} e^{-d_q^2/(4t)}.$$

By Proposition 11.5.9, it suffices to prove the identity with $\mathbf{D}^2$ replaced by the Laplace–Beltrami operator on $\triangle M$. We compute

$$\Delta(e^{-d_q^2/(4t)} h) = (\Delta e^{-d_q^2/(4t)}) h + 2 \sum_i (\partial_{\mathbf{e}_i} e^{-d_q^2/(4t)}) \nabla_{\mathbf{e}_i} h + e^{-d_q^2/(4t)} \Delta h,$$

in an ON-frame $\{\mathbf{e}_i\}$. Here

$$\Delta e^{-d_q^2/(4t)} = \tfrac{1}{\sqrt{g}} \sum_{ij} \partial_i(\sqrt{g} g^{ij} \partial_j e^{-d_q^2/(4t)}) = \left(\frac{d_q^2}{4t^2} - \frac{2n + \partial_{r_q} \ln g}{4t} \right) e^{-d_q^2/(4t)}$$

and

$$2\sum_i \mathbf{e}_i(\partial_{\mathbf{e}_i} e^{-d_q^2/(4t)}) = -\frac{1}{2t}\sum_i \mathbf{e}_i(\partial_{\mathbf{e}_i} d_q^2)e^{-d_q^2/(4t)} = -t^{-1}e^{-d_q^2/(4t)}r_q.$$

Combining these calculations proves the stated formula.   □

Applying $\partial_t - \mathbf{D}^2$ to (12.7), we obtain

$$(\partial_t - \mathbf{D}^2)H_t f$$

$$= \frac{1}{(4\pi)^m}\int_M e^{-d_q^2/(4t)}\Big(\sum_{k=0}^{N} t^{k-m-1}\big(\nabla_{r_q} + \tfrac{1}{4}\partial_{r_q}(\ln g) + k\big)H^k(\dot{p},q)f(q)$$

$$-\sum_{k=1}^{N+1} t^{k-m-1}\mathbf{D}^2 H^{k-1}(\dot{p},q)f(q)\Big)dq.$$

This leads us to the following recursive definition of $H^k(p,q)$. For $p,q \in M$ such that $d(p,q) < \delta$, where δ is the injectivity radius of M, define $\widetilde{H}^k(p,q)$ such that $\widetilde{H}^0(q,q) = I$ and $\widetilde{H}^k(p,q)$ solves

$$\big(\nabla_{r_q} + \tfrac{1}{4}\partial_{r_q}(\ln g) + k\big)\widetilde{H}^k(\dot{p},q)f(q) = \mathbf{D}^2\widetilde{H}^{k-1}(\dot{p},q)f(q) \qquad (12.8)$$

for $k = 0,\ldots,N$ and $f(q) \in (\triangle M)_q$. Here $\widetilde{H}^{-1}(p,q) = 0$. Note that for each k, (12.8) is an ordinary differential equation along the geodesic from q to p, and that for $k \geq 1$, the initial value $\widetilde{H}^k(q,q)f(q) = k^{-1}\mathbf{D}^2\widetilde{H}^{k-1}(q,q)f(q)$ is specified, since $r_q(q) = 0$. Existence theory for ordinary differential equations shows that this uniquely determines maps $H^k(p,q) \in \mathcal{L}(\triangle(T_qM);\triangle(T_pM))$ depending smoothly on $p,q \in M$, at distance $< \delta$.

To extend this construction to general $p,q \in M$, we make a smooth cutoff as follows. Let $\eta \in C^\infty(\mathbf{R})$ be such that $\eta(x) = 1$ for $x < \delta/3$ and $\eta(x) = 0$ for $x > 2\delta/3$. Define

$$H^k(p,q) := \eta(d(p,q))\widetilde{H}^k(p,q),$$

where we understand that $H^k(p,q) = 0$ if $d(p,q) \geq 2\delta/3$.

We have constructed H_t and next compare this ansatz to $e^{t\mathbf{D}^2}$. Let

$$K_t f(p) = \int_M K_t(p,q)f(q)dq := (\partial_t - \mathbf{D}^2)H_t f(p).$$

Here the kernel $K_t(p,q)$ is a smooth function of $p,q \in M$ and $t > 0$, and by construction we have

$$K_t(p,q) = -(4\pi)^{-m}t^{N-m}e^{-d(p,q)^2/(4t)}\mathbf{D}^2 H^N(\dot{p},q)$$

when $d(p,q) < \delta/3$. Consider the difference $H_t f - e^{t\mathbf{D}^2}f$, which satisfies

$$(\partial_t - \mathbf{D}^2)(H_t f - e^{t\mathbf{D}^2}f) = K_t f$$

and initial conditions $\lim_{t \to 0+}(H_t f - e^{t\mathbf{D}^2} f) = 0$. Integration gives

$$H_t f - e^{t\mathbf{D}^2} f = \int_0^t e^{(t-s)\mathbf{D}^2} K_s f\, ds, \tag{12.9}$$

from which we deduce the following.

Proposition 12.3.9 (Trace formula). *We have the formula*

$$\chi(M) = \frac{1}{(4\pi)^m} \int_M \Big(\mathrm{Tr}_{\triangle^{\mathrm{ev}} M}(H^m(q,q)) - \mathrm{Tr}_{\triangle^{\mathrm{od}} M}(H^m(q,q)) \Big) dq$$

for the index of $\mathbf{D} : H^1(M; \triangle^{\mathrm{ev}} M) \mapsto L_2(M; \triangle^{\mathrm{od}} M)$.

Proof. We estimate the trace norm of (12.9). See Example 6.4.8. To estimate the trace norm $\|K_s\|_{\mathcal{L}_1(L_2(M))}$, we factorize into Hilbert–Schmidt operators

$$K_s = (I + \mathbf{D})^{-j} \Big((I + \mathbf{D})^j K_s \Big).$$

We use that the eigenvalues of $\mathbf{D}$ grow in size as $|\lambda_k| = k^{1/n}$, and as a consequence $(I + \mathbf{D})^{-j}$ is a Hilbert–Schmidt operator if $j > m = n/2$. For the second operator and $d(p,q) < \delta/3$, we have $|(I+\mathbf{D})^j K_s(\dot{p}, q)| \lesssim s^{N-m-j}$, whereas for $d(p,q) \geq \delta/3$, we have $|(I + \mathbf{D})^j K_s(\dot{p}, q)| \lesssim e^{-\delta^2/(36s)} s^{-m-j-2}$. This shows that the Hilbert–Schmidt norm of $(I + \mathbf{D})^j K_s$ is bounded by s^{N-m-j} for $0 < s < 1$. Choosing $N > n$ therefore shows that $\|K_s\|_{\mathcal{L}_1(L_2(M))} \lesssim 1$, and in particular,

$$\|H_t - e^{t\mathbf{D}^2}\|_{\mathcal{L}_1(L_2(M))} \leq \int_0^t \|e^{(t-s)\mathbf{D}^2}\|_{\mathcal{L}(L_2(M))} \|K_s\|_{\mathcal{L}_1(L_2(M))} ds \lesssim t \to 0,$$

as $t \to 0^+$, since $\mathbf{D}$ is a skew-adjoint operator, and trace-class operators form an ideal in $\mathcal{L}(L_2)$. Since the trace functional is continuous in the trace norm, we get

$$\chi(M) = \lim_{t \to 0+} \Big(\mathrm{Tr}(e^{t\mathbf{D}^2}|_{L_2(M;\triangle^{\mathrm{ev}} M)}) - \mathrm{Tr}(e^{t\mathbf{D}^2}|_{L_2(M;\triangle^{\mathrm{od}} M)}) \Big)$$

$$= \lim_{t \to 0+} \Big(\mathrm{Tr}(H_t|_{L_2(M;\triangle^{\mathrm{ev}} M)}) - \mathrm{Tr}(H_t|_{L_2(M;\triangle^{\mathrm{od}} M)}) \Big)$$

$$= \frac{1}{(4\pi)^m} \lim_{t \to 0+} \sum_{k=0}^N t^{k-m} \int_M \Big(\mathrm{Tr}_{\triangle^{\mathrm{ev}} M}(H^k(q,q)) - \mathrm{Tr}_{\triangle^{\mathrm{od}} M}(H^k(q,q)) \Big) dq.$$

Since we know that the limit exists, all the terms $0 \leq k < m$ must be zero, and we have proved the stated formula for $\chi(M)$. $\qquad\square$

It remains to compute

$$\mathrm{Tr}_{\triangle^{\mathrm{ev}} M}(H^m(q,q)) - \mathrm{Tr}_{\triangle^{\mathrm{od}} M}(H^m(q,q)).$$

To this end, we fix a normal chart $\mu : x \mapsto \mu(x)$ around q and let $\{e_i\}$ be the associated polar ON-frame from Definition 12.2.4. To handle linear operators in $\mathcal{L}(\wedge M)$, we use the frame

$$\{\mathbf{e}_{s_1}^+ \cdots \mathbf{e}_{s_k}^+ \mathbf{e}_{t_1}^- \cdots \mathbf{e}_{t_l}^-\}_{s_1 < \cdots < s_k,\ t_1 < \cdots t_l,\ 1 \le k,l \le n} \qquad (12.10)$$

generated by the operators of positive and negative left Clifford multiplication by basis vectors, as in Definition 3.4.3. The most useful to us of these basis elements is the volume element

$$T_{2n} := \mathbf{e}_1^+ \mathbf{e}_1^- \mathbf{e}_2^+ \mathbf{e}_2^- \cdots \mathbf{e}_n^+ \mathbf{e}_n^-,$$

which by Example 3.4.8 is the involution operator $T_{2n}(w) = \widehat{w}$.

Proof of Theorem 12.3.1. Consider the recurrence formula

$$\left(\nabla_{r_q} + \tfrac{1}{4}\partial_{r_q}(\ln g) + k\right) H^k(\dot p, q) f(q) = \mathbf{D}^2 H^{k-1}(\dot p, q) f(q).$$

In the polar ON-frame $\{e_i\}$ for TM, with induced frame (12.10) for $\mathcal{L}(\wedge M)$, we have

$$\mathbf{D}^2 h = \sum_i \left(\partial_{e_i} + \frac{1}{4}\sum_{kl}\langle\Gamma_{kl}, e_i\rangle(\mathbf{e}_k^+ \mathbf{e}_l^+ - \mathbf{e}_k^- \mathbf{e}_l^-)\right)$$
$$\cdot \left(\partial_{e_i} h + \frac{1}{4}\sum_{k'l'}\langle\Gamma_{k'l'}, e_i\rangle(\mathbf{e}_{k'}^+ \mathbf{e}_{l'}^+ - \mathbf{e}_{k'}^- \mathbf{e}_{l'}^-)h\right)$$
$$- \sum_i \left(\partial_{\nabla_{e_i} e_i} h + \frac{1}{4}\sum_{kl}\langle\Gamma_{kl}, \nabla_{e_i} e_i\rangle(\mathbf{e}_k^+ \mathbf{e}_l^+ - \mathbf{e}_k^- \mathbf{e}_l^-)h\right)$$
$$- \frac{1}{4}Sh - \frac{1}{8}\sum_{ijkl} R_{ijkl}\mathbf{e}_i^+ \mathbf{e}_j^+ \mathbf{e}_k^- \mathbf{e}_l^- h,$$

for $h = \sum h_s e_s \in C^\infty(M; \wedge M)$, where the ∂_v denote componentwise differentiation $\partial_v g = \sum_s(\partial_v g_s)e_s$. This follows from Proposition 11.5.9, Definition 11.5.6, and Proposition 11.5.3. Note that

$$\mathbf{e}_i \mathbf{e}_j h \mathbf{e}_k \mathbf{e}_l = \mathbf{e}_i^+ \mathbf{e}_j^+ \mathbf{e}_k^- \mathbf{e}_l^- h$$

and

$$\frac{1}{2}[\Gamma_{\mathbf{e}}(v), h] = \frac{1}{4}\sum_{ij}\langle\Gamma_{kl}, v\rangle[\mathbf{e}_k \mathbf{e}_l, h] = \frac{1}{4}\sum_{ij}\langle\Gamma_{kl}, v\rangle(\mathbf{e}_k^+ \mathbf{e}_l^+ - \mathbf{e}_k^- \mathbf{e}_l^-)h,$$

since $wa = a^-(\widehat{w})$ for vectors a according to Example 3.4.7.

Note that

$$\text{Tr}_{\wedge^{\mathrm{ev}} M} H^m(q, q) - \text{Tr}_{\wedge^{\mathrm{od}} M} H^m(q, q) = \text{Tr}(T_{2n} H^m(q, q)),$$

which by Exercise 12.10 equals 2^n times the T_{2n} coordinate of $H^m(q, q)$ in the induced frame (12.10). A key simplifying observation is that in the recurrence where we calculate $H^m(q, q)$ from $H^0(q, q) = 1$, in each of the $m = n/2$ steps we need to multiply by four of the basis operators $\mathbf{e}_s^\pm$ in order to obtain a nonzero T_{2n} coordinate for $H^m(q, q)$. Thus it suffices to use the approximation

$$\mathbf{D}^2 h \approx -\frac{1}{8} \sum_{ijkl} R_{ijkl} \mathbf{e}_i^+ \mathbf{e}_j^+ \mathbf{e}_k^- \mathbf{e}_l^- h,$$

since $\Gamma_{kl}(q) = 0$ by Proposition 12.2.5. We conclude that

$$\mathrm{Tr}_{\triangle^{\mathrm{ev}} M} H^m(q, q) - \mathrm{Tr}_{\triangle^{\mathrm{od}} M} H^m(q, q) = 2^n \Big(T_{2n} \Big(-\frac{1}{8} \sum_{ijkl} R_{ijkl} \mathbf{e}_i^+ \mathbf{e}_j^+ \mathbf{e}_k^- \mathbf{e}_l^- \Big)^m \Big)_\emptyset$$

$$= \frac{(-1)^m}{m! 2^m} \sum_{i_1 j_1 k_1 l_1 \cdots i_m j_m k_m l_m} R_{i_1 j_1 k_1 l_1} \cdots R_{i_m j_m k_m l_m} \cdot$$

$$\cdot \big((\mathbf{e}_1^+ \mathbf{e}_1^- \cdots \mathbf{e}_n^+ \mathbf{e}_n^-)(\mathbf{e}_{i_1}^+ \mathbf{e}_{j_1}^+ \mathbf{e}_{k_1}^- \mathbf{e}_{l_1}^- \cdots \mathbf{e}_{i_m}^+ \mathbf{e}_{j_m}^+ \mathbf{e}_{k_m}^- \mathbf{e}_{l_m}^-) \big)_\emptyset$$

$$= \frac{(-1)^m}{m! 2^m} \sum_{i_1 j_1 k_1 l_1 \cdots i_m j_m k_m l_m} R_{i_1 j_1 k_1 l_1} \cdots R_{i_m j_m k_m l_m} \cdot$$

$$\cdot (-1)^m \big((\mathbf{e}_1^+ \cdots \mathbf{e}_n^+)(\mathbf{e}_1^- \cdots \mathbf{e}_n^-)(\mathbf{e}_{i_1}^+ \mathbf{e}_{j_1}^+ \cdots \mathbf{e}_{i_m}^+ \mathbf{e}_{j_m}^+)(\mathbf{e}_{k_1}^- \mathbf{e}_{l_1}^- \cdots \mathbf{e}_{k_m}^- \mathbf{e}_{l_m}^-) \big)_\emptyset$$

$$= \frac{1}{m! 2^m} \sum_{i_1 j_1 k_1 l_1 \cdots i_m j_m k_m l_m} R_{i_1 j_1 k_1 l_1} \cdots R_{i_m j_m k_m l_m} \cdot$$

$$\cdot \epsilon(i_1, j_1, \ldots, i_m, j_m) \epsilon(k_1, l_1, \ldots, k_m, l_m)$$

$$= \frac{1}{m! 2^m} \sum_{i_1 j_1 \cdots i_m j_m} \langle 2^m R_{i_1 j_1} \wedge \cdots \wedge R_{i_m j_m}, \mathbf{e}_1 \wedge \cdots \wedge \mathbf{e}_n \rangle$$

$$= 2^m \langle \mathrm{Pf}(R), \mathbf{e}_1 \wedge \cdots \wedge \mathbf{e}_n \rangle.$$

With Proposition 12.3.9, this shows that

$$\chi(M) = \frac{1}{(2\pi)^m} \int_M \langle \mathrm{Pf}(R), \mathbf{e}_1 \wedge \cdots \wedge \mathbf{e}_n \rangle dq,$$

which proves the Chern–Gauss–Bonnet theorem. $\qquad\square$

12.4 The Atiyah–Singer Index Theorem

In this section, we prove the following local index theorem for the $\triangle$-Dirac operator.

Theorem 12.4.1 (Atiyah–Singer). *Let $\triangle M$ be a normed spinor bundle of complex dimension 2^m over a compact and closed oriented Riemannian manifold M of even real dimension $n = 2m$, and consider the splitting into chiral subspaces*

$$\triangle M = \triangle^+ M \oplus \triangle^- M,$$

where $\triangle^{\pm}M$ are the ± 1 eigenspaces of the main reflector $w_n = i^{-m}e_{\overline{n}}$. Then the index of $\not{D} : L_2(M; \triangle^+ M) \to L_2(M; \triangle^- M)$ equals

$$\int_M \langle \hat{A}(R), \underline{dp} \rangle,$$

where $\hat{A}(R)$ denotes the A-roof functional of the Riemann curvature operator R. This index is zero when m is odd.

Note that the existence of the subbundles $\triangle^{\pm} M$ locally requires even dimension n, and globally also that the manifold be oriented. We start by defining the A-roof functional $\hat{A}(R)$.

Definition 12.4.2 (A-roof functional). Let $m = n/2$ be an even positive integer and let $p_0(a_1, \ldots, a_m)$ denote the m-homogeneous part in the Taylor expansion of the function

$$f(a_1, \ldots, a_m) = \frac{a_1/2}{\sinh(a_1/2)} \cdots \frac{a_m/2}{\sinh(a_m/2)}.$$

Express this symmetric polynomial p_0 in terms of the elementary symmetric polynomials using

$$p(t_1, \ldots, t_{m/2}) = \sum_{k_1 + 2k_2 + 3k_3 + \cdots + k_{m/2} = m/2} \alpha_{k_1, k_2, \ldots, k_{m/2}} t_1^{k_1} \cdots t_{m/2}^{k_{m/2}},$$

so that p is the unique polynomial such that

$$p(-2(a_1^2 + \cdots + a_m^2), \ldots, 2(-1)^k(a_1^{2k} + \cdots + a_m^{2k}), \ldots) = p_0(a_1, \ldots, a_m).$$

The A-*roof functional* of the Riemann curvature operator R is

$$\hat{A}(R) := \left(\tfrac{1}{2\pi}\right)^m p(\mathrm{Tr}R^2, \mathrm{Tr}R^4, \mathrm{Tr}R^6, \ldots) \in \wedge^n M,$$

where $\mathrm{Tr}R^k := \sum_{i_1 \cdots i_k} R_{i_1 i_2} \wedge R_{i_2 i_3} \wedge \cdots \wedge R_{i_k i_1}$ and p is computed in the commutative algebra $(\wedge^{\mathrm{ev}} M, \wedge)$.

We note that for low-dimensional manifolds, $\hat{A}(R)$ may be simpler to compute recursively as in Example 12.4.7 below.

Lemma 12.4.3. *The quantity $\hat{A}(R)$ is well defined and independent of the ON-frame.*

Proof. The existence and uniqueness of p follow from the theory of symmetric polynomials. It remains to show that $\mathrm{Tr}R^k$ does not depend on the choice of ON-frame. Consider two ON-frames $\{e_i\}$ and $\{\tilde{e}_i\}$, related by $\tilde{e}_i = \sum_j e_j \alpha_{ji}$. By

Proposition 11.3.4, the corresponding curvature bivectors are related by $\tilde{R}_{ij} = \sum_{k,l} \alpha_{ki} R_{kl} \alpha_{lj}$. We get

$$
\sum_{i_1 \cdots i_k} \tilde{R}_{i_1 i_2} \wedge \tilde{R}_{i_2 i_3} \wedge \cdots \wedge \tilde{R}_{i_k i_1}
$$

$$
= \sum_{i_l, j_l, i'_l, j'_l} \alpha_{i'_1 i_1} \alpha_{j'_1 i_2} \alpha_{i'_2 i_2} \alpha_{j'_2 i_3} \cdots \alpha_{i'_k i_k} \alpha_{j'_k i_1} R_{i'_1 j'_1} \wedge R_{i'_2 j'_2} \wedge \cdots \wedge R_{i'_k j'_k}
$$

$$
= \sum_{i'_1 \cdots i'_k} R_{i'_1 i'_2} \wedge R_{i'_2 i'_3} \wedge \cdots \wedge R_{i'_k i'_1},
$$

and in particular, $\hat{A}(R)$ does not depend on the ON-frame. $\qquad\square$

Exercise 12.4.4. Compute the polynomial p in Definition 12.4.2 for manifolds of dimension $n = 4$ and $n = 8$. Write out the index theorem explicitly in these cases and check your calculation with Example 12.4.7.

We now begin the proof of Theorem 12.4.1, which proceeds like the proof of the Chern–Gauss–Bonnet theorem up until Proposition 12.3.9, replacing $\mathbf{D}$ by $\slashed{D}$, $\wedge^{\mathrm{ev}} M$ by $\slashed{\triangle}^+ M$, and $\wedge^{\mathrm{od}} M$ by $\slashed{\triangle}^- M$. We arrive at the index formula

$$
i(\slashed{D} : L_2(M; \slashed{\triangle}^+ M) \to L_2(M; \slashed{\triangle}^- M))
$$
$$
= \frac{1}{(4\pi)^m} \int_M \left(\mathrm{Tr}_{\slashed{\triangle}^+ M}(H^m(q,q)) - \mathrm{Tr}_{\slashed{\triangle}^- M}(H^m(q,q)) \right) dq,
$$

where the $H^k(p,q)$ recursively solve the differential equations

$$
\left(\nabla_{r_q} + \tfrac{1}{4} \partial_{r_q}(\ln g) + k \right) H^k(\dot{p}, q) f(q) = \slashed{D}^2 H^{k-1}(\dot{p}, q) f(q) \tag{12.11}
$$

for $k = 0, 1, \ldots, m$ and p near q, starting from $H^0(q,q) = I$.

To obtain the needed information about $H^m(q,q)$, we fix a normal chart μ around q, and let $\{\mathbf{e}_i\}$ be the associated polar ON-frame. Here and below, we use for i the index set $\{-m, \ldots, -1, 1, \ldots, m\}$ as in Definition 11.6.6, and fix one of the two induced frames $\{\slashed{e}_s\}$ for $\slashed{\triangle} M$, where $s \subset \{1, \ldots, m\}$. We assume that

$$
\mathbf{e}_{\overline{n}} = \mathbf{e}_1 \mathbf{e}_{-1} \cdots \mathbf{e}_m \mathbf{e}_{-m}
$$

is the orientation of M.

Exercise 12.4.5. Modify Exercise 3.4.6 and show that if $T \in \triangle^{\mathrm{ev}} M$, then

$$
\mathrm{Tr}_{\slashed{\triangle}^+ M} T - \mathrm{Tr}_{\slashed{\triangle}^- M} T = (2i)^m T_n,
$$

where T_n is the $\mathbf{e}_{\overline{n}}$ coordinate of T.

From Proposition 11.6.10, Definition 11.5.6, and Proposition 11.6.8, we now get

$$\not{D}^2 h = \sum_i \left(\partial_{\mathbf{e}_i} + \frac{1}{4} \sum_{kl} \langle \Gamma_{kl}, \mathbf{e}_i \rangle \mathbf{e}_k \mathbf{e}_l \right) \left(\partial_{\mathbf{e}_i} h + \frac{1}{4} \sum_{k'l'} \langle \Gamma_{k'l'}, \mathbf{e}_i \rangle \mathbf{e}_{k'} \mathbf{e}_{l'}.h \right)$$
$$- \sum_i \left(\partial_{\nabla_{\mathbf{e}_i} \mathbf{e}_i} h + \frac{1}{4} \sum_{kl} \langle \Gamma_{kl}, \nabla_{\mathbf{e}_i} \mathbf{e}_i \rangle \mathbf{e}_k \mathbf{e}_l.h \right) - \frac{1}{4} Sh,$$

for $h = \sum h_s \not{\phi}_s \in C^\infty(M; \not{\triangle}M)$. As before, ∂_v denotes the componentwise directional derivative in the fixed frame. It is at this stage that the proof of Theorem 12.4.1 becomes more difficult than that of the Chern–Gauss–Bonnet theorem. The reason is that we only need to multiply by n vectors in total in the $n/2$ recursion steps to get a nonvanishing trace difference in Exercise 12.4.5, and so we cannot as before neglect derivative terms in $\not{D}^2$.

To proceed, we Taylor expand each $H^k(p, q)$ around q. More precisely, our setup is the following. Rather than having $H^k(p, q) \in \mathcal{L}((\not{\triangle}M)_q; (\not{\triangle}M)_p)$, since we have a fixed spinor frame $\{\not{\phi}_s\}$, we write $H^k(p, q) \in \triangle \mathbf{R}^n$, which is isomorphic to $\mathcal{L}(\not{\triangle}\mathbf{R}^n)$, since n is even. Moreover, since we have fixed a chart μ, upon identifying x and $p = \mu(x)$ we have multivector fields $H^k(\cdot, q) : D \to \triangle \mathbf{R}^n$ in $D \subset \mathbf{R}^n$. Write

$$H^k(p, q) = \sum_{\beta \geq 0} H^k_\beta(p, q),$$

where $H^k_\beta(p, q)$ is a β-homogeneous $\triangle \mathbf{R}^n$-valued polynomial in the coordinates x of $p = \mu(x)$.

Lemma 12.4.6 (Harmonic oscillator approximation). *Define subspaces*

$$W_i := \triangle^0 \mathbf{R}^n \oplus \triangle^1 \mathbf{R}^n \oplus \cdots \oplus \triangle^i \mathbf{R}^n.$$

We have $H^k_\beta(p, q) \in W_{2k+\beta}$ and the recurrence formula

$$(k + \beta) H^k_\beta = D_2 H^{k-1}_{\beta+2} + D_1 H^{k-1}_\beta + D_0 H^{k-1}_{\beta-2} \quad \mod W_{2k+\beta-1}, \qquad (12.12)$$

with $H^0_0 = 1$, $H^{-1}_\beta = H^k_{-1} = H^k_{-2} = 0$, where

$$D_2 := \Delta = \sum_i \partial_i^2,$$

$$D_1 := \frac{1}{2} \sum_{ij} x_i R^0_{ij} \partial_j,$$

$$D_0 := -\frac{1}{16} \sum_{ijk} x_i R^0_{ij} R^0_{jk} x_k,$$

and $R^0_{ij} := \frac{1}{2} \sum_{kl} R_{ijkl}(q) e_{kl} \in \triangle^2 \mathbf{R}^n$.

Furthermore, omitting the term D_1 in (12.12) yields the same $H^k_\beta(p, q) \in W_{2k+\beta} \mod W_{2k+\beta-1}$.

These calculations use the Clifford product, but it is only the top exterior product, as in Proposition 3.1.9, which will not produce a $W_{2k+\beta-1}$ error term.

Proof. (i) From Proposition 12.2.5, we get the Taylor approximation

$$\frac{1}{4}\sum_{kl}\langle\Gamma_{kl},\mathbf{e}_i\rangle\mathbf{e}_{kl} = \frac{1}{8}\sum_{jkl}(x_j R_{jikl}(q)+O(x^2;\triangle^0))\mathbf{e}_{kl} = -\frac{1}{4}\sum_j R_{ij}^0 x_j + O(x^2;\triangle^2),$$

where we use $O(x^j;\triangle^k)$ to denote $\triangle^k\mathbf{R}^n$-valued error terms of order x^j. The recurrence formula (12.11) thus reads

$$\left(\sum_i x_i\partial_i - \frac{1}{4}\sum_{ij}x_i R_{ij}^0 x_j + O(x^3;\triangle^2)+O(x^2)+k\right)H^k$$

$$=\sum_i\left(\partial_i - \frac{1}{4}\sum_j R_{ij}^0 x_j + O(x^2;\triangle^2)\right)\left(\partial_i - \frac{1}{4}\sum_{j'}R_{ij'}^0 x_{j'}+O(x^2;\triangle^2)\right)H^{k-1}$$

$$-\sum_{ii'}\langle\nabla_{\mathbf{e}_i}\mathbf{e}_i,\mathbf{e}_{i'}\rangle\left(\partial_{i'}-\frac{1}{4}\sum_j R_{i'j}^0 x_j + O(x^2;\triangle^2)\right)H^{k-1}-\frac{1}{4}SH^{k-1},$$

where $\mathbf{e}_i = e_i + O(x^2;\triangle^1)$. We note on the left-hand side that $\sum_{ij}x_i x_j R_{ij}^0 = 0$, and on the right-hand side that $\partial_i(R_{ij}^0 x_j H^{k-1}) = R_{ij}^0 x_j \partial_i H^{k-1}$, since $R_{ii}^0 = 0$.

Now fix k,β and assume $H_{\beta'}^{k-1}\in W_{2k+\beta'-2}$ for all β' and $H_{\beta'}^k\in W_{2k+\beta'}$ for $\beta'<\beta$. Computing the β-homogeneous part, modulo $W_{2k+\beta-1}$, of the above recurrence formula gives (12.12).

(ii) To show that D_1 may be omitted, we compute the commutators

$$[D_1,D_2] = \frac{1}{2}\sum_{ijk}[x_i,\partial_k^2]R_{ij}^0\partial_j = -\sum_{ij}\partial_i R_{ij}^0\partial_j = 0,$$

since $R_{ij}^0 = -R_{ji}^0$, and

$$-32[D_1,D_0] = \sum_{i'j'ijk}x_{i'}[R_{i'j'}^0\partial_{j'},x_i R_{ij}^0 R_{jk}^0 x_k]$$

$$=\sum_{i'j'ijk}x_{i'}\left(R_{i'j'}^0[\partial_{j'},x_i x_k]R_{ij}^0 R_{jk}^0 + [R_{i'j'}^0,R_{ij}^0 R_{jk}^0]x_i x_k\partial_{j'}\right)$$

$$=\sum_{i'j'ijk}x_{i'}\left([R_{i'k}^0,R_{ij}^0 R_{jk}^0]x_i + [R_{i'j'}^0,R_{ij}^0 R_{jk}^0]x_i x_k\partial_{j'}\right).$$

Note that $[D_1,D_0]$ is a multiplier by a W_4-valued function, since

$$[R_{ij}^0,R_{kl}^0]\in W_2$$

by Proposition 3.1.9. Consequently, in the computation of H_β^k from $H_{\beta-2}^{k-2}$ with $D_1 D_0$, we can write $D_1 D_0 H_{\beta-2}^{k-2} = (D_0 D_1 + [D_1,D_0])H_{\beta-2}^{k-2}$, where the commutator can be omitted, since it gives only a $W_{2k+\beta-2}$ term as $H_{\beta-2}^{k-2}\in W_{2k+\beta-6}$.

We conclude that in the computation of H_β^k from H_β^0, we can omit any term containing D_1, since it may be commuted to the right, where $D_1 H_\beta^0 = 0$. Therefore H_β^k can be computed recursively as

$$(k + \beta)H_\beta^k = D_2 H_{\beta+2}^{k-1} + D_0 H_{\beta-2}^{k-1}$$

for all $k \geq 0$. $\square$

Note that after the approximations in Lemma 12.4.6, the remaining operator $D_2 + D_0$ resembles the harmonic oscillator from Example 6.3.7 if we diagonalize the symmetric matrix $\{R_{ij}^0 \wedge R_{jk}^0\}_{ik}$.

Example 12.4.7 (Low-dimensional M). Before finding an explicit formula for the solution H_0^m to the recurrence

$$(\beta + k)H_\beta^k = D_2 H_{\beta+2}^{k-1} + D_0 H_{\beta-2}^{k-1} \quad \text{mod } W_{2k+\beta-1},$$

in the general case, we observe that the only H_β^k that can be nonzero modulo $W_{2k+\beta-1}$ are

$$H_0^0 = I, \quad H_2^1, \quad H_0^2, H_4^2, \quad H_2^3, H_6^3, \quad H_0^4, H_4^4, H_8^4, \quad H_2^5, H_6^5, H_{10}^5, \quad \ldots.$$

This shows that the index of $\not{D}$ vanishes when m is odd, that is, when $n \equiv 2$ mod 4, since $H_0^m = 0$ if m is odd. Let us also compute this index in the case of a four-dimensional manifold. We have

$$H_0^2 = \frac{1}{6} D_2 D_0 1 = -\frac{1}{96} \sum_{ijkl} (\partial_l^2 x_i x_k) R_{ij}^0 R_{jk}^0 = -\frac{1}{48} \sum_{jl} R_{lj}^0 R_{jl}^0.$$

Thus the index of $\not{D} : H^1(M; \mathcal{A}^+ M) \to L_2(M; \mathcal{A}^- M)$ is

$$\frac{1}{(4\pi)^2} \int_M 2^2 i^2 \left(\frac{-1}{48}\right) \sum_{ij} \langle R_{ij}^0 R_{ji}^0, e_1 e_{-1} e_2 e_{-2}\rangle dp = \frac{1}{192\pi^2} \int_M \sum_{ij} \langle R_{ij} \wedge R_{ji}, \underline{dp}\rangle.$$

To obtain the explicit formula for the index in higher dimensions, we first consider the following model case, in which we replace $R_{ij}^0 \in \Delta^2 \mathbf{R}^n$ by the scalar antisymmetric matrix

$$A := \begin{bmatrix} 0 & a_1 & \cdots & 0 & 0 \\ -a_1 & 0 & \cdots & 0 & 0 \\ \vdots & \vdots & \ddots & \vdots & \vdots \\ 0 & 0 & \cdots & 0 & a_m \\ 0 & 0 & \cdots & -a_m & 0 \end{bmatrix}, \tag{12.13}$$

in an orthogonal splitting as in Proposition 4.3.6.

Proof of Theorem 12.4.1. (i) Consider the recurrence

$$(\beta + k)H_\beta^k = D_2 H_{\beta+2}^{k-1} + D_0 H_{\beta-2}^{k-1}$$

for β-homogeneous polynomials $H_\beta^k : \mathbf{R}^n \to \mathbf{R}$ with $H_0^0 = 1$ and $H_\beta^0 = 0$ when $\beta \geq 1$, where

$$D_2 := \Delta = \sum_i \partial_i^2,$$

$$D_0 := -\frac{1}{16} \sum_{ijk} x_i A_{ij} A_{jk} x_k,$$

with $A = (A_{ij})$ the matrix (12.13). In this case, we have

$$-\sum_{ijk} x_i A_{ij} A_{jk} x_k = a_1^2(x_1^2 + x_{-1}^2) + \cdots + a_m^2(x_m^2 + x_{-m}^2) = \sum_j a_j^2 x_j^2,$$

with $a_{-j} := a_j$. To start with, assume that $\operatorname{Re} a_j = 0$ and consider the harmonic oscillator

$$\partial_t f = \sum_j (\partial_j + (a_j/4)^2 x_j^2) f$$

for a scalar function $f(t, x)$ and $t > 0$, $x \in \mathbf{R}^n$. With $\omega = ia_j/4$ in Example 6.3.7, $j = 1, \ldots, m$, we obtain from Mehler's formula the explicit solution

$$f(t, x) = e^{t(D_2 + D_0)} f(x)$$

$$= \int_{\mathbf{R}^n} \prod_j \sqrt{\frac{\gamma(ia_j t)}{4\pi t}} \exp\left(-\frac{\gamma(ia_j t)}{4t}\left((x_j^2 + y_j^2)\cos(2a_j t) - 2x_j y_j\right)\right) f(0, y) dy,$$

where $\gamma(z) := (z/2)/\sinh(z/2)$. Alternatively, we can proceed as in (12.7) and make an ansatz

$$f(t, x) = \int_{\mathbf{R}^n} (4\pi t)^{-m} e^{-|x-y|^2/(4t)} \sum_k t^k H^k(x, y) f(y) dy.$$

The equality of these two kernels at $x = y = 0$ shows that

$$\sum_k t^k H^k(0, 0) = \prod_j \sqrt{\gamma(ia_j t)} = \prod_{j>0} \gamma(ia_j t).$$

Equating the t^m terms yields $H^m(0, 0) = i^m p_0(a_1, \ldots, a_m)$, where p_0 is the polynomial from Definition 12.4.2. This coincides with the result for H_0^m from the recurrence $(\beta + k)H_\beta^k = D_2 H_{\beta+2}^{k-1} + D_0 H_{\beta-2}^{k-1}$.

(ii) By Definition 12.4.2, we have

$$H^m(0, 0) = i^m p(\operatorname{Tr} A^2, \operatorname{Tr} A^4, \ldots).$$

By analytic continuation, this continues to hold for all $a_j \in \mathbf{C}$. Furthermore, it continues to hold for general $A \in \underline{SO}(\mathbf{R}^n)$, by changing basis to an orthogonal splitting for A as in Proposition 4.3.6 and using the invariance of the trace functional. Similarly, using instead the commutative algebra $(\wedge^{ev}\mathbf{R}^n, \wedge)$, it is clear that the recurrence (12.12) with $D_1 = 0$ will result in

$$H^m(q, q) = i^m p(\mathrm{Tr}R^2, \mathrm{Tr}R^4, \ldots).$$

Using Exercise 12.4.5, we obtain that the index of

$$\not{D} : L_2(M; \not{\triangle}^+ M) \to L_2(M; \not{\triangle}^- M)$$

is

$$\frac{1}{(4\pi)^m} \int_M (2i)^m i^m \langle p(\mathrm{Tr}R^2, \mathrm{Tr}R^4, \ldots), \underline{dq} \rangle.$$

Since we have shown in Exercise 12.4.7 that the index vanishes when m is odd, this completes the proof of the Atiyah–Singer index theorem. $\qquad\square$

To quote a famous mathematician: I think I'll stop here.

12.5 Comments and References

12.3 The Gauss–Bonnet theorem goes back to Gauss, who, however, never published it. Bonnet first published a special case of the theorem in 1848. The full Chern–Gauss–Bonnet theorem for a general compact manifold was first proved by Chern in 1945.

12.4 The index theorem for general elliptic differential operators was proved by Atiyah and Singer in 1963, and is regarded as one of the great landmarks of twentieth-century mathematics. A main special case of this general index theorem is the index theorem for the $\not{\triangle}$-Dirac operator. In fact, this operator was rediscovered by Atiyah and Singer in their work on the index theorem.

The early proofs of the index theorem used different methods. The now dominant heat equation method, which we use, originates in the works of Atiyah, Bott, and Patodi [3] and Gilkey. Standard references for the index theory of Dirac operators include the books by Gilkey [43], Berline, Getzler, and Vergne [20] and Lawson and Michelsohn [63]. Further treatments of the index theory can be found in Taylor [92] and Gilbert and Murray [42].

The proofs of the index theorems given here follow the book [96] by Yanlin Yu, and they do not rely on the theory of pseudodifferential operators. Some minor variations on the setup from [96] include our use of the pointwise constructed polar ON-frame from Definition 12.2.4 and our use of skew-adjoint Dirac operators rather than self-adjoint Dirac operators. By a Wick-type rotation argument it is straightforward to see that this does not affect the index of the operator. A reference for results on differential equations that we use, including Weyl's law and existence results for ordinary differential equations, is Taylor [91, 92]. The local indices $\mathrm{Pf}(R)$ and $\hat{A}(R)$, the integrands appearing in the Chern–Gauss–Bonnet and Atiyah–Singer index theorems, are examples of what are called characteristic classes, which are certain polynomials in the curvature bivectors. A reference for the theory of symmetric polynomials, used in Definition 12.4.2, is Nicholson [73].

.

Bibliography

[1] AHLFORS, L. V. Möbius transformations in R^n expressed through 2×2 matrices of clifford numbers. *Complex Variables Theory Appl. 5*, 2-4 (1986), 215–224.

[2] ARNOLD, D., FALK, R., AND WINTHER, R. Finite element exterior calculus, homological techniques, and applications. *Acta Numer. 255*, 15 (2006), 1–155.

[3] ATIYAH, M., BOTT, R., AND PATODI, V. K. On the heat equation and the index theorem. *Invent. Math. 19* (1973), 279–330.

[4] AUSCHER, P., AXELSSON, A., AND HOFMANN, S. Functional calculus of Dirac operators and complex perturbations of Neumann and Dirichlet problems. *J. Funct. Anal. 255*, 2 (2008), 374–448.

[5] AUSCHER, P., AXELSSON, A., AND MCINTOSH, A. Solvability of elliptic systems with square integrable boundary data. *Ark. Mat. 48* (2010), 253–287.

[6] AUSCHER, P., HOFMANN, S., LACEY, M., MCINTOSH, A., AND TCHAMITCHIAN, P. The solution of the Kato square root problem for second order elliptic operators on $\mathbf{R}^n$. *Ann. of Math. (2) 156*, 2 (2002), 633–654.

[7] AXELSSON, A. Oblique and normal transmission problems for Dirac operators with strongly Lipschitz interfaces. *Comm. Partial Differential Equations 28*, 11-12 (2003), 1911–1941.

[8] AXELSSON, A. *Transmission problems for Dirac's and Maxwell's equations with Lipschitz interfaces*. PhD thesis, The Australian National University, 2003. Available at
https://openresearch-repository.anu.edu.au/handle/1885/46056.

[9] AXELSSON, A. Transmission problems and boundary operator algebras. *Integral Equations Operator Theory 50*, 2 (2004), 147–164.

[10] AXELSSON, A. Transmission problems for Maxwell's equations with weakly Lipschitz interfaces. *Math. Methods Appl. Sci. 29*, 6 (2006), 665–714.

© Springer Nature Switzerland AG 2019

A. Rosén, *Geometric Multivector Analysis*, Birkhäuser Advanced Texts Basler Lehrbücher,
https://doi.org/10.1007/978-3-030-31411-8

[11] AXELSSON, A., GROGNARD, R., HOGAN, J., AND MCINTOSH, A. Harmonic analysis of Dirac operators on Lipschitz domains. In _Clifford analysis and its applications (Prague, 2000)_, vol. 25 of _NATO Sci. Ser. II Math. Phys. Chem._ Kluwer Acad. Publ., Dordrecht, 2001, pp. 231–246.

[12] AXELSSON, A., KEITH, S., AND MCINTOSH, A. Quadratic estimates and functional calculi of perturbed Dirac operators. _Invent. Math. 163_, 3 (2006), 455–497.

[13] AXELSSON, A., KOU, K., AND QIAN, T. Hilbert transforms and the Cauchy integral in euclidean spaces. _Studia Math. 193_, 2 (2009), 161–187.

[14] AXELSSON, A., AND MCINTOSH, A. Hodge decompositions on weakly Lipschitz domains. In _Advances in analysis and geometry_, Trends Math. Birkh'auser, Basel, 2004, pp. 3–29.

[15] AXLER, S. Down with determinants! _Amer. Math. Monthly 102_, 2 (1995), 139–154.

[16] AXLER, S., BOURDON, P., AND RAMEY, W. _Harmonic function theory._ No. 137 in Graduate Texts in Mathematics. Springer-verlag, 1992.

[17] BANDARA, L., MCINTOSH, A., AND ROSÉN, A. Riesz continuity of the Atiyah-Singer Dirac operator under perturbations of the metric. _Math. Ann. 370_, 1-2 (2018), 863–915.

[18] BANDARA, L., AND ROSÉN, A. Riesz continuity of the Atiyah-Singer Dirac operator under perturbations of local boundary conditions. To appear in Communications in Partial Differential Equations, 2019, DOI: 10.1080/03605302.2019.1611847.

[19] BENN, I., AND TUCKER, R. _An introduction to spinors and geometry with applications in physics._ Adam Hilger, Ltd., 1987.

[20] BERLINE, N., GETZLER, E., AND VERGNE, M. _Heat kernels and Dirac operators._ No. 298 in Grundlehren der Mathematischen Wissenschaften. Springer-verlag, 1992.

[21] BOTT, R., AND TU, L. _Differential forms in algebraic topology._ No. 82 in Graduate Texts in Mathematics. Springer-Verlag, 1982.

[22] BOURGUIGNON, J.-P. Spinors, Dirac operators, and changes of metrics. In _Differential geometry: geometry in mathematical physics and related topics (Los Angeles, CA, 1990)_, vol. 54 of _Proc. Sympos. Pure Math._ Amer. Math. Soc., 1993, pp. 41–44.

[23] BRACKX, F., DELANGHE, R., AND SOMMEN, F. _Clifford Analysis._ No. 76 in Research Notes in Mathematics. Pitman, 1982.

[24] BRÖCKER, T., AND TOM DIECK, T. _Representations of compact Lie groups._ Graduate Texts in Mathematics. Springer-Verlag, 1985.

[25] CARTAN, E. Sur certaines expressions différentielles et le problème de Pfaff. *Ann. Sci. Ecole Norm. Sup. 3*, 16 (1899), 239–332.

[26] CARTAN, E. *The theory of spinors*. The M.I.T. Press, Cambridge, Mass., 1967.

[27] CLIFFORD, W. Applications of Grassmann's Extensive Algebra. *Amer. J. Math. 1*, 4 (1878), 350–358.

[28] COIFMAN, R. R., McINTOSH, A., AND MEYER, Y. L'intégrale de Cauchy définit un opérateur borné sur L^2 pour les courbes lipschitziennes. *Ann. of Math. (2) 116*, 2 (1982), 361–387.

[29] COLTON, D., AND KRESS, R. *Integral equation methods in scattering theory*, first ed. John Wiley & Sons, New York, 1983.

[30] COLTON, D., AND KRESS, R. *Inverse acoustic and electromagnetic scattering theory*, second edition ed. Springer-Verlag, Berlin, 1998.

[31] COSTABEL, M., AND McINTOSH, A. On Bogovskiĭ and regularized Poincaré integral operators for the de Rham complexes on Lipschitz domains. *Math. Z. 265*, 2 (2010), 297–320.

[32] DARLING, R. *Differential forms and connections*. Cambridge University Press, 1994.

[33] DELANGHE, R., SOMMEN, F., AND SOUVCEK, V. *Clifford algebra and spinor-valued functions. A function theory for the Dirac operator*. Mathematics and its Applications. Kluwer Academic Publishers Group, 1992.

[34] DIEUDONNÉ, J. The tragedy of Grassmann. *Linear and multilinear algebra 8*, 1 (1979/80), 1–14.

[35] EVANS, L. *Partial differential equations*, vol. 19 of *Graduate Studies in Mathematics*. American Mathematical Society, 1998.

[36] EVANS, L., AND GARIEPY, R. *Measure theory and fine properties of functions*. Studies in Advanced Mathematics. CRC Press, 1992.

[37] FEARNLEY-SANDER, D. Hermann Grassmann and the creation of linear algebra. *Amer. Math. Monthly 86*, 10 (1979), 809–817.

[38] FEDERER, H. *Geometric measure theory*. Die Grundlehren der mathematischen Wissenschaften, Band 153. Springer-Verlag, 1969.

[39] FLANDERS, H. Liouville's theorem on conformal mapping. *J. Math. Mech. 15* (1966), 157–161.

[40] FULTON, W. *Algebraic topology*. No. 153 in Graduate Texts in Mathematics. Springer-verlag, 1985.

[41] FULTON, W., AND HARRIS, J. *Representation theory. A first course*. No. 129 in Graduate Texts in Mathematics. Springer-verlag, 1991.

[42] GILBERT, J., AND MURRAY, M. *Clifford algebras and Dirac operators in harmonic analysis.* Cambridge Studies in Advanced Mathematics. Cambridge University Press, 1991.

[43] GILKEY, P. *Invariance theory, the heat equation, and the Atiyah-Singer index theorem.* No. 11 in Mathematics Lecture Series. Publish or Perish, Inc., 1984.

[44] GRASSMANN, H. *Die Lineale Ausdehnungslehre, ein neuer Zweig der Mathematik.* 1844.

[45] GRASSMANN, H. *Die Ausdehnungslehre: Vollständig und in strenger Form bearbeitet.* 1864.

[46] GREUB, W. *Multilinear algebra. Second edition.* Universitext. Springer-Verlag, 1864.

[47] GRIFFITHS, P., AND HARRIS, J. *Principles of algebraic geometry.* Pure and Applied Mathematics. Wiley-Interscience, 1978.

[48] GRISVARD, P. *Elliptic problems in nonsmooth domains.* Monographs and Studies in Mathematics. Pitman, 1985.

[49] GROVE, L. *Classical groups and geometric algebra.* No. 39 in Graduate Studies in Mathematics. American Mathematical Society, 2002.

[50] HARVEY, F. R. *Spinors and calibrations.* No. 9 in Perspectives in Mathematics. Academic Press, Inc., 1990.

[51] HELSING, J. Solving integral equations on piecewise smooth boundaries using the RCIP method: a tutorial. Available at https://arxiv.org/abs/1207.6737v9.

[52] HELSING, J., AND KARLSSON, A. On a Helmholtz transmission problem in planar domains with corners. *J. Comput. Phys. 371* (2018), 315–332.

[53] HELSING, J., AND KARLSSON, A. Physical-density integral equation methods for scattering from multi-dielectric cylinders. *J. Comput. Phys. 387* (2019), 14–29.

[54] HERTRICH-JEROMIN, U. *Introduction to Möbius differential geometry.* No. 300 in London mathematical society lecture notes series. Cambridge university press, 2003.

[55] HESTENES, D. *Space-time algebra.* Gordon and Breach, 1966.

[56] HESTENES, D. *New foundations for classical mechanics.* No. 99 in Fundamental Theories of Physics. Kluwer Academic Publishers Group, 1999.

[57] HESTENES, D., AND SOBCZYK, G. *Clifford algebra to geometric calculus. A unified language for mathematics and physics.* Fundamental Theories of Physics. D. Reidel Publishing Co., 1984.

[58] HITCHIN, N. The Dirac operator. In *Invitations to geometry and topology*, Oxf. Grad. Texts Math. Oxford Univ. Press, Oxford, 2002, pp. 208–232.

[59] HLADIK, J. *Spinors in physics*. Graduate Texts in Contemporary Physics. Springer-Verlag, 1999.

[60] JANCEWICZ, B. *Multivectors and Clifford algebra in electrodynamic*. No. 11 in Mathematics Lecture Series. World Scientific Publishing Co., Inc., 1988.

[61] KATO, T. *Perturbation theory for linear operators*, second ed. Springer-Verlag, Berlin, 1976. Grundlehren der Mathematischen Wissenschaften, Band 132.

[62] KRESS, R. *Linear integral equations*. No. 82 in Applied Mathematical Sciences. Springer-Verlag, New York, 1999.

[63] LAWSON, H. B., J., AND MICHELSOHN, M.-L. *Spin geometry*. No. 38 in Princeton Mathematical Series. Princeton University Press, 1989.

[64] LOUNESTO, P. *Clifford algebras and spinors*. London Mathematical Society Lecture Note Series. Cambridge University Press, 2001.

[65] MADSEN, I., AND TORNEHAVE, J. *From calculus to cohomology. de Rham cohomology and characteristic classes*. No. 11 in Mathematics Lecture Series. Cambridge University Press, 1997.

[66] MCINTOSH, A. Clifford algebras and the higher-dimensional Cauchy integral. In *Approximation and function spaces (Warsaw, 1986)*, vol. 22 of *Banach Center Publ.* PWN, Warsaw, 1989, pp. 253–267.

[67] MCINTOSH, A., AND MITREA, M. Clifford algebras and Maxwell's equations in Lipschitz domains. *Math. Methods Appl. Sci. 22*, 18 (1999), 1599–1620.

[68] MCINTOSH, A., AND MONNIAUX, S. Hodge-Dirac, Hodge-Laplacian and Hodge-Stokes operators in L^p spaces on Lipschitz domains. *Rev. Mat. Iberoam. 34*, 4 (2018), 1711–1753.

[69] MEYER, Y. *Wavelets and operators*. No. 37 in Cambridge Studies in Advanced Mathematics. Cambridge University Press, 1992.

[70] MITREA, D., MITREA, M., AND MONNIAUX, S. The Poisson problem for the exterior derivative operator with Dirichlet boundary condition in nonsmooth domains. *Commun. Pure Appl. Anal. 7*, 6 (2008), 1295–1333.

[71] MITREA, M. *Clifford Wavelets, Singular Integrals and Hardy Spaces*. No. 1575 in Lecture Notes in Mathematics. Springer, 1994.

[72] MORREY, C.B., J. *Multiple integrals in the calculus of variations*. No. 130 in Die Grundlehren der mathematischen Wissenschaften. Springer-Verlag, 1966.

[73] NICHOLSON, W. K. *Introduction to abstract algebra*. Wiley-Interscience. John Wiley & Sons, 2007.

[74] PICARD, R. An elementary proof for a compact imbedding result in general-ized electromagnetic theory. *Math. Z. 187*, 2 (1984), 151–164.

[75] PORTEOUS, I. *Clifford algebras and the classical groups.* No. 50 in Cambridge Studies in Advanced Mathematics. Cambridge University Press, 1995.

[76] PRYDE, A. J. The five lemma for Banach spaces. *Proc. Amer. Math. Soc. 65*, 1 (1977), 37–43.

[77] RESETNJAK, J. G. Liouville's conformal mapping theorem under minimal regularity hypotheses. *Sibirsk. Mat. 8* (1967), 835–840.

[78] RIESZ, M. *Clifford numbers and spinors. With the author's private lectures to E. Folke Bolinder.* No. 54 in Fundamental Theories of Physics. Kluwer Academic Publishers Group, 1993.

[79] RINDLER, W. *Relativity. Special, general, and cosmological.* Oxford University Press, 2006.

[80] ROSÉN, A. Boosting the Maxwell double layer potential using a right spin factor. To appear in Integral Equations and Operator Theory.

[81] ROSÉN, A. Fredholm theory, singular integrals and Tb theorems. Unpublished lecture notes from 2011, available at
http://www.math.chalmers.se/~rosenan/FST.html.

[82] ROSÉN, A. Layer potentials beyond singular integral operators. *Publ. Mat. 57*, 2 (2013), 429–454.

[83] ROSÉN, A. Square function and maximal function estimates for operators beyond divergence form equations. *J. Evol. Equ. 13*, 3 (2013), 651–674.

[84] ROSÉN, A. A spin integral equation for electromagnetic and acoustic scat-tering. *Appl. Anal. 96*, 13 (2017), 2250–2266.

[85] SCHWARZ, G. *Hodge decomposition - a method for solving boundary value problems.* No. 1607 in Lecture Notes in Mathematics. Springer-Verlag, 1995.

[86] SOMMEN, F. Spingroups and spherical means. In *Clifford algebras and their applications in mathematical physics (Canterbury, 1985)*, vol. 183 of *NATO Adv. Sci. Inst. Ser. C Math. Phys. Sci.* Reidel, Dordrecht, 1986, pp. 149–158.

[87] SOMMEN, F. Spingroups and spherical means. II. In *Proceedings of the 14th winter school on abstract analysis (Srni, 1986)*, no. 14. Rend. Circ. Mat. Palermo (2) Suppl., 1987, pp. 157–177.

[88] SOMMEN, F. Spingroups and spherical means. III. In *Proceedings of the Winter School on Geometry and Physics (Srni, 1988)*, no. 21. Rend. Circ. Mat. Palermo (2) Suppl., 1989, pp. 295–323.

[89] STEIN, E., AND WEISS, G. On the theory of harmonic functions of several variables. I. The theory of H^p-spaces. *Acta Math. 103* (1960), 25–62.

[90] TAUBES, C. *Differential geometry. Bundles, connections, metrics and curvature.* No. 23 in Oxford Graduate Texts in Mathematics. Oxford University Press, 2011.

[91] TAYLOR, M. *Partial differential equations. I. Basic theory.* No. 115 in Applied Mathematical Sciences. Springer-Verlag, 1996.

[92] TAYLOR, M. *Partial differential equations. II. Qualitative studies of linear equations.* No. 116 in Applied Mathematical Sciences. Springer-Verlag, 1996.

[93] THALLER, B. *The Dirac equation.* Texts and Monographs in Physics. Springer-Verlag, 1992.

[94] VERCHOTA, G. Layer potentials and regularity for the Dirichlet problem for Laplace's equation in Lipschitz domains. *J. Funct. Anal. 59*, 3 (1984), 572–611.

[95] WATSON, G. N. *A Treatise on the Theory of Bessel Functions.* Cambridge University Press, New York, 1944.

[96] YU, Y. *The index theorem and the heat equation method.* No. 2 in Nankai Tracts in Mathematics. World Scientific Publishing Co., 2001.

Index

© Springer Nature Switzerland AG 2019

A. Rosén, *Geometric Multivector Analysis*, Birkhäuser Advanced Texts Basler Lehrbücher,

https://doi.org/10.1007/978-3-030-31411-8

Printed in the United States
By Bookmasters